Lecture Notes in Computer Science   14768

Founding Editors

Gerhard Goos
Juris Hartmanis

AF166369

The series Lecture Notes in Computer Science (LNCS), including its subseries Lecture Notes in Artificial Intelligence (LNAI) and Lecture Notes in Bioinformatics (LNBI), has established itself as a medium for the publication of new developments in computer science and information technology research, teaching, and education.

LNCS enjoys close cooperation with the computer science R & D community, the series counts many renowned academics among its volume editors and paper authors, and collaborates with prestigious societies. Its mission is to serve this international community by providing an invaluable service, mainly focused on the publication of conference and workshop proceedings and postproceedings. LNCS commenced publication in 1973.

Hiroyuki Kajimoto · Pedro Lopes ·
Claudio Pacchierotti · Cagatay Basdogan ·
Monica Gori · Betty Lemaire-Semail ·
Maud Marchal

Editors

Haptics:
Understanding Touch;
Technology and Systems;
Applications and Interaction

14th International Conference on Human Haptic Sensing
and Touch Enabled Computer Applications, EuroHaptics 2024
Lille, France, June 30 – July 3, 2024
Proceedings, Part I

 Springer

Editors
Hiroyuki Kajimoto ⓘ
The University of Electro-Communications
Tokyo, Japan

Pedro Lopes ⓘ
University of Chicago
Chicago, IL, USA

Claudio Pacchierotti ⓘ
CNRS
Rennes, France

Cagatay Basdogan ⓘ
Koc University
Istanbul, Türkiye

Monica Gori ⓘ
Italian Institute of Technology
Genova, Italy

Betty Lemaire-Semail ⓘ
University of Lille
Lille, France

Maud Marchal ⓘ
University of Rennes
Rennes, France

ISSN 0302-9743   ISSN 1611-3349 (electronic)
Lecture Notes in Computer Science
ISBN 978-3-031-70057-6   ISBN 978-3-031-70058-3 (eBook)
https://doi.org/10.1007/978-3-031-70058-3

Preface

Eurohaptics 2024 brings together researchers from different domains and industrial partners to explore the complex interactions between touch, perception, and technology. This year, out of 142 regular paper submissions, 81 were accepted based on the evaluations of at least three reviewers per paper through a single-blind review process. To maintain a single-track oral presentation format, only 37 papers with the highest reviewer scores were selected for oral presentation while 44 papers were presented in poster format, allowing for dedicated interaction time with the presenters throughout the conference. Additionally, 63 work-in-progress posters and 65 hands-on demonstrations showcasing haptic technology were featured at Eurohaptics 2024. The accepted papers are grouped into three categories according to their thematic focus:

1. **Understanding Touch:** The human sense of touch, one of our most vital sensory modalities, forms the cornerstone of our interaction with the world. From the gentle caress of a loved one's hand to the firm grip of a tool, touch allows us to perceive textures, pressures, temperatures, and vibrations. The papers in this category focus on the intricacies of human touch sensing, exploring the underlying physiological mechanisms and the fundamental processing that governs our tactile perception. It's important to note that haptics, the science of touch, encompasses not only the sensation perceived by the hand but also extends to the entire body, including complex interactions such as proprioception and kinesthesia.
2. **Technology and Systems:** Advances in technology have enabled the design of devices and systems that aim to replicate and augment the human sense of touch. From kinesthetic devices that provide force feedback to tactile displays that render realistic sensations, these technologies are revolutionizing human-machine interaction. The papers in this category explore the design principles and engineering innovations behind such devices and systems.
3. **Applications and Interaction:** The integration of haptic technology into real and virtual environments has opened up new frontiers in multi-modal interaction between humans and devices. From medical simulators that allow surgeons to practice complex procedures to immersive virtual reality experiences that simulate physical sensations, the applications of haptics are vast and diverse. The papers in this category examine the multifaceted interactions between humans and haptic-enabled devices, exploring their potential impact on fields such as medicine, rehabilitation, education, data visualization, automotive technology, and remote communication.

We extend our sincere gratitude to the associate editors and reviewers for their diligent work in the paper selection and review process. We hope that the attendees enjoy the diverse and engaging conference program.

May 2024

Cagatay Basdogan
Monica Gori
Claudio Pacchierotti
Betty Lemaire-Semail

Message from the Conference Chair

The 14th Eurohaptics is the European international conference focused on the latest research in the domain of haptics. We were delighted to host Eurohaptics 2024 as fully in-person and enjoyed seeing, learning, sharing, and hanging out with our vibrant community.

This extraordinary event took place in Lille, France, June 30th – July 3rd, 2024, at the University campus. As capital of Flanders, Lille's rich cultural heritage provided a picturesque backdrop to our conference, offering a perfect blend of intellectual exchange and cultural immersion. After a stimulating day of conference activities, attendees could enjoy the city of Lille or sit with friends and colleagues to taste the typical food of the North of France.

Eurohaptics 2024 was a special event that brought us all back together after the challenging pandemic years to present our innovative research, experience new advances in haptics, and catch up with what's happening with our community. We are pleased and appreciative of our research community that contributed a tremendous number of exciting paper presentations, workshops, tutorials, research demos, posters and works-in-progress. We are excited to announce that this year, we organized a dedicated "Art/Design" track, providing a unique opportunity to explore the intersection of haptics with creativity and design. Furthermore, we committed to strengthening our ties with industrial partners. Eurohaptics 2024 featured job dating sessions and an industry forum, facilitating meaningful connections between academia and industry, and fostering collaborative opportunities that drive haptic technology forward. Finally, we had three amazing haptics pioneers lined up for the keynote talks as well as one industry keynote talk.

These are challenging financial times for many corporations, so we would especially like to thank our sponsors who with their generous donations helped this conference happen, while promoting diversity, equity, inclusion, and accessibility, including supporting those with financial barriers. We would like to thank the Eurohaptics society for its continuous support for the conference.

Welcome to the proceedings of Eurohaptics 2024.

Frédéric Giraud
Maud Marchal

Organization

General Chairs

Frédéric Giraud   University of Lille, France
Maud Marchal   Univ. Rennes, INSA/IRISA, France

Program Committee Members

Claudio Pacchierotti   CNRS – IRISA, France
Hiroyuki Kajimoto   University of Electro-Communications, Japan
Pedro Lopes   University of Chicago, USA

Editorial Board Members

Cagatay Basdogan   Koç University, Turkey
Monica Gori   Italian Institute of Technology, Italy
Claudio Pacchierotti   CNRS – IRISA, France
Betty Semail   University of Lille, France

Workshops Chairs

David Gueorguiev   CNRS – ISIR, France
Dasha Kolesnyk   University of Twente, Netherlands

Demonstration Chairs

Matthieu Rupin   HAP2U, France
Masaya Takasaki   Saitama University, Japan

Work-in-Progress Chairs

Charles Hudin   CEA LIST, France
Sabrina Panëels   CEA LIST, France

Industry and Sponsorship Chairs

Dan Shor   Innovobot, Germany
Christophe Giraud-Audine   Arts & Métiers Sciences and Technology, France

Communication Chair

Mounia Ziat   Bentley University, USA

Associate Editors

Basdogan Cagatay   Koc University, Istanbul, Turkey
Farkhatdinov Ildar   Queen Mary University of London, UK
Gori Monica   Italian Institute of Technology, Italy
Gueorguiev David   CNRS – ISIR, France
Hashtrudi-Zaad Keyvan   Queen's University, Canada
Ho Hsin-Ni   Kyushu University, Japan
Jeon Seokhee   Kyung Hee University, Republic of Korea
Kim Jin Ryong   University of Texas at Dallas, USA
Kuroki Scinob   Nippon Telegraph and Telephone Corporation,
                              Japan
Leonardis Daniele   Scuola Superiore Sant'Anna, Italy
Levesque Vincent   École de Technologie Supérieure, Canada
Lisini Baldi Tommaso   University of Siena, Italy
Malvezzi Monica   University of Siena, Italy, Istituto Italiano di
                              Tecnologia (IIT), Italy
McIntyre Sarah   Linköping University, Sweden
Metzger Anna   Bournemouth University, UK
Okamoto Shogo   Tokyo Metropolitan University, Japan
Pacchierotti Claudio   CNRS-IRISA, Rennes, France
Park Gunhyuk   Gwangju Institute of Science and Technology,
                              Republic of Korea
Saal Hannes   University of Sheffield, UK
Semail Betty   University of Lille, France
Smeets Jeroen   VU Amsterdam, Netherlands
Steinbach Eckehard   TU Munich, Germany
Tanaka Yoshihiro   Nagoya Institute of Technology, Japan
Van Beek Femke   Eindhoven University of Technology, Netherlands
Wang Dangxiao   State Key Lab of Virtual Reality Technology and
                              Systems, Beihang University, China
Ziat Mounia   Bentley University, USA

Reviewers

Abbasimoshaei Alireza
Abdlkarim Diar
Abdulali Arsen
Ablart Damien
Ackerley Rochelle
Adeyemi Ayoade
Aggravi Marco
Aghababaei Ramin
Ahmetovic Dragan
Alabbas Ali
Alberto Parmiggiani
Alea Mark Daniel
Aliabbasi Easa
Altamirano Cabrera Miguel
Anwar Eisa
Arcangeli Dorine
Arif Ahmed
Arikan Belkis Ezgi
Armand Mehran
Arnold Gabriel
Awan Mudassir
Azañón Elena
Bailly Gilles
Balaji Ananta Narayanan
Ban Yuki
Banerjee Premankur
Barbareschi Giulia
Basile Valerio
Batmaz Anil Ufuk
Beckerle Philipp
Ben Dhiab Ayoub
Benallegue Mehdi
Bennewitz Roland
Bennewitz Roland
Bernard Corentin
Bhardwaj Amit
Bhardwaj Ayush
Bianchi Andrea
Bigué Jean-Philippe
Birznieks Ingvars
Blount Hannah
Bo Valerio
Bontula Anisha

Boucaud Fabien
Bouchigny Sylvain
Bouchihan Hélie
Bouzbib Elodie
Brahimaj Detjon
Braun Doris
Brayda Luca
Brayda Luca
Brock Anke
Brogi Chiara
Browder Jonathan
Buckingham Gavin
Caarls Wouter
Cabaret Pierre-Antoine
Cacucciolo Vito
Cacucciolo Vito
Calisto Francisco Maria
Camardella Cristian
Camardella Cristian
Campus Claudio
Carbon Claus-Christian
Carolina Tammurello
Carolina Tammurello
Casado-Palacios Maria
Cataldo Antonio
Cavdan Müge
Cecamore Matteo
Celebi Bora
Cespedes Gomez Nathalia
Cha Youngsu
Chaichaowarat Ronnapee
Chandrasekar Srinivasan
Chase Elyse
Chee Keith
Chen Yonghua
Chen Chao
Cheng Zelei
Chinello Francesco
Chinello Francesco
Cho Byung Jin
Choi Dong-Soo
Choi Seungmoon
Choi Hyouk Ryeol

Chu Yingguang
Chu Yingguang
Cleland Luke
Coe Patrick
Coelho Lara
Colgate J. Edward
Colley Mark
Córdova Bulens David
Cornelio Patricia
Corniani Giulia
Corniani Giulia
Cosgun Akansel
Costes Antoine
Cox Samuel
Cretu Ana-Maria
D'aurizio Nicole
Dagli Smit
Dalsgaard Tor-Salve
Damian Dana
Danieau Fabien
Daunizeau Thomas
Davide Esposito
Degraen Donald
Delbos Benjamin
Delhaye Benoit
Di Campli San Vito Patrizia
Ding Xueting
Dione Mariama
Dometios Athanasios
Dong Gaofeng
Dosen Strahinja
Dragusanu Mihai
Drewing Knut
Driller Karina
Dunkelberger Nathan
Duvernoy Basil
Duvernoy Basil
Duvernoy Basil
Edmondson Laura
Eid Mohamad
Eleonora Montagnani
Elizondo Sonia
Ercan Samet Mert
Erkat Orhan Batuhan
Evangelou George

Evreinov Grigori
Evreinov Grigori
Fairhurst Merle
Fairhurst Merle
Fan Zhuzhi
Farooq Ahmed
Fartook Ori
Faucheu Jenny
Favrot Jeanne
Ferdous Md Javedul
Feuchtner Tiare
Filingeri Davide
Fink Paul
Fishel Jeremy
Fitter Naomi
Fontana De Vargas Mauricio
Foo Esther
Fortin Pascal
Fortin Pascal
Fraisse Philippe
Franco Leonardo
Freeman Euan
Frier William
Friesen Rebecca
Fruchard Bruno
Fruchard Bruno
Gabardi Massimiliano
Gallo Simone
Gather Mark
Gatti Elia
Georgiou Orestis
Georgiou Theodoros
Gerling Gregory
Ghosh Avik
Gilles Benjamin
Gilles Benjamin
Giordano Marcello
Giraud Frederic
Giraud-Audine Christophe
Giudici Gabriele
Goguey Alix
Göksel Orcun
Goktepe Nedim
Gomez Hernandez Pedro
Götzelmann Timo

Gourishetti Ravali
Gülecyüz Basak
Guo Hanzhe
Guo Xingwei
Guo Miaoxian
Hachisu Taku
Haghighi Osgouei Reza
Hagura Nobuhiro
Han Amy Kyungwon
Harders Matthias
Hartcher-O'brien Jess
Hasegawa Shoichi
Hatira Amal
He Dazhong
Hecquard Jeanne
Hernández Damián
Hesse Constanze
Hidaka Souta
Hirao Yutaro
Hirota Koichi
Hojatmadani Mehdi
Hojatmadani Mehdi
Holschuh Brad
Honrales Daniel
Horie Arata
Howard Thomas
Howard Thomas
Howard Thomas
Hu Yaoping
Hudin Charles
Huisman Gijs
Huloux Nicolas
Hwang Chiwoong
Hwang Inwook
Ichihashi Sosuke
Inoue Seki
Irisarri Josu
Ishikawa Yuri
Ishikawa Yuri
Ishizuka Hiroki
Iyer Abhilash
Jager Edwin
Jain Neera
Janko Marco
Jarocka Ewa

Jeschke Michaela
Jiang Chao
Johansson Roland
Jones Lynette
Ju Yulan
Kabdyshev Nurlan
Kaci Anis
Kakaraparthi Vishnu
Kameoka Takayuki
Kaneko Seitaro
Kang Jeonggoo
Kangas Jari
Kangas Jari
Kao Anika
Kappassov Zhanat
Kappers Astrid
Kappers Astrid M. L.
Katircilar Didem
Kawazoe Anzu
Kern Thorsten
Khalilianmotamed Bonab Ali
Khamis Mohamed
Kianzad Soheil
Kianzad Soheil
Kim Sang-Youn
Kim Dong-Geun
Kim Jinwook
Kim Sang-Youn
Kim Myung Jin
Kim Junwoo
Kim Giryeon
Klatzky Roberta L.
Kolesnyk Dasha
Kollannur Sandeep Zechariah George
Kommuri Krishna Dheeraj
Komninos Andreas
Krieger Kathrin
Krishnappa Babu Pradeep Raj
Kuang Lisheng
Kuling Irene
Kumar Paras
Kuntz Alan
Kuroda Yoshihiro
Kurogi Tadatoshi
Lacôte Inès

Lacôte Inès
Lambert Scott
Lavenant Suliac
Le Thanh-Loan
Lécuyer Anatole
Lee Hyosang
Lee Jungeun
Lee Dajin
Lei Zhenhong (Brad)
Lelevé Arnaud
Leonardis Daniele
Ley-Flores Judith
Li Ao
Li Bingxu
Li Zijun
Li Yixuan
Liang Eddy Zexin
Liao Yi-Chi
Lim Soo-Chul
Lin Lisa Pui Yee
Lin Hongnan
Lineykin Simon
Liu Tianyi
Liu Yiming
Liu Siwen
Liu Qian
Locharulu Vinay
Long Benjamin
Lorenzo Landolfi
Lucia Schiatti
Luzhnica Granit
Lylykangas Jani
M. Manivannan
Macé Marc
Maclean Karon
Maeda Tomosuke
Maggioni Emanuela
Magnusson Charlotte
Maiello Guido
Maiero Jens
Majumdar Swarnamouli
Makino Yasutoshi
Makino Yasutoshi
Malacria Sylvain
Maldonado Jaime

Malik Hassam
Mallick Chandrama
Manjulalayam Rajendran Rajashree
Marchal Maud
Mariabianca Amadeo
Marshall Andrew
Martina Riberto
Matsubayashi Atsushi
Matsumoto Mitsuharu
Mazursky Alex
Meli Leonardo
Memeo Mariacarla
Miller Luke
Miller Luke
Mini Thulasi Rishika
Mirza Vishal
Mizoguchi Izumi
Mizushima Daisuke
Moesgen Tim
Mohammed Shahed
Montano Murillo Roberto
Moon Hyungpil
Moore Warren
Morisaki Tao
Morosi Federico
Moscatelli Alessandro
Mukhopadhyay Mayukh
Mulot Lendy
Nagano Hikaru
Nagi Saad
Naish Michael
Nakamura Takuto
Nakatani Masashi
Nam Saekwang
Nault Emilyann
Ng Kevin
Nicolau Hugo
Nisky Ilana
Nockenberg Lars
Norman Daniel
Norman Farley
Normand Erwan
Norouzi Nahal
O'Malley Marcia
O'Modhrain Sile

Oakley Ian
Oh Seungjae
Ohtsuka Satoshi
Omarali Bukeikhan
Oron-Gilad Tal
Otaduy Miguel
Otaran Ata
Ouari Mondher
Overvliet Krista
Pacchierotti Claudio
Palmer Jasmin
Panëels Sabrina
Pantera Lucie
Paolocci Gianluca
Parisi David
Park Wanjoo
Park Yong-Lae
Parry Ross
Paschall Courtnie
Pawluk Dianne
Peiris Roshan
Peng Danyang
Perquin Marlou
Pezent Evan
Pietrzak Thomas
Pittera Dario
Plaisier Myrthe
Plaisier Myrthe A.
Plante Jean-Sebastien
Pontreau Marion
Pozzi Maria
Pruks Vitalii
Puppala Sai
Qiu Yujuan
R. Kermani Mehrdad
Raiano Luigi
Rajanna Vijay
Rakhsha Ramtin
Rakkolainen Ismo
Rakotondrabe Micky
Ramasamy Priyanka
Rantala Jussi
Rassmus-Gröhn Kirsten
Raza Ahsan
Raza Ahsan

Reardon Gregory
Regimbal Juliette
Richard Grégoire
Roberts Roberta
Rohou–Claquin Baptiste
Romeo Katerine
Rossa Carlos
Routray Prasanna Kumar
Ruan Cihan
Sadeghi Aval Shahr Mohammad
Saga Satoshi
Saint-Aubert Justine
Saint-Bauzel Ludovic
Samengo Ines
Samur Evren
Sand Antti
Santhosh Joseph
Sarac Mine
Sato Katsunari
Saviano Giuseppe
Savoye Yann
Sawada Hideyuki
Sawayama Masataka
Sazara Cem
Schneider Oliver
Schuhler Guillaume
Schuwerk Clemens
Secciani Nicola
Seifi Hasti
Seim Caitlyn
Seita Daniel
Semail Betty
Semenzi Ivan
Serra Federica
Setti Walter
Shafique Shehzaib
Shao Yitian
Shao Yitian
Shields Rachael
Shim Grace
Shinoda Hiroyuki
Shultz Craig
Simner Julia
Singh Richa
Singhal Yatharth

Singhal Yatharth
Singhal Anshul
Sood Arpit
Spagnoletti Giovanni
Speicher Marco
Spiers Adam
Sridhar Varun
Stellmacher Carolin
Su Zhe
Sun Qingyu
Suzuishi Yosuke
Suzuki Shun
Sylaiou Stella
Szymkowski Maciej
Takagi Atsushi
Tame Luigi
Tan Hong
Tanabe Takeshi
Tavakoli Mahdi
Teng Shan-Yuan
Terao Masahiko
Theivendran Karthikan
Tong Qianqian
Tong Cui
Topp Sven
Toscani Matteo
Troisi Danilo
Tummala Neeli
Turco Enrico
Turco Enrico
Tursynbek Iliyas
Tymms Chase
Ujitoko Yusuke
Umehara Rodan
Ushiyama Keigo
Valigi Maria Cristina
Van Laake Lucas Carolus
Vardar Yasemin
Vibol Yem
Vickery Richard
Vigni Francesco
Villani Alberto
Visentin Francesco
Vitali Helene

Vlachos Christoforos
Vollert Jan
Vyas Preeti
Wang Run
Wang Haokun
Wang Yun
Wang Hsueh-Cheng
Watanabe Junji
Watkins Roger
Weber Gerhard
Weda Judith
Wessberg Johan
Wiedemann Oliver
Wiertlewski Michael
Wijntjes Maarten
Wilson Graham
Witzel Christoph
Xia Pingjun
Xiao Yu
Xiao Shuangshuang
Xu Jiayi
Xu Jiayi
Yang Zhengbao
Yang Zhengbao
Yang Tae-Heon
Yannier Nesra
Yasmin Shamima
Yokosaka Takumi
Yokosaka Takumi
Yoo Yongjae
Youssef Fady
Zanchi Silvia
Zárate Juan
Zhang Jun
Zhang Dan
Zhao Siyan
Zheng Yilei
Zhou Ziliang
Zhou Jianshu
Zhu Jihong
Zhu Kening
Ziemke Tom
Zook Zane
Zopf Regine

Sponsors

Contents – Part I

Technology and Systems

Applications and Interaction

Contents – Part II

Technology and Systems

Applications and Interaction

Understanding Touch

Human Identification Performance of Vibrotactile Stimuli Applied on the Torso Along Azimuth or Elevation

Junwoo Kim[1]([✉])(iD), Jaejun Park[1]([✉])(iD), Chaeyong Park[1]([✉])(iD),
Junseok Park[2]([✉])(iD), and Seungmoon Choi[1]([✉])(iD)

[1] Pohang University of Science and Technology, Pohang, Republic of Korea
kjw8515@postech.ac.kr, jjpark17@postech.ac.kr, pcy8201@postech.ac.kr,
choism@postech.ac.kr
[2] Electronics and Telecommunications Research Institute, Daejeon, Republic of Korea
parkjs@etri.re.kr

Abstract. This study investigates the human recognition performance of vibrotactile stimuli applied on the torso along azimuth or elevation under the context of tactile egocentric directional cueing. We conducted two absolute identification experiments for azimuth and elevation, respectively, using real tactile stimuli generated by physical actuators and illusory stimuli rendered by funneling illusion. For both azimuth and elevation, the recognition accuracies were very high, over 95.6%, when only real stimuli were used to indicate 6–8 directions. However, combining the real stimuli with the illusory ones to double the spatial resolution resulted in significantly lower accuracies between 74.1% and 77.8%. The estimated information transfer (IT) values also remained very similar. Using identical methods, our results quantify the human identification performance of torso-distributed tactile stimuli for azimuth or elevation. The detailed results provide general guidelines for designing torso-based tactile systems to enhance spatial awareness and navigation.

Keywords: Torso · Vibrotactile Stimuli · Tactile Illusion · Directional Cue

1 Introduction

Humans naturally associate a tactile stimulus applied on the body, especially the torso, with the egocentric orientation directing outward from the body center through the stimulation point [5]. This affordance is instrumental for wearable directional interfaces that are easy and effective to use; see a review in Sect. 2.

In this paper, we use identical methods to evaluate the human recognition performance of vibrotactile stimuli applied on the torso at different egocentric azimuths or elevations. In study 1, we quantified the identification performance of eight vibrotactile stimuli at different azimuths around the torso at three heights. This standard azimuth mapping was extended to double the angular resolution

H. Kajimoto et al. (Eds.): EuroHaptics 2024, LNCS 14768, pp. 3–15, 2025.
https://doi.org/10.1007/978-3-031-70058-3_1

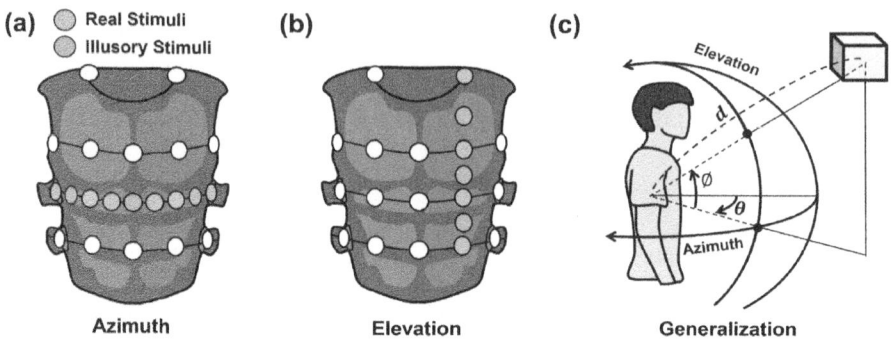

Fig. 1. Real and illusory vibrotactile stimuli used in our research (a and b) with the general definitions of azimuth θ, elevation ϕ, and distance d (c).

using illusory vibrotactile stimuli. A point between two neighbor vibration points was indicated by activating the two vibrations, imitating the tactile funneling illusion (static phantom sensation) [20]. This technique using illusory sensations allowed us to assess 16 angular spots around the torso for azimuth recognition without using more actuators (Fig. 1(a)). The performance measures used were identification accuracy and information transfer (IT). Study 2 followed essentially the same procedure to examine the human vibrotactile recognition performance of elevation on the torso (Fig. 1(b)). The experimental results of the two studies allowed us to evaluate and objectively compare the human identification performance of real and illusory tactile stimuli applied on the torso along the azimuth or elevation direction, with implications for the design of effective tactile directional cueing methods.

All experiments in this paper were approved by the Institutional Review Board of POSTECH (PIRB-2022-E036).

2   Related Work

The past 20 years have seen extensive research interests in wearable haptic interfaces delivering tactile directional cues. Associated research began with a belt equipped with 6–8 tactors around the waist for directional navigation [11,28], way-point navigation [8], and car navigation [1], also for blind users [9]. Belt-type interfaces enclosing other body parts were also developed, such as bands with 4, 6, or 8 tactors around the neck [23], arms [3], and legs [18]. Most of these studies activate the actuator at the specific body position, connecting the body center to the target direction [2].

Additionally, tactile funneling illusions (phantom sensations) were used to effectively increase the angular resolution using a small number of tactors. For example, six tactors combined with phantom sensation rendering were used in an ankle band to represent 12 directions [18] and in a wrist band to indicate any continuous angle [22]. Liao et al. demonstrated that illusory tactile cues with an average directional resolution of 15.35° could effectively guide directions to users [18]. Salazar et al. reported that a 49.3% reduction in hand trajectory error resulted from continuous vibrotactile directional guidance [22].

Some studies investigated the performance of vibrotactile localization. Chen et al. measured the vibrotactile identification performance for a 3×3 tactor array on the wrist [4]. Cholewiak et al. examined the effects of the number and locations of vibratory stimuli on the abdomen [6] and arm [7]. In particular, the experiments in [6] used belts with 6, 8, or 12 tactors. The vibration recognition accuracy for azimuth was 92% with 8 tactors and 74% with 12 tactors.

The above studies were concerned with directional information through the egocentric azimuth spanning the body's horizontal plane. Recent studies have expanded the research to conveying 3D spatial information to contribute to spatial awareness [14,17] and navigation [15,21]. Other studies proposed methods for representing 3D directional information using grid-shaped tactile actuators on the head [16], head and waist [26], and torso [19]. Specifically, Tawa et al. [26] designed a system using two vibrotactile belts with eight tactors worn on the head and waist, respectively, and evaluated the user perception performance for orientation, height, and depth. The identification accuracy was 54% for 12 azimuth stimuli and 45% for four elevation stimuli within a 45° range. Understanding the human recognition performance for both azimuth and elevation is crucial for designing accurate 2D-direction notification methods, but the scientific knowledge available is insufficient yet.

The torso is ideal for delivering a range of spatial stimuli owing to its large size, making it effective for transmitting directional cues. The recent releases of several commercial haptic vests [12,24,27,30] substantially improve the potential of utilizing the torso as a communication channel. In these circumstances, this research assesses the human performance of recognizing vibrotactile stimuli applied on the torso for designating the 2D azimuth or elevation directions.

3 Study 1: Azimuth Identification

We conducted two perceptual experiments to assess the human identification performance of vibrotactile stimuli applied across the torso at evenly-spaced azimuths. One experiment (REL) used eight real vibrations. The other one (R+I) combined the eight real vibrations with an additional eight illusory vibrations using funneling illusion, resulting in 16 vibrations. In both experiments, we evaluated the identification performance using accuracy and information transfer.

3.1 Methods

Participants. Twelve participants (six males and six females) without somatosensory disorders took part in each experiment (age 23.3 ± 4.1 years in REL and 22.2 ± 3.2 in R+I). Before the experiment, they understood the purposes and procedure of the study based on a written document and verbal explanations, and signed an informed consent form. They were paid at KRW 20,000 per hour.

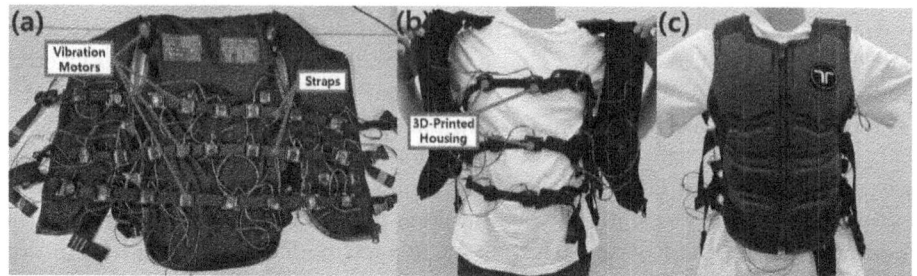

Fig. 2. A customized haptic suit.

Apparatus. We customized a commercial haptic vest (bHaptics, TactSuit X40) to accurately control vibration locations on the torso against various individual body shapes and sizes (Fig. 2). The original vest had all vibration motors affixed to its inner side, which did not allow us to control the stimulated body spots exactly for different users. We took these motors out, put them into 3D-printed housings, and attached eight motors each to three belts using length-adjustable straps. We also added one additional motor to each shoulder. As a result, the vest allowed vertical and horizontal repositioning of the 26 motors, ensuring a customized fit for each individual.

For each participant, four vibration motors on the three belts were initially aligned at the navel, center of the back, and right and left centers of the waist. The other motors were evenly distributed between each pair of motors. The three belts were vertically positioned at the navel, solar plexus, and upper chest while maintaining similar distances between two adjacent belts. The vest was tightly fastened to ensure stable contact between the motors and the torso.

Experimental Conditions. The vibration stimulation positions on the torso are specified in Fig. 3. In REL, the eight tactors were evenly distributed around the torso. The vibrations had an amplitude of 10 G and a frequency of 61 Hz, which enabled strong perception [10]. Each stimulus lasted for 3 s: 1-s vibration, 1-s rest, and 1-s vibration.[1] The vibration vertical positions were grouped as the upper chest (TOP), solar plexus (MID), and navel (BTM).

In R+I, illusory stimuli were presented using a tactile illusion rendering algorithm in [13], where the intensity of the illusory stimulus, A_I, was calculated using the intensities of the two neighbor real vibrations, $A_{R,1}$ and $A_{R,2}$, as follows:

$$A_I^2 = A_{R,1}^2 + A_{R,2}^2. \tag{1}$$

To achieve $A_I = 10$ G, we set $A_{R,1} = A_{R,2} = 7$ G. This design led to testing 16 equi-azimuth locations around the torso as in Fig. 3(b).

---

[1] In pilot experiments, this pattern seemed more effective, e.g. than one 3-s vibration, in catching and maintaining participants' attention.

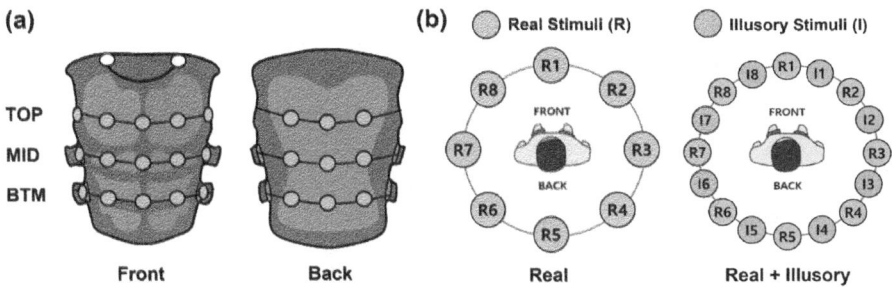

Fig. 3. Stimulus positions for the azimuth identification experiments.

Procedure. Participants wore lightweight clothes and then put on the custom haptic vest. They also wore noise-canceling headphones that played white noise to block external sounds. They used a graphical user interface (GUI) shown on a touchscreen during the experiment, which visualized buttons associated with the stimulus locations similar to Fig. 3(b). The GUI did not include any labels or color codes on the buttons.

Both experiments had a within-subjects design. Each experiment included three sessions for the three belt positions: TOP, MID, and BTM. Their order was counter-balanced across participants using a Latin square. Each session consisted of five sub-sessions: familiarization 1 → practice 1 → familiarization 2 → practice 2 → main. In the familiarization sub-sessions, participants experienced each stimulus many times for two minutes by selecting the button for the stimulus location on the GUI. In the practice sub-sessions, participants were presented with a randomly chosen stimulus and asked to select the GUI button that corresponded to the perceived stimulus location among the eight buttons in REL or sixteen in R+I. After each response, correct answer feedback was provided so that participants could adjust their identification criteria. The main sub-session had the same procedure as the practice sub-session, but correct answer feedback was not provided. Each stimulus was repeated 10 times in REL and 8 times in R+I in random order. Participants took a 2-min break before each main sub-session and after each session. Each experiment was finished in 60 min for REL and 90 min for R+I, respectively.

Data Analysis. Using the obtained data, we computed the identification accuracy as the average accuracy over all stimuli and participants per condition. We also estimated the IT from the pooled data using the standard formula in [25]:

$$IT_{est} = \sum_{i=1}^{k} \sum_{j=1}^{k} \left(\frac{n_{i,j}}{n} \right) \log_2 \left(\frac{n_{i,j} \cdot n}{n_{i,*} \cdot n_{*,j}} \right), \tag{2}$$

where k is the number of stimuli used in each experiment, n is the total number of trials, $n_{i,j}$ is the number of responses to stimulus j for stimulus i, $n_{i,*}$ and

$n_{*,j}$ are computed by $n_{i,*} = \sum_{j=1}^{k} n_{i,j}$ and $n_{*,j} = \sum_{i=1}^{k} n_{i,j}$. The term $2^{IT_{est}}$ represents the number of patterns that can be perfectly identified.

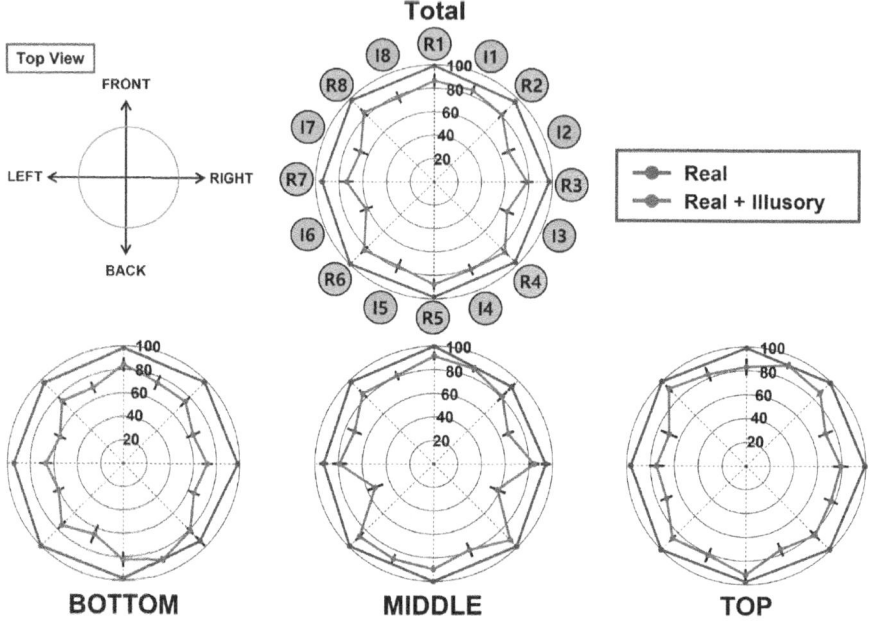

Fig. 4. Tactile azimuth identification accuracies with standard errors (Study 1).

3.2 Results

The azimuth identification accuracies are visualized in Fig. 4. The average accuracy for REL was very high at 97.7%. The accuracy was not affected significantly by the stimulus azimuth or vertical position (two-way repeated-measures ANOVA; $F(7, 77) = 1.88$, $p = 0.0841$; $F(2, 22) = 2.83$, $p = 0.0806$)[2].

When both the real and illusory stimuli were presented (R+I), the average accuracy dropped to 77.8%. A two-way repeated-measures ANOVA showed that the stimulus azimuth and vertical position were significant for the accuracy ($F(15, 165) = 4.83$, $p < 0.0001$; $F(2, 22) = 6.04$, $p = 0.0081$), but their interaction term was not ($F(30, 330) = 1.20$, $p = 0.2185$). The accuracies per vertical position were 81.3, 79.0, and 72.5% for TOP, MID, and BTM, with TOP and MID significantly higher than BTM by Tukey's HSD tests. Additionally, the result of multiple comparisons regarding azimuth indicated that I2, I3, I6, and I7, all illusory stimuli presented on the left and right sides of the torso, exhibited significantly lower accuracies than the others. Another two-way ANOVA revealed

---

[2] The participant gender was insignificant in any experiment in this paper.

that the stimulus type (real vs. illusory) and the vertical position were significant for the recognition accuracy ($F(1, 11) = 65.3$, $p < 0.0001$; $F(2, 22) = 6.04$, $p = 0.0081$). Real stimuli demonstrated a significantly higher average accuracy at 82.4% than virtual ones at 73.3% by Tukey's HSD test. Finally, the estimated information transfer, IT_{est}, was 2.83 bits (out of 3 bits) for REL and 2.91 (out of 4 bits) for R+I. These values were computed by (2) from all pooled data of the three vertical positions. They correspond to the numbers of perfectly identifiable stimuli of 7.13 and 7.52, respectively.

3.3   Discussion

Effects of Stimulus Vertical Position. The average azimuth identification accuracy was as high as 97.7% regardless of the stimulus vertical position when employing only eight real vibrations. However, the average accuracy significantly decreased to 77.8% when the stimuli included another eight illusory stimuli, while demonstrating vertical position dependence. The stimuli applied around the chest (TOP) and the solar plexus (MID) led to better identification performance than the waist (BTM). It could be because the former two body parts include bones right underneath the skin, which is advantageous for vibration propagation, whereas the latter has thicker and softer skin.

Real vs. Real + Illusory. The average identification accuracy of the eight real stimuli was 97.7% in REL but was noticeably lower at 77.8% in R+I. A reason could be that participants had 8 choices to select in REL but 16 choices in R+I, which included the same 8 real stimuli with the 8 more illusory stimuli.

In R+I, the real tactile stimuli yielded a higher average identification accuracy (82.4%) than the illusory stimuli (73.3%). One potential explanation is that the illusory stimuli elicit the positional sensation less clearly. Another possibility, although not tested in this study, is that the real stimuli might have been better aligned with the body's cardinal axes. Conducting another experiment where the positions of all real and illusory stimuli are interchanged could provide more definite results.

The four illusory stimuli (I2, I3, I6, and I7) positioned on the left and right sides of the torso resulted in significantly lower identification accuracies than the others. The vibration motors were evenly spaced in length on our belts. However, the stimuli on the left and right sides of the torso were denser in azimuth than those on the front or back due to the torso's elliptical shape. This angular density difference may account for the lower recognition accuracies of I2, I3, I6, and I7.

The results of IT_{est} provide another view. When the stimuli increased from 8 (real) to 16 (real + illusory), the number of perfectly recognizable azimuths increased from 7.13 (2.83 bits) to 7.52 (2.91 bits). These values amount to an increase of approximately 0.4 azimuths, suggesting that adding the illusory stimuli was inefficient in enhancing information transmission capacity.

In summary, incorporating real and illusory tactile stimuli in the azimuth space barely improves the information transmission, but it decreases the average recognition accuracy. For tasks requiring accurate recognition, like route navigation, relying solely on real stimuli is likely more effective.

4   Study 2: Elevation Identification

Study 2 investigated the human recognition performance of vibrotactile stimuli stimulating the torso at different elevations. Similar to study 1, study 2 included two identification experiments: one using only real stimuli (REL) and the other using real and illusory stimuli (R+I).

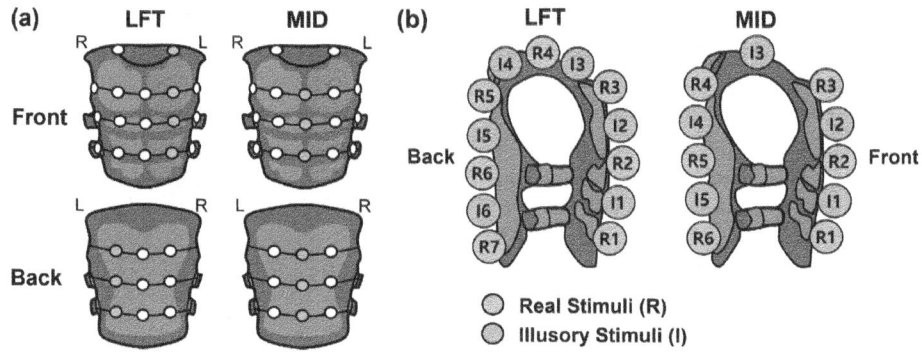

Fig. 5. Stimulus positions for the elevation identification experiments.

4.1   Methods

The methods common to Study 1 are omitted for conciseness.

Participants. The two experiments involved 12 participants each (6 males and 6 females; ages 22.8 ± 3.4 and 23.3 ± 2.0 years).

Experimental Conditions. One experiment REL used six-to-seven elevations evenly spaced on the torso (Fig. 5). The other one R+I also included illusory stimuli, each at the middle point between two real stimuli. Each experiment considered two horizontal positions of the stimuli: the center (MID) and left (LFT) of the torso. Note that LFT had a motor on the shoulder, but MID had none around the neck. The right body side was excluded, as we assumed symmetry in perception. Participants were instructed to use the GUI with their right hands. Each experimental condition was repeated ten times in both experiments.

4.2   Results

The elevation identification accuracies are shown in Fig. 6. The average accuracies for LFT and MID in REL were very high at 95.6% and 98.1%, respectively. A one-way repeated-measures ANOVA showed the elevation was significant for LFT ($F(6,66) = 2.640$, $p = 0.0235$), but it was not for MID ($F(5,55) = 1.976$, $p = 0.0966$)[3]. According to Tukey's HSD tests for LFT, the real stimulus R1 had a lower average accuracy than R2 and R3.

In R+I, the average identification accuracy decreased to 76.9% and 74.1% for LFT and MID, respectively. A one-way repeated-measures ANOVA indicated the significance of elevation for both vertical positions ($F(12,132) = 3.730$, $p < 0.0001$ on LFT; $F(10,110) = 5.222$, $p < 0.0001$ on MID). Tukey's tests showed that the stimuli positioned in the upper back (I5, R5 in LFT; R4, I4 in MID) had significantly lower accuracies than other stimuli (R2 in LFT; R2, I2, R3, I3, I5 in MID). Additionally, the real stimulus R1 had a significantly lower accuracy than R2 and R4 in LFT. In addition, one-way ANOVAs with an independent variable of stimulus type showed no significant differences between real and illusory stimuli ($F(1,11) = 0.17$, $p = 0.687$ for LFT; $F(1,11) = 0.03$, $p = 0.877$ for MID).

Finally, in REL, IT_{est} was 2.56 bits (out of 2.81) for LFT and 2.45 bits (out of 2.59) for MID. In R+I, IT_{est} was 2.67 bits (out of 3.70) for LFT and 2.39 bits (out of 3.46) for MID. These values correspond to 5.90 and 5.46 perfectly identifiable stimuli for REL and 6.34 and 5.24 stimuli for R+I, respectively.

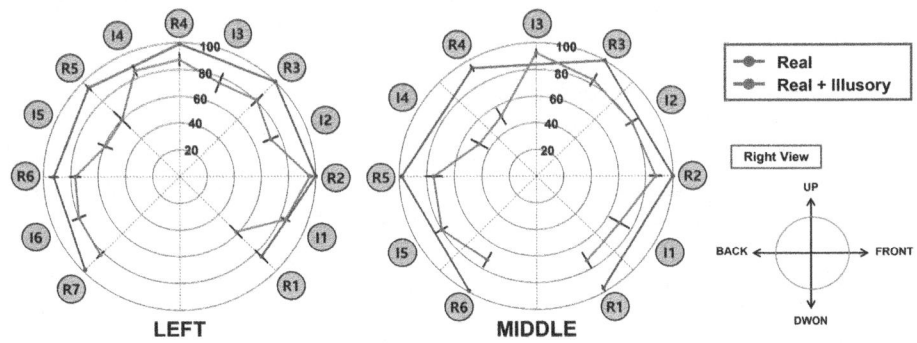

Fig. 6. Tactile elevation identification accuracies with standard errors (Study 2).

4.3   Discussion

Similar to the previous azimuth study, the average accuracies for REL were high (95.6% for LFT and 98.1% for MID), but they considerably decreased to 76.9%

---

[3] Since LFT and MID used different numbers of elevations, we present separate ANOVA results in this section.

and 74.1% in R+I. This accuracy reduction can be due to the approximately doubled number of stimuli or the addition of the illusory stimuli.

This study unveiled two torso areas on which tactile recognition is particularly inaccurate. First, in LFT of REL, the lowest tactile stimulus, R1, exhibited a significantly lower accuracy than the others. This part on the left-bottom stomach has thick and soft skin with no bones underneath, and it appears to make the identification of tactile stimuli on it difficult. Second, the stimuli positioned on the upper back in R+I exhibited particularly low average accuracies ranging from 48% to 60%. A potential reason is that the lower upper back has low tactile acuity [29], which may induce low recognition ability.

Interestingly, there were no significant differences in average accuracy between the real and illusory stimuli in R+I, whereas the opposite was true for azimuth in Study 1. It can be due to the substantially lower accuracies of both real and illusory stimuli applied on the upper back.

The IT_{est} results offer valuable insights about the elevation perception. In REL, the number of stimuli perfectly identified was 5.90 (2.56 bits) for LFT and 5.46 (2.45 bits) for MID. For R+I, the numbers were 6.34 (2.67 bits) and 5.24 (2.39 bits). Even though R+I presented more stimuli, IT_{est} was almost unchanged, indicating no practical merits in using additional illusory stimuli for elevation representation. Consequently, REL should be prioritized for precise information delivery using elevations.

5   General Discussion

5.1   Summary of Results

We measured the human identification performance of vibrotactile stimuli applied evenly on the torso along azimuth or elevation under the context of egocentric directional cueing. We considered two cases: one using only real stimuli and the other combining illusory stimuli attempting to increase the spatial resolution. The primary findings are summarized below:

1. Using only evenly-spaced real tactile stimuli, eight for azimuth and six-to-seven for elevation, allows for near-perfect recognition, regardless of their specific locations.
2. Combining the real tactile stimuli with the illusory stimuli rendered using phantom sensations slightly improves the information transfer capacity. However, the added benefit is insignificant compared with the lowered recognition accuracy of 74.1–77.8%.
3. A few body parts show considerably low recognition accuracies. They are the leftmost and rightmost parts of the torso for illusory tactile stimuli in azimuth and the upper back for both real and illusory stimuli in elevation.

These findings provide general guidelines for applications that require accurate identification of vibrotactile stimuli applied on the torso, also highlighting the need to consider the curvature and sensitivity differences across different regions.

5.2   Comparisons with Previous Studies

For a similar identification task of azimuth, vibration belts around the waist with 8 and 12 tactors resulted in 92% and 74% accuracy, respectively [6]. Here, the 8-tactor case is consistent with our result, although slightly lower. The accuracy of the 12-tactor case is similar to that of our condition of the 8 real and 8 illusory stimuli. Thus, repeating our experiment with 16 real stimuli may not improve the identification accuracy further, but it needs to be verified. Besides, the identification accuracy of 12 azimuths was 54% when expressed using static phantom sensations simultaneously by head and waist belts with 8 tactors each [26].

We were unable to find the previous studies that provide tactile elevation identification results that can be directly compared with ours.

5.3   Future Work

To improve tactile identification accuracy further, we can position vibration motors on the torso based on the somatosensory sensitivity rather than at equal spacing as tested in this study. Additionally, we plan to estimate the human recognition performance of vibrotactile stimuli applied to the torso when their positions vary along both azimuth and elevation. These results are expected to allow us to design and validate an optimal egocentric directional tactile cueing scheme for 3D applications involving spatial awareness and navigation.

Acknowledgments. This work was supported by a grant 23ZS1200 from ETRI and a Mid-Career Researcher Program (2022R1A2C2091161) of NRF.

References

1. Asif, A., Boll, S.: Where to turn my car? Comparison of a tactile display and a conventional car navigation system under high load condition. In: Proceedings of the 2nd International Conference on Automotive User Interfaces and Interactive Vehicular Applications, pp. 64–71 (2010)
2. Bajpai, A., Powell, J.C., Young, A.J., Mazumdar, A.: Enhancing physical human evasion of moving threats using tactile cues. Trans. Haptics **13**(1), 32–37 (2019)
3. Baldi, T.L., Scheggi, S., Aggravi, M., Prattichizzo, D.: Haptic guidance in dynamic environments using optimal reciprocal collision avoidance. Robot. Autom. Lett. **3**(1), 265–272 (2017)
4. Chen, H.-Y., Santos, J., Graves, M., Kim, K., Tan, H.Z.: Tactor localization at the wrist. In: Ferre, M. (ed.) EuroHaptics 2008. LNCS, vol. 5024, pp. 209–218. Springer, Heidelberg (2008). https://doi.org/10.1007/978-3-540-69057-3_25
5. Choi, S., Kuchenbecker, K.J.: Vibrotactile display: perception, technology, and applications. Proc. IEEE **101**(9), 2093–2104 (2013)
6. Cholewiak, R.W., Brill, J.C., Schwab, A.: Vibrotactile localization on the abdomen: effects of place and space. Percept. Psychophys. **66**(6), 970–987 (2004)
7. Cholewiak, R.W., Collins, A.A.: Vibrotactile localization on the arm: effects of place, space, and age. Percept. Psychophys. **65**(7), 1058–1077 (2003)
8. Van Erp, J.B.F., Van Veen, H.A.H.C., Jansen, C., Dobbins, T.: Waypoint navigation with a vibrotactile waist belt. Trans. Appl. Percept. **2**(2), 106–117 (2005)

9. Flores, G., Kurniawan, S., Manduchi, R., Martinson, E., Morales, L.M., Sisbot, E.A.: Vibrotactile guidance for wayfinding of blind walkers. Trans. Haptics **8**(3), 306–317 (2015)
10. García-Valle, G., Arranz-Paraíso, S., Serrano-Pedraza, I., Ferre, M.: Estimation of torso vibrotactile thresholds using eccentric rotating mass motors. Trans. Haptics **14**(3), 538–550 (2020)
11. Heuten, W., Henze, N., Boll, S., Pielot, M.: Tactile wayfinder: a non-visual support system for wayfinding. In: Proceedings of the 5th Nordic Conference on Human-Computer Interaction: Building Bridges, pp. 172–181 (2008)
12. bHaptics Inc.: Next generation full body haptic suit - bhaptics tactsuit (2024). https://www.bhaptics.com/. Accessed 07 Jan 2024
13. Israr, A., Poupyrev, I.: Tactile brush: drawing on skin with a tactile grid display. In: Proceedings of the SIGCHI Conference on Human Factors in Computing Systems, pp. 2019–2028 (2011)
14. de Jesus Oliveira, V.A., Brayda, L., Nedel, L., Maciel, A.: Designing a vibrotactile head-mounted display for spatial awareness in 3D spaces. Trans. Vis. Comput. Graph. **23**(4), 1409–1417 (2017)
15. Katzschmann, R.K., Araki, B., Rus, D.: Safe local navigation for visually impaired users with a time-of-flight and haptic feedback device. Trans. Neural Syst. Rehabil. Eng. **26**(3), 583–593 (2018)
16. Kaul, O.B., Rohs, M.: HapticHead: a spherical vibrotactile grid around the head for 3D guidance in virtual and augmented reality. In: Proceedings of the SIGCHI Conference on Human Factors in Computing Systems, pp. 3729–3740 (2017)
17. Lee, J.Y.F., Rajeev, N., Bhojan, A.: Goldeye: enhanced spatial awareness for the visually impaired using mixed reality and vibrotactile feedback. In: Proceedings of the 3rd ACM International Conference on Multimedia in Asia. ACM, New York (2022)
18. Liao, Z., Luces, J.V.S., Hirata, Y.: Human navigation using phantom tactile sensation based vibrotactile feedback. Robot. Autom. Lett. **5**(4), 5732–5739 (2020)
19. Monica, R., Aleotti, J.: Improving virtual reality navigation tasks using a haptic vest and upper body tracking. Displays **78**, 102417 (2023)
20. Park, G., Choi, S.: Tactile information transmission by 2D stationary phantom sensations. In: Proceedings of the SIGCHI Conference on Human Factors in Computing Systems, pp. 1–12. no. 258. ACM (2018). https://doi.org/10.1145/3173574.3173832
21. Pascher, M., Franzen, T., Kronhardt, K., Gruenefeld, U., Schneegass, S., Gerken, J.: HaptiX: vibrotactile haptic feedback for communication of 3D directional cues. In: Proceedings of the Extended Abstracts SIGCHI Conference on Human Factors in Computing Systems, pp. 1–7 (2023)
22. Salazar, J., Okabe, K., Hirata, Y.: Path-following guidance using phantom sensation based vibrotactile cues around the wrist. Robot. Autom. Lett. **3**(3), 2485–2492 (2018)
23. Schaack, S., Chernyshov, G., Ragozin, K., Tag, B., Peiris, R., Kunze, K.: Haptic collar: vibrotactile feedback around the neck for guidance applications. In: Proceedings of the 10th Augmented Human International Conference, pp. 1–4 (2019)
24. Skinetic: Skinetic by actronika (2024). https://www.skinetic.actronika.com/. Accessed 26 Jan 2024
25. Tan, H.Z., Choi, S., Lau, F.W., Abnousi, F.: Methodology for maximizing information transmission of haptic devices: a survey. Proc. IEEE **108**(6), 945–965 (2020)

26. Tawa, S., Nagano, H., Tazaki, Y., Yokokohji, Y.: Three-dimensional position presentation via head and waist vibrotactile arrays. Trans. Haptics **17**(3), 319–333 (2024)
27. Teslasuit: Teslasuit | meet our haptic VR suit and glove with force feedback (2024). https://teslasuit.io/. Accessed 07 Jan 2024
28. Tsukada, K., Yasumura, M.: ActiveBelt: belt-type wearable tactile display for directional navigation. In: Davies, N., Mynatt, E.D., Siio, I. (eds.) UbiComp 2004. LNCS, vol. 3205, pp. 384–399. Springer, Heidelberg (2004). https://doi.org/10.1007/978-3-540-30119-6_23
29. Weinstein, S.: Intensive and extensive aspects of tactile sensitivity as a function of body part, sex and laterality. In: The First International Symposium on the Skin Senses (1968)
30. Woojer: Woojer - born to feel - woojer vest 3 (2024). https://www.woojer.com/. Accessed 26 Jan 2024

Utilizing Absence of Pacinian Corpuscles in the Forehead for Amplitude-Modulated Tactile Presentation

Yuma Akiba$^{(\boxtimes)}$, Shota Nakayama$^{(\boxtimes)}$, Keigo Ushiyama$^{(\boxtimes)}$, Izumi Mizoguchi$^{(\boxtimes)}$, and Hiroyuki Kajimoto$^{(\boxtimes)}$

The University of Electro-Communications, 1-5-1 Chofugaoka, Chofu, Tokyo, Japan
{yuma.akiba,nakayama,ushiyama,mizoguchi,kajimoto}@kaji-lab.jp

Abstract. This study proposes a method for tactile presentation to the forehead, focusing on the low-frequency band with a small linear resonant actuator. We utilized the absence of the Pacinian corpuscles, responsible for sensing vibrations around 200 Hz in the forehead, and amplitude modulation to compensate for the challenge of the limited frequency range of linear resonant actuators. The amplitude modulation was achieved by around 200-Hz carrier wave around the actuator's resonant frequency. In two experiments, we investigated the efficacy of amplitude modulation in representing low-frequency vibrations (Experiment 1) and the quality of vibration (Experiment 2) on the forehead. We found that participants could clearly feel the original low-frequency vibration on the forehead compared to other body locations.

Keywords: Amplitude modulation · Forehead · Linear resonant actuator · Vibration

1 Introduction

The increasing demand for virtual reality (VR) experiences, fueled by the widespread adoption of head-mounted displays (HMDs), poses a significant challenge: the lack of tactile information in many widely used VR applications, which is needed to enrich the sense of presence. Researchers have been exploring wearable tactile presentation devices to address this challenge [1–3]. However, the time-consuming nature of wearing many hand-worn devices prompts the investigation of integrating a tactile presentation device into HMD [4–7]. The proposed tactile presentation targets the "forehead" area, where the HMD and skin interact, offering feedback on the sensation when a user's fingertips touch an object in the VR environment [4].

Integrating a small tactile presentation device into HMD is necessary for compact and concise tactile presentation, but when using a widely employed linear resonant actuator (LRA), a specific issue related to the forehead emerges. While the LRA is a compact yet powerful vibration device, the available frequency range is narrow around the resonant frequency (typically around 200 Hz) since LRAs generate vibrations using mechanical

H. Kajimoto et al. (Eds.): EuroHaptics 2024, LNCS 14768, pp. 16–28, 2025.
https://doi.org/10.1007/978-3-031-70058-3_2

resonance phenomena. On the flip side, the forehead is a unique part of the body, and it is known that there are no Pacinian corpuscles that sense vibrations at around 200 Hz [8–10]. Consequently, presenting tactile sensations on the forehead using a small LRA is challenging. Myles and Kerb [11] also reported that vibration above 150 Hz becomes uncomfortable. In fact, most previously proposed tactile presentation devices for HMD were large and primarily delivered shock or pressure sensations [12–15].

While there has been research on compact tactile presentation devices capable of delivering vibrations in the low-frequency range [16, 17], they are not suitable for long-term use. The device created by Manabe et al. [16] employed bonded permanent magnets that repel each other, leading to a potential divergence over time, indicating a lack of structural resilience. In contrast, the device developed by Liu et al. [17] featured a robust design, though the risk of skin burns with extended usage remained a concern.

On another note, a technique for presenting low-frequency sensations by modulating high-frequency vibrations has been known for hand tactile presentation. Specifically, amplitude modulation (AM) is employed to represent the low-frequency signal to be presented as an envelope wave using a carrier wave [18, 19]. However, although the frequency of the envelope can be perceived, the high-frequency component of the carrier wave is also sensed, making the sensation noisy, and it is known that the tactile sensation differs from genuine low-frequency sensations [20].

This study proposes a novel method for delivering the *pure* low-frequency sensation (i.e., people sense only low-frequency sensation) to the forehead using a small LRA by utilizing the forehead haptic characteristic: *the absence of Pacinian corpuscles*. Although this can be a disadvantage for haptic presentation in general, we leveraged this feature to present the low-frequency vibration by employing amplitude modulation as previously suggested for hand tactile presentation. Namely, even though high frequency (around 200 Hz) is used as a carrier wave, the pure low-frequency vibration should be more clearly perceived since the forehead hardly senses higher frequencies.

The objective of this study is to investigate whether amplitude modulation with a carrier wave at around 200 Hz—the typical resonant frequency of LRAs—can effectively depict low-frequency vibrations on the forehead, ranging from a few Hz to several tens of Hz. Additionally, the high-frequency component will not be picked up by the afferents in the forehead because there are no fast-adapting type II neurons innervating Pacinian corpuscles in the forehead, so it should be easier to convey low-frequency signals through the forehead as compared to other body locations.

Two experiments were conducted to achieve these objectives. The first experiment evaluated the frequency perception induced by amplitude-modulated waves on tactile sensation using the method of adjustment. Sinusoidal vibrations were presented to the fingerpads of the non-dominant hand while sinusoidal vibrations or amplitude-modulated vibrations were presented to the index fingerpad or forearm of the dominant hand or forehead. The perceived frequencies were adjusted by the non-dominant hand to match amplitude-modulated vibration. The adjustment was performed by pressing the arrow keys on a keyboard using fingers other than the thumb and index finger of the non-dominant hand. In the second experiment, we evaluated the subjective quality of tactile perception when exposed to amplitude-modulated waves. We anticipated that the high-frequency component around 200 Hz, constituting the modulated wave, should be strongly perceived at the fingerpad and forearm due to the presence of Pacinian corpuscles. In contrast, on the forehead, the high-frequency component is not perceived, and the low-frequency component—the original signal—is expected to be perceived more distinctly.

2 Apparatus

The device producing the amplitude-modulated vibration signals was a small LRA (VG0640001D, Vybronics), as depicted in Fig. 1a. This device, with a diameter of 6 mm, was attached to the center of the forehead, index fingerpad, and ventral part of forearm for experiments, as illustrated in Fig. 1b. The choice of the index fingerpad was based on its common usage for tactile presentation due to its high spatial resolution of tactile sensation. The forearm was selected as a site where Pacinian corpuscles are housed but with relatively lower spatial resolution than the index fingerpad. Essentially, the forearm was chosen to investigate whether the results of the forehead experiment stemmed from low tactile receptor density or the absence of Pacinian corpuscles.

To compare sinusoidal and amplitude-modulated vibration signals in Experiment 1, we also used a Haptuator [21] (TL002–14-A, Tactile Labs, Fig. 1a) to present simple sinusoidal vibrations. Sinusoidal wave signals were generated by a custom program (Processing), and the signals were input to the vibration actuator through an audio interface (UltraLite mk5, MOTU) and a stereo audio amplifier (FX-AUDIO-, FX202A/FX-36A PRO, North Flat Japan). In contrast, the amplitude-modulated vibration signals were generated by a function generator (DG812, RIGOL) and amplified by the audio amplifier. The amplitude modulation wave in this study is expressed by the following equationdisple

$$g(t) = A \sin(2\pi f_e t) \sin(2\pi f_c t), \tag{1}$$

where, f_e represents the frequency of the original signal, set in this study between 2 Hz and 32 Hz. f_c stands for the carrier frequency of amplitude modulation, utilizing the resonant frequency of the LRA (210 Hz). Through amplitude modulation, the original signal's frequency is theoretically altered by the carrier wave, and the signal $g(t)$ contains frequencies $f_c \pm f_e$.

Before conducting experiments, we confirmed whether the vibration actuators (LRA and the Haptuator) with wave signals used in two experiments could output perceivable intensity. The vibrations were measured by an accelerometer (BMX055, BOSCH), and the measurement was conducted by one author. For measurement, the LRA was driven by an amplitude-modulated signal, and the Haptuator was driven by a sinusoidal wave. The vibration amplitudes were measured with a constant signal amplitude where participants could clearly perceive vibration at 2, 4, 8, 16, and 32 Hz. The measurements were conducted on the same body locations as in the two experiments (see Table 1).

The result is illustrated in Fig. 2. The result indicates an almost consistent acceleration at the index fingerpad, forearm, and forehead. The condition of the Haptuator at 2 Hz has its own actuator limit and acceleration was decreased. Based on this result, the amplifier volume used for experiments remained constant across all participants throughout the two experiments and was not individually adjusted.

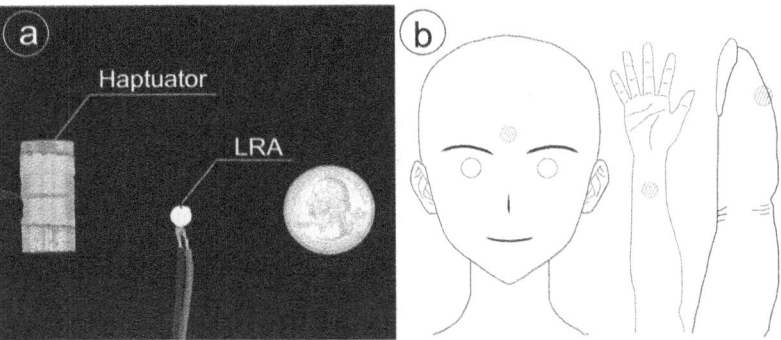

Fig. 1. Experimental conditions: (a) Apparatus used in experiments. Left: TL002–14-A, Tactile Labs, Middle: VG0640001D, Vybronics. (b) Body site, center of the forehead; ventral part of forearm; and index fingerpad. The shaded circles were the presentation sites.

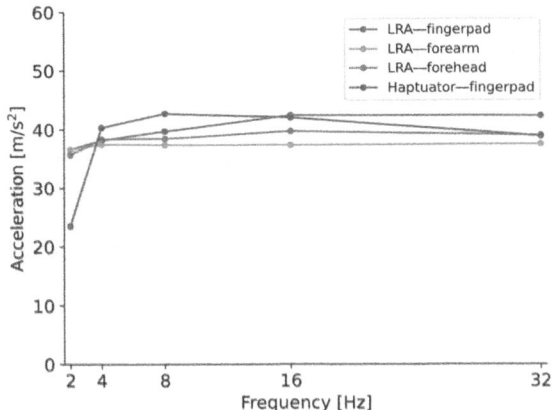

Fig. 2. Result of acceleration measurements at each frequency used in two experiments. Amplitude modulation signals were applied to LRA, and sinusoidal signals were applied to the Haptuator.

3 Experiment 1

3.1 Experimental Procedure

In this experiment, we measured how people perceive the frequency of amplitude-modulated waves when applied to the index fingerpad, forearm, and forehead. We recruited 11 participants aged 22–24 from our university. All experiments were conducted with the approval of the ethics committee of The University of Electro-Communications (No. 23076).

We asked participants to adjust the frequency of sinusoidal vibration using the Haptuator on the non-dominant hand so that it would match the frequency of vibration perceived in the other location. The stimulation conditions are summarized in Table 1. A summary figure corresponding to Table 1 is shown in Fig. 3. We presented amplitude-modulated vibrations with the small LRA at the index fingerpad on the dominant-hand side (Finger condition), forearm on the dominant-hand side (Arm condition), and fore-head (Head condition). The LRA was secured to the body with surgical tapes (Transpore, 3M). We also presented sinusoidal vibrations with the Haptuator. It was grasped with the index finger and thumb on the dominant-hand side (Haptuator condition). The condition, where the Haptuator driven by a sinusoidal wave was used to explore frequency perception instead of the LRA, was added as a baseline to check if participants could answer the perceived frequency in the experiment setup. Therefore, two Haptuators (i.e., one in the dominant hand and one in the non-dominant hand) were used in the Haptuator condition.

The frequency was adjusted by pressing the arrow keys on the keyboard using fingers other than the thumb and index finger of the non-dominant hand as illustrated in Fig. 4. Pressing the up-arrow button increased the frequency by 1 Hz, and pressing the down-arrow button decreased it by 1 Hz. When reaching the desired frequency, the user finalized it by pressing Enter. Participants regulated their visual focus by keeping their eyes closed, except when necessary for locating or manipulating buttons.

There were five stimulation frequencies, which were 2, 4, 8, 16, and 32 Hz. They were modulated by a 210-Hz (resonant frequency of the LRA) carrier wave to drive the LRA. The duration of stimulation was unlimited, and stimulation continued until the participants had finished adjusting the frequency. Participants adjusted the vibration frequency of the Haptuator twice. The adjustment was started from 1 Hz or 300 Hz for each trial. The order of the stimulation frequency (2, 4, 8, 16, and 32 Hz) and the starting frequency (1 Hz or 300 Hz) was randomized. The order of the condition (Finger, Arm, Head, Haptuator) was also randomized. The average frequency (starting from 1 Hz and 300 Hz) was calculated as a result of each stimulation frequency. There were no limits on adjustment time, and it was not measured. Participants wore headphones and listened to white noise to mask bone conduction and ambient noise.

Prior to the measurement at each stimulation frequency, participants conducted two practice trials in the same way as the measurement. At the end of the experiment, participants were also able to provide open comments about the experiment in general. Participants were not given feedback on how accurate their answers were.

Table 1. Experimental conditions

Condition	Vibrator	Body site	Signal	Frequency [Hz] (original signal)
Finger	LRA	Fingerpad	Amplitude-modulated wave	2,4,8,16,32
Arm	LRA	Forearm	Amplitude-modulated wave	2,4,8,16,32
Head	LRA	Forehead	Amplitude-modulated wave	2,4,8,16,32
Haptuator	Haptuator	Fingerpads	Sinusoidal wave	2,4,8,16,32

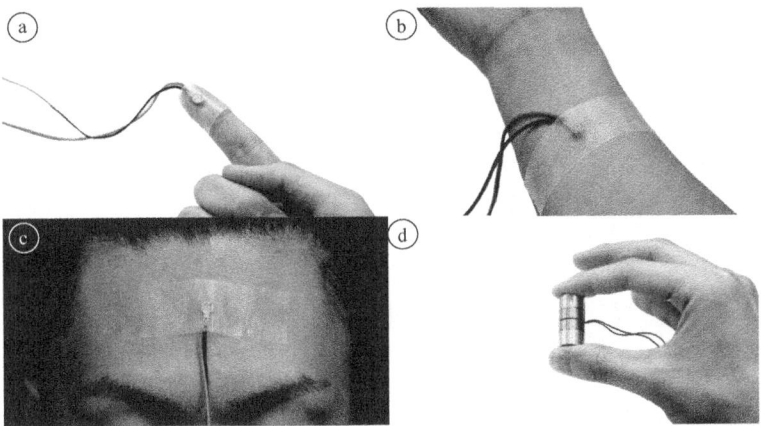

Fig. 3. A summary figure corresponding to Table 1. (a) Finger, (b) Arm, (c) Head, and (d) Haptuator. The LRA was secured to the body with surgical tapes (Transpore, 3M). The Haptuator was grasped with the index finger and thumb. This shows an example of a right-handed participant. The non-dominant hand should have the Haptuator in all conditions, and the frequency of it should be adjusted.

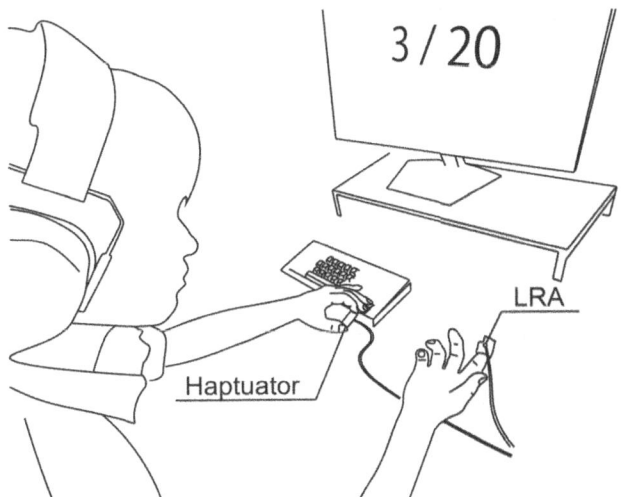

Fig. 4. An example scene of the experiment with a right-handed participant (Finger condition). The monitor showed the number of times remaining in the experiment, but participants were instructed to keep their eyes closed except when operating the buttons as much as possible.

3.2 Result

The perceived frequency results for each set frequency and each site are illustrated in Fig. 5. For each stimulation frequency condition (2, 4, 8, 16, and 32 Hz), the Friedman test indicated the main effects for the 4 Hz ($p < 0.001$), 8 Hz ($p < 0.01$), and 16 Hz ($p < 0.001$) conditions. Multiple comparisons with Bonferroni correction revealed 5%

significant differences for the Arm condition and the Head condition at 4 Hz (p = 0.046), the Arm condition and the Head condition at 16 Hz (p = 0.023), and the Arm condition and the Haptuator condition at 16 Hz (p = 0.035).

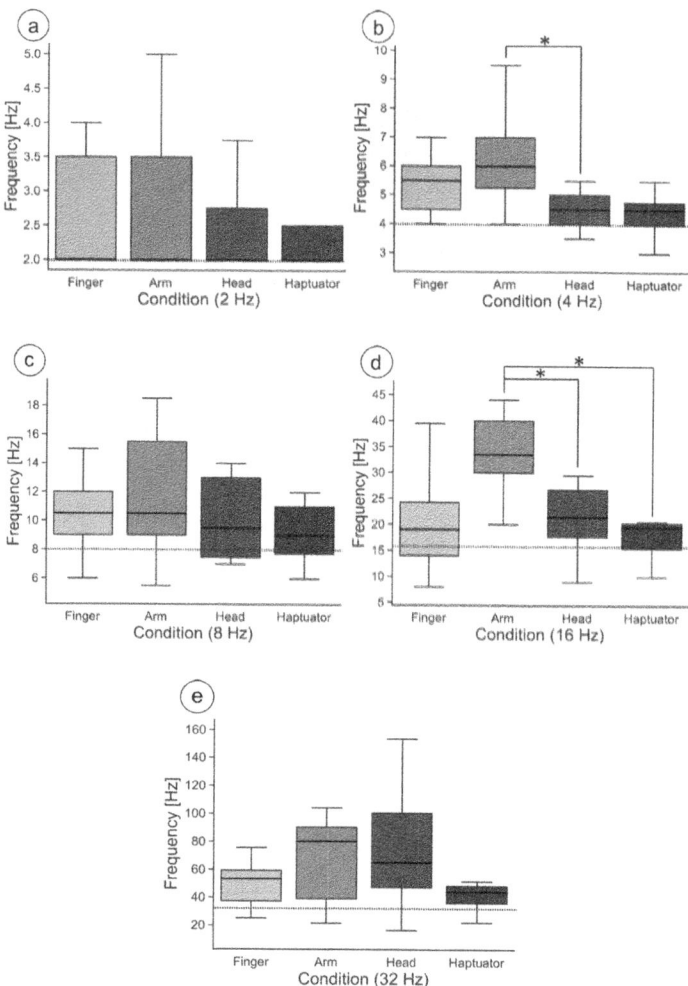

Fig. 5. Perceived frequency results. The dotted lines are the reference vibrotactile stimulus. The green-colored graphs show the group that was stimulated with amplitude-modulated vibration using LRA. The purple-colored graph shows that it was stimulated with sinusoidal vibration using the Haptuator set as the baseline. (a) 2 Hz, (b) 4 Hz, (c) 8 Hz, (d) 16 Hz, and (e) 32 Hz (*: p < 0.05)

Following the experiment, some comments were obtained such as "Lower frequencies were easier to match", "Matching frequencies with the forehead was easier than with the forearm", and "Matching frequencies with the Haptuator (sinusoidal wave) was the easiest". As other comments, some participants also commented, "The task of adjusting frequencies itself was challenging" and "Starting with a higher trial frequency made it harder to match than starting with a lower frequency".

4 Experiment 2

4.1 Experimental Procedure

In this experiment, we examined how people perceived the quality of sensations of amplitude-modulated vibration on their index fingerpad, forearm, and forehead. The purpose of this experiment was to qualitatively assess whether the participants purely felt the low frequencies. Ten participants aged 22–24 took part in this experiment (eight of them also joined Experiment 1). We applied the same vibratory stimulation as in Experiment 1 (i.e., Finger, Arm, and Head condition in Table 1. Five amplitude-modulated vibration conditions: 2, 4, 8, 16, and 32 Hz. The carrier frequency was 210 Hz) using the same devices. The duration of stimulation was 10 s. Participants then evaluated the tactile sensations using a 9-point Likert scale from 1 to 9 for the following attributes: (A) speed (slow-fast), (B) smoothness (bumpy-smooth), (C) hardness (soft-hard), (D) strongness (weak-strong), (E) distinctness (vague-distinct), (F) weight (light-heavy), (G) clarity (dull-clear), and (H) preference (dislike-like) of vibration [22, 23]. Participants wore headphones and listened to white noise to mask auditory cues. The order of the amplitude modulation frequency and the presenting body site was randomized. At the end of the experiment, participants were also able to provide open comments about their sensations and the experiment in general.

4.2 Result

The results are displayed in Fig. 6. It depicts the answers regarding the quality of vibration at different frequencies and sites. The Friedman test revealed the main effects for distinctness ($p < 0.05$) and clarity ($p < 0.01$) at 2 Hz, distinctness ($p < 0.05$) and clarity ($p < 0.05$) at 4 Hz, distinctness ($p < 0.001$) and clarity ($p < 0.001$) at 8 Hz, and smoothness ($p < 0.01$), distinctness ($p < 0.01$), and clarity ($p < 0.001$) at 16 Hz. For multiple comparisons with Bonferroni correction, 5% significant differences were observed for forearm and forehead at 2-Hz distinctness ($p = 0.041$), forearm and forehead at 2-Hz clarity ($p = 0.026$), forearm and forehead at 4-Hz distinctness ($p = 0.044$), fingerpad and forehead ($p = 0.017$), forearm and forehead at 4-Hz clarity ($p = 0.026$), fingerpad and forehead ($p = 0.026$), forearm and forehead at 8-Hz distinctness ($p = 0.017$), fingerpad and forehead ($p = 0.036$), forearm and forehead at 8-Hz clarity ($p = 0.025$), forearm and forehead at 16-Hz smoothness ($p = 0.048$), fingerpad and forehead ($p = 0.03$), forearm and forehead at 16-Hz distinctness ($p = 0.039$), and fingerpad and forehead ($p = 0.026$), forearm and forehead at 16-Hz clarity ($p = 0.026$).

After the experiment, participants commented, "The tactile sensation felt on the forehead seemed quite different from that on the forearm and fingerpad". Additionally, they remarked, "On the forehead, it felt softer and smoother," and they perceived, "The higher frequency felt faster in speed." They found, "The fingerpad and forearm had similar vibration sensations."

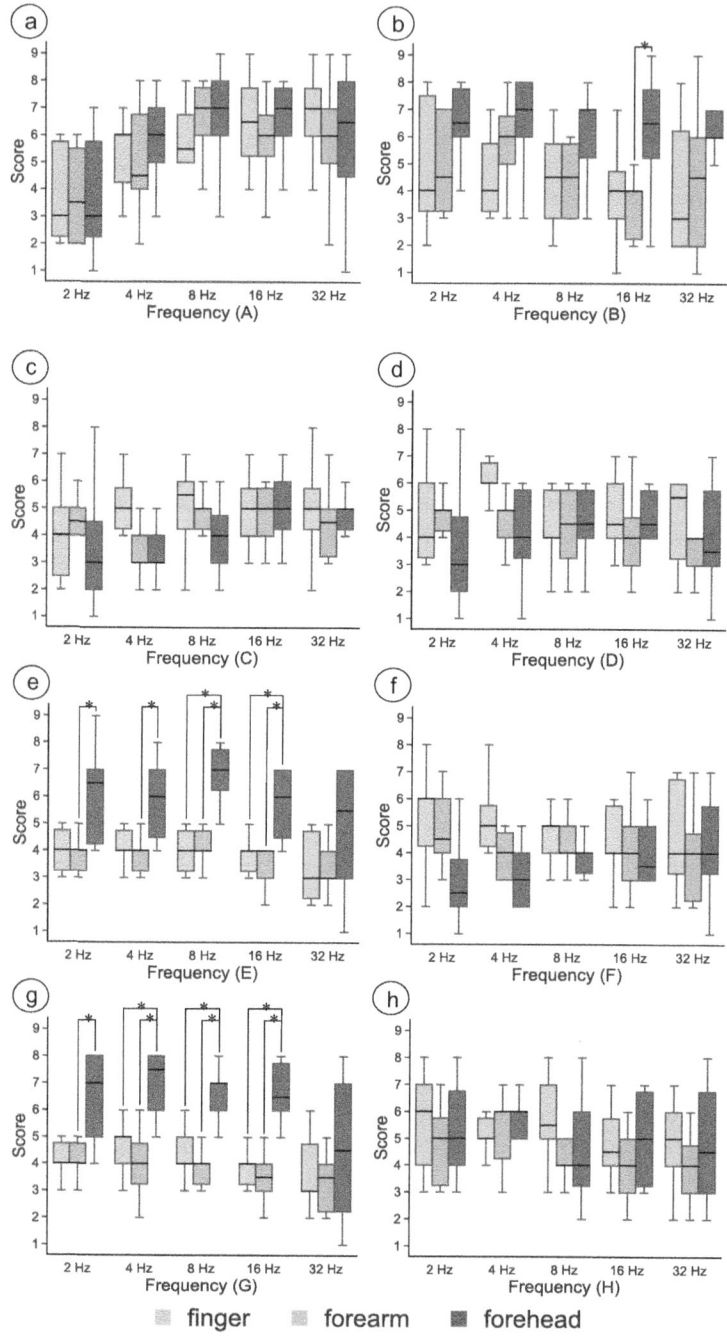

Fig. 6. Results of vibration quality responding with adjective pairs. (a) A: speed (slow-fast), (b) B: smoothness (bumpy-smooth), (c) C: hardness (soft-hard), (d) D: strongness (weak-strong), (e) E: distinctness (vague-distinct), (f) F: weight (light-heavy), (g) G: clarity (dull-clear), and (h) H: preference (dislike-like) (*: $p < 0.05$)

5   Discussion

The result of Experiment 1 indicated that the accuracy (i.e., the deviation from the expected frequency) in perceiving vibrations on the index fingerpad and forehead was comparable. This finding is relevant in light of previous studies that have shown that the envelope of amplitude-modulated waves can be identified in the fingerpad when the signal wave is below 50 Hz [24].

Frequency perception showed a slight increase overall. For example, the 4 Hz amplitude-modulated wave was scored around a 5 to 6 Hz sinusoidal wave for the fingerpad and forearm. The 8 Hz amplitude-modulated wave was scored around a 10 Hz sinusoidal wave for the fingerpad and forearm, and so on. The reason why the perceived frequency values exceeded the predicted frequencies overall could be due to the inclusion of the high default frequency (i.e., the carrier frequency, 210 Hz). Moreover, it is possible that the fact that there was a starting frequency of 300 Hz in the experimental task affected the results of the experiment. In the experiment, there were two adjustment patterns: one from 1 Hz and the other from 300 Hz. Among these, the pattern of adjusting from 300 Hz required holding the button down from 2 Hz to 32 Hz, which might have been more difficult than adjusting from 1 Hz. Furthermore, the observation that several participants mentioned in their comments, "Starting with a higher trial frequency made it harder to match than starting with a lower frequency" indicates that it might have been difficult to adjust from a higher frequency.

Interestingly, the forearm demonstrated lower frequency perception accuracy than the other locations overall. The superior accuracy in the index fingerpad compared to the forearm may stem from the index fingerpad's higher receptor density [25]. The discussion revolves around the higher accuracy of the forehead compared to the forearm in frequency perception. The vibration signal modulated by amplitude modulation with a carrier wave around 200 Hz was perceived as noisy on the forearm, accompanied by the perception of the carrier wave itself. In contrast, because the Pacinian corpuscles, sensitive to vibrations around 200 Hz, are missing, the forehead perceived a higher percentage of the original data than the forearms, resulting in a higher frequency perception accuracy.

Notably, the result of the forehead for 32 Hz has a much larger deviation. This challenge might also be attributed to the absence of Pacinian corpuscles in the forehead. Fast-adapting I type neurons (FA I), slowly-adapting I type neurons (SA I), and slowly-adapting II (SA II) likely played a role in perceiving frequency because there are no fast-adapting II type neurons (FA II) innervating Pacinian corpuscles in the forehead. Among them, SA I and SA II are primarily involved in perceiving frequency in the forehead since the majority of the afferents in the face are slowly-adapting neurons [8]. The responsiveness of SA I and SA II is nearly identical, being more responsive to vibrations below 15 Hz. At frequencies higher than about 20 Hz, the perceived frequency thresholds increase linearly [26]. These could be attributed to the fact that 32 Hz vibrations were difficult to perceive in the forehead. On the flip side, hairy skin such as the face and forearms have hair follicle afferents that are sensitive to around 30 Hz in the epidermis [27]. Considering that the face has a high density of hair follicle afferents sensitive to around 30 Hz among them [28, 29], it is possible that 32 Hz was easily perceived due to hair follicle afferents, but the results of the experiment did not suggest this. Additionally, the significance of afferent types is debatable, as spike pattern low-frequencies can activate Pacinian corpuscles [30]. However, it is possible that the low-frequency vibration in this experiment was perceived as the sinusoidal vibration rather than the spike-pattern vibration in the forehead, which awaits further clarification.

It is also possible that bone conduction occurred when the vibration was applied to the forehead, making it easier to adjust the frequency. However, bone conduction might have been masked by white noise from the headphones. No one commented that bone conduction made it easier to adjust the frequency in the forehead, and the human forehead is comparatively known to have a higher threshold of bone-conducted sounds than the condyle and vertex under white noise conditions [31].

In Experiment 2, the forehead consistently outscored the fingerpad and forearm in terms of smoothness, clarity, and distinctness. This suggests that the forehead exhibited reduced sensitivity to noisy vibrations induced by an amplitude-modulated carrier wave and helped capture vibrations more faithfully to the original stimulus. The participants' comments underscored the unique tactile experience on the forehead.

These experiments demonstrated that the participants could vividly feel the envelope of amplitude-modulated vibration (i.e., modulation frequency) at the forehead more than at other body locations as it involves the absence of Pacinian corpuscles. This result emphasizes the viability of employing a relatively compact LRA for low-frequency stimulation on the forehead through amplitude modulation.

Current research has several limitations. Foremost among them is our incomplete understanding of perceptual dynamics for frequencies below 2 Hz. When we tested the amplitude-modulated 1-Hz signal, the resultant forehead vibrations lacked the anticipated smoothness. Although Experiment 1 was conducted by adjusting frequencies, experiments on frequency discrimination perception should also be conducted to verify the ability to present frequencies. Furthermore, the mechanism under the demodulation of amplitude-modulated vibration signals on the forehead remains elusive. Insufficient contact between the vibrator and the skin could result in a mechanism in which the vibrator intermittently "taps" the skin, where the amplitude modulation signal is demodulated. However, in the present situation, such inadequate contact is unlikely to happen since the actuator was taped to the forehead. Consequently, demodulation likely happened at the receptor or neural activity level or central integration. Lastly, although our investigation centered on the forehead, the potential contact points of an HMD include the cheeks, temples, crown, and back of the head. Whether our method can be applied to such body sites should be further investigated. It would also be useful to run the detection threshold measurement on the forehead and fingerpad using the amplitude-modulated wave.

6 Conclusion

This study investigated the feasibility of implementing low-frequency vibration representation on the forehead through amplitude modulation using a high-frequency carrier wave to integrate it into an HMD. We explored frequency perception on the forehead, fingerpad, and forearm for modulation frequencies ranging from 2 Hz to 32 Hz. The results indicate that, within the 2 Hz to 16 Hz range, the forehead and fingerpad exhibited similar frequency perception capability, and participants perceived frequencies more accurately at the forehead than at the forearm. Including the comments from the participants, it is suggested that the forehead could extract pure low-frequency vibrations differently from the fingerpad and forearm.

Future research aims to validate this approach in other HMD-contacting areas. In addition, small LRAs will be integrated into the HMD and used to represent haptic events in the VR experience. In this way, we will continue to develop and apply the technology to be integrated into HMD in the future.

Acknowledgements. This research was supported by JSPS KAKENHI Grant Number JP20K20627.

References

1. Hinchet, R., Vechev, V., Shea, H., Hilliges, O.: DextrES: wearable haptic feedback for grasping in VR via a thin form-factor electrostatic brake. In: Proceedings of the ACM Symposium on User Interface Software and Technology (UIST), pp. 901–912 (2018)
2. Choi, I., Hawkes, E.W., Christensen, D.L., Ploch, C.J., Follmer, S.: Wolverine: a wearable haptic interface for grasping in virtual reality. In: Proceedings of the IEEE/RSJ International Conference on Intelligent Robots and Systems (IROS), pp. 986–993 (2016)
3. Zhang, Z.-Y., Chen, H.-X., Wang, S.-H., Tsai, H.-R.: ELAXO: rendering versatile resistive force feedback for fingers grasping and twisting. In: Proceedings of the ACM Symposium on User Interface Software and Technology (UIST), pp. 1–14 (2022)
4. Kameoka, T., Kajimoto, H.: Design of suction-type tactile presentation mechanism to be embedded in HMD. Front. Virtual Real. **3**, 894873 (2022)
5. Peiris, R.L., Peng, W., Chen, Z., Chan, L., Minamizawa, K.: ThermoVR: exploring integrated thermal haptic feedback with head mounted displays. In: Proceedings of the ACM SIGCHI Conference on Human Factors in Computing Systems, pp. 5452–5456 (2017)
6. Shen, V., Shultz, C., Harrison, C.: Mouth haptics in VR using a headset ultrasound phased array. In: Proceedings of the ACM SIGCHI Conference on Human Factors in Computing Systems, pp. 1–14 (2022)
7. Ranasinghe, N., et al.: Season traveler: multisensory narration for enhancing the virtual reality experience. In: Proceedings of the ACM SIGCHI Conference on Human Factors in Computing Systems, pp. 1–13 (2018)
8. Johansson, R.S., Trulsson, M., Olsson, K.Å., Westberg, K.-G.: Mechanoreceptor activity from the human face and oral mucosa. Exp. Brain Res. **72**, 204–208 (1988)
9. Siemionow, M.Z., Gharb, B.B., Rampazzo, A.: The Face as a Sensory Organ, pp. 11–23. In: Siemionow, M. (eds.) The Know-How of Face Transplantation. Springer, London (2011). https://doi.org/10.1007/978-0-85729-253-7_2
10. Barlow, S.M.: Mechanical frequency detection thresholds in the human face. Exp. Neurol. **96**, 253–261 (1987)
11. Myles, K., Kalb, J.T.: Guidelines for Head Tactile Communication: Defense Technical Information Center, report no. ARL-TR-5116 (2010)
12. Rietzler, M., et al.: VaiR: simulating 3D airflows in virtual reality. In: Proceedings of the ACM SIGCHI Conference on Human Factors in Computing Systems, pp. 5669–5677 (2017)
13. Tsai, H.-R., Chen, B.-Y.: ElastImpact: 2.5D multilevel instant impact using elasticity on head-mounted displays. In: Proceedings of the ACM Symposium on User Interface Software and Technology (UIST), pp. 429–437 (2019)
14. Chang, H.-Y., et al.: FacePush: introducing normal force on face with head-mounted displays. In: Proceedings of the ACM Symposium on User Interface Software and Technology (UIST), pp. 927–935 (2018)

15. Liu, S.-H., et al.: HeadBlaster: a wearable approach to simulating motion perception using head-mounted air propulsion jets. ACM Trans. Graph. (TOG) **39**(4), 84–91 (2020)

16. Manabe, M., Ushiyama, K., Takahashi, A., Kajimoto, H.: Energy efficient wearable vibrotactile transducer utilizing the leakage magnetic flux of repelling magnets. In: Proceedings of the IEEE Conference on Virtual Reality and 3D User Interfaces Abstracts and Workshops (VRW), pp. 599–600 (2023)

17. Liu, Y., et al.: Skin-integrated haptic interfaces enabled by scalable mechanical actuators for virtual reality. IEEE Internet Things J. **10**(1), 653–663 (2022)

18. Weisenberger, J.M.: Sensitivity to amplitude-modulated vibrotactile signals. J. Acoust. Soc. Am. **80**(6), 1707–1715 (1986)

19. Park, G., Choi, S.: Perceptual space of amplitude-modulated vibrotactile stimuli. In: Proceedings of the IEEE World Haptics Conference (WHC), pp. 59–64 (2011)

20. Kim, T., Shim, Y.A., Lee, G.: Heterogeneous stroke: using unique vibration cues to improve the wrist-worn spatiotemporal tactile display. In: Proceedings of the ACM SIGCHI Conference on Human Factors in Computing Systems, pp. 1–12 (2021)

21. Yao, H.-Y., Hayward, V.: Design and analysis of a recoil-type vibrotactile transducer. J. Acoust. Soc. Am. **128**(2), 619–627 (2010)

22. Hwang, I., Choi, S.: Perceptual space and adjective rating of sinusoidal vibrations perceived via mobile device. In: Proceedings of the IEEE Haptics Symposium (HAPTICS), pp. 1–8 (2010)

23. Park, G., et al.: Tactile effect design and evaluation for virtual buttons on a mobile device touchscreen. In: Proceedings of the ACM International Conference on Human Computer Interaction with Mobile Devices and Services, pp. 11–20 (2011)

24. Makino, Y., Maeno, T., Shinoda, H.: Perceptual characteristic of multi-spectral vibrations beyond the human perceivable frequency range. In: Proceedings of the IEEE World Haptics Conference (WHC), pp. 439–443 (2011)

25. Johansson, R.S., Vallbo, A.B.: Tactile sensibility in the human hand: relative and absolute densities of four types of mechanoreceptive units in glabrous skin. J. Physiol. **286**(1), 283–300 (1979)

26. Toma, S., Nakajima, Y.: Response characteristics of cutaneous mechanoreceptors to vibratory stimuli in human glabrous skin. Neurosci. Lett. **195**(1), 61–63 (1995)

27. Schmidt, D., Schlee, G., Milani, T.L., Germano, A.M.C.: Vertical contact forces affect vibration perception in human hairy skin. PeerJ **11**, e15952 (2023)

28. Szabo, G.: The regional anatomy of the human integument with special reference to the distribution of hair follicles, sweat glands and melanocytes. Philos. Trans. R. Soc. Lond. B Biol. Sci. **252**(779), 447–485 (1967)

29. Otberg, N., et al.: Variations of hair follicle size and distribution in different body sites. J. Invest. Dermatol. (JID) **122**(1), 14–19 (2004)

30. Birznieks, I., et al.: Tactile sensory channels over-ruled by frequency decoding system that utilizes spike pattern regardless of receptor type. Elife **8**, e46510 (2019)

31. McBride, M., Letowski, T., Tran, P.: Bone conduction reception: head sensitivity mapping. Ergonomics **51**(5), 702–718 (2008)

Optimizing Haptic Feedback in Virtual Reality: The Role of Vibration and Tangential Forces in Enhancing Grasp Response and Weight Perception

Yunxiu Xu[✉][ID], Siyu Wang[ID], and Shoichi Hasegawa[ID]

Tokyo Institute of Technology, Yokohama, Japan
{yunxiu,siw131,hasevr}@haselab.net

Abstract. This study explored haptic feedback methods that do not rely on tangential forces, aiming to address the miniaturization challenges of haptic feedback devices in current virtual reality environments. We employed a six-degree-of-freedom haptic interface to construct a virtual grasping scene that simulates the stick-slip phenomenon. By comparing user grip response speed and force adjustment capability under different conditions: no haptic feedback, vibration feedback, and vibration with tangential force feedback, we collected data on users' grip responses under different haptic conditions. The results demonstrate that vibration feedback can significantly improve users' grip response speed without tangential forces. However, without tangential forces, users find it hard to differentiate responses based on the weight of objects. This indicates that while vibration feedback can simplify the design of haptic devices and enhance response speed, tangential forces are still essential for accurate weight perception. The theoretical and practical significance of this research lies in providing a new direction for haptic feedback devices in virtual reality.

Keywords: Haptics · Virtual Reality · Grip Force Adjustment

1 Introduction

In the realm of virtual reality (VR), haptic feedback technology plays a crucial role in enhancing user interaction by simulating touch sensations. This technology enables users to interact with and manipulate three-dimensional objects in virtual environments more realistically. Despite its potential, a significant challenge lies in the miniaturization of haptic devices, particularly those designed for hand-haptic feedback. Current solutions often rely on mechanical structures with movable parts, which can be bulky and limit the range of tactile sensations they can provide.

Research by Flanagan et al. [4] showed that the human body has reflexes that finely adjust grip force in response to changes in load force to prevent

© The Author(s), under exclusive license to Springer Nature Switzerland AG 2025
H. Kajimoto et al. (Eds.): EuroHaptics 2024, LNCS 14768, pp. 29–42, 2025.
https://doi.org/10.1007/978-3-031-70058-3_3

objects from slipping or being damaged. Wiertlewski et al. [14] explored how vibrations caused by sliding affect the grasping reflex, indicating that vibrational signals play a significant role in modulating grip force to compensate for unexpected object sliding. These studies highlight the importance of vibration in human tactile perception and suggest that using vibration instead of other types of force feedback could potentially reduce the size of devices. Konyo et al. [8] demonstrated that the sensation of friction can be conveyed through vibration without needing tangential forces at the fingertips. We aim to explore further haptic interfaces capable of generating tangential forces to investigate whether the coupling mechanism between grip force and load force remains after the removal of tangential forces, as well as the role of tangential forces in the dexterous manipulation of virtual objects. In our previous study [15], we investigated a tactile device for dexterous manipulation that relies on pressure and vibration, omitting tangential forces. This approach resulted in a reduction in the size of fingertip tactile devices. However, the effects of eliminating tangential forces on object grasping within our environment have not yet been investigated.

Our goal is to enable users to perform dexterous manipulations in virtual reality, especially to change the contact points of grasped objects through actions such as rolling and sliding [5], making the interaction more realistic. Therefore, identifying the minimal haptic elements that help dexterous manipulation is a worthwhile research topic. In other words, finding the types of haptic feedback that are less crucial during the object-grasping process is a valuable subject of study.

In this study, we utilized a 6-degree-of-freedom (DOF) haptic interface to construct a virtual object grasping scenario in a virtual environment that replicates the stick-slip phenomenon. Participants were asked to repeatedly perform "Regrasping" actions on virtual objects in their hands, which involved brief sliding movements in a controlled environment. This task requires changing the current grasping posture and quickly transmit to another more advantageous posture. We compared three experimental conditions: no haptic feedback, vibration feedback, and vibration combined with tangential force feedback. We observed the characteristics of grip force changes during re-grasping, dividing the force changes into three stages and recording the users' reaction times based on these stages.

Our results indicate that the presence of both types of haptic conditions significantly improved users' grasping response speeds compared to visual feedback only, but there were no significant differences between having tangential force feedback and not. Regarding the grasping force after reflex, our findings suggest that most users could only distinguish the grasping force for objects of different weights when tangential force was present. Our results demonstrate that the coupling mechanism between grip force and load force remains intact with vibration alone, even with the removal of tangential force. However, our result also suggests that simply removing tangential force may not be sufficient for users to perceive the different weights, indicating the need for further investigation.

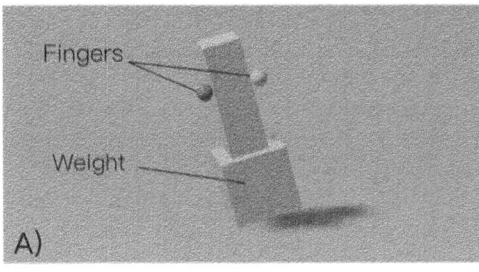
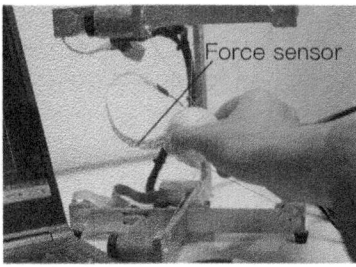

Fig. 1. Overview of the Experimental System: A) The user's view during the experiment. The mass of the virtual object is concentrated in the cube at the bottom, creating a "hammer"-like structure to keep it vertical, preventing undesired rotation slip. B) The device is a haptic interface equipped with a force sensor.

2 Related Work

Our research is based on the grip force applied by the fingers during object grasping, which can respond to sudden changes in load force. Coal et al. [3] indicated that the grip force began increasing in all subjects from 60 to 90 ms. Also, human can dynamically adjust their grip force based on the feedback of the load force [4]. Based on these basic adjustment principles, the tactile signals at the fingertips provide information on contact events, such as the contact timing, location, and direction of contact forces. This information is used to monitor task progress and achieve grasp stability, considered crucial for dexterous manipulation [6]. Moreover, distributed vibration stimuli on the fingers can unconsciously control human grip force [10].

Regarding the research on using limited tactile feedback as a substitute for traditional force feedback, some studies [12] utilized skin deformation feedback instead of traditional force feedback, showing that skin deformation could convey force information effectively. Konyo et al. [8] compared vibration friction against the friction generated by force feedback devices, indicating that tangential forces are not necessary to convey a sense of friction, simplifying the device and allowing for its combination with force feedback devices. We will further investigate whether tactile feedback replicating the friction sensation can evoke a grip response to sudden changes in load force.

Additionally, research by King et al. [7] has shown that, without tactile feedback, both new users and experts apply a grip force far greater than the minimum required to complete a task, indicating that continuous haptic feedback is necessary to maintain optimal grip strength. We will explore whether vibration alone can replicate this phenomenon.

Studies also suggest that vibration tactile feedback can help dexterously manipulate objects. A classic scenario of sudden load change is when an object unexpectedly falls out of grasp. Walker et al. [13] designed a scenario where the virtual floor suddenly drops, requiring users to prevent the object from moving. The results showed that when visual feedback was disabled, vibration feedback

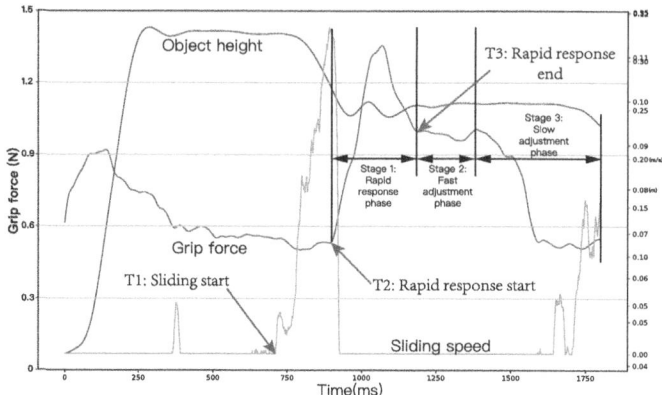

Fig. 2. The typical grip force adjustment process: Stage 1: Due to the sudden change in grip force caused by sliding, users begin to adjust their grip force to deal with unexpected sliding. Stage 2: To grip the object more comfortably, users will adjust their grip strength quickly, seeking the optimal grip. Stage 3: Users might release or continue manipulating the object once the grip is stable. T1 moment: To adapt to sliding, users gradually reduce their grip force, at which point the object begins to slide. T2 moment: Users perceive the sliding of the object and quickly increase their grip force in response, attempting to stop the sliding. T3 moment: Once the sliding is controlled, the grip force will stop increasing and tend to reduce. This figure also shows three stages of grip force adjustment:

significantly positively impacted manipulation. Li et al. [9] designed a virtual task of grasping and holding fragile objects, proving that dual-modal feedback (sliding vibration and squeezing feedback) is significantly better than any single-modal feedback. However, their feedback signals were not applied to the hands, and we aim to verify if the results differ when stimulating mechanoreceptors on the hand.

Wiertlewski et al. [14] demonstrated that vibration characteristics were used by the central nervous system to regulate the grasping reflex and suggested that larger tangential forces and vibration power improve response speed. However, in their system, tangential forces were always present, and the scenario of vibration only has not been explored. Okamoto et al. [11] added vibration cues before sliding can enhance the speed of humans' grip adjustment response. However, in the real world, vibration cues synchronous with stick-slip occurrences, occurring simultaneously as sliding. We need further validation for physics-based manipulation.

3   Experiment

Participants were invited to take part in the experiment involving the grasping and sliding of objects in a virtual scene. This included a series of tasks requiring them to achieve controlled sliding of objects in their hands under different haptic conditions, collecting data on their reaction time and grip force.

3.1 Participants

Eight participants (six male and two female) joined the experiment after providing informed consent. The participants ranged in age from 24 to 33, with one being left-hand dominant. The experimental procedure received approval from the Tokyo Institute of Technology Review Board.

3.2 System Set-Up

For the hardware, we utilized a 6-DoF haptic interface equipped with pressure sensors, following the hardware configuration described in the work by Balandra et al. [1]. This setup allowed users to manipulate a virtual coupling in the virtual environment, controlling its position and rotation. Two spherical haptic pointers are linked to two device positions in the virtual environment. These pointers move in accordance with the device's grip position and rotation while the distance between the two pointers decreases proportionally to the applied pressure, enabling them to exert pressure on virtual objects to achieve grasping and releasing. The force feedback to the grip begins to appear when the user starts to touch virtual objects. The system could provide users with tangential forces in any direction and vibration feedback. When users grasp a virtual object or when the virtual object slips within their hands due to gravity acting on the object, a vertically downward tangential force is applied to the surface of the users' fingertips. Our haptic interface drives motors to pull the strings, providing users with a tangential force on their fingers in the corresponding direction. Users typically utilize this information to directly perceive the object's weight. When the object slides or when users shake the object while holding it, they will also experience greater tangential forces. The system setup is shown in Fig. 1.

For software, we employed the Springhead[1] physics engine, which calculates the forces experienced by the virtual coupling and outputs them directly on the haptic interface. During the manipulation of virtual objects, the physics engine detected the moment when static friction transitioned to dynamic friction between the coupling and the objects. At this moment, a 120 Hz damped sinusoidal wave proportional to the contact force was output to create stick-slip tactile sensations. Similarly, during dynamic friction, a continuous damped sinusoidal wave proportional to the sliding velocity and contact force was output to create sliding vibratory sensations. The example of vibration signal is shown in Fig. 3B. The system's update rate was 1000 Hz.

3.3 Experimental Procedure

Our experiment focused on three haptic conditions: 0: no haptic feedback, 1: tangential force and vibration, 2: vibration only, and two mass conditions of

---

[1] https://springhead.info/.

virtual object: mass = 150, mass = 100. We avoided a significant difference in mass to prevent participants from subjectively perceiving the difference. Participants could always see the virtual scene on the screen, meaning visual was always available.

Before the experiment began, participants were given free play time under different conditions until they could successfully control the sliding of virtual objects to minimize the learning effect on the results. Each individual's free play time varies because this task is relatively challenging even in the real world, and different users have different operating habits. However, we strictly require users to complete the task with "haptic on" skillfully. When "haptic off", they must demonstrate the corresponding grasping reflex before ending the free play time, even if they cannot successfully complete the task. This requirement aims to minimize learning effects in subsequent experiments.

The specific experiment consisted of six trials, combining the three haptic conditions with the two mass conditions, and the order is condition 1 mass 100 g, condition 1 mass 150 g, condition 2 mass 100 g, condition 2 mass 150 g, condition 0 mass 100 g, condition 0 mass 150 g. We later arranged the condition 0 (no haptic feedback) trials to avoid the practice's effects. Because the 'Re-Grasp' task is an innate ability in real life that does not require extensive practice to accomplish, we provided participants with sufficient practice time before the experiment to familiarize themselves with the system. To streamline the experimental process, we fixed the order of the conditions, gradually increasing the difficulty level from the presence of haptic feedback to its absence.

During the experiment, participants were instructed to induce slippage by gradually and intentionally reducing their grip force, but making the object completely leave the fingers was prohibited. Furthermore, we allowed participants to attempt to minimize the slip by using a small release duration, but such attempts were not considered successful sliding. In the experiments under each condition, the trial concludes when the user successfully 're-grasps' the object 5 times. However, in the 'no haptic' condition, many users struggle to complete the experiment. Therefore, if the object is observed slipping and the user exhibits a reflexive tightening of the fingertips, it is considered a successful completion even if they fail to grasp the object.

Participants were not informed of the experimental conditions and were instructed to follow the specified order. An experimental trial was considered complete when at least three effective slides (the object should not leave the virtual finger or drop on the ground) were observed. The system recorded data throughout, including output from the force sensors, the vertical position of the virtual object, the relative speed between the virtual coupling and the object, and the output of the haptic interface.

4 Results

The typical sliding process under condition 1 is shown as Fig. 3.

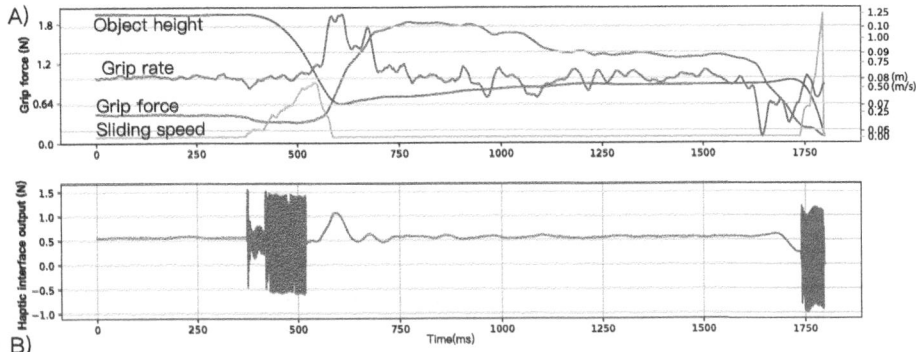

Fig. 3. In the typical sliding process under condition 1 (tangential force plus vibration) when grasping an object around 50 g, the object height shown in the figure above (A) represents the object's vertical position. At around 500 ms, the object begins to slide but does not touch the ground. During this process, to control the slide, the grip force suddenly increases; simultaneously, the sliding speed maintains a non-zero value throughout the sliding process. The grip rate—the derivative of grip force—helps us accurately locate the specific moment of the sudden force change. The figure below (B) shows the output of the haptic interface in the upward vertical direction, including the overlapped damped sinusoidal produced during the sliding process and the tangential force generated by the object's weight and the sliding.

4.1   Grip Force Response Latency

The initial analysis included ANOVA, which revealed significant differences between the conditions. The ANOVA results indicated that the effect of condition on grip response time was statistically significant ($F(2, 234) = 84.83$, $p < 2.10e-28$). This result suggests that at least one group's mean differs significantly from the others. Then, we employ the Tukey HSD multiple comparison test and independent sample t-tests to find the impact of tangential force and vibration on response time. We analyzed the grip response time following sliding under different haptic conditions to see the significance of applying tangential forces and vibratory haptic feedback on fingers. We will collect the time of T2–T1 shown in Fig. 2. The overall and individual result is shown in Fig. 4 and Fig. 5.

In this research, we compared latency differences between three conditions (labeled 0, 1, 2), In our experiments, we required participants to perform 're-grasp' five times under each condition, which should have yielded 240 data. During subsequent processing, 3 data were eliminated from the analysis because the speed of finger opening was too fast, resulting in a lack of sliding friction. This resulted in the analysis of 237 valid samples across three haptic conditions and two weights. Our analysis aims to evaluate the impact of the presence or absence of vibration on grip force adjustment.

According to the results of the Tukey HSD multiple comparison test, we have found significant differences between group 0 and group 1, with a mean difference of -87.4211 ($p < 0.001$, confidence interval: -106.0696 to -68.7726),

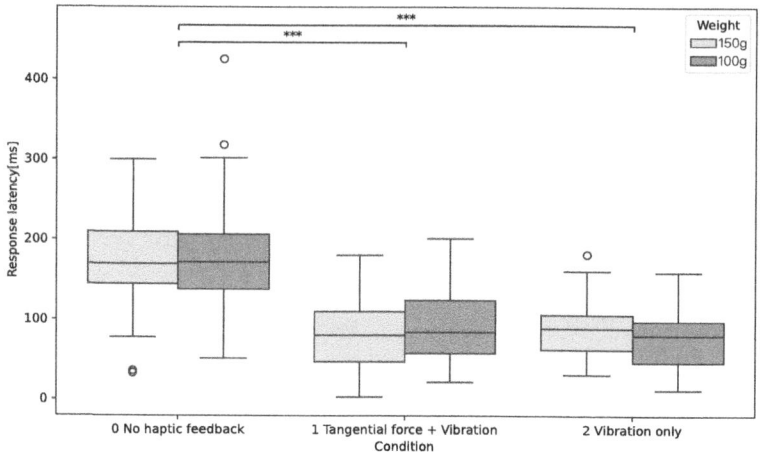

Fig. 4. The response latency after object sliding for all individuals under three conditions (0, 1, 2) shows significant differences between condition 0 and condition 1 and also between condition 0 and condition 2. However, the results between condition 1 and 2 are similar and do not show a significant difference. Additionally, the data range for condition 0 is wider. There is also no significant difference between weights 150 and 100.

and between group 0 and group 2, with a mean difference of -91.5215 (p < 0.001, confidence interval: -110.2877 to -72.7554). This indicates that the grip force adjustment latency in group 0 is significantly lower than in groups 1 and 2. However, the mean difference between groups 1 and 2 was -4.1004 (p = 0.8623, confidence interval: -22.7489 to 14.548), showing no significant difference between these two groups in grip force response time.

Further independent sample t-test results were consistent with the findings of the Tukey HSD test, revealing significant differences between group 0 and group 1 (t = 9.951, p < 0.001) and between group 0 and group 2 (t = 10.842, p < 0.001). In contrast, the comparison between groups 1 and 2 (t = 0.648, p = 0.518) did not reach statistical significance.

For condition 0, regardless of whether the mass was 150 g (median response time: 169.53 ms, SD = 58.37 ms) or 100 g (median response time: 171.49 ms, SD = 72.00 ms), the reaction times for participants were longer in the absence of vibration. In contrast, under conditions 1 and 2, where vibration was present, the median response times significantly decreased, regardless of the mass. In condition 1, the median response time was 79.82 ms (SD = 43.74 ms) for a mass of 150 g and 83.69 ms (SD = 43.80 ms) for a mass of 100 g. Similarly, in condition 2, the median response time was 88.3 ms (SD = 35.00 ms) for a mass of 150 g and 79.50 ms (SD = 36.30 ms) for a mass of 100 g. These results suggest that the presence of vibration alone can significantly reduce grip force response time, irrespective of the object's mass.

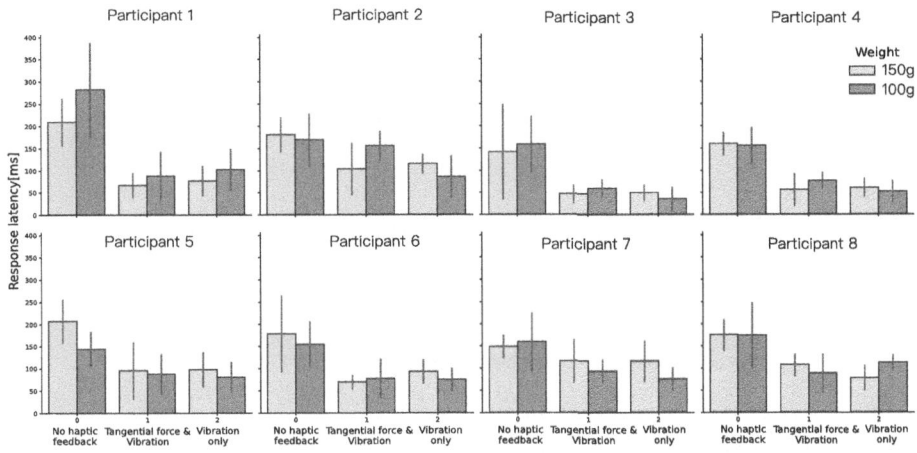

Fig. 5. The response latency after object sliding for each individual. Generally, the latency under condition 0 is greater than that under conditions 1 and 2.

Moreover, we observed the variability in grip force adjustment response across conditions by comparing the minimum and maximum values. The maximum values in condition 0 reached 300.13 ms and 425.31 ms, much higher than in other conditions, indicating more significant variability in response time without vibration. The maximum values in conditions 1 and 2 were relatively lower, suggesting that users might tend to make predictions of sliding or respond slowly without tactile feedback.

Employing average-based analysis allows for comparing overall differences between various conditions. Subsequently, we attempt to conduct individual data-based analyses to uncover potential outliers and observe individual variations. Regarding individual user data, everyone demonstrated noticeably slower responses under the condition without haptic feedback than those with haptic feedback.

4.2   Grip Force After Sliding

To investigate whether users have different behaviors toward virtual objects of different weights under specific haptic conditions, we collected the gripping force of users after the object slid. Under each condition, there is no significant difference between weights of 150 and 100 ($P > 0.05$). Therefore, we will focus on analyzing each individual's performance and try to find if individual differences can be observed. The result of an individual is shown as Fig. 6. We conducted independent sample t-tests on the data of each user under a specific tactile condition. We will collect the grip force of T3 shown in Fig. 2. The results showed that under condition 1 (tangential force and vibration), 6 out of 8 users displayed significant differences ($P < 0.05$) in their data, while two users did not show significant differences ($P > 0.05$). In contrast, under condition 0 (no haptic

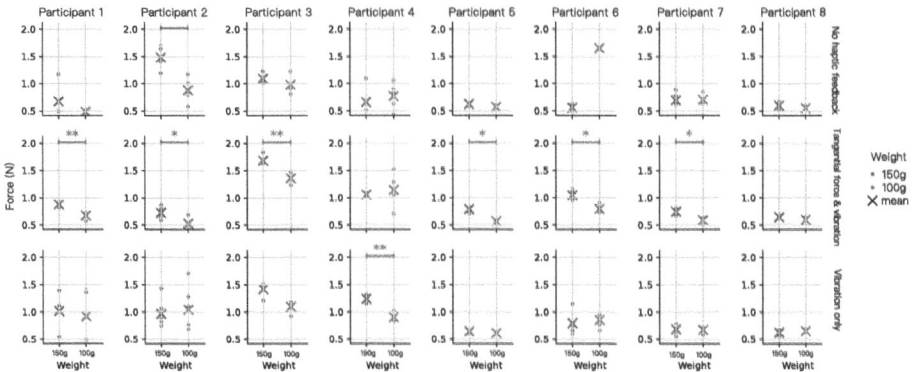

Fig. 6. During object sliding, the gripping force has a process of initial increase then decrease. We collected grip force data at the end of the rapid response phase under various weight conditions and presented them in scatter plots. In this chart, each column represents data from an individual participant, and in rows, with each row representing a condition (no haptic feedback, tangential force & vibration, vibration only).

feedback) and condition 2 (vibration only), only one user's data in each condition showed significant differences.

To explore the correlation between the force magnitudes under two weight conditions, we calculated the ratio of the mean grip force under two weights for each person under one tactile condition. For clarity in plotting, we subtracted one from the ratio, making the graph more understandable, shown as Fig. 7. The ideal ratio would be $(150\,\mathrm{g}/100\,\mathrm{g}) - 1 = 0.5$. The results indicated that under condition 1, 6 people had ratios greater than 0.2, while under condition 0, 2 people did, and under condition 2, 2 people did. Additionally, condition 1 had one negative value, whereas conditions 0 and 2 had two and three negative values, respectively. These negative values indicate that some users almost responded opposite to the two weights. Overall, the ratios under condition 1 were closer to the ideal state than conditions 0 and 2, and condition 2 showed a relatively stable phenomenon compared to condition 0, without more significant variances.

5   Discussion

5.1   Grip Force Response Latency

From the results, it is clear that in the absence of haptic feedback (condition 0), latency is the longest, significantly higher than in conditions with haptic feedback (condition 1 and condition 2). However, there was no significant difference between condition 1 and condition 2. This indicates that haptic feedback can significantly enhance the response speed in the "Re-Grasp" task compared to relying solely on visual cues. More importantly, vibration feedback alone can improve users' grasping reflex speed, even when removing tangential forces. This

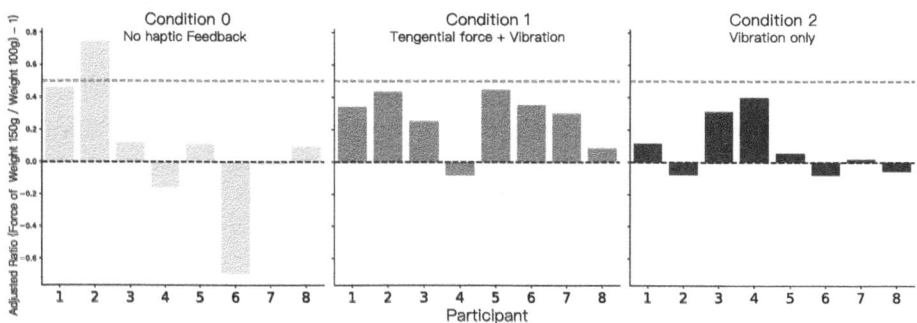

Fig. 7. The ratio of the average grasping force at the end of the rapid response phase for two different weights (150 g and 100 g) under various haptic conditions. For each condition, we calculate the ratio of the average grasping force for weights of 150 g and 100 g, respectively. Then, we obtain the data presented in the chart by subtracting one from this ratio. Bars closer to the dashed line at the top indicate that the force ratio under is closer to the ideal value of 1.5. If the ratio is negative, it indicates that participants exert more force on the lighter object, demonstrating that they can't distinguish objects of different weights.

is likely because vibration feedback is sufficient to trigger a rapid response that bypasses visual processing.

During this phase, the average response delay under conditions 0 and 1 is about 80 ms. It is believed that a Long-Latency Reflex (LLR) [16] may occur, which is a reaction to the perception of sliding events, typically appearing 50 to 70 ms after the disturbance. Following is the delay from the electromyographic (EMG) signals to the muscle-generating response force, which spans approximately 30 to 100 ms [2]. This series of reactions constitutes the entire response process. A voluntary response may be required for condition 0, which is visual only and may cause a slow reaction. Furthermore, by comparing the minimum and maximum values, we can observe the variability in grip force adjustment response under different conditions. The condition without haptic feedback (condition 0) shows the most significant variance and numerous outliers. The presence of these outliers is not difficult to explain; sometimes, users might react slower due to not noticing the object slipping. However, many outliers show a faster response speed than in conditions with haptic feedback. These outliers might have originated from very slight slips. To improve success rates, some users remember the amount of force used during previous slips and then attempt to proactively predict the object's falling behavior, managing to "accidentally" succeed by quickly controlling the opening and closing of their fingers. The lack of haptic feedback makes this strategy more appealing in condition 0, as success rates are generally lower. Moreover, there is no significant difference between different weights, as the weight does not affect the speed of generating haptic signals.

Okamoto et al. [11] showed that grasping responses are significantly slower in conditions with only vibration feedback than those with tangential forces, which seems to conflict with our findings. However, the experimental conditions between the two studies are different. In their experiment, the vibration occurred before the object slipped, which helped identify the neural mechanisms related to tangential forces. In contrast, in our conditions, vibration occurred synchronously with the slip, serving as the primary source of information for users to detect object displacement.

5.2   Grip Force After Sliding

From the observations of Fig. 6, we note significant individual differences in the perception of force. Despite this, under condition 1, most individuals could still differentiate between two weights without being explicitly informed of the weight difference. However, the data from the two participants did not show significant differences. Specifically, participant 4, when dealing with the lighter weight, may have experienced finger fatigue due to too many attempts without adequate rest, affecting the accuracy of the data. Participant 8, perceiving the task as challenging, adopted a strategy that caused only a slight displacement of the object, resulting in low grip strength values and subtle variations. However, the overall data suggest that tangential force enables users to exhibit different grasping responses when handling objects of different weights. From Fig. 7, it is evident that under condition 1, the ratio of grip force more closely aligns with the ideal value. In contrast, under conditions 0 and 2, observing different responses from users towards objects of varying weights is challenging, implying that vibration feedback alone is ineffective in distinguishing between different weights.

Wiertlewski et al. [14] proposed a correlation between the power of vibration generated during sliding and the subsequent peak in gripping force. Our analysis suggests that haptic signals mainly originate from the stick-slip phenomenon in our experimental setup. Due to the effect of static friction, objects do not always slip after being successfully grasped, even if the grip force is reduced within a specific range. Zangrandi et al. [16] mentioned the slip process includes stick, partial slip, and complete slip phases. However, the partial slip is hard to display without tangential force under only vibration and no haptic feedback conditions. Without partial slip or sliding, it becomes difficult to perceive differences in weight. Moreover, our experimental setup allows for the complete elimination of tangential force in a vibration-only condition, differing from the conditions of previous experiments. This preliminarily indicates that in manipulation environments based on physics, we cannot conclusively state that tangential force is unnecessary for enabling users to distinguish weights through haptics.

5.3   Experimental Design Impact

The study had a relatively small number of participants. In the Grip Force Response Latency experiment, results showed no significant fluctuations or inconsistencies, suggesting that increasing participant numbers would unlikely change the conclusions. However, in the Grip Force After Sliding experiment, some differences were observed between the tangential force condition and the other two conditions. Increasing participant numbers might make these differences more statistically significant.

Although participants were given sufficient practice time and the experimental process was streamlined by fixing the condition order, order effects may not have been entirely eliminated. Counterbalancing the condition order across participants would have helped control for potential order effects.

6   Future Work

We will focus on validating the correlation between different weights and vibration power to find if it will help to distinguish different weights without tangential force. Also, we aim to explore the potential of fingertip haptic devices lacking tangential force feedback, seeking the minimal tactile feedback necessary for achieving the most flexible and intuitive grasp.

7   Conclusion

In this study, we utilized a 6-DoF haptic interface to explore the effectiveness of vibration feedback in virtual object grasping, finding that vibration feedback significantly improves grasping response speed even without tangential forces, highlighting its importance in simplifying the design of haptic devices. However, tangential force is still essential for distinguishing object weight in virtual environments. Future research will delve into optimizing vibration cues and exploring the minimal haptic feedback required for physics-based manipulation.

Acknowledgments. This work was supported by JST, the establishment of university fellowships towards the creation of science technology innovation, Grant Number JPMJFS2112, and JSPS KAKENHI Grant Number 23H03432.

References

1. Balandra, A., Gruppelaar, V., Mitake, H., Hasegawa, S.: Enabling two finger virtual grasping on a single grasp 6-DOF interface, by using just one force sensor. In: 2017 IEEE World Haptics Conference, pp. 382–387. IEEE (2017)
2. Cavanagh, P.R., Komi, P.V.: Electromechanical delay in human skeletal muscle under concentric and eccentric contractions. Eur. J. Appl. Physiol. **42**, 159–163 (1979)
3. Cole, K.J., Abbs, J.H.: Grip force adjustments evoked by load force perturbations of a grasped object. J. Neurophysiol. **60**(4), 1513–1522 (1988)
4. Flanagan, J.R., Tresilian, J.R.: Grip-load force coupling: a general control strategy for transporting objects. J. Exp. Psychol. Hum. Percept. Perform. **20**(5), 944 (1994)

5. Han, L., Trinkle, J.C.: Dextrous manipulation by rolling and finger gaiting. In: Proceedings of the 1998 IEEE International Conference on Robotics and Automation, vol. 1, pp. 730–735. IEEE (1998)
6. Johansson, R.S., Flanagan, J.R.: Coding and use of tactile signals from the fingertips in object manipulation tasks. Nat. Rev. Neurosci. **10**(5), 345–359 (2009)
7. King, C.H., et al.: Tactile feedback induces reduced grasping force in robot-assisted surgery. IEEE Trans. Haptics **2**(2), 103–110 (2009)
8. Konyo, M., Yamada, H., Okamoto, S., Tadokoro, S.: Alternative display of friction represented by tactile stimulation without tangential force. In: Ferre, M. (ed.) EuroHaptics 2008. LNCS, vol. 5024, pp. 619–629. Springer, Heidelberg (2008). https://doi.org/10.1007/978-3-540-69057-3_79
9. Li, K., Brown, J.D.: Dual-modality haptic feedback improves dexterous task execution with virtual EMG-controlled gripper. IEEE Trans. Haptics **16**(4), 816–825 (2023)
10. Masataka, N., Konyo, M., Maeno, T., Tadokoro, S.: Reflective grasp force control of humans induced by distributed vibration stimuli on finger skin with ICPF actuators. In: Proceedings of the 2006 IEEE International Conference on Robotics and Automation, pp. 3899–3904. IEEE (2006)
11. Okamoto, S., Wiertlewski, M., Hayward, V.: Anticipatory vibrotactile cueing facilitates grip force adjustment during perturbative loading. IEEE Trans. Haptics **9**(2), 233–242 (2016)
12. Quek, Z.F., Schorr, S.B., Nisky, I., Provancher, W.R., Okamura, A.M.: Sensory substitution and augmentation using 3-degree-of-freedom skin deformation feedback. IEEE Trans. Haptics **8**(2), 209–221 (2015)
13. Walker, J.M., Blank, A.A., Shewokis, P.A., O'Malley, M.K.: Tactile feedback of object slip improves performance in a grasp and hold task. In: 2014 IEEE Haptics Symposium, pp. 461–466. IEEE (2014)
14. Wiertlewski, M., Endo, S., Wing, A.M., Hayward, V.: Slip-induced vibration influences the grip reflex: a pilot study. In: 2013 World Haptics Conference, pp. 627–632. IEEE (2013)
15. Xu, Y., Wang, S., Hasegawa, S.: Realistic dexterous manipulation of virtual objects with physics-based haptic rendering. In: ACM SIGGRAPH 2023 Emerging Technologies, pp. 1–2 (2023)
16. Zangrandi, A., D'Alonzo, M., Cipriani, C., Di Pino, G.: Neurophysiology of slip sensation and grip reaction: insights for hand prosthesis control of slippage. J. Neurophysiol. **126**(2), 477–492 (2021)

Audiovisual-Haptic Simultaneity Perception Across the Body for Multisensory Applications

Jiwan Lee(iD), Gyeore Yun(iD), and Seungmoon Choi(✉)(iD)

Pohang University of Science and Technology, Pohang 37673, Republic of Korea
jiwan95@postech.ac.kr, ykre0827@postech.ac.kr, choism@postech.ac.kr

Abstract. This paper explores human performance on simultaneity judgments between audiovisual and haptic stimuli across the body for use in future real-time haptic applications. Three representative body sites, the torso, fingertip, and foot, were stimulated with vibration, and a media clip was used as an audiovisual stimulus. The results showed that: (1) the timing delay that humans can tolerate in the audiovisual leading-haptic following case was 55 ms at the fingertip, 65 ms at the chest, and 45 ms at the foot, and these values were significantly different from each other, (2) the regression curves shifted toward the haptic-leading direction from the chest down to the foot, showing significantly different points of subjective synchrony (PSS) between the body sites but similar window widths of simultaneity, (3) the PSSs were obtained between 20 and 40 ms in the haptic leading-audiovisual following case, and (4) a significant asymmetry was observed in the curves between haptic-leading and audiovisual-leading stimuli, with a higher temporal sensitivity in the audiovisual-leading case. We expect our results can provide essential information for multisensory applications.

Keywords: Temporal synchrony perception · Simultaneity judgment · Multiple body sites · Vibrotactile feedback · Viewing experiences

1 Introduction

Media is in a transition period, evolving into a new form that appropriately stimulates various sensory modalities in addition to the visual and auditory systems. In this situation, haptic feedback is most demanded to enhance the viewing experience and foster a sense of participation in the media [5]. Recent studies affirm the positive impact of haptic feedback on both immersion and enjoyment [24,25], and full-body haptics covering multiple body sites are accelerating the proliferation of haptic feedback [10]. Here, ensuring the seamless real-time delivery of haptic feedback is paramount for viewers, particularly as the majority now experience media content through online streaming rather than recorded logs; representative instances include eSports spectatorship. However, owing to the computational time involved in generating haptic stimuli based on audiovisual

H. Kajimoto et al. (Eds.): EuroHaptics 2024, LNCS 14768, pp. 43–55, 2025.
https://doi.org/10.1007/978-3-031-70058-3_4

information from content, these stimuli inevitably lag behind their audiovisual counterparts. To guarantee effective haptic stimulation, it is crucial to investigate the minimum thresholds above which timing delays between audiovisual and haptic stimuli become noticeable to human observers.

Such temporal synchrony perception can be quantified by the paradigm of simultaneity judgment (SJ), which has been investigated over decades in a vast body of research [18]. This experimental paradigm yields two important variables, the window of simultaneity (WS) and the point of subjective simultaneity (PSS), based on 'simultaneous' responses for various stimulus-onset asynchronies (SOAs) [23]. The former represents a certain tolerable SOA range beyond human detection, and the latter corresponds to the SOA value at which the probability of synchronous responses is the highest. Both are essential information for haptic designers to provide synchronous multisensory experiences. The same variables can also be obtained with the temporal order judgment paradigm [23], but it requires more cognitive activity to determine the temporal order of stimuli [2]. Some adopt even a ternary SJ task that combines the two tasks, as the two tasks have been proven to produce different results with different decision criteria [18].

Studies on synchrony perception have steadily expanded their measurements by examining human performance in increasingly diverse setups [18]. They initially looked at temporal perception within one sensory modality and later across two different modalities. Their stimuli were designed to be as simple as possible, such as a flashing light bulb, beep, brief touch or vibration, for most accurate measurements. Results of tolerable delay were usually less than 50 ms with such stimuli [9]. Higher thresholds were observed for intersensory perception than intrasensory, and the lowest thresholds were found for auditory-haptic perception among visual-auditory, auditory-haptic, and visual-haptic [18]. Meanwhile, recent studies tested more complex stimuli, such as speech and musical instrument recordings [19,20]. Increased stimulus complexity did not shift PSS systematically, but did so on WS; high complexity resulted in poor perceptual performance compared to the conventional stimuli [20]. In comparison, PSS was affected by observer-stimulus distance [16] or body sites [4,14]. These studies showed a tendency that the farther away a haptic stimulus is from the body, the earlier it needed to be presented than an auditory or visual stimulus to be perceived as synchronous. On the other hand, WS was not affected by such factors.

Despite such important discoveries, more extensive scientific knowledge is required for haptic designers due to two main reasons. First, most studies involving haptic stimulation used stimuli presented only to the fingertips. Only a few studies considered different body sites [4,14], despite the emerging trend of full-body haptics [11,13]. Future investigations should encompass other representative body sites like the torso. Second, the previous results were based on relatively simple stimuli to precisely measure human performance. In particular, the conventional stimuli involve one or two sensory channels about timing and are a far cry from actual complex multisensory stimuli used in applications.

This paper investigated the temporal synchrony perception between audiovisual and haptic stimuli across multiple body sites under the context of multisensory applications. Using a gameplay recording and a vibration effect as stimuli, human responses were collected by stimulating representative areas of the hand, torso, and foot under the SJ paradigm. Psychometric functions for simultaneity perception were estimated from the garnered data for each body site and then compared with each other. Our results showed that the thresholds were relatively high over 50 ms and the effect of body site was significant in all performance variables except WS. Additionally, significant asymmetry was found between haptic-leading and audiovisual-leading stimuli.

Fig. 1. Haptic actuators attached to three body sites: (A) left index fingertip, (B) upper chest, and (C) foot. The actuators were attached using elastic straps.

2   Methods

The study was approved by the Institutional Review Board at POSTECH under an application No. PIRB-2022-E039.

2.1   Participants

Eighteen volunteers (11 males and 7 females; aged 21–30 with a mean of 25.0) participated in this experiment. They were all right-handed and had normal or corrected-to-normal sensory abilities. All participants gave their informed consent prior to the experiment. They were paid USD 15 per hour of participation.

2.2   Stimuli

All stimuli combined audiovisual and haptic stimuli. They were produced in the form of a multimedia file with three sound channels, two for sound and one for vibration. The video was displayed full-screen on a 32-in. monitor using the VLC media player. The haptic stimulus was presented with a vibration actuator (HapCoil–One, Tactile Labs) attached to the body site using an elastic

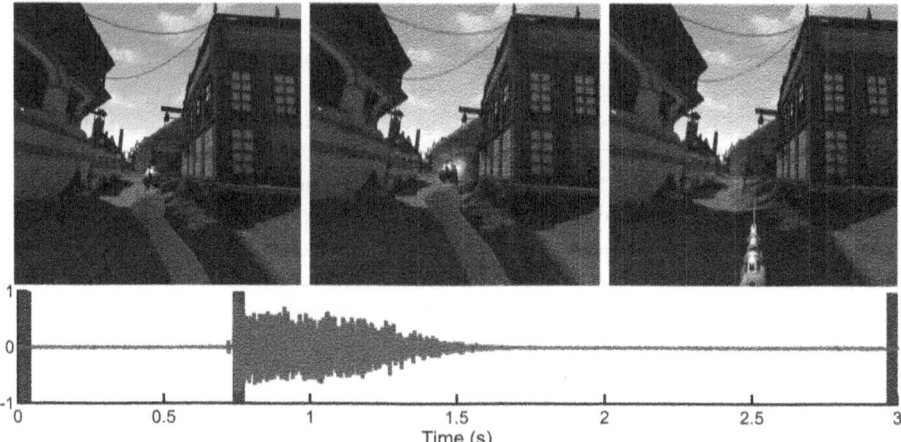

Fig. 2. Screenshots of the video (top) and the sound waveform (bottom). An image and the sound data in a shaded area are paired from left to right. The center image is the first frame from when the gunshot occurred.

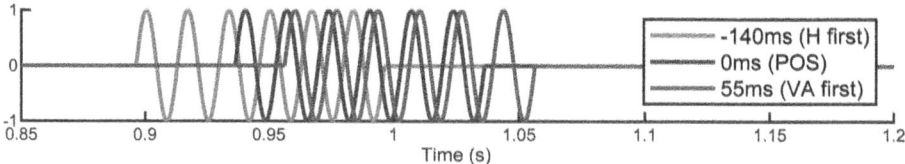

Fig. 3. Haptic stimulus examples with different SOAs. The blue waveform shows the point of objective simultaneity (POS) with no physical delay. (Color figure online)

Velcro belt, as shown in Fig. 1. The sound was delivered through noise-canceling headphones (QuietComfort 35 II, Bose) while blocking audible noises from the vibration actuator. The three stimuli were measured using an ambient light sensor (SEN080605, YwRobot), a free space binaural microphone (3Dio), and an accelerometer (8765A250M5, Kistler) through a data acquisition card (USB-6251, National Instruments) with an acquisition resolution of 50 ns. We attached the sensors to the monitor screen, headphones, and haptic actuator and measured the end-to-end timing delays of the visual-sound and sound-haptic pairs multiple times. As in [6,12], the timing delays between actually generated stimuli were adjusted to be sufficiently short (less than 1–2 ms) for both pairs, representing the reference of synchronous stimuli. We made other stimuli by temporally shifting the haptic stimulus by a desired SOA from on the reference.

The audiovisual stimulus was taken from a recording of Dead and Buried II, a typical FPS game from Meta. It was a 3-s long video including a gunshot occurring at 750 ms. The gunshot appeared visually as a bright circular flash of 48 ms with a visual angle of ~2° in the center of the screen. It also made a sound that peaked sharply at a sound pressure level of 72 dB and then decreased slowly. The sound was about 800 ms, much longer than the flash, with its energy focused mostly in the early 550 ms. The video had the full HD image resolution with an aspect ratio of 1.0 and a frame rate of 24 fps. Fig. 2 shows an example of the audiovisual stimuli.

The haptic stimulus was an 80-Hz sinusoidal vibration of 100 ms, a brief feedback suitable for gunshot. It was delivered to three body sites: fingertip, upper chest, and foot instep, representing the hand, torso, and foot. The stimulus intensities were 3.9 G, 28.5 G, and 15.7 G, respectively. These parameters allowed the stimulus to be clearly and similarly perceived across the body sites using the haptic actuator. The stimulus intensities were determined through a pilot experiment with six participants. The actual vibration duration was 115.2 ms including a residual vibration of 15.2 ms. Fig. 3 shows some stimulus examples.

2.3   Procedure

Participants sat upright in a chair in front of a table with a computer. Their upper bodies were positioned about 57 cm away from the monitor, which displayed the visual stimulus. They wore one layer of thin clothing to ensure clear haptic stimulus transmission. The actuators were attached tightly to the three body sites of each person using the straps at the beginning of the experiment, as shown in Fig. 1. The actuator positions remained fixed throughout the experiment.

The experiment followed the method of constant stimuli to obtain a psychometric function of intersensory synchrony. Ten SOAs between the audiovisual and haptic stimuli were used, ranging from −240 ms to 120 ms with 40 ms intervals. Participants responded to 100 stimuli (10 SOAs × 10 repetitions) for each body site in a within-subject design. The experiment consisted of three sessions, and each session used one body site. The order of the sessions was completely counterbalanced across the participants by presenting them in every possible order. The order of stimuli was randomized in each session. Before the first session, ten randomly selected stimuli out of those used in the experiment were presented for familiarization. Participants rested for 3 min after each session. The experiment took an hour on average.

In each trial, participants perceived a set of audiovisual and haptic stimuli once. Then, they made a simultaneity judgment, that is, answered whether they felt the audiovisual and haptic stimuli simultaneously. The only possible responses were yes and no. No correct answer feedback was provided to avoid biasing the results [8].

2.4   Data Analysis

We pooled the data of the simultaneous responses across all participants for each body site. Then, an asymmetrical bell-shaped model was fit to the data using two cumulative Gaussians functions, which allows us to model different slopes

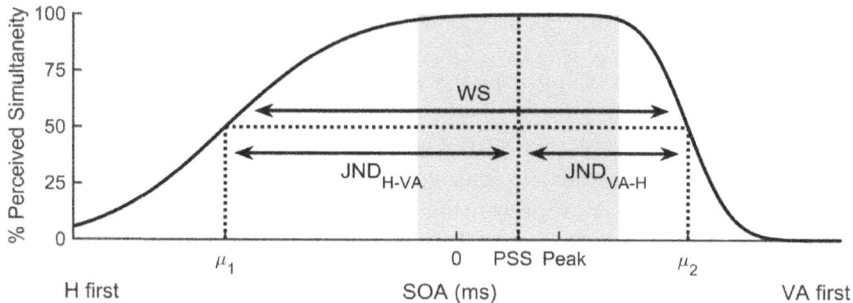

Fig. 4. Example of the asymmetrical model fitted to SJ responses. The shaded area indicates the interval of SOAs with near 100% simultaneity, and the PSS is calculated as the midpoint of the interval instead of the peak.

in the two halves [15, 17], such that

$$\Phi(t; \mu_1, \sigma_1, \mu_2, \sigma_2) = \frac{1}{2} \left(\text{erf} \left(\frac{t - \mu_1}{\sqrt{2}\sigma_1} \right) - \text{erf} \left(\frac{t - \mu_2}{\sqrt{2}\sigma_2} \right) \right), \tag{1}$$

where t is the multisensory SOA, and erf($\cdot$) denotes the Gaussian error function with two parameters of mean μ and standard deviation σ. The two means, μ_1 and μ_2, represent the decision points at which people begin to feel from 'simultaneous' to 'asynchronous' for haptic-leading and audiovisual-leading stimuli, respectively. The two standard deviations, σ_1 and σ_2, represent the noises therein. The best-fit parameters were estimated by the Levenberg-Marquardt non-linear least squares algorithm.

Important variables for simultaneity perception were derived from the estimated parameters by referring to [23]. The WS was set as the time interval between 50% percentiles of each curve, and its length was $\mu_2 - \mu_1$. The PSS was first determined as the time at which the curve had the peak value. However, when the curve had multiple SOAs above 98% ($\sim$3 σ) and thus formed a plateau, the midpoint of the plateau was used as the PSS instead. The PSS calculated in this way better represents the part perceived simultaneously than the peak, as illustrated in Fig. 4. Additionally, just-noticeable difference (JND) values, $\text{JND}_{\text{H-VA}}$ and $\text{JND}_{\text{VA-H}}$, were the differences between the respective two means from the PSS.

3   Results

The collected data fit the psychometric model well with low root mean squared errors (RMSEs) for all three body sites. The regression curves are shown in Fig. 5, and their RMSEs were 0.02, 0.05, and 0.04 from (A) to (C). The estimated curve parameters and the variables derived from them are shown in Table 1. Important results obtained from the two tables are summarized below:

Table 1. Estimated parameters of regression curves for the three body sites. Estimates are for the best fits to the pooled data, and 95% CIs are for the individual fits of the participants.

	Chest		Fingertip		Foot	
	Est.	95% CI	Est.	95% CI	Est.	95% CI
μ_1	−127.78	−142.04 −115.30	−141.80	−157.60 −130.70	−147.57	−162.63 −134.35
σ_1	37.31	15.09 28.52	36.41	16.08 31.89	38.61	18.52 32.10
μ_2	65.76	58.52 76.00	55.37	44.75 68.72	43.88	34.72 54.00
σ_2	25.77	10.33 25.02	28.97	10.73 22.32	29.11	12.14 25.31
WS	193.54	181.28 210.58	197.17	185.23 216.55	191.45	175.12 210.59
PSS	−19.49	−35.69 −16.30	−35.79	−47.07 −25.56	−42.41	−54.58 −36.64
$\text{JND}_{\text{H-VA}}$	108.29	90.14 115.21	106.01	97.03 118.64	105.16	91.42 114.35
$\text{JND}_{\text{VA-H}}$	85.25	84.74 101.77	91.16	82.59 103.51	86.29	80.46 99.49

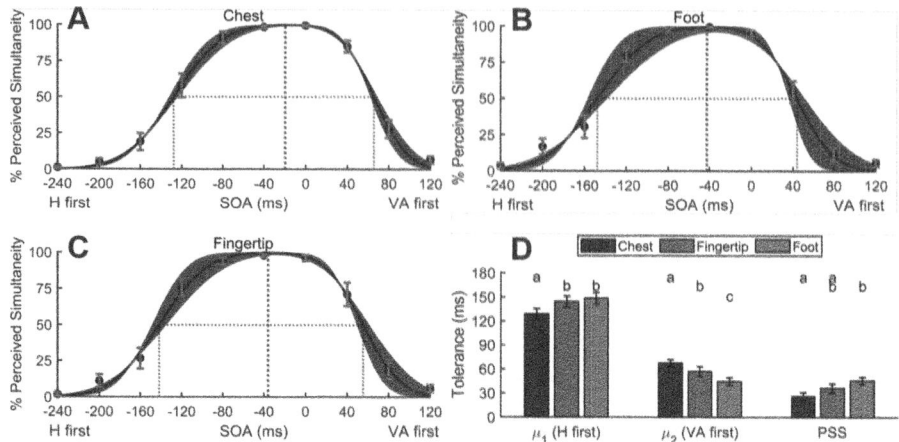

Fig. 5. Regression curves of simultaneity judgment (A)–(C) and the degrees of tolerance (D) for the three body sites. The shaded areas represent the curves within the 95% confidence intervals (CIs) for the participants' individual parameters. The bars grouped with the same letter indicate that they did not show significant differences by the Tukey tests.

1. μ_2 represents the 50%-threshold within which the audiovisual and haptic stimuli are perceived synchronous when the audiovisual stimulus precedes the haptic stimulus. In Table 1, $\mu_2 = 65.76, 55.37$, and 43.88 ms in the order of the chest, fingertip, and foot. The body site had a significant effect on μ_2 (one-way repeated-measures ANOVA; $F(2, 34) = 14.33, p < 0.001$). The chest had a significantly higher μ_2 than the fingertip and foot (Tukey test; $p < 0.05$ and $p < 0.001$), and the fingertip than the foot (p = 0.018).
2. μ_1 is the 50%-threshold within which the audiovisual and haptic stimuli are perceived synchronous for the haptic-preceding cases. $\mu_1 = -127.78, -141.80$, and -147.57 ms for the chest, fingertip, and foot, respectively. The body site had a significant effect on μ_1 ($F(2, 34) = 11.61, p < 0.001$). The chest had a significantly different μ_1 from the fingertip and foot ($p < 0.05$ and $p < 0.001$).
3. WS, the time delay interval during which the audiovisual and haptic stimuli are perceived to occur simultaneously, were 193.54, 197.17, and 191.45 ms for the chest, fingertip, and foot. The body site had no significant effects on WS ($F(2, 34) = 0.81, p = 0.453$).
4. PSS, the point of subjective simultaneity, was $-19.49, -35.79$, and -42.41 ms for the chest, fingertip, and foot, indicating a bias to haptic-leading stimuli. The body site was significant for PSS ($F(2, 34) = 4.47, p = 0.019$). The chest had a significantly different PSS from the foot ($p = 0.014$). The chest's PSS was closest to the absolute simultaneity of 0 ms.
5. σ_1 and σ_2 are the noises involved in simultaneity perception for haptic-leading and audiovisual-leading stimuli, respectively. σ_1 was 37.31, 36.41, and 38.61 ms for the chest, fingertip, and foot, and σ_2 was 25.77, 28.97, and 29.11 ms, respectively. The body site did not significantly affect either

σ_1 ($F(2,34) = 0.41, p = 0.670$) or σ_2 ($F(2,34) = 0.23, p = 0.799$). The stimulus order (audiovisual or haptic leading) had a significant effect on σ, i.e., $\sigma_2 < \sigma_1$ (two-way repeated-measures ANOVA also including the body site as an independent factor; $F(1,17) = 7.10, p = 0.016$).

6. JNDs are the minimum differences from PSS beyond which audiovisual and haptic stimuli begin to feel asynchronous. JND_{H-VA} was 108.29, 106.01, 105.16 ms for the chest, fingertip, and foot, and JND_{VA-H} was 85.25, 91.16, 86.29 ms for the same order. The body site did not have significant effects on either JND_{H-VA} ($F(2,34) = 0.46, p = 0.637$) and JND_{VA-H} ($F(2,34) = 0.30, p = 0.739$). The stimulus order had a significant effect ($F(1,17) = 9.10, p = 0.008$), so $JND_{VA-H} < JND_{H-VA}$.

4 Discussion

4.1 Summary of Results

We investigated the human temporal synchrony perception for multisensory media experiences between audiovisual and haptic information. Particularly, we measured on three representative body sites: chest, fingertip, and foot according to the paradigm of simultaneity judgment task. A media clip and appropriate vibration signal were used as stimuli, which are much more complex than the stimuli used in previous studies. The resulting simultaneous responses were regressed well to a typical bell-shaped model with adequate goodness-of-fit despite the high stimulus complexity. The regression curves showed a significant asymmetry between audiovisual-leading and haptic-leading stimuli, with a higher temporal sensitivity in the audiovisual-leading case.

The 50% percentile and even the 16% percentile of the curves for the audiovisual-leading stimuli, actually important and practical values for haptic design, were less than 90 ms. This appeared similarly across all three body sites. However, as the body site was located further away from the head, the curve appeared to shift to the left, toward the haptic-leading SOAs, thereby decreasing the tolerable audiovisual-haptic delay. This dependence on the body site was significant.

4.2 Multiple Body Sites

This study explored the synchrony perception on the three body sites. The regression curves moved to the left with the body site in the order of chest, fingertip, and foot. All parameters except WS were significantly affected by the body site. The tolerable delay with the audiovisual-leading stimuli for the three body sites ranged from 80 ms to 90 ms at the 16% percentile and from 45 ms to 65 ms at the 50% percentile.

There exist related studies for hands and feet [4,14]. According to [4], haptic sensations in the lower body had to occur significantly earlier than those in the hand to be perceived as maximally synchronous with vision. Although their results were obtained from a special group of amputees, their regression curves are consistent with ours between the foot and fingertip. Another study [14] examined such extremities specifically with different body postures within the haptic modality. Here again, haptic stimulation had to be delivered from the foot before the hand to be perceived in synchrony in the hand-foot combined condition.

It appears that our results are the first about the chest. When considering the chest together, we can come up with one theory: to feel synchrony, people should receive haptic feedback earlier the farther the stimulated body site is located from the head. This hypothesis is consistent with the fact that the farther the body site is from the head, the more time it physically takes for the signals received by the receptors to reach the brain, where information is actually processed [3]. This also can be seen in [16] with the distance between a person and a stimulus. This hypothesis certainly needs to be confirmed with more diverse stimuli and body sites.

4.3   Stimulus Complexity

Our stimuli contain information about the timing of a specific event in three sensory modalities, of which visual and auditory stimuli are from media content. The stimuli are usually at least five times longer than the simple stimuli used in the previous studies, providing irrelevant information such as the sky and buildings. Like ours, there have been some attempts to measure the synchrony perception with stimuli recorded from the real world [1,19,20]. Most of such studies have controlled the SOAs between visuals and sounds, and they typically used music or musical instruments, words, and hammer strikes as stimuli. Their specific measurements vary depending on the stimuli used, but overall, the results for music and speech clearly showed poor performance with high JND values over 150 ms and correspondingly wide integration windows due to the stimulus complexity compared to simple stimuli [1,20]. In comparison, our results range from around 110 ms for audiovisual-leading stimuli to around 200 ms for haptic-leading ones. These numbers are close to the results for real-world stimuli, given that those for simple stimuli were usually less than 50 ms [7,9,22]. One recent study [6] showed the results between ours and those of the simple stimuli, with moderately simple stimuli cued by changing color of a cube but with a background in a virtual environment and haptic feedback. PSS could vary across stimuli but does not appear to vary with stimulus complexity [20].

There could be timing delays between visual and auditory elements within media content. Such inherent delays are typically adjusted to be unnoticed when media content is created. We doubt that the audiovisual delays can have a significant impact on the synchrony perception with other modalities if they are difficult to perceive. However, the delays may become perceptually pronounced as the distance between the observer and the media clip changes [16], which may suddenly affect the synchrony perception.

4.4   Asymmetry Between Audiovisual and Haptic-Leading Stimuli

Applying an asymmetric bell-shaped model to the SJ responses allowed us to see differences in results depending on which sensory information comes first from the estimated or derived variables. In the results, a significant difference in sensitivity for synchrony perception was found between audiovisual-leading and haptic-leading stimuli, with smaller JND and standard deviation for audiovisual-leading stimuli. Such difference can be seen that two different decision criteria are involved [18], or some may explain this phenomenon through ecological validity as in [17]. However, we expect that this is due to the tendency for humans to have higher perceptual sensitivity in familiar or natural cases, whereas in the opposite situations there is greater uncertainty in the responses, leading to lower sensitivity. In fact, people often notice that the audiovisual video is followed by additionally authored haptic feedback. Relatedly, there are some studies where this presumed cause may apply [14,21]. As with our results for the audiovisual-leading stimuli, better performance was achieved on the synchrony-related task when the limbs were uncrossed than when crossed [14] and when object movement followed gravity rather than when it did not [21].

4.5   Future Work

Our future work could be to expand the results using more diverse parameters of haptic stimulation. Also, we can recruit participants of different ages, or try more different types of media content for generalization.

Acknowledgments. This work was supported by a mid-career researcher grant (2022-R1A2C2091161) from the National Research Foundation of Korea.

References

1. Altinsoy, M., Blauert, J., Treier, C.: Inter-modal effects of non-simultaneous stimulus presentation. In: Proceedings of the 7th International Congress on Acoustics (2001)
2. Binder, M.: Neural correlates of audiovisual temporal processing-comparison of temporal order and simultaneity judgments. Neuroscience **300**, 432–447 (2015). https://doi.org/10.1016/j.neuroscience.2015.05.011
3. Carr, C.E.: Processing of temporal information in the brain. Annu. Rev. Neurosci. **16**(1), 223–243 (1993). https://doi.org/10.1146/annurev.ne.16.030193.001255
4. Christie, B.P., Graczyk, E.L., Charkhkar, H., Tyler, D.J., Triolo, R.J.: Visuotactile synchrony of stimulation-induced sensation and natural somatosensation. J. Neural Eng. **16**(3), 036025 (2019). https://doi.org/10.1088/1741-2552/ab154c
5. Covaci, A., Zou, L., Tal, I., Muntean, G.M., Ghinea, G.: Is multimedia multisensorial?-A review of mulsemedia systems. ACM Comput. Surv. **51**(5), 1–35 (2018). https://doi.org/10.1145/3233774
6. Di Luca, M., Mahnan, A.: Perceptual limits of visual-haptic simultaneity in virtual reality interactions. In: IEEE World Haptics Conference (WHC), pp. 67–72. IEEE (2019). https://doi.org/10.1109/WHC.2019.8816173

7. Fujisaki, W., Nishida, S.: Audio-tactile superiority over visuo-tactile and audio-visual combinations in the temporal resolution of synchrony perception. Exp. Brain Res. **198**, 245–259 (2009). https://doi.org/10.1007/s00221-009-1870-x

8. Gengel, R.W., Hirsh, I.J.: Temporal order: the effect of single versus repeated presentations, practice, and verbal feedback. Percept. Psychophys. **7**(4), 209–211 (1970). https://doi.org/10.3758/BF03209360

9. Hirsh, I.J., Sherrick, C.E., Jr.: Perceived order in different sense modalities. J. Exp. Psychol. **62**(5), 423 (1961). https://doi.org/10.1037/h0045283

10. Konishi, Y., Hanamitsu, N., Outram, B., Minamizawa, K., Mizuguchi, T., Sato, A.: Synesthesia suit: the full body immersive experience. In: ACM SIGGRAPH 2016 VR Village, p. 1 (2016). https://doi.org/10.1145/2929490.2932629

11. Lee, D., Oh, S., Choi, S., You, B.J.: Vibrotactile metaphor of physical interaction using body-penetrating phantom sensations: stepping on a virtual object. In: IEEE World Haptics Conference (WHC), pp. 367–372. IEEE (2021). https://doi.org/10.1109/whc49131.2021.9517144

12. Park, C., Choi, S.: Perceptual simultaneity between vibrotactile and impact stimuli. In: IEEE World Haptics Conference (WHC), pp. 148–155. IEEE (2023). https://doi.org/10.1109/whc56415.2023.10224459

13. Park, J., Kim, J., Han, S., Park, C., Park, J., Choi, S.: Information transfer of full-body vibrotactile stimuli: an initial study with one to three sequential vibrations. In: IEEE World Haptics Conference (WHC), pp. 41–47. IEEE (2023). https://doi.org/10.1109/WHC56415.2023.10224391

14. Schicke, T., Röder, B.: Spatial remapping of touch: confusion of perceived stimulus order across hand and foot. Proc. Natl. Acad. Sci. **103**(31), 11808–11813 (2006). https://doi.org/10.1073/pnas.0601486103

15. Stevenson, R.A., Wallace, M.T.: Multisensory temporal integration: task and stimulus dependencies. Exp. Brain Res. **227**, 249–261 (2013). https://doi.org/10.1007/s00221-013-3507-3

16. Stone, J., et al.: When is now? Perception of simultaneity. Proc. Roy. Soc. Lond. Ser. B: Biol. Sci. **268**(1462), 31–38 (2001). https://doi.org/10.1098/rspb.2000.1326

17. Van Eijk, R.L., Kohlrausch, A., Juola, J.F., van de Par, S.: Audiovisual synchrony and temporal order judgments: Effects of experimental method and stimulus type. Percept. Psychophys. **70**, 955–968 (2008). https://doi.org/10.3758/pp.70.6.955

18. Vatakis, A., Balcı, F., Di Luca, M., Correa, Á.: Timing and Time Perception: Procedures, Measures, & Applications. Brill (2018). https://doi.org/10.1163/9789004280205

19. Vatakis, A., Navarra, J., Soto-Faraco, S., Spence, C.: Audiovisual temporal adaptation of speech: temporal order versus simultaneity judgments. Exp. Brain Res. **185**, 521–529 (2008). https://doi.org/10.1007/s00221-007-1168-9

20. Vatakis, A., Spence, C.: Audiovisual synchrony perception for music, speech, and object actions. Brain Res. **1111**(1), 134–142 (2006). https://doi.org/10.1016/j.brainres.2006.05.078

21. Vatakis, A., Spence, C.: Enhanced audiovisual temporal sensitivity when viewing videos that appropriately depict the effect of gravity on object movement. In: Vatakis, A., Esposito, A., Giagkou, M., Cummins, F., Papadelis, G. (eds.) Multidisciplinary Aspects of Time and Time Perception. LNCS (LNAI), vol. 6789, pp. 116–124. Springer, Heidelberg (2011). https://doi.org/10.1007/978-3-642-21478-3_10

22. Vogels, I.M.: Detection of temporal delays in visual-haptic interfaces. Hum. Factors **46**(1), 118–134 (2004). https://doi.org/10.1518/hfes.46.1.118.30394

23. Vroomen, J., Keetels, M.: Perception of intersensory synchrony: a tutorial review. Atten. Percept. Psychophys. **72**(4), 871–884 (2010). https://doi.org/10.3758/APP.72.4.871
24. Yun, G., Lee, H., Han, S., Choi, S.: Improving viewing experiences of first-person shooter gameplays with automatically-generated motion effects. In: Proceedings of the 2021 CHI Conference on Human Factors in Computing Systems, pp. 1–14 (2021). https://doi.org/10.1145/3411764.3445358
25. Yun, G., Mun, M., Lee, J., Kim, D.G., Tan, H.Z., Choi, S.: Generating real-time, selective, and multimodal haptic effects from sound for gaming experience enhancement. In: Proceedings of the 2023 CHI Conference on Human Factors in Computing Systems, pp. 1–17 (2023). https://doi.org/10.1145/3544548.3580787

Apparent Thermal Motion on the Forearm

Tim Moesgen[1]([✉]) [iD], Hsin-Ni Ho[2] [iD], and Yu Xiao[1] [iD]

[1] Aalto University, Espoo, Finland
tim.moesgen@aalto.fi
[2] Kyushu University, Fukuoka, Japan

Abstract. The concept of Apparent Tactile Motion (ATM) has been extensively studied in the field of haptics, allowing people to perceive a sense of dynamic motion through tactile stimuli such as vibrations, tapping or mid-air stimuli. However, there is a lack of research on whether a similar perception of motion can be achieved using thermal stimuli. As prior research suggests that particularly the stimuli onset asynchrony (SOA) of two stimuli is a significant contributor to the perception of motion, in this study, we examine different SOAs between two warm stimuli on the forearm in order to induce a sensation of motion. Our results indicate that the sensation of motion can be achieved on the forearm with SOAs close to the signal duration. We further found a negative correlation between SOAs and the perception of speed and report findings of participants' perceptions of motion through drawings. With our study, we strengthen the understanding of dynamic thermal feedback through apparent thermal motion that may lead to the development of lighter and more sustainable wearable thermal devices.

Keywords: Apparent motion · Thermal feedback · Thermal illusion

1 Introduction

Virtual reality (VR) experiences become increasingly immersive by not only producing high definition visual and auditory content but also stimulating the sense of touch and temperature through haptic interfaces [30]. Growing research further explores the rendering of not only static but dynamic experiences caused by thermal [9] or tactile stimulation [11]. Dynamic tactile feedback allows to create more nuanced tactile and thermal sensations to increase users' sense of immersion and realism in, for instance, multisensory VR experiences that simulate heat, rain, and winds [8] or for achieving dynamic user awareness of objects in the virtual surroundings [29].

The illusionary perception of motion has been examined through sensory phenomena such as the apparent tactile motion (ATM) [2,31] or the cutaneous rabbit illusion [5]. Both illusions use successive signals to create sensations of motion. While the cutaneous rabbit requires a series of closely positioned actuators, apparent tactile motion can elicit a perception of movement between two or more actuators placed at greater distances from each other, even across hands

H. Kajimoto et al. (Eds.): EuroHaptics 2024, LNCS 14768, pp. 56–68, 2025.
https://doi.org/10.1007/978-3-031-70058-3_5

[22]. ATM generates a continuous sensation on the skin between two stimuli which can potentially result in a reduced number of actuators and may lead to lighter and more energy-efficient wearable haptic interfaces or hand-held devices. To achieve ATM between two or more successive tactile signals applied to the skin, it is crucial to determine the right Stimuli Onset Asynchrony (SOA) time. SOA refers to the point in time when the subsequent signal activates, hence, it describes a delay in the actuator's activation. If this time is too large, an individual would perceive two discrete signals. Conversely, if the time is too small, the signals would be perceived as simultaneous. Only when the SOA timing is optimal can a person perceive a continuous motion between two tactile signals. Moreover, the SOA is dependent on the stimulus duration (DOS), hence an increased DOS leads to an increased SOA [24].

Research on apparent tactile motion can already be dated back to the beginning of the 19th century [2,31]. Previous studies primarily concentrated on utilizing vibrotactile feedback to explore motion sensation. However, as research in the field has expanded, more recent studies have introduced mid-air haptics as another tactile modality to produce motion sensations [21,26]. To elicit the perception of apparent motion with vibrotactile [2,22,31] or mechanical tapping stimulation [13,15] prior work has shown that particularly stimulus duration (DOS) and SOA timing show a significant effect on the perception of continuous tactile motion. Israr and Poupyrev [10] showed that, in addition, frequency of vibration signals might impact motion perception. In terms of body locations, Chu et al. [4] developed a MotionRing that produces an illusionary motion sensation through an array of vibration motors around the head. Hands as body site were examined by [7,12,22]. Pittera et al. [22] report a successful motion perception between two hands of a participant while participants in [7] perceived motion traveling from hand to the hand of another person. Takeda and colleagues [26] investigated ATM of air stimuli at cheek and ear. Perceiving motion remained achievable even when utilizing different devices with different actuator vibration frequencies, for instance, from a game controller to a smart watch on the wristband [12]. In terms of speed perception, Lacôte and colleagues [15] investigated the discrimination of velocity changes conveyed during apparent tactile motion and found that there was no significant discrimination of speed levels for different SOAs.

Since the prior research has focused specifically on apparent motion illusion induced by tactile stimuli through vibration or pressure [4,7,10,22,26,32], only very limited research has explored if motion illusions are achievable with thermal signals as well [3,6]. As an example, consider the possibility of VR enabling users to experience dynamic sensations such as hot steam or cold wind moving across their bodies. A key challenge compared to tactile stimulation here is that human spatial resolution of thermal perception is rather poor [27,28], so the question is if people would be able to distinguish different locations of traveling stimuli.

Prior thermal perception studies have investigated the effects of different SOAs and temperatures on thermal motion perception intermanually and on the face [3,6]. Gongora and colleagues [6] were able to produce a motion sensation with both warm and cold stimuli between two hands of a person. They also found that cold stimuli are more sensitive for motion quality than warm signals. Chen [3] found that participants were able to indicate the correct direction of motion with an accuracy of 70%. Nevertheless, there remains a notable gap in current research regarding the feasibility of generating thermal motion on other body sites, such as the arm (see Fig. 1), which is frequently targeted by haptic sleeves [33] or gloves [20]. Ongoing research of such devices make the arms an suitable area for generating motion sensations. For example, haptic and thermal stimuli can travel along the length of the arms, either from distal to proximal or vice versa.

To understand whether and how a motion illusion can be produced with thermal stimuli on the forearm, we conducted an experiment with ten participants and investigated (1) What is the optimal SOA to perceive the motion of a thermal stimulus produced by two Peltier elements on the forearm? Moreover, we are interested in the perception of velocity and ask (2) Does SOA influence the perception of speed for thermal stimulation?

This research contributes to the understanding of motion illusions induced by thermal feedback both for the scientific and engineering community. By exploring the relationship between thermal stimuli and perceived motion, our study contributes to advancing the apparent motion phenomenon beyond the tactile modality.

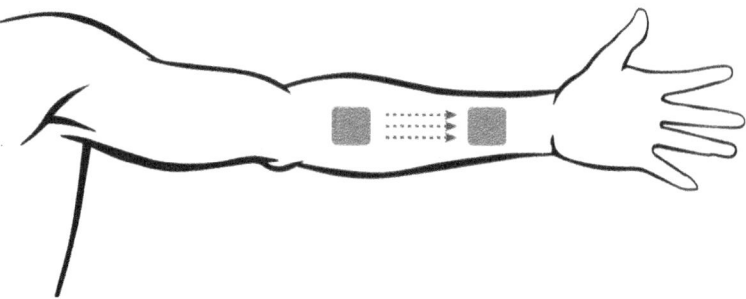

Fig. 1. Is it possible to achieve a sensation of motion across the forearm by modulating the stimulus onset asynchrony of successive thermal signals from two Peltiers attached to the arm?

2 Experiment

2.1 Participants

Ten participants (five female, five male) took part in the experimental study. Their age ranged from 19 to 41 (mean age: 29). Participants were recruited through a public call. No participant reported any abnormalities in their sense of touch or temperature on the forearms used for thermal evaluation. At the

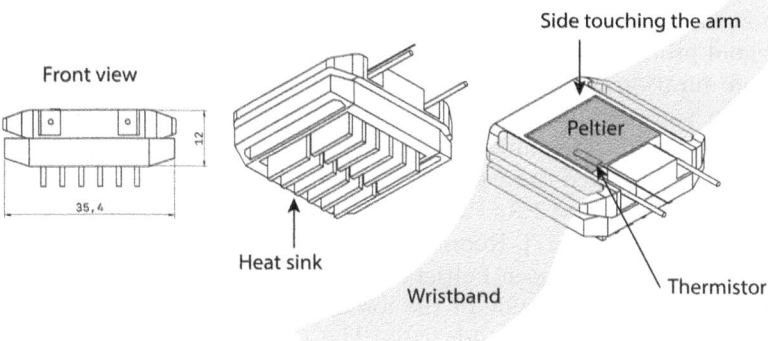

Fig. 2. Custom designed and 3D-printed case for Peltier component and heat sink, holes left and right allow to attach wristband. See in use in Fig. 5.

beginning of the experiment, participants received general information about the study context and procedure and were asked to sign a consent form. After study completion we handed them a 20€ restaurant gift card. In accordance with the Declaration of Helsinki this study was confirmed by our university's ethics committee.

2.2 Apparatus

To apply the thermal stimuli we used two bracelets each containing a Peltier cell (20 × 20 mm, CUI Devices CP20251, maximum heat output 80 °C) attached to a metal heat sink for heat dissipation. A custom designed and 3D printed case (PLA material with heat resistance of ca. 60 °C) with elastic wristbands help to attach the module to the arm (see Fig. 2). A thermistor (TEWA Sensors LLC, TT6-10KC8-9-25) placed between the skin and the Peltier module measured the temperature applied to the skin. The Peltier cells were controlled with an Arduino Uno microcontroller.

2.3 Procedure and stimuli

Based on [10] we define the concept of motion as a thermal sensation that

a) presents continuous motion,
b) has a clear start and end,
c) is perceived as a single unit that cannot be subdivided, and
d) can move with varying speed.

We presented this definition to the participants. To familiarize participants with the dynamic thermal signals on the arm and check if participants were able to perceive the two distinct spots we provided two test signals to the forearm. One of the signals had a large SOA (i.e., 12 s) indicating two discrete signals while one signal produced a hypothetical continuous motion sensation (i.e., 6 s).

The signal duration was set to 8 s. This DOS was chosen since in our pilot study it took a considerable longer time to perceive a thermal signal on the forearm than compared to the fingertip. For the SOAs, we opted for five intervals centered around the selected duration: 4 s, 6 s, 8 s, 10 s, 12 s (see Fig. 3). The rate of temperature change was at 1.8 s/°C, with a target temperature of 39 °C which is below the pain threshold [17]. Room temperature was kept at 20 °C for every participant. The spacing between Peltiers was 10 cm (see Fig. 5). We placed the Peltiers centrally on the underside of the forearm which is less hairy and more sensitive compared to the outer side [16]. The overall experiment comprised of 25 trials per participant (5 SOAs x 5 repetitions) (see Fig. 4) and lasted approximately 1.5 h. The order of trials was counterbalanced. The stimuli lasted either 12 s (with SOA of 4 s), 14 s (SOA: 6 s), 16 s (SOA: 8 s), 18 s (SOA: 10 s), or 20 s (SOA: 12 s). We allowed a break of a few minutes after a set of five trials.

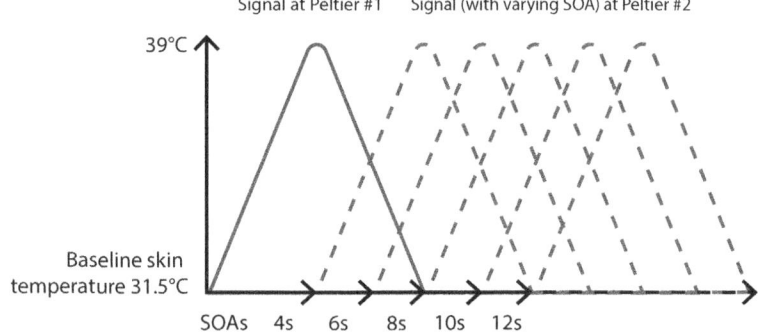

Fig. 3. Different SOAs between first and second thermal signal; each signal lasted for a fixed duration of 8 s

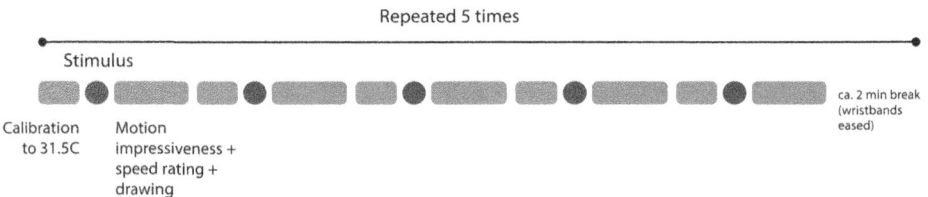

Fig. 4. Study procedure: 1. calibration phase, 2. stimulus (consists of two thermal signals with a different SOA) and 3. participant ratings and drawing plus break

To ensure a perceptual distinction between stimuli it must be ensured that a normal skin temperature is regained [23]. Based on our pilot tests on the arm and hand, we observed that this process takes a considerably higher amount of time on the forearm compared to, for instance, the fingertips (some participants described that heat is "lingering" after a stimulus is applied). This makes it more challenging to repeat trials and display multiple, successive stimuli on the arm. To ensure thermal recovery after each trial, we allowed a time period of 30 s for the skin to return to a normal temperature state, allowing thermal recovery. During that time the participant gave their ratings and drew their motion sensation. Secondly, after every set of five trials, the wristband was eased, and participants were instructed to rub their skin where the stimuli were applied. Lastly, before initiating each new trial, we calibrated the Peltiers so that the skin reaches a baseline skin temperature of 31.5 °C (normal skin temperature). Following this calibration phase, a sound was played as an indicator of the stimulus to start. After each trial participants were asked to rate the motion impressiveness from 0 to 3:

'0' indicated that no motion was felt,

'1' that motion was weak or vague,

'2' that motion is definitely present but not so clear, and

'3' indicated impressive and continuous motion from one stimulating point to the other (based on [13]).

In addition, we instructed the participant after each trial to rate the perception of speed from '0' (no motion), '1' (slow), '2' (medium), and '3' (fast). Lastly, we asked the participant to draw their perceptions on a tablet that displayed a forearm (1:1 scale) to indicate further features of the sensation such as intensities or directions. We offered suggestions for visualizing the sensations (e.g. colours or arrows) but allowed the participant to create their drawings freely. Mapping sensations in body outlines is a common research tool that allows to capture complex bodily experiences [1]. This is done by asking participants to visualize their sensations in a body map. Body mapping as a tool can, for instance,

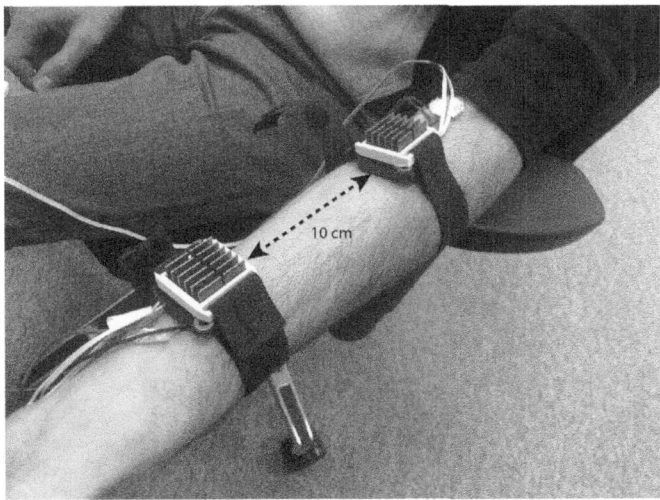

Fig. 5. Study setup of Peltiers attached with a wristband to the forearm

assist researchers to make sense of subjective bodily experiences that cannot necessarily be expressed through words or numbers.

3   Results

3.1   Ratings

We calculated the mean ratings for motion impressiveness and speed (ranging from 0 to 3) for each SOA provided by the participants and conducted a one-way ANOVA test to prove statistical signifance. We used Tukey HSD post hoc test to identify exactly which groups differ from each other. Figure 6 depicts participants' ratings on motion impressiveness and Fig. 7 speed as a function of SOA. For continuous motion induced by thermal signals of 8 s duration, the optimal SOA value (highest mean for motion impressive) is 6 s, with a corresponding mean rating of 2.22 (SD = 1.00), closely followed by 8 s (mean impressiveness rating of 1.97; SD = 1.09) and 10 s (mean = 1.6; SD = 1.09). SOA of 12 s received the worst ratings of 1.04 (SD = 1.19). A quadratic trend line was fitted to the plot ($R^2 = 0.97$). We can theorize that as the values of SOA become smaller than 4 s or larger than 12 s, participants are likely to perceive a decrease in the continuity of motion. A one-way ANOVA revealed that SOA timings have a significant effect on the motion impressiveness ($F(4, 242) = 7.8, p < 0.01$). The post hoc pairwise comparisons using Tukey HSD test (no violation against homogeneity of variances with $p > 0.05$) show that there are significant differences between 4 s and 12 s, 6 s and 12 s, 8 s and 12 s, and 6 s and 10 s ($p < 0.005$). Taken

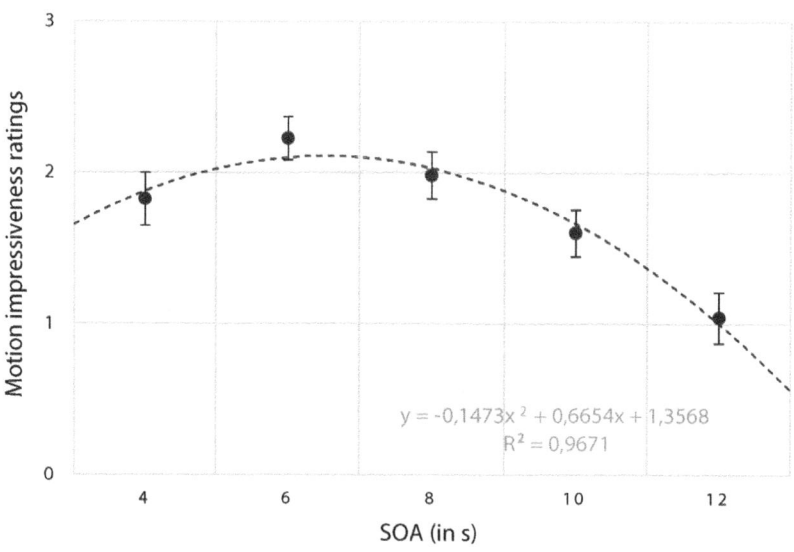

Fig. 6. Mean ratings of motion impressiveness

together, these results suggest that a higher SOA such as 12 s cause significantly less impressive motion sensations than the smaller SOAs.

For speed perception, findings reveal another noticeable trend. As SOAs decrease, there is an observable increase in the perception of speed (see linear trend line in Fig. 7). This implies that despite a fixed duration shorter SOAs may induce a feeling of faster motion. We identified a significant effect of SOA speed perception with one-way ANOVA ($F(4, 242) = 9.061, p < 0.001$). With our data meeting the assumption of homogeneity of variance ($p > 0.05$), Tukey HSD post hoc test confirms statistical differences between 4 s and 12 s, 6 s and 12 s, 8 s and 12 s, and 10 s and 12 s ($p < 0.005$).

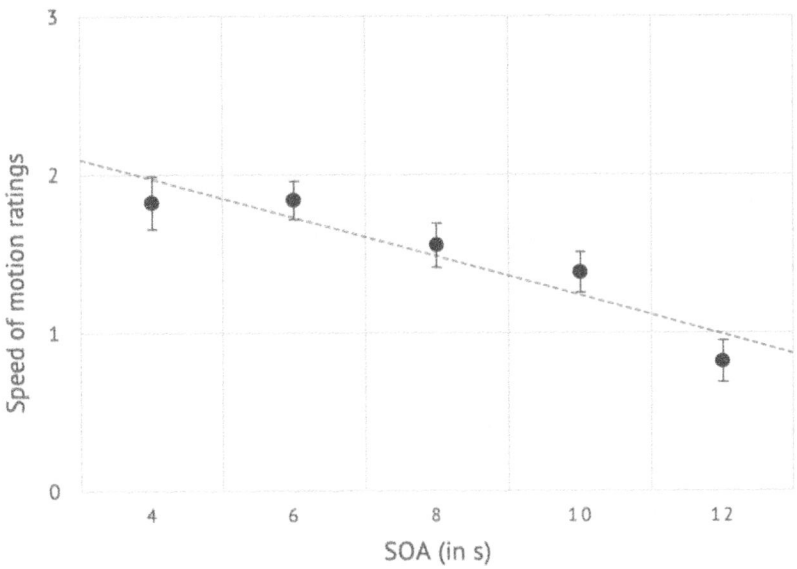

Fig. 7. Mean ratings of speed of motion

3.2 Drawings

In analyzing participants' sketches, we overlaid all drawings and generated a heat map illustrating the occurrence frequency of depicted forearm sections (see Fig. 8). Additionally, we represented commonly noted sensations such as distinctions in intensity across different areas. These visual representations align with the ratings of motion impressiveness, revealing that 12 s were perceived as less continuous motion compared to SOAs of 6 s or 8 s. Interestingly, participants indicated that when they perceived motion, it was not in the form of a slender line moving from one point to another. Instead, it manifested as 'spatial expansions' (P10), encompassing broader areas on the forearm and between the two

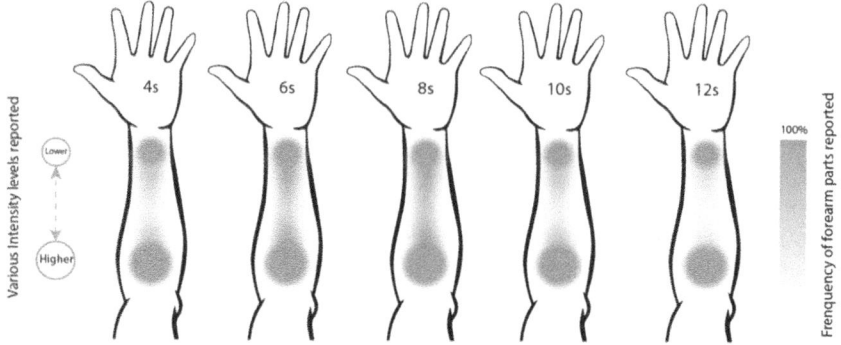

Fig. 8. Patterns of participants' sensations on the forearm produced by different SOAs

Peltiers. This sensation has been discovered in several illustrations such as in the depicted drawing in Fig. 9a. Often the signal at the wrist was perceived as less intense than the one closer to elbow. This was often visualized by a smaller area covered at the wrist (see in Fig. 9b). Different perceptions of intensity on the forearm aligns with research on thermal sensation that showed a reduced sensitivity along the arm from proximal to distal [25].

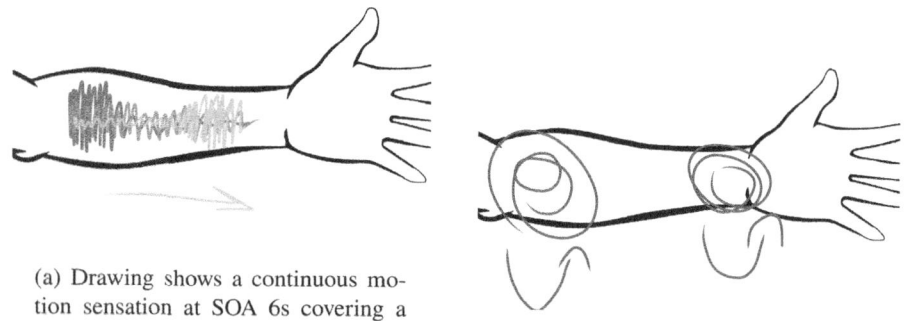

(a) Drawing shows a continuous motion sensation at SOA 6s covering a larger area of the forearm. The colors here represent the order of sensations starting from red via orange to yellow (P7).

(b) Drawing shows the sensation of two discrete signals at SOA: 12s (P10). Arrows indicate that thermal signals flare up and disappear again.

Fig. 9. Two participant drawings of thermal motion perception

4 Discussion and Conclusion

In this study, we investigated if apparent thermal motion is perceptible on the forearm and what is the optimal SOA timing to perceive clear and continuous motion. Moreover, we examined if SOA would affect the perception of speed and how people perceived motion through drawings.

To our knowledge, this is the first study to examine apparent thermal motion across the forearm. Despite spatial summation and a moderate spatial acuity of thermal stimuli in human perception [27] participants were able to perceive a spatio-temporal change in the thermal signals and reported most continuous sensation of motion on the forearm with SOAs ranging around 6 s at a DOS of 8 s. This discovery is in approximate alignment with the assertion by Gongora et al. [6], suggesting that a 40% overlap in time results in more continuous motion sensations. In our study, with an ideal SOA timing of 6 s the degree of overlap is at 25% of the DOS (8 s), while SOAs of 4 s or 8 s of SOA did not show significantly less motion continuity. Furthermore, we report a significant change in speed perception showing that lower SOAs lead to the perception of faster motion. This may allow the control of speed perception through SOA selection. However, there is a trade-off effect as in very small SOAs (<4 s) according to our hypothesis may produce a less continuous motion sensation at the same time. The speed perception task was of subjective nature due to time constraints in the overall experiment time but may be realized as discrimination task in future studies such as in [15].

The varied distribution of thermal receptors across the body [18], and differences even within various regions of the arm [25] may affect the perception of thermal signals and motion. Participants reported that the wrist region often felt less intense than the sensations produced closer to the elbow. When designing thermal motion across body regions different receptor densities and sensitivity levels need to be taken into consideration. Moreover, participants noted an increasing difficulty of discerning and assessing thermal signals over time. We recommend to further mitigate this obstacle by allowing sufficient time for the skin's thermal recovery.

Moving forward, we aim to investigate the impact of varying temperatures (both cold and warm), signal durations, distances and directions (distal to proximal) on the perception of apparent thermal motion on the arm. We also like to investigate the outer, more hairy side of the forearm that is less sensitive to thermal stimuli. We like to highlight that incorporating the drawing task into the experimental procedure proved beneficial, facilitating participants to express and discuss their thermal sensations with the experimenter. This qualitative approach helped us to interpret and understand participants perceptions. Hence, a mixed method approach of quantitative and qualitative data collection tools will support researchers to understand thermal perceptions and the design of future thermal experiences [19]. However, we want to stress that, similar to thermal perception and thermal comfort [14], drawings and verbal communication can be greatly shaped by a participant's cultural background.

The observable apparent thermal motion illusion in our study has the potential to be used for interactive thermal experiences that guide, for instance, the user's awareness with motion feedback to certain body regions or to render dynamic VR experiences such as hot steam traveling across the body. Furthermore, utilizing perceptual illusions can reduce the number of energy-intensive Peltier elements in a wearable device and, hence, contributes to a more sustainable advancement in thermal interfaces.

Acknowledgments. We would like to thank all study participants and particularly Yuhan Tseng and Esa Vikberg for supporting in the hardware development. We also thank Ramyah Gowrishankar for advising in the study design. This study has received funding from the European Union Horizon Europe Research and Innovation Programme under Grant Agreement No 101070533 and is part of EMIL (European Media Immersion Lab).

References

1. Anne Cochrane, K., et al.: Body maps: a generative tool for soma-based design. In: Sixteenth International Conference on Tangible, Embedded, and Embodied Interaction, pp. 1–14 (2022). https://doi.org/10.1145/3490149.3502262
2. Burtt, H.E.: Tactual illusions of movement. J. Exp. Psychol. **2**(5), 371 (1917). https://doi.org/10.1037/h0074614
3. Chen, Z., Peiris, R.L., Minamizawa, K.: A thermal pattern design for providing dynamic thermal feedback on the face with head mounted displays. In: Proceedings of the Eleventh International Conference on Tangible, Embedded, and Embodied Interaction, pp. 381–388 (2017). https://doi.org/10.1145/3024969.3025060
4. Chu, S.Y., Cheng, Y.T., Lin, S.C., Huang, Y.W., Chen, Y., Chen, M.Y.: Motion-ring: creating illusory tactile motion around the head using 360 vibrotactile headbands. In: The 34th Annual ACM Symposium on User Interface Software and Technology, pp. 724–731 (2021). https://doi.org/10.1145/3472749.3474781
5. Geldard, F.A., Sherrick, C.E.: The cutaneous "rabbit": a perceptual illusion. Science **178**(4057), 178–179 (1972). https://doi.org/10.1126/science.178.4057.178
6. Gongora, D., Peiris, R.L., Minamizawa, K.: Towards intermanual apparent motion of thermal pulses. In: Adjunct Proceedings of the 30th Annual ACM Symposium on User Interface Software and Technology, pp. 143–145 (2017). https://doi.org/10.1145/3131785.3131814
7. Hachisu, T., Suzuki, K.: Tactile apparent motion through human-human physical touch. In: Haptics: Science, Technology, and Applications: 11th International Conference, EuroHaptics 2018, Pisa, Italy, June 13–16, 2018, Proceedings, Part I 11, pp. 163–174. Springer (2018). https://doi.org/10.1007/978-3-319-93445-7_15
8. Han, P.H., et al.: Haptic around: multiple tactile sensations for immersive environment and interaction in virtual reality. In: Proceedings of the 24th ACM Symposium on Virtual Reality Software and Technology, pp. 1–10 (2018). https://doi.org/10.1145/3281505.3281507
9. Ho, H.N., Sato, K., Kuroki, S., Watanabe, J., Maeno, T., Nishida, S.: Physical-perceptual correspondence for dynamic thermal stimulation. IEEE Trans. Haptics **10**(1), 84–93 (2016). https://doi.org/10.1109/TOH.2016.2583424
10. Israr, A., Poupyrev, I.: Tactile brush: drawing on skin with a tactile grid display. In: Proceedings of the SIGCHI Conference on Human Factors in Computing Systems, pp. 2019–2028 (2011). https://doi.org/10.1145/1978942.1979235
11. Israr, A., Zhao, S., Schwemler, Z., Fritz, A.: Stereohaptics toolkit for dynamic tactile experiences. In: HCI International 2019–Late Breaking Papers: 21st HCI International Conference, HCII 2019, Orlando, FL, USA, July 26–31, 2019, Proceedings 21, pp. 217–232. Springer (2019). https://doi.org/10.1007/978-3-030-30033-3_17

12. Kasaei, S., Levesque, V.: Effect of vibration frequency mismatch on apparent tactile motion. In: 2022 IEEE Haptics Symposium (HAPTICS), pp. 1–6 (2022). 10.1109/HAPTICS52432.2022.9765602
13. Kirman, J.H.: Tactile apparent movement: the effects of interstimulus onset interval and stimulus duration. Percept. Psychophys. **15**(1), 1–6 (1974). https://doi.org/10.3758/BF03205819
14. Knez, I., Thorsson, S.: Influences of culture and environmental attitude on thermal, emotional and perceptual evaluations of a public square. Int. J. Biometeorol. **50**, 258–268 (2006). https://doi.org/10.1007/s00484-006-0024-0
15. Lacôte, I., Gueorguiev, D., Pacchierotti, C., Babel, M., Marchal, M.: Speed discrimination in the apparent haptic motion illusion. In: International Conference on Human Haptic Sensing and Touch Enabled Computer Applications, pp. 48–56. Springer (2022)
16. Lee, D.K., McGillis, S.L., Greenspan, J.D.: Somatotopic localization of thermal stimuli: I. A comparison of within-versus across-dermatomal separation of innocuous thermal stimuli. Somatosens. Motor Res. **13**(1), 67–71 (1996). https://doi.org/10.3109/08990229609028913
17. Lue, Y.J., Wang, H.H., Cheng, K.I., Chen, C.H., Lu, Y.M.: Thermal pain tolerance and pain rating in normal subjects: gender and age effects. Eur. J. Pain **22**(6), 1035–1042 (2018). https://doi.org/10.1002/ejp.1188
18. Luo, M., Wang, Z., Zhang, H., Arens, E., Filingeri, D., Jin, L., Ghahramani, A., Chen, W., He, Y., Si, B.: High-density thermal sensitivity maps of the human body. Build. Environ. **167**, 106435 (2020). https://doi.org/10.1016/j.buildenv.2019.106435
19. Moesgen, T., Gowrishankar, R., Xiao, Y.: Designing beyond hot and cold - exploring full-body heat experiences in sauna. In: Eighteenth International Conference on Tangible, Embedded, and Embodied Interaction (TEI '24), pp. 163–174 (2024)
20. Perret, J., Vander Poorten, E.: Touching virtual reality: a review of haptic gloves. In: ACTUATOR 2018; 16th International Conference on New Actuators, pp. 1–5. VDE (2018)
21. Pittera, D., Georgiou, O., Frier, W.: 'I see where this is going': a psychophysical study of directional mid-air haptics and apparent tactile motion. IEEE Trans. Haptics **16**, 322–333 (2023). https://doi.org/10.1109/TOH.2023.3280263
22. Pittera, D., Obrist, M., Israr, A.: Hand-to-hand: an intermanual illusion of movement. In: Proceedings of the 19th ACM International Conference on Multimodal Interaction, pp. 73–81 (2017). https://doi.org/10.1145/3136755.3136777
23. Shani, Y.Y., Lineykin, S.: Thermal cues composed of sequences of pulses for transferring data via a haptic thermal interface. Bioengineering **10**(10), 1156 (2023). https://doi.org/10.3390/bioengineering10101156
24. Sherrick, C.E., Rogers, R.: Apparent haptic movement. Percept. Psychophys. **1**(3), 175–180 (1966). https://doi.org/10.3758/bf03210054
25. Singhal, A., Jones, L.: Space-time dependencies and thermal perception. In: Haptics: Perception, Devices, Control, and Applications: 10th International Conference, EuroHaptics 2016, London, UK, July 4-7, 2016, Proceedings, Part I 10, pp. 291–302. Springer (2016). https://doi.org/10.1007/978-3-319-42321-0_27
26. Takeda, T., Niijima, A., Mukouchi, T., Satou, T.: Creating illusion of wind blowing with air vortex-induced apparent tactile motion. In: Extended Abstracts of the 2020 CHI Conference on Human Factors in Computing Systems, pp. 1–7 (2020). https://doi.org/10.1145/3334480.3382811
27. Taus, R.H., Stevens, J.C., Marks, L.E.: Spatial localization of warmth. Percept. Psychophys. **17**(2), 194–196 (1975). https://doi.org/10.3758/BF03203885

28. Tewell, J., Bird, J., Buchanan, G.R.: The heat is on: a temperature display for conveying affective feedback. In: Proceedings of the 2017 CHI Conference on Human Factors in Computing Systems, pp. 1756–1767 (2017). https://doi.org/10.1145/3025453.3025844

29. Valkov, D., Linsen, L.: Vibro-tactile feedback for real-world awareness in immersive virtual environments. In: 2019 IEEE Conference on Virtual Reality and 3D User Interfaces (VR), pp. 340–349. IEEE (2019). https://doi.org/10.1109/VR.2019.8798036

30. Wee, C., Yap, K.M., Lim, W.N.: Haptic interfaces for virtual reality: challenges and research directions. IEEE Access **9**, 112145–112162 (2021). https://doi.org/10.1109/ACCESS.2021.3103598

31. Whitchurch, A.K.: The illusory perception of movement on the skin. Am. J. Psychol. **32**(4), 472–489 (1921). https://doi.org/10.2307/1413769

32. Wu, W., Culbertson, H.: Wearable haptic pneumatic device for creating the illusion of lateral motion on the arm. In: 2019 IEEE World Haptics Conference (WHC), pp. 193–198. IEEE (2019). https://doi.org/10.1109/WHC.2019.8816170

33. Zhu, M., et al.: Pneusleeve: In-fabric multimodal actuation and sensing in a soft, compact, and expressive haptic sleeve. In: Proceedings of the 2020 CHI Conference on Human Factors in Computing Systems, pp. 1–12 (2020). https://doi.org/10.1145/3313831.3376333

Surface Tactile Presentation to the Palm Using an Aerial Ultrasound Tactile Display

Naoki Kishi$^{(\boxtimes)}$ ⓘ, Atsushi Matsubayashi ⓘ, Yasutoshi Makino ⓘ, and Hiroyuki Shinoda ⓘ

The University of Tokyo, Tokyo, Japan
`kishi@hapis.k.u-tokyo.ac.jp`

Abstract. In tactile interaction in a virtual space, tactile presentation to the palm over a wider plane is required. A highly uniform tactile stimulus is required to present a stationary flat surface on the palm, and it is necessary to identify the parameters of tactile presentation that improve uniformity. Generating ultrasound pressure distribution over a surface area improves the uniformity of tactile stimulus, whereas the perceived intensity decreases as the presentation area expands because the sound pressure is dispersed. In this study, we aimed to present a stronger surface stimulus by using spatiotemporal modulation (STM) to rapidly move the ultrasound focus within the presentation area and thus distribute the concentrated sound pressure over the entire area. We showed that tactile presentation of a square area as a simple shape can be achieved using STM by adjusting the focal path and the number of path passes per unit of time, thereby improving the perceived intensity while maintaining spatial uniformity and temporal constancy compared to acoustic field control.

Keywords: Aerial ultrasound tactile presentation · STM · Spatial uniformity · Temporal constancy

1 Introduction

In recent years, research on tactile sensing and tactile presentation on skin surfaces has progressed, leading to the realization of applications that use these technologies. Intuitive and immersive actions and interactions with others in a virtual space can be improved by projecting 3D images onto the virtual space (as in a 3D display) and overlaying tactile sensations.

Aerial ultrasound tactile presentation is one of the methods of tactile presentation when touching 3D images. This method provides tactile sensations by stimulating the fingers and palms with ultrasound waves [2,6]. One of the features of this method is that it allows tactile sensations to be presented without wearing a device on the body. The method of providing tactile sensations by wearing a device on the hand has the disadvantages that the feeling of contact between the hand and the device interferes with the tactile sensation when touching the 3D image, and that the device blocks the field of view, which obstructs

H. Kajimoto et al. (Eds.): EuroHaptics 2024, LNCS 14768, pp. 69–81, 2025.
https://doi.org/10.1007/978-3-031-70058-3_6

the 3D image projection. Therefore, aerial ultrasound tactile presentation is more suitable for tactile presentation when touching 3D images. An ultrasound transducer array is controlled to align the ultrasound phase at a point in the air, which increases the pressure at that point and presents a stimulus to the skin surface. However, when interacting with 3D images, it is necessary to present not only point stimulus by focusing ultrasound, but also surface contact sensation. For example, when a cube is placed on the palm of a hand, as in Fig. 1(a), it is necessary to present the stimulus on the entire base of the cube. In this case, spatial uniformity and temporal constancy of the tactile stimulus are essential. Spatial uniformity means that the tactile stimulus is presented over the entire area and no rapid changes in intensity are perceived within this area. Temporal constancy means that the stimulus is always presented with the same intensity throughout the entire area, regardless of time, without the sensation that the stimulus is moving within the presentation area.

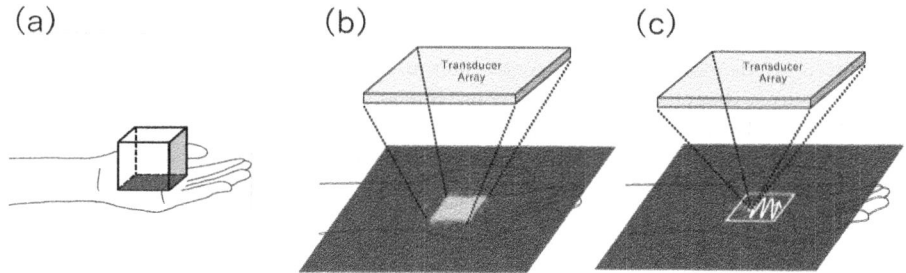

Fig. 1. (a) The drawing shows a cube on the palm of a hand. The blue area is the area where the cube and palm are in contact. (b) Tactile presentation to a surface area by sound field control. (c) Tactile presentation to a surface area by spatiotemporal modulation (STM).

The tactile stimulation of a certain surface area can be achieved by controlling the ultrasound transducers to generate a sound field that allows tactile presentation of a specific shape, as shown in Fig. 1(b). The generation of target sound pressure distributions enables tactile presentation on a surface area, such as tactile interaction with a liquid in a virtual space [9] and tactile interaction with a 3D shape [7,11]. This method can present spatially uniform and temporally constant tactile stimulus over the entire area. However, the perceived intensity is reduced compared with that of a single focus stimulus because the sound pressure is distributed over the entire area. Therefore, we investigated the tactile presentation of a surface area by fast focus shift using spatiotemporal modulation (STM) [16] as shown in Fig. 1(c). STM is a method of presenting spatially varying tactile stimulus by moving an ultrasound focus along a determined path. Compared with the sound field control method, the perceived intensity is enhanced because the sound pressure concentrated at the focal point permeates the entire area. In a related study of tactile presentation of a surface area to the palm, Jang et al. [8] showed tactile interaction with a 3D shape using STM. They showed that the tactile presentation of the surface area that constitutes the 3D shape by using STM improved the accuracy of shape and boundary recognition. Additionally, Barreiro et al. [1] realized the sensation of touching a gaseous fluid medium in a virtual space with the palm of the hand in such a way that it followed the pressure distribution of the fluid medium. They implemented an algorithm to find the optimal focus movement path of the STM based on the given target sound pressure distribution, and demonstrated that a tactile presentation that reproduces the target sound pressure distribution can be performed. In these studies, tactile presentation methods using STM were investigated to improve the shape and boundary recognition accuracy and to reproduce the target sound pressure distribution. However, spatial uniformity and temporal constancy have not been verified, and the parameters of the focus movement that improve these evaluation indices need to be clarified.

In this study, we applied tactile presentation to a surface area by STM in order to realize tactile presentation to a surface area with high spatial uniformity and temporal constancy. In the case of tactile presentation by moving the focal point, if the update frequency of the movement is low, the distance moved per step becomes larger, so that a discrete set of stimulus points is perceived on the skin of the palm. Meanwhile, in the case of tactile presentation by focus movement, the change in the focal point position is not noticed on the skin of the palm if the movement is sufficiently fast. However, the slower the speed of movement, the more the movement is noticed, and hence, the provided stimulus is no longer perceived as temporally constant. Stimulus uniformity could be maintained even with a focus movement that does not require a rapid phase shift, such as tracing only the edge of the area. Thus, we searched for parameters of focus movement that would increase the perceived intensity while maintaining spatial uniformity and temporal constancy. We performed tactile presentation using a square-shaped surface area as an example of a simple shape. In the user study, we searched for the focal movement paths and number of path passes per unit of time that would enable tactile presentation with high perceptual intensity while retaining the same degree of spatial uniformity and temporal constancy when generating pressure distributions in a surface area.

2   User Study

We conducted user studies to show that tactile presentation by STM improves the stimulus intensity while retaining spatial uniformity and temporal constancy compared with sound field reconstruction through phase optimization. Two experiments were carried out, the first to determine the effect of the focal movement path, and the second to determine the effect of the different number of path passes per unit of time.

2.1   User Study 1: Effect of Focus Movement Path

First, we conducted an experiment to determine the focal movement path that could provide high spatial uniformity, temporal constancy, and tactile stimulus intensity.

(a)   (b)

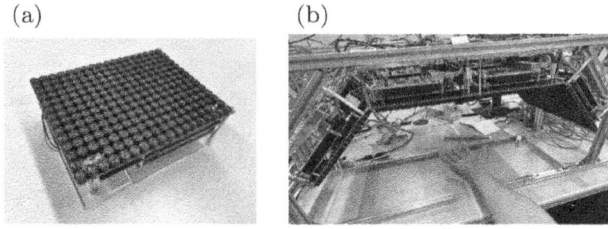

Fig. 2. (a) Airborne ultrasound tactile display (AUTD). (b) User study setup.

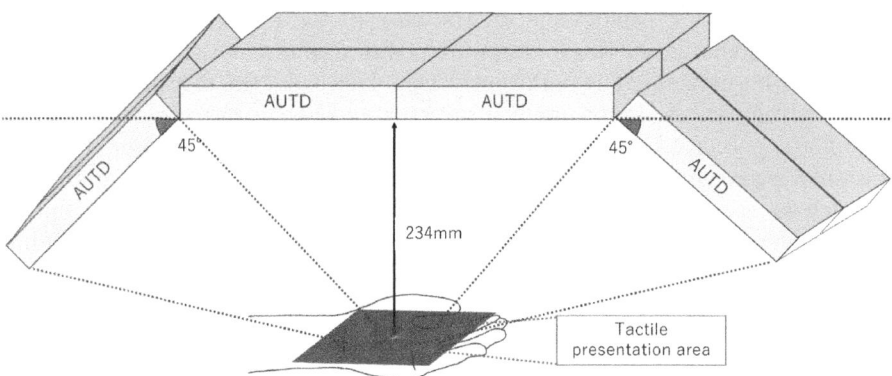

Fig. 3. Overview of the user study system. A hand was fixed to the bottom center of the system, where 8 AUTDs were set up, with the palm up, and the surface tactile sensation was presented on the palm.

Implementation. For tactile presentation, airborne ultrasound tactile displays (AUTDs) [15] were used in this study (Fig. 2(a)). Each AUTD consisted of 249 ultrasound transducers. We implemented a tactile presentation system with two rows of four AUTDs arranged in an arch shape, as shown in Fig. 2(b). Figure 3 exhibits a front view of the system. The participants held the back of their hands down to touch the table and fixed their palms at the height of the plane to which the focus moved, and tactile presentations were realized on the surface area of their palms.

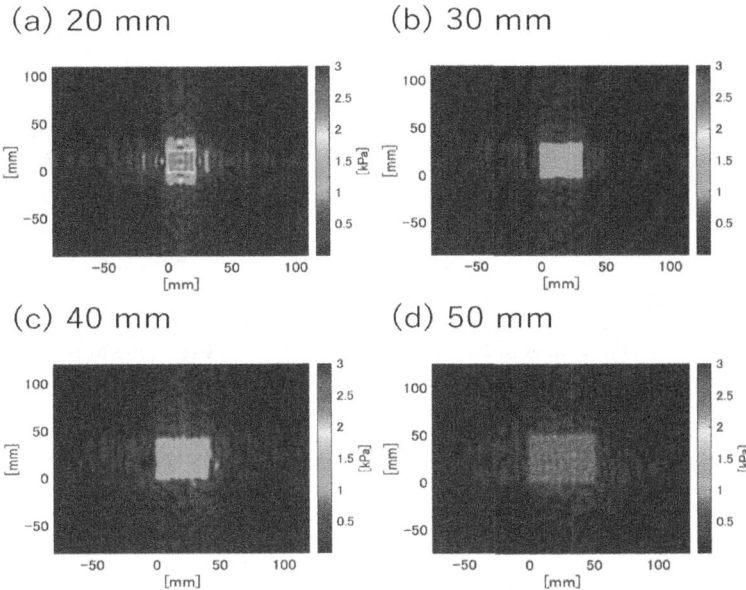

Fig. 4. Sound pressure distribution in the surface area.

For square-shaped sound field control, we used the method of Matsubayashi et al. [13]. Figure 4 shows the simulation of sound pressure distribution in a square region with four different sizes (20, 30, 40, and 50 mm on a side). In all cases, (a)–(d), a uniform sound pressure distribution was generated in the center of a square area of 200 mm per side. This square area was divided into fine square grids of 1 mm per side, and a target sound pressure value was set on the points of each grid in the surface area where tactile presentation was performed. The amplitude of sound pressure was determined so that the sound pressure distribution within the surface area would be constant. As a result, the sound pressure was set to (a) 2.28 kPa at 20 mm, (b) 1.425 kPa at 30 mm, (c) 1.14 kPa at 40 mm, and (d) 0.798 kPa at 50 mm. Outside of the surface area for tactile presentation, the amplitude of sound pressure on each grid point was set to 0. Then, a nonlinear least-squares optimization called the Levenberg-Marquardt method [10,12] was used to generate the sound pressure distribution. Figure 4 shows that, the sound pressure inside the area for tactile presentation is sufficiently higher than the pressure outside the area. In fact, the sound pressure outside the area is almost zero. Furthermore, the sound pressure value decreases as the area of tactile presentation expands.

Four different paths were used in the experiment, as shown in Fig. 5(a)–(d). In this experiment, the output of all AUTD transducers was always maximized in all trials. Figure 6 presents the simulation of the sound pressure distribution of a single ultrasound focus. Because transducers are arranged in long horizontal arrays, it has an elliptical shape with a long diameter of approximately 10 mm

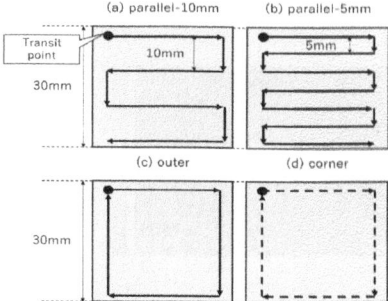

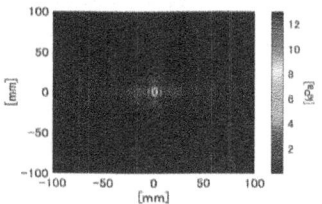

Fig. 5. Focus movement paths in the STM. (a) Path repeating a parallel shift along the edge followed by a 10 mm vertical shift. (b) The same path as in (a) but with vertical shift of 5 mm. (c) Path that moves along the perimeter of the area. (d) Path that moves only to the vertices of the region in sequence.

Fig. 6. Simulation of sound pressure distribution at a single focus moved in STM.

and a short diameter of 5 mm, and it was generated on each of the points on the path. As the phase shift of the AUTD transducers takes place in discrete time, the focal points actually travel in sequence at equally spaced discrete points along the path. Based on the fact that the maximum frequency of the AUTD phase switching is approximately 40 kHz, the transit points of the ultrasound focus set on the path were defined according to the number of times the focal point moves along the entire path per second, such that the focal point moved to the next transit point at a frequency of 40 kHz. Hereafter, this frequency is referred to as the 'STM frequency.' For example, if the focal point moved along the path 200 times per second (STM frequency = 200 Hz), the number of transit points per round is 40000/200 = 200 because the transit points are switched 40,000 times per second. We designed the movement paths shown in Fig. 5(a) 'parallel-10 mm' and (b) 'parallel-5 mm' based on the idea that tactile stimulus of uniform intensity over the entire region can be effectively presented by STM along a path such that the focus moves all over the area. Focal point movement begins at the vertex of the square. Starting from a point on a square edge, the focal point moves parallel along the edge orthogonal to that edge until it reaches the opposite edge, then moves along the opposite edge by 10 mm in (a) and by 5 mm in (b). These movements are repeated until the focus passes through all vertices. This is the path of travel for one round, and after reaching the end of the path, the focus returns to the start point and moves along the path repeatedly. The movement path in Fig. 5(a) was designed with a width of 10 mm to avoid overlapping the passing area of the elliptical ultrasound focal point. In Fig. 5(b), the width is halved to 5 mm so that the high-pressure region of the focal point can move around the area universally. The 'outer' path in Fig. 5(c) is the movement path along the perimeter of the tactile presentation area. The 'corner' path shown in Fig. 5(d) is a path that passes only through the

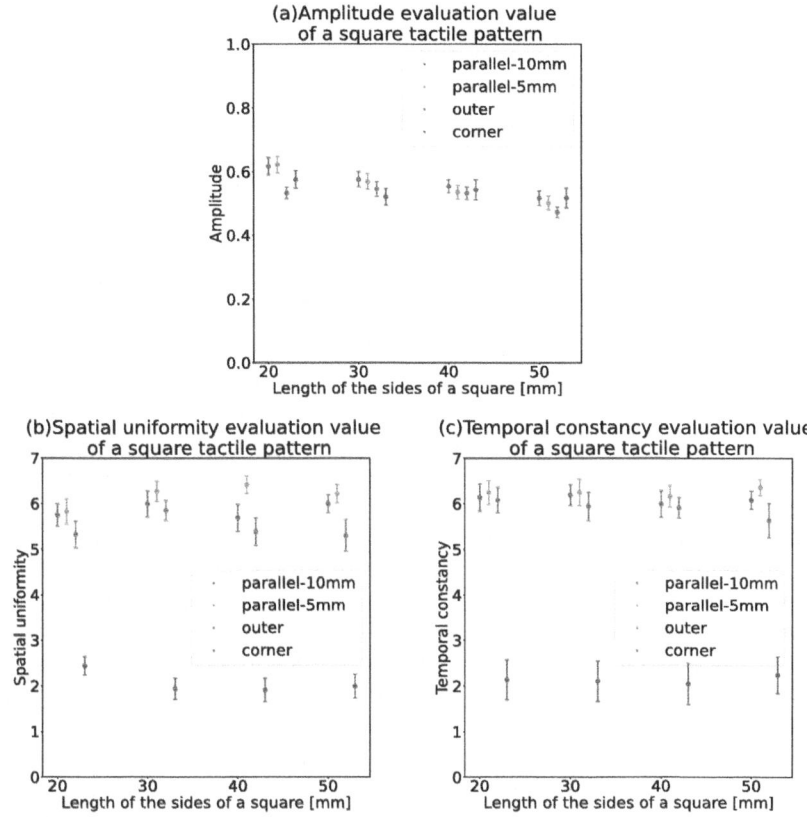

Fig. 7. Average evaluation of stimulus intensity, spatial uniformity, and temporal constancy for each focal movement path. The plots are shown when the length of a side of the square is 20, 30, 40, and 50 mm. Each plot is shifted so that the standard error bars do not overlap.

vertices of the tactile presentation area. These paths have been used in shape recognition by STM. We designed these two paths in the expectation that the tactile presentation of a surface stimulus would be sufficiently uniform if only the edge was presented without filling the interior.

Procedure. This section describes the flow of the evaluation trials for comparison by focal movement path. First, the participant was provided with a reference tactile pattern, which was the square-area sound field generated by phase optimization. An amplitude modulation of a sine pulse of 150 Hz was applied. Next, the participant was presented with the tactile pattern to be stimulated by the STM using the movement path to be evaluated. The STM frequency was fixed at 200 Hz for the parallel-5 mm, parallel-10 mm, and outer paths, and at 10 Hz for the corner path. 200 Hz accords with the frequency at which pacinian corpuscles are most likely to be sensitive. The reason why the frequency was set to 10 Hz only for the corner path was to make it easier for the participant to perceive the difference in temporal constancy. The participant was asked to adjust the perceived intensity of the evaluation tactile pattern so that the intensity was the same as that of the reference tactile pattern. The intensity values were set up in 101 steps between 0 and 1, and the intensity value was set to 0.5 when the first evaluation tactile pattern was presented. This value corresponds to how many times the ultrasound amplitude emitted from each transducer should be multiplied by the maximum amplitude. For example, if the intensity value was 0.5 after adjusting the perceptual intensity, it means that the evaluation pattern was rated to have the same perceptual intensity as the reference pattern when the ultrasound amplitude of the evaluation pattern was 0.5 times the maximum amplitude. After adjusting for the perceived intensity, the participant was asked to give a 7-point rating of spatial uniformity and temporal constancy. In both evaluation cases, 7 was the highest rating and 1 was the lowest one, and a comparative evaluation was made with the reference tactile pattern as 7. The spatial uniformity was defined as 7 when the perception was uniform without gaps as much as reference tactile patterns, and 1 when it was totally not uniform. The evaluation value was lowered as unevenness or gaps were perceived. The temporal constancy was defined as 7 when the tactile stimulus was perceived as stationary, and 1 when a sensation of movement or vibration was perceived significantly. The evaluation value was lowered as the sensation of tactile stimulus movement on the palm or temporal vibration in intensity was perceived. As the ultrasound focus movement by STM generates driving noise, we set a limit on the phase change within the range that did not affect the focus movement to reduce the noise [14]. As the driving noise could be heard even when suppressed, the participant worked while hearing white noise in order to avoid being affected by the noise.

Results. We conducted experiments on four different sizes of square areas (20, 30, 40, and 50 mm per side), and showed the relationship between the focal movement path and the uniformity and intensity of the stimulus. For each situation, a total of 36 trials were performed by 12 participants, and the average of the evaluation values was calculated. Figure 7 shows the average evaluation of stimulus intensity, spatial uniformity, and temporal constancy for each focal movement path. The bars at each point in the graph indicate the standard error.

Figure 7(a) shows that the average intensity adjustment value was within the range of 0.47–0.63, regardless of the edge length of the square area or the focal movement path. In other words, for any evaluation tactile pattern, the perceived intensity of the reference pattern was less than 63% of the maximum output of that being evaluated. Thus, for any focal movement path, tactile presentation to the surface area by STM could provide greater stimulus intensity than the method using sound pressure distribution generation. In addition, the average value of strength adjustment tended to decrease as the length of the sides of the square area increased. In the STM, as the path length increases, the stimulus time per point decreases, and the intensity is expected to decrease. The fall range was smaller than the surface sound pressure distribution presentation. This phenomenon is called temporal integration. When tactile stimulus is presented to pacinian corpuscles, it is known that the detection threshold decreases as the duration of the stimulus increases [5]. As for the movement paths, there was no significant difference in the average intensity adjustment for square areas of any side length.

It would be expected that longer path lengths in STM would decrease the perceived intensity because the stimulus duration per point would be shorter, but this was not the case. As the intensity of the surface sound field presented as the reference pattern decreases when the square size increases, the intensity values may not have varied.

The spatial uniformity and temporal constancy showed significant differences between the corner path and the other three paths from Fig. 8(b) and (c). The difference in spatial uniformity between the outer path, where only edges were stimulated, and the parallel paths, where the interior is also stimulated, were large when the side length of the square area was 40 and 50 mm. We observed that the effect of an increase in the area of the interior that could not be stimulated became more apparent as the length of the edges increased. Additionally, between the parallel-5 mm and parallel-10 mm paths, the parallel-5 mm path always had the higher evaluation value. Spatial uniformity is considered to be higher if the receptors to which the tactile stimulus is presented are more densely clustered. Based on the movement path of tactile stimulus, its density is expected to be higher in the order of parallel-5 mm, parallel-10 mm, and outer paths. Among these paths, the density is supposed to be particularly lower in the outer one. These differences in density affect the differences in spatial uniformity.

In parallel-10 mm, parallel-5 mm, and outer paths, the intensity adjustment value was less than 0.63, which is far lower than 1, and the spatial uniformity and temporal constancy were the highest at approximately 6. The spatial uniformity was particularly different from that of the outer path when the square area was large. Particularly, the STM tactile presentation method with a parallel-5 mm focal movement path enabled surface tactile presentation with greater perceived intensity while keeping the highest spatial uniformity and temporal constancy compared with the same square surface area for sound pressure distribution generation. It was also found that outer maintained spatial uniformity and temporal constancy comparable to parallel-5 mm and parallel-10 mm, even with shorter path lengths, that is, with hardware processing savings.

2.2   User Study 2: Effect of STM Frequency

Next, we conducted experiments to find an STM frequency with high spatial uniformity, temporal constancy, and tactile stimulus intensity. As mentioned above, STM frequency was defined as the number of times the focal point moves along the entire path per second. The STM frequency was varied using the tactile presentation method of STM with a parallel-5 mm focal movement path, which was shown to be the most effective in the former experiment. In the experiment,

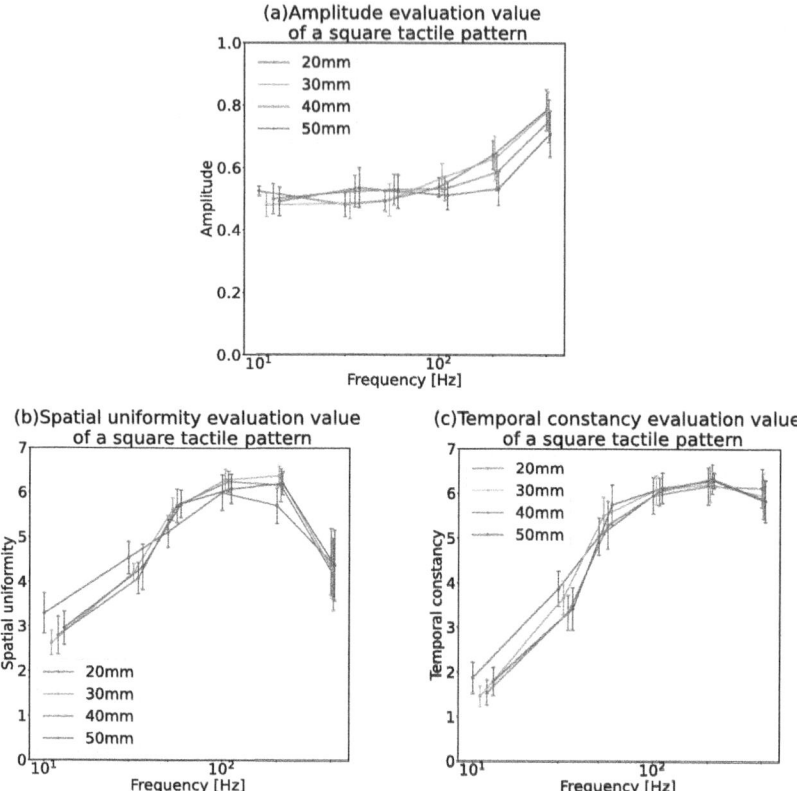

Fig. 8. Average evaluation of stimulus intensity, spatial uniformity, and temporal constancy for each STM frequency. The plots correspond to STM frequencies of 10, 30, 50, 100, 200, and 400 Hz. Each plot is shifted so that the standard error bars do not overlap.

we performed tactile presentations of the same four square areas as those in user study 1 at six different frequencies (10, 30, 50, 100, 200, and 400 Hz), and evaluated the intensity adjustment value, spatial uniformity, and temporal constancy by comparing them with the reference method. For each condition, a total of 24 trials were performed by 8 participants, and the average of the evaluation values was calculated.

Figure 8 is a logarithmic graph that represents the average evaluation of stimulus intensity, spatial uniformity, and temporal constancy at each frequency. The bars at each point in the graph represent the standard error.

Figure 8(a) shows that the intensity adjustment values were within the range of 0.48–0.57 for STM frequencies in the range of 10–100 Hz. The intensity adjustment value increased as the STM frequency increased from 200 to 400 Hz, and the intensity adjustment value at 400 Hz was approximately 0.71–0.79. This means that the stimulus intensity achievable by the STM surface tactile presentation was maintained higher until the STM frequency reached 100 Hz, and decreased when the STM frequency exceeded 100 Hz. Additionally, at STM frequencies of 200 and 400 Hz, the strength adjustment value tended to be smaller as the side length of the square area increased, especially between 30 and 40 mm, and between 40 and 50 mm. Meanwhile, at lower STM frequencies, the intensity adjustment values tended to increase with the side length of the square region. Furthermore, in 24 of the 96 trials with the STM frequency set to 400 Hz, the participant attempted to tune the intensity adjustment value to a value higher than 1. This behavior was concentrated in some participants, and it can be said that the intensity of the reference tactile pattern was greater in the trials. In the graph in Fig. 8(a), the intensity adjustment value in the above 24 trials was set to 1.

When the STM frequency increased to 400 Hz, the increase in phase change width was supposed to have decreased the output power. Moreover, we considered that the sound pressure of the reference tactile pattern decreased as the proposed area became larger, and the intensity to be aligned decreased. Additionally, as shown in Fig. 8(c), the temporal constancy decreased with the side length of the square areas when the STM frequency was low. Therefore, in that case, the perceived intensity of the tactile stimulus was lowered due to the time when it was weak, and the intensity adjustment value was raised.

Figure 8(b) and (c) show a peak at 100–200 Hz. The spatial uniformity and temporal constancy were particularly low at low STM frequencies. This might be due to the slow focus speed, which made the sensation of the focus movement perceptible in the palms, thus decreasing the spatial uniformity and temporal constancy. When the STM frequency was 100–200 Hz, the spatial uniformity and temporal constancy were close to 6, indicating that tactile presentation of a surface area by STM did not significantly impair spatial uniformity and temporal constancy compared to a method of sound pressure distribution generation.

In the case of the STM frequency of 100 Hz and the parallel-5 mm movement path, the tactile presentation on a square plane of 50 mm per side showed sufficient perceived intensity, and the movement speed of the ultrasound focus was 60 m/s. It was faster than the 5–8 m/s movement speeds at which the peak perceived intensity was recorded in the study by Frier et al. [3] Moreover, in the study by Frier et al. [4], the perceived intensity was greater when the transit point was 10 points, while in the case of the STM frequency of 100 Hz and the parallel-5 mm movement path, the number of transit points was 400. In the proposed method, the transit points were set densely within the plane, so the distance between each other was short. When tactile stimulus was presented to the points close to each other, the same receptors might have been stimulated. As a result, it is considered that the same receptors are stimulated for a sufficiently long time even if the stimulation time at each transit point is short, thus showing a different trend between movement speed, number of transit points, and perceived intensity from that in the previous studies.

Based on the above, we showed that the tactile presentation of a surface area using STM with a parallel-5 mm focal movement path could be achieved with a larger stimulus intensity with an STM frequency in the range of 100–200 Hz, without losing spatial uniformity and temporal constancy, compared to the method using the square area sound field generated using phase optimization.

3 Conclusion

In this study, we proposed a method for tactile presentation of a surface area on the palm. Given that the perceived intensity was reduced when using the sound field control method, we performed tactile presentation of the surface area using STM. To realize tactile presentation of a stationary flat surface area, we searched for a tactile presentation method capable of retaining spatial uniformity and temporal constancy. In the user study, a tactile presentation using STM with a focal movement path over a square area was designed and compared to the generation of a sound pressure distribution over the entire presentation area. We verified whether it was possible to improve the stimulus intensity while keeping spatial uniformity and temporal constancy. As a result, in the tactile stimulus presentation using the parallel-5 mm focal movement path, when the STM frequency was 100–200 Hz, the stimulus intensity could be increased without significantly losing spatial uniformity and temporal constancy compared to the generation of sound pressure distribution over the entire presentation area. On the other hand, if the performance of the hardware that presents the tactile sensation improves, the perceived intensity might not decrease even if the phase switching width in the STM is increased. In this experiment, we found that only STM at high frequencies could provide a uniform tactile presentation over a surface area. However, the tactile sensation when touching an object is essentially a pressure sensation without vibration, and it is difficult to realize such a sensation using this method. Future work includes the development of a tactile presentation method for a flat surface that reduces the sensation of vibration perceived by the palm while maintaining a high intensity of the stimulus.

References

1. Barreiro, H., Sinclair, S., Otaduy, M.A.: Path routing optimization for STM ultrasound rendering. IEEE Trans. Haptics **13**(1), 45–51 (2020). https://doi.org/10.1109/TOH.2019.2963647
2. Carter, T., Seah, S.A., Long, B., Drinkwater, B., Subramanian, S.: Ultrahaptics: multi-point mid-air haptic feedback for touch surfaces. In: UIST 2013 - Proceedings of the 26th Annual ACM Symposium on User Interface Software and Technology, pp. 505–514 (2013). https://doi.org/10.1145/2501988.2502018
3. Frier, W., et al.: Using spatiotemporal modulation to draw tactile patterns in mid-air. In: Prattichizzo, D., Shinoda, H., Tan, H.Z., Ruffaldi, E., Frisoli, A. (eds.) EuroHaptics 2018. LNCS, vol. 10893, pp. 270–281. Springer, Cham (2018). https://doi.org/10.1007/978-3-319-93445-7_24
4. Frier, W., Pittera, D., Ablart, D., Obrist, M., Subramanian, S.: Sampling strategy for ultrasonic mid-air haptics. In: Conference on Human Factors in Computing Systems - Proceedings, pp. 1–11 (2019). https://doi.org/10.1145/3290605.3300351

5. Gescheider, G., Bolanowski, S., Verrillo, R.: Some characteristics of tactile channels. Behav. Brain Res. **148**(1), 35–40 (2004). https://doi.org/10.1016/S0166-4328(03)00177-3
6. Hoshi, T., Takahashi, M., Iwamoto, T., Shinoda, H.: Noncontact tactile display based on radiation pressure of airborne ultrasound. IEEE Trans. Haptics **3**(3), 155–165 (2010). https://doi.org/10.1109/TOH.2010.4
7. Inoue, S., Makino, Y., Shinoda, H.: Active touch perception produced by airborne ultrasonic haptic hologram. IEEE World Haptics Conf. WHC **2015**, 362–367 (2015). https://doi.org/10.1109/WHC.2015.7177739
8. Jang, J., Frier, W., Park, J.: Multimodal volume data exploration through mid-air haptics. In: Proceedings - 2022 IEEE International Symposium on Mixed and Augmented Reality, ISMAR 2022 **D**, pp. 243–251 (2022).https://doi.org/10.1109/ISMAR55827.2022.00039
9. Jang, J., Park, J.: SPH fluid tactile rendering for ultrasonic mid-air haptics. IEEE Trans. Haptics **13**(1), 116–122 (2020). https://doi.org/10.1109/TOH.2020.2966605
10. Levenberg, K.: A method for the solution of certain non-linear problems in least squares. Q. Appl. Math. **2**(2), 164–168 (1944)
11. Long, B., Seah, S.A., Carter, T., Subramanian, S.: Rendering volumetric haptic shapes in mid-air using ultrasound. ACM Trans. Graph. **33**(6), 1 (2014). https://doi.org/10.1145/2661229.2661257
12. Marquardt, D.W.: An algorithm for least-squares estimation of nonlinear parameters. J. Soc. Ind. Appl. Math. **11**(2), 431–441 (1963). https://doi.org/10.1137/0111030
13. Matsubayashi, A., Makino, Y., Shinoda, H.: Accurate control of sound field amplitude for ultrasound haptic rendering using the Levenberg-Marquardt method. In: IEEE Haptics Symposium, HAPTICS **2022-March**, pp. 1–6 (2022). https://doi.org/10.1109/HAPTICS52432.2022.9765564
14. Suzuki, S., Fujiwara, M., Makino, Y., Shinoda, H.: Reducing amplitude fluctuation by gradual phase shift in midair ultrasound haptics. IEEE Trans. Haptics **13**(1), 87–93 (2020). https://doi.org/10.1109/TOH.2020.2965946
15. Suzuki, S., Inoue, S., Fujiwara, M., Makino, Y., Shinoda, H.: AUTD3: scalable airborne ultrasound tactile display. IEEE Trans. Haptics **14**(4), 740–749 (2021). https://doi.org/10.1109/TOH.2021.3069976
16. Takahashi, R., Hasegawa, K., Shinoda, H.: Tactile stimulation by repetitive lateral movement of midair ultrasound focus. IEEE Trans. Haptics **13**(2), 334–342 (2020). https://doi.org/10.1109/TOH.2019.2946136

Humans Terminate Their Haptic Explorations According to an Interplay of Task Demands and Motor Effort

Michaela Jeschke[1]([⊠]) [ID], Anna Metzger[2] [ID], and Knut Drewing[1] [ID]

[1] Justus-Liebig University, 35390 Giessen, Germany
Michaela.Jeschke@psychol.uni-giessen.de
[2] Bournemouth University, Poole BH12 5BB, UK

Abstract. Haptic exploration is an inherently active process by which humans gather sensory information through physical contact with objects. It has been proposed that humans generally optimize their exploration behavior to improve perception. We hypothesized that the duration of haptic explorations is the result of an optimal interplay of sensory and predictive processes, also taking costs such as motor effort into account. We assessed exploration duration and task performance in a two-alternative forced-choice spatial frequency discrimination task under varying conditions of task demand and motor effort. We manipulated task demands by varying the discriminability of virtual grating stimuli and manipulated motor effort by implementing forces counteracting the participants' movements while switching between stimuli. Participants were instructed to switch between stimuli after each swipe movement. Results revealed that higher task demands lead to higher numbers of exploratory movements (i.e. longer exploration duration), likely reflecting a compensatory mechanism that enables participants to attain a certain level of task performance. However, this effect is reduced when motor effort is increased; while low and medium task demands yield similar numbers of movements regardless of related motor effort, higher demands are not associated with increased numbers of movements when the required motor effort is high. In conclusion, the extent to which increased task demands are compensated via the extension of an exploration seems to depend on the motor costs that the agent is confronted with.

Keywords: Haptic exploration · behavioral optimization · motor control

1 Introduction

Humans gather sensory information about objects, materials, or textures via active touch, i.e. by haptic exploration. Previous research has illustrated the adaptive nature of such explorations with various examples: In texture perception for instance, humans systematically use different scanning velocity patterns depending on the perceptual task [1] or they gradually adjust their movement direction to the orientation of grating textures over the course of an exploration [2]; or in curvature perception, participants update

H. Kajimoto et al. (Eds.): EuroHaptics 2024, LNCS 14768, pp. 82–93, 2025.
https://doi.org/10.1007/978-3-031-70058-3_7

their contact force according to the present surface curvature [3]. The adjustments of those exploration parameters serve as optimization mechanisms, increasing task performance [2, 4, 5] and efficiency of explorations [6]. Another parameter that might be subject to optimization is the duration of haptic explorations, i.e. the time individuals dedicate to touching and perceiving objects. The amount of time spent on an exploration can drastically differ across situations and objectives: Humans might be rather quick when checking for keys in their pocket or lifting the milk package to check whether it is empty, though explorations might take longer when they do woodwork and investigate the smoothness of edges, or when surgeons inspect tissues and organs during procedures.

Humans typically conduct serial exploration with repetitive movements, i.e. multiple indentations for deformable objects or swiping and rubbing movements for rough surfaces [7]. The percept then results from the integration of the sensory information gained during each of these single movements [8]. Prolonged exploration, i.e. the extension of the exploration over space and time, was shown to increase perceptual reliability [9–13] (up to a saturation point). This is consistent with the Maximum Likelihood Estimation (MLE) model of optimal integration of information [14]. More recent (modified Kalman filter-) models also take memory limitations into account [10, 15], and hence are able to explain why the increase in the percept's reliability in a prolonged exploration is overestimated by the MLE model [16].

Various factors have been previously shown to produce or affect optimization behavior (e.g., adjustment of contact forces and movement orientation) during haptic explorations, such as the (expected) task difficulty [5], the availability of prior information [17, 18], or motivational factors related to the individual relevance of the task and goal [19]. However, it is not yet clear how these factors interact in determining the time after which individuals decide to terminate their exploration. Given that the spatiotemporal extension of a haptic exploration increases the differential sensitivity of an agent, we could predict that the exploration would be extended in response to increased task demands to maintain a certain level of task performance. As the benefits of the extension of the exploration come at a cost of additional motor effort, we could also predict that it has a counteracting effect on the exploration length.

It is widely established in fields such as ecology and economics that costs and benefits are central determinants of behavior, combined to form a utility function that can guide choices (e.g., [20]). This principle has been demonstrated to apply to motor control as well, with motor effort being represented as costs which humans seek to minimize and weigh up against potential positive outcomes [21]: Motor control has been shown to be an optimal process derived from the maximization of the weighted differences between anticipated rewards and motor efforts, i.e. energetic expenditure, with examples such as walking, flying, or reaching [22]. Extending this concept to the gathering of haptic information, we speculate that haptic explorations are similarly governed by closed-loop and open-loop optimization processes following the principles of established utility frameworks for decision and motor control with regards to cost-benefit discounting. More concretely, when being confronted with higher motor costs, humans should terminate their exploratory behavior earlier than when being confronted with lower motor costs. In line with this, humans typically stop their exploration before reaching the abovementioned saturation point of task performance [11, 12, 15]: Motor costs appear to be a plausible reason why humans would display this premature termination behavior. Here we assessed the exploration duration as a function of task demands and the motor effort required to explore an object.

In our experiment, participants in each trial explored two virtual gratings (rendered by a force-feedback device) and decided which one had higher spatial frequency. Exploration duration can be operationalized in multiple ways, e.g. as the overall time spent on the exploration, the length of movement trajectories, or as the number of (repetitive) movement segments that are being executed. Here, we operationalized duration as the number of individual swiping movements over the stimuli (= swipes). Participants were instructed to switch after every swipe, which yields maximum perceptual performance by minimizing "memory loss" [16] and allows for thorough control of motor effort. Motor effort was manipulated by implementing forces that counteract the participants' movements when switching between the virtual stimuli. Task demand was manipulated by varying the discriminability of the stimuli. We hypothesized that with increasing task demands, the exploration duration would increase, resembling a compensatory mechanism. Further, we expected increased motor efforts to reduce the expected utility of an exploration extension, leading to earlier termination of the exploration when compared to lower motor efforts.

2 Methods

2.1 Participants

Previous studies that investigated optimization processes in exploratory movement control oftentimes reported large effect sizes (e.g., [2, 17]). A sample size calculation for a large effect (f = .40), a power of 80% and an alpha of 5% yields a sample of 12 for a three-level within-participant factor of a repeated measures ANOVA (G*Power, [23]). In the present study, 12 right-handed students from Justus-Liebig University Giessen participated (8 female, 4 male, age 19–35 years, mean: 24.89 years). None of them reported motor or sensory impairments. All participants were naïve to the purpose of the experiment, provided written informed consent, and received financial compensation (8€/hour). The experiment was approved by the local ethics committee LEK FB06 and conducted in accordance with 2013 Declaration of Helsinki, except for preregistration.

2.2 Setup and Stimuli

The setup (Fig. 1a) consisted of a Geomagic Touch™ haptic force feedback device (spatial resolution: ~ 0.055 mm, temporal resolution: 1000 Hz). This was embedded in a virtual-reality environment to guide the participants through the experiment while preventing them from seeing their actual interaction with the haptic device. For this, we used the HTC Vive Pro Eye virtual head-mounted-display (HMD; 1440×1600 pixels per eye, 90 Hz; HTC Corp., Xindian, New Taipei, Taiwan). The experiment was implemented and programmed using Unity (Version 2021.2.7f1; Unity Technologies, Inc., San Francisco, CA, USA), SteamVR (Version 2.1.9), the Unity Experiment Framework package (Version 2.2.1, [24]) and the Haptics Direct Unity Plugin (Version 1.0;

3D Systems). It was run on a custom-built desktop PC (Intel Core i9-12900KF CPU at 3.2 GHz, 64 GB RAM, Dual NVidia GeForce RTX3080 GPU).

The visual scene (Fig. 1b) consisted of the two virtual stimuli, two response buttons above the stimuli (grey cubes), and the bright green stylus that participants used to interact with the objects. The stimuli bases were rendered in white while the ridges were invisible. As soon as participants moved the pen over the ridge area, the stylus became invisible as well. Thus, no information on movement velocity or the spatial frequency of the stimuli was visually revealed to them via movement characteristics of the stylus. A semi-transparent pink ellipsoid object was permanently placed between the two stimuli to remind participants of the prescribed switching movement trajectory (arc-like motion over the ellipsoid) and prevent them from carelessly moving too close to the stimuli during the transportation phase. The virtual stimuli were simulated by the force feedback device by applying reaction forces ($\overrightarrow{F_P}$) as a function of the 3D-position of the end effector. The force magnitude is directly proportional to the indentation depth (i_p) of the virtual stimulus and its' spring constant (K), i.e. its' stiffness. The direction of the indentation is computed by Unity's built-in 3d physics engine (Nvidia PhysX 3.4). The force direction is normal to the texture's surface at the contact point ($\vec{n}_P$):

$$\overrightarrow{F_P} = \vec{n}_P \times \left|\overrightarrow{F_P}\right|, \left|\overrightarrow{F_P}\right| = K \times i_P \tag{1}$$

The virtual grating stimuli all consisted of a $10 \times 4 \times 0.5$ cm rectangular cuboid and $0.1 \times 40 \times 0.05$ mm cylindrical segments (= ridges) on their top side (Fig. 1c). The distance between the centers of adjacent ridges (= period) defined the spatial frequency and differed between stimuli. Amplitude and ridge width was constant. With constant stimulus sizes, higher spatial frequencies would always be characterized by higher absolute number of ridges. To partly decouple the spatial frequency from the absolute number of ridges, we varied the length of the ridged stimulus area between trials. The length of the ridge area could comprise ca. 7 cm, 5.5 cm, and 4.5 cm (slightly varying depending on the respective period), with each size appearing equally often in randomized order. At the front and back ends of each stimulus, there were areas without ridges: 2.5×4 cm at the front end, being the starting area, and $0.5, 2,$ or 3×4 cm at the back end, depending on the length of the ridged area. The stimulus set comprised two reference stimuli with periods of 10.84 mm and 13.38 mm. For each demand-level, reference stimuli were paired with comparison stimuli of either $\pm/-$ 2.54 mm period (high demand), $\pm/-$ 3.81 mm (medium demand), or $\pm/-$ 4.45 mm (low demand), resulting in 4 stimulus pairs for each demand level. The stimulus pairs were selected after examination of psychometric functions derived from a pilot-study (N = 5): their differences corresponded to discrimination performances of about 65%, 75%, and 85%. Throughout the experiment, brown noise was presented via the headset's headphones additional white noise was played from a speaker directly above the force feedback device.

2.3 Design and Procedure

The experimental design was a 3 (Demand) $\times$ 3 (Motor Effort) - design, with $3 \times 3 \times 4$ (stimulus pairs) $\times$ 6 (repetitions) = 216 trials in total. The experiment was split in two blocks. Each block contained 3 repetitions of the trial types and the trial order

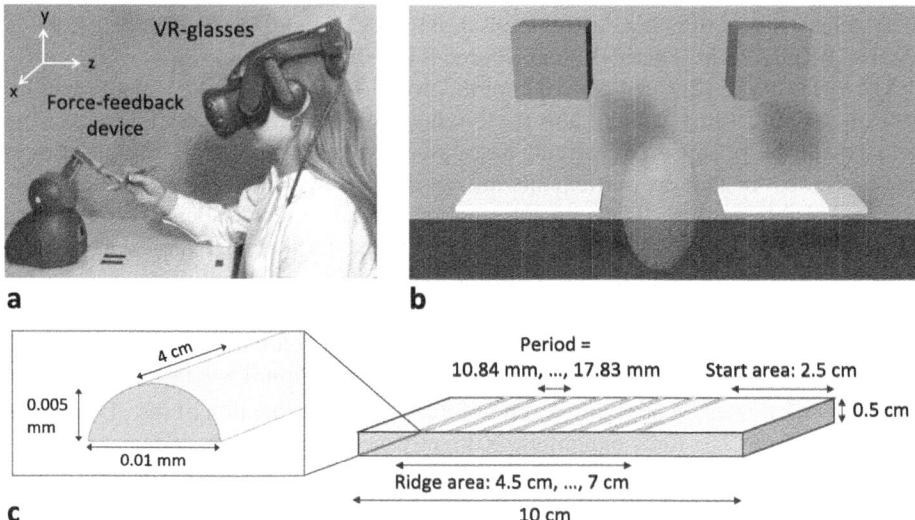

Fig. 1. a, setup. b, visual scene (before participant touches the stimulus for the first time in a trial), yellow area indicates start stimulus. c, Schematic depiction of virtual stimulus.

was randomized within each block. There was a break of 15 min between the blocks and a break of 5 min in the middle of each block. All in all, the session took about 2 h. In each trial, participants had to complete a two-alternative forced choice (2AFC) discrimination task, in which they had to successively explore two grating stimuli and decide which one had a higher spatial frequency. To ensure a reasonable level of effort in the completion of the task, we provided participants with a monetary incentive (i.e. they received points for correct responses, which could amount to max. Total 8.64 €). After every 10 trials, the number of accumulated reward points was displayed (3 s), to not give immediate performance feedback after each trial. The starting area of the respective stimulus on which participants were supposed to begin the exploration was highlighted in bright yellow before each trial until they first touched it (Fig. 1b). Exploration began equally often on the left and right stimulus, in randomized order. This was also the case for the position of the reference stimulus. As indicated via the starting area, exploration always began at the "outer" area of a stimulus towards the "inner" area (with regards to the visual scene). Participants were instructed to switch to the other stimulus after one swipe. Here again, exploration started at the outer area and ended in the inner area. Thus, the switching movement was sufficiently long to allow for proper manipulation of motor effort. Participants were free to switch as often as desired. To log their decision, they had to touch the virtual button (cube) above the respective stimulus that they perceived to have the higher spatial frequency, triggering the next trial to begin. We manipulated „motor effort" by implementing forces counteracting the participants' movement when switching from one stimulus to the other. Thus, when a participant finished one swipe, lifted the stylus up, and moved to the other stimulus, the device exerted a constant force Fx of either 1.7 N (high effort), 0.85 N, (medium) or 0 N (baseline) along the x-axis of the device in the opposite direction of the movement vector. Thus, participants had to put a constant additional effort into moving from one stimulus end area to the other stimulus start area. The force was only active > 1.5 cm above and next to the stimuli, to ensure that the sensory perception during exploration was not directly affected by e.g. increased muscular effort. The onsets and offsets of the counterforce were not abrupt; the force linearly increased/decreased from zero to the respective value and vice versa by 3.6N/s. Participants were instructed that they should never touch the stimuli while switching. To standardize the switching movement, they were additionally instructed to make an arc-like motion and avoid the semitransparent pink capsule object between the two stimuli. This way, the switching movement comprised a minimum distance of ~22 cm.

Before each session, participants had a familiarization phase for the HMD and the force-feedback device and a subsequent training phase of about 10 min (10 practice trials) to practice the movement coordination. During training and experimental trials, we aimed to keep scanning velocities relatively constant between and especially within participants at around 100–120 mm/s to avoid any potential confounds [11, 25–27]. This target velocity has been observed to occur naturally and was proven feasible during piloting. While moving over the stimulus, the velocity was continuously tracked and averaged over the last previous 250 ms for smoothing. Whenever the average value exceeded or fell below the threshold of 180 mm/s or 60 mm/s respectively, an acoustic warning signal was played (low pitch tone or high-pitched beep tone, duration 300 ms). We kept the criterion relatively liberal, enabling participants to mainly focus on the task rather than on the movement execution. Consequently, the training phase also helped with automatizing the prescribed velocity. It additionally gave a rough orientation on how much force participants should exert during stimulus contact. This was verbally instructed by the experimenter, who received visual feedback on the current reaction forces of the device (desired range: approx. 0.4–1 N). After the experiment, participants filled out a brief questionnaire to check whether they noticed any behavioral changes in reaction to the motor effort manipulations.

2.4 Data Analysis

Raw data of individual observers is available at https://doi.org/10.5281/zenodo.103 70635. Data analysis was performed using MATLAB (Version R2020b). Raw data comprised the positions of the end effector represented as Cartesian coordinates in three-dimensional space, the movement velocities, and the participants' responses. The number of swipes was derived from the position of the end effector, i.e. the number of times that the effector changed from one stimulus area to the other stimulus area +1. We compared the number of swipes using a two-way repeated measures analysis of variance (ANOVA) with the within-participant factors Task Demand and Motor Effort. Likewise, we analyzed the average response accuracy. As a manipulation check, we also assessed whether there were systematic differences in the average movement velocities between the different conditions, conducting a similar ANOVA as the previous ones. Whenever the assumption of sphericity was violated, the p-values of the respective ANOVA were Greenhouse-Geisser adjusted [28]. The assumption of normality (tested with the Shapiro-Wilk test) was not violated (all p > .07).

3 Results

The average numbers of exploratory movements can be seen in Fig. 2a. A two-way repeated measures ANOVA with the within-participants factors Task Demand (low, medium, high) and Motor Effort (baseline, medium, high) showed that a higher number of movements was executed when task demands were increased, $F(2,22) = 4.65$, $p = .021$, $\eta_p^2 = 0.30$, confirmed by a linear trend, $t(22) = 2.9$, $p = .008$. There was no significant main effect of Motor Effort, $F(2,22) = 1.79$, $p = .19$, $\eta_p^2 = 0.14$, but an interaction between the two factors, $F(4,44) = 3.76$, $p = .01$, $\eta_p^2 = 0.26$. Bonferroni-corrected post-hoc tests (36 pairwise comparisons) revealed that for Baseline Motor Effort, more movements were executed in the High Demand condition than in the Low Demand condition, $t(11) = 4.54$, $p = .001$, $d = 0.83$. For High Motor Effort, this was not the case ($p > .99$). For High Demand, in line with that, more movements were executed in the Baseline Effort condition than in the High Effort condition, $t(11) = 3.65$, $p = .022$, $d = 0.65$. Trivially, more movements were executed for Baseline Effort/High Demand than for High Effort/Low Demand, $t(11) = 3.47$, $p = .035$, $d = 0.73$. For Low Demand, the amount of movements did not differ regardless of the Effort level, all $p > .99$ (same for all other 33 comparisons). In summary, participants extended their explorations in reaction to higher task demands, but only when motor effort was low. Average accuracies entered a similar repeated measures ANOVA (Fig. 2b), which revealed a main effect of

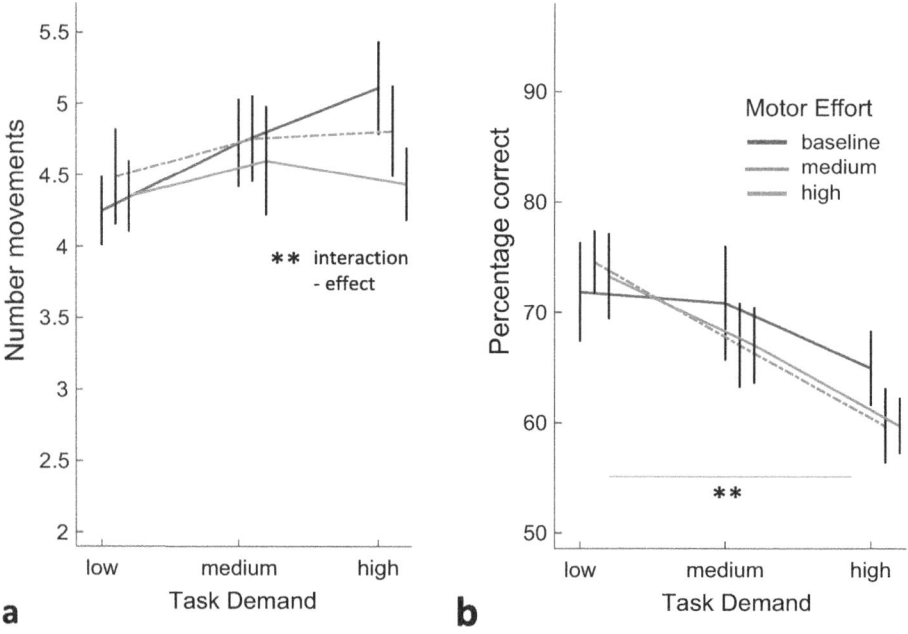

Fig. 2. a, Exploration duration (number of swiping movements) as a function of task demands separately for all motor effort conditions. b, Task performance (average percentage of correct responses) as a function of task demands separately for all motor effort conditions. Error bars represent the standard error of the mean.

Task Demand, $F(2,22) = 13.01, p = <.001, \eta_p^2 = 0.54$, with accuracies decreasing with increasing task demand, confirmed by a linear trend, $t(22) = 5.08, p = <.001$. No main effect of Motor Effort, $F(2,22) = 0.34, p = .072, \eta_p^2 = .03$, and no interaction effect, $F(4,44) = 0.78, p = 0.55, \eta_p^2 = 0.07$, was found. As intended, participants kept their movement velocities rather constant across conditions, resulting in no significant main effect of Task Demand on the average velocities, $F(2,22) = 0.12, p = 0.89, \eta_p^2 = 0.01$, no main effect of Motor Effort, $F(2,22) = 2.58, p = .1, \eta_p^2 = 0.19$, and no interaction effect, $F(4,44) = 0.57, p = .69, \eta_p^2 = .05$.

4   Discussion

The current study provides evidence that humans adjust the duration of their haptic exploration as a function of both task demands and the motor efforts that are associated with the execution of the exploration. When being confronted with higher task demands, participants extended their exploration by executing more exploratory movements. This likely constitutes an adaptation mechanism serving perception/task performance–given that previous literature showed that extension of exploration generally leads to an increased sensitivity of the agent [9–12]. The effect though seems to strongly depend on the motor effort participants must exert during the execution of their exploration, as it disappears when the associated motor effort is high. While participants devoted the same number of movements to less demanding trials, they did not extend their exploration during the more demanding trials anymore.

Furthermore, higher task demands were associated with reduced accuracies, demonstrating successful manipulation of task demands. However, one could expect an interaction effect with motor effort here as well, with stronger effects of task demands when motor effort is increased, resulting from the lack of the observed adaptation behavior. This what not the case though, possibly due to other, e.g. attentional compensatory mechanisms.

With this study, we hope to provide a first cornerstone for a comprehensive understanding of how humans adapt their haptic exploration strategy with respect to the extension of the exploration. In light of our findings, we theorize that humans adjust the duration of their exploration according to the principles of an expectancy × value framework (see e.g. [29]). Herein, the utility of the extension of an exploration would be derived from the expected task performance, which in turn consists of the agent's perceptual ability and the task demands at hand, combined with the value that is associated with a correct task performance (e.g. monetary reward, positive consequences). Thus, higher task demands would be compensated with extended explorations due to expected task performance differences. Additionally, aligning with motor control literature [22], motor costs would be discounted with the expected utility. The accumulated costs increased with increasing motor efforts, while the (intrinsic) reward stayed constant. Consequently, the expected utility of an exploration extension would be reduced due to the changed ratio of (expected) motor costs and reward, leading to earlier termination of exploration behavior.

Even though it seems plausible that the observed effect of the effort manipulation is a result of cost-benefit-weighting, there are more banal alternative explanations. One would be the mere exhaustion of the participant, i.e. a ceiling effect due to physical restrictions of their bodies. This might produce an absolute limit of exploration duration, only manifesting itself in higher task demand trials due to the necessary exploration durations being longer. More extensive assessments could rule this out, but while the progressing fatigue over time might have some moderating effect, it does not seem likely that it is the main driving force for the observed behavior. Participants did not report that they noticed immense exhaustion of their arm or changes in their behavior. One might still wonder why the motor effort manipulation affects the behavior when task demands are high but does not affect it when demands are low. A possible explanation could involve the expected utility of the exploration segments/swipes. As task demands increase, the exploration duration needed for a consistent task performance also increases. Hence, when exploring a very demanding stimulus pair, each swipe yields a lower repetition gain, i.e. benefit regarding expected task performance, than when exploring a less demanding pair. Thus, the expected utility of each swipe would be lower for high task demands then for low demands. As a consequence, with the reward (value) staying constant and the utility of each swipe being lower, the increased motor efforts (= increased accumulated costs) would carry higher weights for high demand trials than for low demand trials, thus reducing the expected utility of an exploration extension more. In other words: For high demand trials, one can renounce one or two swipes as a measure for cost reduction without affecting the task performance too drastically. This might possibly not be the case for the low demand swipes, as the impact of e.g. just one swipe less might be immense. Obtaining the individual perceptual performances of each participant as a function of instructed exploration duration would help elucidating on that matter.

Please note that the artificial manipulation of motor effort did not only have unspecific binary effects; it could have been possible that the manipulation is perceived as unnatural and disruptive, so that participants virtually cease to explore as soon as any additional effort is introduced. However, when motor effort was only increased by one level (= medium effort), the participants' behavior only slightly deviated from the baseline level. That is, the impact of the effort- manipulation is likely not dichotomous; we can rather expect a gradual progression of the effect's magnitude, just as we would expect it for the real life. This underlines the feasibility of the experimental procedure. Further, one might speculate that sensory consequences of the counterforce in medium- and high motor effort conditions could have tampered with perceptual performance due to masking effects or conflicts with proper sensory input. However, as the average accuracy for low demand trials does not substantially differ between the baseline and high effort condition, this seems improbable. Finally, an evident peculiarity of the present study is the restricted exploration scheme. Investigating unrestricted explorations could offer valuable insights: One might expect less switching as a reaction to increased efforts and possibly more swipes as a consequence [4]. Still, the driving mechanisms would be the same, i.e. cost reduction and cost-benefit weighting; but could be assessed from a different viewpoint.

For future studies, a more thorough examination of the influence of task demands on the duration of haptic explorations might provide interesting insights. Data of the baseline effort condition in the current study suggests a linear relationship between exploration duration and the three implemented levels of task demands. However, it can be expected that exploration duration would at some demand level reach a saturation point, in line with the observed saturation point of task performance as a function of instructed exploration duration [10], and because trivially, humans would not explore infinitely long. It is likely as well that the relationship is not linear when task demands are closer to saturation: Task demands might become so high that people terminate earlier again, as the expected utility-gain from an extension could be perceived as insufficient. As every additional movement is associated with costs but would yield only extremely limited information gain due to the very high task demand, the marginal increase in expected task performance might not provide sufficient "incentive", resulting in premature termination of exploration. Additionally, it would be necessary to confirm the derived conclusions in a more naturalistic setting. In this context, one could also tackle the question whether the artificial motor effort manipulation introduces additional cognitive load; although participants did not report notable disturbances in the questionnaire, the perceived unnaturalness of the effort manipulation might yield subtle effects.

Ultimately, we plan to model the trade-off between motor effort and task performance and aim to be able to predict natural exploration duration by also taking motivational influences into account, as these were proven to be crucial in the motor control literature [21]. We expect this to provide valuable insights: Dissecting the prediction and valuation processes that take place during haptic explorations is worthwhile for both practical and theoretical reasons, as the derived conclusions can inform the development of more efficient and ergonomic human-computer interfaces, improve design choices for haptic experiences, and enhance our understanding of the intricacies of human perception and action.

Acknowledgments. We would like to thank Marai Söhngen, Kimberly Glas and Lars Hagenmeier for their help with data collection. This research was supported by the Deutsche Forschungsgemeinschaft (DFG, German Research Foundation) – project number 222641018 – SFB/TRR 135, A5.

Disclosure of Interests. The authors have no competing interests to declare that are relevant to the content of this article.

References

1. Tanaka, Y., Bergmann Tiest, W.M., Kappers, A.M.L., Sano, A.: Contact force and scanning velocity during active roughness perception. PLOS ONE (2014). https://doi.org/10.1371/journal.pone.0093363
2. Lezkan, A., Drewing, K.: Interdependences between finger movement direction and haptic perception of oriented textures. PloS one (2018). https://doi.org/10.1371/journal.pone.0208988
3. Weiss, E.J., Flanders, M.: Somatosensory comparison during haptic tracing. Cerebral cortex (New York, N.Y.: 1991) (2011). https://doi.org/10.1093/cercor/bhq110
4. Metzger, A., Drewing, K.: Switching between objects improves precision in haptic perception of softness. In: Nisky, I., Hartcher-O'Brien, J., Wiertlewski, M., Smeets, J. (eds.) Haptics: Science, Technology, Applications. EuroHaptics 2020. LNCS, vol. 12272, pp. 69–77. Springer, Cham (2020). https://doi.org/10.1007/978-3-030-58147-3_8

5. Kaim, L., Drewing, K.: Exploratory strategies in haptic softness discrimination are tuned to achieve high levels of task performance. IEEE transactions on haptics (2011). https://doi.org/10.1109/TOH.2011.19

6. Jeschke, M., Drewing, K.: Prior static visual information on material properties increases the efficiency of a subsequent haptic exploration. J. Vis. (2023). https://doi.org/10.1167/jov.23.9.4921

7. Lederman, S.J., Klatzky, R.L.: Hand movements: a window into haptic object recognition. Cogn. Psychol. (1987).https://doi.org/10.1016/0010-0285(87)90008-9

8. Metzger, A., Lezkan, A., Drewing, K.: Integration of serial sensory information in haptic perception of softness. J. Exp. Psychol. Hum. Percept. Perform. (2018). https://doi.org/10.1037/xhp0000466

9. Drewing, K., Lezkan, A., Ludwig, S.: Texture discrimination in active touch: effects of the extension of the exploration and their exploitation. In: 2011 IEEE World Haptics Conference. IEEE (2011). https://doi.org/10.1109/whc.2011.5945488

10. Lezkan, A., Drewing, K.: Processing of haptic texture information over sequential exploration movements. Atten. Percept. Psychophys. (2018). https://doi.org/10.3758/s13414-017-1426-2

11. Louw, S., Kappers, A.M.L., Koenderink, J.J.: Haptic detection of sine-wave gratings. Perception (2005). https://doi.org/10.1068/p5425

12. Giachritsis, C.D., Wing, A.M., Lovell, P.G.: The role of spatial integration in the perception of surface orientation with active touch. Atten. Percept. Psychophys. (2009).https://doi.org/10.3758/APP.71.7.1628

13. Metzger, A., Drewing, K.: The longer the first stimulus is explored in softness discrimination the longer it can be compared to the second one. In: 2017 IEEE World Haptics Conference (WHC). IEEE (2017). https://doi.org/10.1109/whc.2017.7989852

14. Ernst, M.O., Bülthoff, H.H.: Merging the senses into a robust percept. Trends Cogn. Sci. (2004). https://doi.org/10.1016/j.tics.2004.02.002

15. Metzger, A., Drewing, K.: A Kalman filter model for predicting discrimination performance in free and restricted haptic explorations. In: 2021 IEEE World Haptics Conference (WHC). IEEE (2021). https://doi.org/10.1109/whc49131.2021.9517242

16. Lezkan, A., Drewing, K.: Unequal but fair? Weights in the serial integration of haptic texture information. In: Auvray, M., Duriez, C. (eds.) Haptics: Neuroscience, Devices, Modeling, and Applications. EuroHaptics 2014. LNCS, vol. 8618, pp. 386–392. Springer, Berlin, Heidelberg (2014). https://doi.org/10.1007/978-3-662-44193-0_48

17. Jeschke, M., Zöller, A.C., Drewing, K.: Influence of prior visual information on exploratory movement direction in texture perception. In: Seifi, H., et al. (eds.) Haptics: Science, Technology, Applications. EuroHaptics 2022. LNCS, vol. 13235, pp. 30–38. Springer, Cham (2022). https://doi.org/10.1007/978-3-031-06249-0_4

18. Zoeller, A.C., Lezkan, A., Paulun, V.C., Fleming, R.W., Drewing, K.: Integration of prior knowledge during haptic exploration depends on information type. J. Vis. (2019). https://doi.org/10.1167/19.4.20

19. Lezkan, A., Metzger, A., Drewing, K.: Active haptic exploration of softness: indentation force is systematically related to prediction, sensation and motivation. Front. Integr. Neurosci. (2018). https://doi.org/10.3389/fnint.2018.00059

20. Kahneman, D., Tversky, A.: Prospect Theory: An Analysis of Decision Under Risk. In: Handbook of the Fundamentals of Financial Decision Making, pp. 99–127. WORLD SCIENTIFIC (2013)

21. Rigoux, L., Guigon, E.: A model of reward- and effort-based optimal decision making and motor control. PLoS Comput. Biol. (2012). https://doi.org/10.1371/journal.pcbi.1002716

22. Shadmehr, R., Huang, H.J., Ahmed, A.A.: A representation of effort in decision-making and motor control. Current Biol. CB (2016). https://doi.org/10.1016/j.cub.2016.05.065

23. Faul, F., Erdfelder, E., Buchner, A., Lang, A.-G.: Statistical power analyses using G*Power 3.1: tests for correlation and regression analyses. Behav. Res. (2009). https://doi.org/10.3758/BRM.41.4.1149

24. Brookes, J., Warburton, M., Alghadier, M., Mon-Williams, M., Mushtaq, F.: Studying human behavior with virtual reality: the unity experiment framework. Behav. Res. (2020).https://doi.org/10.3758/s13428-019-01242-0

25. Gamzu, E., Ahissar, E.: Importance of temporal cues for tactile spatial- frequency discrimination. J. Neurosci. Off. J. Soci. Neurosci. (2001).https://doi.org/10.1523/JNEUROSCI.21-18-07416.2001

26. Smith, A.M., Gosselin, G., Houde, B.: Deployment of fingertip forces in tactile exploration. Exp. Brain Res. (2002). https://doi.org/10.1007/s00221-002-1240-4

27. Boundy-Singer, Z.M., Saal, H.P., Bensmaia, S.J.: Speed invariance of tactile texture perception. J. Neurophysiol. (2017). https://doi.org/10.1152/jn.00161.2017

28. Greenhouse, S.W., Geisser, S.: On methods in the analysis of profile data. Psychometrika (1959). https://doi.org/10.1007/BF02289823

29. Tolman, E.C.: Principles of performance. Psychol. Rev. (1955). https://doi.org/10.1037/h0049079

The TIP Benchmark: A Tactile Image-Based Psychophysics-Inspired Benchmark for Artificial Tactile Sensors

Tianyi Liu[✉] and Benjamin Ward-Cherrier

University of Bristol, Bristol, UK
{tianyi.liu,b.ward-cherrier}@bristol.ac.uk

Abstract. We introduce a comprehensive benchmarking method, the TIP benchmark, to assess the spatial acuity of tactile sensors. The TIP benchmark is made up of 4 stages, in which data output from a tactile sensor performing a psychophysics-inspired task is converted into images. The Structural Similarity Index Measure (SSIM) and support vector machine (SVM) classifier are then used to evaluate sensor performance across 4 metrics, representing sensor accuracy (Accuracy, Distance), stability (IQR) and generalisability (Margin). The TIP benchmark is validated to determine an ideal indentation depth and evaluate noise degradation on a tactile task, and is then employed for a grid search hardware optimization of a neuromorphic tactile sensor (9 configurations, 2 hardware design parameters). Sensors with shorter, denser internal pins are shown as having greater spatial acuity on a grating orientation discrimination task, demonstrating the TIP benchmark's potential to quantitatively compare tactile sensors, paving the way for establishing unified methods for hardware design and benchmarking for real-world applications.

Keywords: Tactile sensing · Benchmarking

1 Introduction

Tactile perception is crucial for humans to interact with their environment and detect textures, shapes, and pressure. Improvements in artificial tactile sensors are ongoing, and the demand for high-sensitivity tactile sensors is increasing in fields such as robotics, prosthetics, and virtual reality. However, a standardized, universally applicable method for assessing the spatial sensitivity of tactile sensors is still lacking, posing a significant challenge for the development and comparison of existing tactile sensors.

While some tactile sensor evaluation methods are available [5, 22], their applicability and reliability are limited. The challenge in creating a universal benchmark for tactile sensors specifically stems from the large variety of transduction methods and underlying algorithms for tactile perception. We aim to address this

© The Author(s), under exclusive license to Springer Nature Switzerland AG 2025
H. Kajimoto et al. (Eds.): EuroHaptics 2024, LNCS 14768, pp. 94–106, 2025.
https://doi.org/10.1007/978-3-031-70058-3_8

challenge by treating tactile data as images and leveraging structural similarity
index measure (SSIM), an image-based comparison metric.

As well as aiding in comparing existing tactile sensors and optimizing sensor
hardware for tactile tasks, a standardized evaluation method for tactile sensors
could help mitigate safety risks in tactile sensing applications. In robotics for
instance, inaccurate tactile feedback can damage objects and equipment, whereas
erroneous tactile sensor feedback in prosthetics might cause discomfort or pain to
users. A universal benchmark for tactile sensors, mirroring recent developments
in benchmarks for grasping and manipulation [1,14] would have benefits in both
industry and academia, enabling the most appropriate transduction systems to
be selected for a given task.

Therefore, the aims of this paper are as follows:

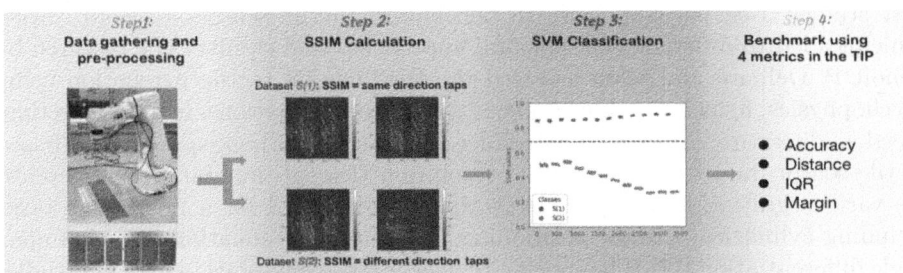

Fig. 1. The TIP (Tactile Image-based Psychophysics-inspired) Benchmark comprises
4 steps. From left to right: Data gathering on a psychophysics inspired tactile task and
data pre-processing (conversion to images, specifically heatmaps, and Gaussian filtering), SSIM calculation, SVM classification and final benchmark based on 4 performance
metrics.

- Establish a universal evaluation method for spatial accuracy that is applicable to all tactile sensors, leveraging an image-based comparison metric and
  inspired by psychophysics (Fig. 1).
- Apply the benchmark to determine the noise degradation of a neuromorphic
  tactile sensor on a given task, and to optimize the sensor's design through a
  grid search over 2 hardware parameters.

Our TIP benchmark uses 4 metrics to evaluate spatial acuity across 3 areas:
accuracy, sensor stability and generalisability. This approach shows promise for
the straight-forward comparison, benchmarking, and selection of tactile sensors,
ultimately contributing to improved performance and safety in applications such
as robotics, prosthetics, and virtual reality.

2   Background and Related Work

In fields like computer vision, manipulation, and grasping, benchmarks such as
ImageNet [18], the YCB Object Set [1], and the Cornell Grasping Dataset [9]

have catalyzed key advancements by providing standard metrics and challenges for evaluation. These benchmarks enable direct comparisons between approaches and spur innovation by defining clear objectives. Inspired by these successes, our goal is to establish a similar benchmarking method in tactile sensing.

The aim of biomimetic tactile perception is to match human tactile sensitivity. To achieve that task, a variety of tactile sensors have been developed based on different transduction methods such as piezoresistive [20], capacitive [6], vision-based [11] or optical event-based [25]. The variety of underlying technologies in tactile sensors and their associated perception algorithms has made the process of establishing a universal benchmark particularly challenging. Existing studies have tended to focus on one-off evaluation methods that apply to limited sensor hardware types or specific tactile tasks. Psychophysical experiments on human touch provide key benchmarks for evaluating artificial tactile sensors. Although past studies used psychophysics to benchmark tactile sensors, sensors' varied principles led to different experimental approaches and results. The research by Benoit P. Delhaye and team assessed the BioTac's [5] tactile perception using psychophysics, focusing on four aspects: punctate indentations, motion direction, speed, and texture. Wang et al. introduce HSVTac [22], a high-speed vision-based tactile sensor, and tested its spatial resolution through its ability to differentiate various grating textures. The Tactip was evaluated for a number of tasks including cylindrical surface positioning [12], gap discrimination [13], and apex angle differentiation [17]. The specificity of evaluation methods in existing studies underscores a prevailing trend towards assessments that are narrowly focused on particular sensor technologies, often rendering these benchmarks less applicable across diverse tactile sensing methods.

Although artificial tactile sensors produce data in a wide range of formats (analog signals [20], digital readings [6], images [11], events [25]...), they can all be converted to an image-based visual representation with some minor preprocessing. We thus propose an image-based benchmarking method which capitalizes on existing computer vision methods and is independent of each sensor's individual transduction principle. In particular, we seek to exploit existing image-based quality metrics to compare data from different trials of a tactile task. Focusing on the grating orientation discrimination task, we employ the Structural Similarity Index (SSIM) [23] as our primary image quality metric due to its superior alignment with human visual perception, in contrast to traditional metrics like Peak Signal-to-Noise Ratio (PSNR) [8]. SSIM's effectiveness in capturing and analyzing image features makes it an ideal choice for evaluating tactile data representations. Here we evaluate our method on the Neurotac [25], a sensor which produces event-based data output, to demonstrate the generalisability of our image-based approach.

We also take inspiration for this benchmark from clinical tactile perception evaluation in humans, which is generally performed through psychophysical tests like two-point discrimination [4], grating orientation resolution [3], edge detection [16], and texture detection [15]. Gratings are widely used for evaluating tactile spatial acuity in both clinical and non-clinical populations [3, 19, 21].

In our project, grating orientation discrimination has been selected because of the inherent challenges in methods like two-point discrimination, which include maintaining consistent criteria, the difficulty in establishing a clear threshold for a "two-point" response and significant variability in results across and within subjects.

3 Methods

3.1 Hardware

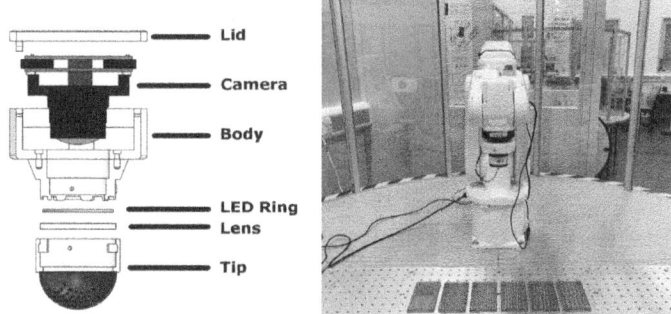

Fig. 2. Experimental setup. Left: Internal structure of the NeuroTac sensor [25] used to evaluate our benchmark. Right: The NeuroTac tactile sensor is attached to a 6-dof industrial robot arm (ABB, IRB120) and taps vertically downwards on 12 3d-printed gratings.

The NeuroTac [25] is an event-based optical tactile sensor(Fig. 2, Left panel). Its design integrates the TacTip's [24] biomimetic hardware with an event-based camera (iniVation, DVXplorer) with 640 * 480 pixel resolution for power-efficient and effective information representation. The NeuroTac outputs event-based data in the form of precisely timed "on" (luminance increased above a set threshold) or "off" (luminance decreased below a set threshold) events at each pixel. Despite NeuroTac not directly outputting optical images, its event-based data is dynamically captured can be processed into static heatmap images (as described in Sect. 3.3), making it compatible with our image-based TIP benchmark methodology. This adaptability highlights the potential of the TIP benchmark to be applied not just to image-based tactile sensors but to sensors with a wide range of transduction methods.

3.2 Experimental Setup

Our experimental setup mirrors human psychophysical experiments [10], in which participants assess two sequences of tapped gratings: $S(1)$, with both gratings aligned longitudinally to the finger, and $S(2)$, where the first is longitudinal and the second is perpendicular to the finger's length. We 3D printed 12 gratings on a resin printer (Photon Mono X, Anycubic), with gap sizes ranging from 0.0 (smooth resin surface) to 5.5 mm in 0.5 mm increments (Fig. 2, Right panel). The NeuroTac is mounted on a 6-degree-of-freedom robotic arm (ABB, IRB20) and vertically taps each grating at 8 mm/s, then rotates 90°C and taps again, repeating this process for each of the 12 gratings for 9 trials per depth in Experiment 1 (noise degradation) and 20 trials per sensor in Experiment 2 (hardware optimization).

We therefore gather a total of 1728 samples for Experiment 1 (4 indentation depths, 12 gratings, 2 grating orientations, 2 event types and 9 trials) and 8460 samples for Experiment 2 (9 sensors, 12 gratings, 2 grating orientations, 2 event types and 20 trials).

3.3 Data Processing

Heatmap Generation and Preprocessing. The NeuroTac stores its data in the Address-Event Representation (AER) format [2]. In the first step of the TIP benchmarking method, we convert the data into images by creating a heatmap of events for each trial. These heatmaps collapse the time dimension and illustrate the sensor output's spatial distribution during a trial, with color intensity indicating event count at each pixel. Heatmaps are cropped to focus on the taxels of the NeuroTac sensor, and a Gaussian filter ($\sigma = 2$) is applied to filter noise from the data. This heatmap generation process can be adapted to any tactile sensor by interpreting taxels as pixels, thus enabling a standardized visualization of tactile events across different sensor technologies.

Structural Similarity Index Measure (SSIM) Calculation. The Structural Similarity Index Measure (SSIM) [23] assesses image similarity, reflecting human visual perception's focus on structural information, a key factor in image quality. Here, we apply it pairwise to our pre-processed tactile sensor data to determine similarity between trials.

For each grating, SSIM comparisons are conducted within distinct trials of orientation $\theta_1 = 0°$, forming dataset $S(1)$. Subsequently, we compare each heatmap at orientation $\theta_1 = 0°$ with each heatmap at orientation $\theta_2 = 90°$, forming dataset $S(2)$. The threshold on SSIM values to classify between $S(1)$ and $S(2)$ is set using a Support Vector Machine classifier.

Support Vector Machine (SVM) Classification. The Support Vector Machine (SVM) classifier [7] is memory efficient, resilient to outliers due to its focus on margin maximization and able to assign varying weights to classes to address data imbalance. These advantages make SVM an ideal method classify betweeen the SSIM values determined previously in $S(1)$ (same orientation gratings) and $S(2)$ (different orientation gratings).

Metrics Used for TIP Benchmark. Given our objective of comparing and evaluating the spatial acuity of tactile sensors, a key evaluation metric is the **Accuracy** of the SVM classification results. However, to enable performance discrimination for sensors that report similar accuracy, we introduce three additional metrics—**Distance, IQR**, and **Margin** described below. These metrics are applied sequentially during benchmarking in the following order: Accuracy, Distance, IQR and Margin.

Accuracy. Accuracy ($\mathcal{A}$) is a key metric for assessing classifiers. Its formula is:

$$\mathcal{A} = \frac{TP + TN}{TP + TN + FP + FN} \tag{1}$$

where TP is True Positives, TN True Negatives, FP False Positives, and FN False Negatives. A higher $\mathcal{A}$ value is preferable, indicating a larger ratio of accurate predictions. Accuracy is the key metric and the first to be assessed as it tends to be the primary goal of any tactile task.

Distance. The average distance between Datasets $S(1)$ and $S(2)$ for all gratings, denoted as $\mathcal{D}$, is calculated as follows:

$$\mathcal{D} = \frac{1}{N} \sum_{i=1}^{N} \frac{1}{M} \sum_{j=1}^{M} ||X_{S(1)}^{(i,j)} - X_{S(2)}^{(i,j)}|| \tag{2}$$

where N is the number of gratings, M is the number of samples per grating, and $X_{S(1)}^{(i,j)}$ and $X_{S(2)}^{(i,j)}$ represent the SSIM values of the two datasets $S(1)$ and $S(2)$ (j-th sample in the i-th grating) respectively. A higher value of $\mathcal{D}$ is preferred as it reflects the discriminative ability of the tactile sensor to different types of tactile stimuli. A larger average distance indicates that the features extracted from different tactile stimuli are more distinct, leading to more robust performance for instance when noise is introduced. The Distance metric allows us to distinguish performance between sets with 100% accuracy on a task.

Interquartile Range (IQR). The average interquartile range of a dataset is calculated as follows:

$$\overline{IQR}_X = \frac{1}{N} \sum_{i=1}^{N} \left[Q_3(X^{(i)}) - Q_1(X^{(i)}) \right] \tag{3}$$

In Eq. 3, X represents a generic dataset, which in our experiments refers to either $S(1)$ or $S(2)$. Here, i denotes the index of a sample within the dataset containing N samples. $Q_1(X)$ and $Q_3(X)$ represent the first and third quartiles of the dataset X, respectively. The average of $\overline{IQR}_{S(1)}$ and $\overline{IQR}_{S(2)}$, denoted as $\mathcal{I}$, is then calculated as:

$$\mathcal{I} = \frac{\overline{IQR}_{S(1)} + \overline{IQR}_{S(2)}}{2} \tag{4}$$

In SVM binary classification, the Interquartile Range (IQR) indicates the dispersion of data within each class. A smaller $\mathcal{I}$ indicates that the responses are more consistent and tightly grouped. It is used here as an indication of the stability of the sensor's response.

Margin. The margin $\mathcal{M}$ of the SVM model is calculated from the SVM weight vector w as Eq. 5. Greater values of $\mathcal{M}$ are preferred, as a larger margin implies stronger model generalization. This is important for ensuring that the tactile sensor and classifier can perform well across a variety of different, potentially unseen tactile stimuli.

$$\mathcal{M} = \frac{2.0}{||w||} \tag{5}$$

We use Min-Max Normalization to adjust the values of all four metrics into the range of [0.1, 1]. This avoids zero values and ensures each metric is fairly compared on the same scale. We use the Inverted IQR ($\mathcal{I}_{inverted-norm} = 1 - 0.9\,\mathcal{I}_{norm}$), to ensure a consistent evaluation criterion for all metrics. This approach aligns with the notion that a larger value for all metrics in the TIP benchmark indicate better performance of the tactile sensor.

$\mathcal{A}$ is generally considered the most critical metric due to its direct reflection of the sensor's reliability in accurately classifying stimuli. The importance of $\mathcal{D}$ and $\mathcal{I}_{inverted-norm}$ varies based on application needs: Distance is crucial if noise could be introduced, while IQR is key for ensuring consistency in sensor responses in precision-dependent scenarios. We consider $\mathcal{M}$ the least critical metric, however it could become important in environments requiring strong generalization to handle a wide range of stimuli. Thus, the TIP benchmark allows for comparison along three different aspects of sensor performance: accuracy (Accuracy and Distance metrics), stability (IQR metric) and generalisability (Margin metric).

4 Results

4.1 Inspection of Data

Gaussian Filtering. Gaussian filtering, as shown in Fig. 3, significantly boosts the Structural Similarity Index Measure (SSIM) values and accentuates the difference between datasets $S(1)$ (same orientation) and $S(2)$ (different orientation). SSIM values stay fairly constant for $S(1)$ (same orientation contacts) as grating gap size increases. In contrast, SSIM values for $S(2)$ decrease noticeably with larger grating gaps. When orientations vary, the difference becomes more distinct with larger grating periods, mirroring human tactile perception trends.

Combining Event Types. As discussed in Sect. 3.1, the NeuroTac sensor captures two types of data: "on events" and "off events", leading to heatmaps that are visually similar. Our analysis indicates that using only "on" or "off" events often

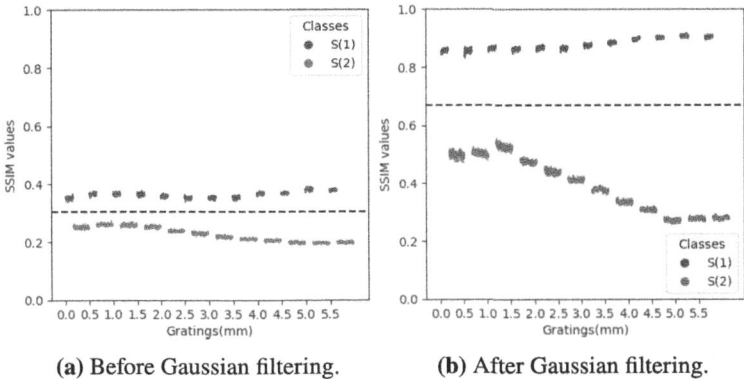

(a) Before Gaussian filtering.   **(b)** After Gaussian filtering.

Fig. 3. SSIM values calculated across all pairwise combinations $S(1)$ (same orientation) and $S(2)$ (different orientation) for all grating gap sizes. The threshold calculated by SVM is shown as a horizontal black dotted line.

yields comparable SSIM results. However, by combining "on" and "off" events into a single heatmap, we obtain slightly increased accuracy on the grating orientation task with 0–7 mm gratings, likely caused by some averaging out of noise internal to the sensor (Fig. 4).

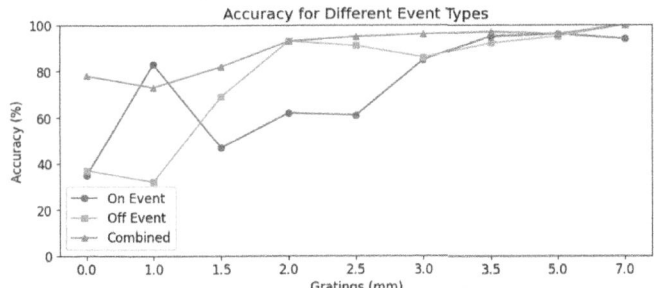

Fig. 4. Accuracy of SVM classification of datasets $S(1)$ (same orientation) and $S(2)$ (different orientation) after the pre-processing and SSIM stages, for "on" events, "off" events, and combined "on and off" events.

4.2 Experiment 1: Using the TIP Benchmark to Determine Ideal Indentation Depth and Noise Degradation

In this experiment, the TIP benchmark is applied to identify the preferred indentation depth on a grating orientation task, and following that to evaluate how the sensor degrades with noise in its indentation location.

Identifying the Indentation Depth with Highest Spatial Accuracy.
The NeuroTac tip taps the gratings at four distinct indentation depths
$\{D_i|0, 1, 2, 3\,\text{mm}\}$. At $D_{0\,\text{mm}}$ the NeuroTac tip makes contact with the surface
but does not apply a downwards force.

The metrics of the TIP benchmark are calculated as described in Sect. 3.3 for
each indentation depth. As indentation increases, the performance of the tactile
sensor increases (Fig. 5, Left panel) as shown by Accuracy and Distance metrics
improving with deeper indentations.

Grating orientations are therefore better discriminated at larger indentation
depths for this tactile sensor. However the NeuroTac sensor got lightly damaged
at $D_{3\,\text{mm}}$ during the 10th trial. Thus, the most recommended indentation depth
is $D_{2\,\text{mm}}$ as a balance between performance and sensor durability.

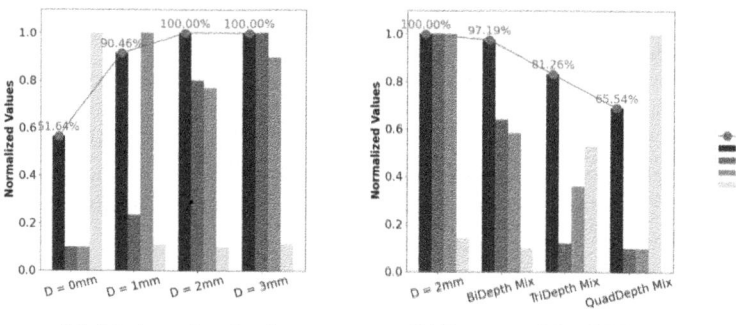

(a) 4 indentation depths.   (b) $D_{2\text{mm}}$ and the 3 Mixtures.

Fig. 5. Accuracy (red line) and normalized 4 metrics of the TIP benchmark (bar chart)
for indentation depths ranging from 0–3 mm (Left) and mixed datasets described in
Sect. 4.2 (Right).

Assessing Noise Degradation in Indentation Depth

- BiDepth Mix: Mixing the heatmaps from $D_{2\,\text{mm}}$ and $D_{3\,\text{mm}}$.
- TriDepth Mix: Mixing the heatmaps from $D_{2\,\text{mm}}$, $D_{3\,\text{mm}}$, and $D_{1\,\text{mm}}$.
- QuadDepth Mix: Mixing the heatmaps from $D_{2\,\text{mm}}$, $D_{3\,\text{mm}}$, $D_{1\,\text{mm}}$ and $D_{0\,\text{mm}}$.

By mixing datasets with different indentation depths, we can see that performance on the TIP benchmark degrades the more indentation depths we add
to the mix (Fig. 5, right panel). This is expected, reflecting the degradation of
performance with added noise in indentation depth. However, the Margin metric
increases with additional datasets, indicating enhanced model generalization.

Mixing heatmaps from varied depths decreases similarity in same-orientation
heatmaps and increases it in different orientations. Despite a reduction in accuracy, distance, and IQR, due to increased noise, the model's improved generalization at QuadDepth Mix could make it more robust to unseen data. The TIP

benchmark thus provides a nuanced way of describing how the addition of noise along a particular dimension affects sensor performance on a tactile task.

4.3 Experiment 2: Using the TIP Benchmark to Optimize Tactile Sensor Hardware

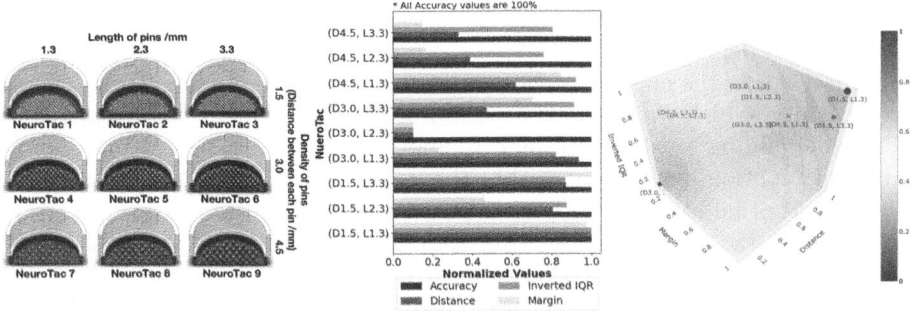

Fig. 6. Hardware and optimization results. Left: Sensors vary along 2 hardware design parameters: pin length (increasing left to right) and pin density (decreasing top to bottom). Middle: Individual metrics of the TIP benchmark for each of the 9 NeuroTac sensors tested. Right: The Distance, IQR and Margin vs pin length and pin density.

This experiment validates the TIP benchmark for a hardware optimization task, by comparing the spatial acuity of 9 NeuroTac sensors with variations only in the sensors' pin length and density, selected via a grid search over both parameters (Fig. 6, left panel). The previously explored optimum indentation depth of $D_{2\,mm}$ was used for all samples in this experiment.

The middle panel of Fig. 6 presents the relationship between our hardware design parameters (pin density and length) and the 4 metrics of the TIP benchmark. NeuroTac 1 is the best overall performer with the highest values in all 4 metrics. Despite all NeuroTacs achieving 100% classification accuracy, the additional TIP benchmark metrics enable us to distinguish more detailed performance levels for each sensor (Fig. 6, right panel). Increasing pin density generally increases the distance as it is linked to the spatial resolution of the sensor. However, the pin length's impact on classification performance is more ambiguous. We hypothesize that during taps, longer pins lead to larger bending angles, but the bending direction of the pins could also become more random. However, medium-length pins display worse performance here. This could be linked to the lack of strict control procedures in the manufacture of the sensors, leading to variations in additional uncontrolled parameters such as gel quantity.

4.4 Discussion

The TIP benchmark has been validated for identifying the optimal indentation depth and assessing noise impact on tactile tasks. It's further used in a grid search for hardware optimization of a neuromorphic tactile sensor, examining 9 configurations and 2 design parameters. The findings indicate that sensors with shorter, denser pins excel in spatial resolution during grating orientation tasks, showcasing TIP's capability to quantitatively evaluate tactile sensors.

By using psychophysics experiments [10], we can effectively compare technological sensory capabilities with human performance. For instance here we demonstrated that the NeuroTac can outperform humans on a grating orientation task in controlled environments (but performance degrades with uncertainty in indentation depth).

Limitations of the current version of the TIP Benchmark include its focus on spatial acuity, which overlooks the temporal dimension of tactile data, potentially undervaluing sensors with high temporal resolution. The grating orientation discrimination experiment within the TIP benchmark can be expanded to include perceptual aspects such as texture sensitivity and force discrimination. The use of data pre-processing (Gaussian filtering) and classification (SVM) algorithms, while inevitable to establish a performance metric, will likely also influence the outcome of our evaluation. Although we attempted to restrict this influence through straightforward methods with low computational complexity, we plan to explore the effect of these algorithms on benchmark outcomes in future through a comparative analysis of different algorithms. TIP's heatmap generation process is designed to be used with all types of tactile sensors by interpreting taxels as pixels. In future works, we will also apply the benchmark to a larger variety of tactile sensors based on different transduction technologies, to demonstrate its applicability across the field of tactile sensing.

The TIP benchmark enables a standardized comparison of tactile sensor spatial acuity. It could be extended and applied to varying tactile sensor technologies to help identify strengths and areas for improvement, guiding the development of more versatile and effective tactile sensors.

5   Conclusion

This project introduces the TIP Benchmark for evaluating the spatial acuity of tactile sensors using an image comparison metric (SSIM), which employs four metrics—**Accuracy**, **Distance**, **IQR**, and **Margin**. The method has been validated on three separate tasks: establishing ideal indentation depth, testing noise degradation and optimizing hardware design on a grating orientation task. The TIP benchmark enables a comparative analysis of the spatial acuity of existing tactile sensors and could help with designing novel highly performing, safe and reliable tactile sensors for real-world applications.

Acknowledgments. We sincerely thank George Brayshaw for his help with revising the manuscript and providing feedback on writing.

References

1. Calli, B., Singh, A., Walsman, A., Srinivasa, S., Abbeel, P., Dollar, A.M.: The YCB object and model set: towards common benchmarks for manipulation research. In: International Conference on Advanced Robotics, pp. 510–517. IEEE (2015)
2. Chan, V., Liu, S.C., van Schaik, A.: AER EAR: a matched silicon cochlea pair with address event representation interface. IEEE Trans. Circuits Syst. I Regul. Pap. **54**(1), 48–59 (2007)
3. Craig, J.C.: Grating orientation as a measure of tactile spatial acuity. Somatosens. Motor Res. **16**(3), 197–206 (1999)
4. Dellon, A.L., Mackinnon, S.E., Crosby, P.M.: Reliability of two-point discrimination measurements. J. Hand Surg. **12**(5), 693–696 (1987)
5. Fishel, J.A.: Design and use of a biomimetic tactile microvibration sensor with human-like sensitivity and its application in texture discrimination using Bayesian exploration. Ph.D. thesis, University of Southern California (2012)
6. Gray, B.L., Fearing, R.S.: A surface micromachined microtactile sensor array. In: IEEE International Conference on Robotics and Automation, vol. 1, pp. 1–6 (1996)
7. Hearst, M.A., Dumais, S.T., Osuna, E., Platt, J., Scholkopf, B.: Support vector machines. IEEE Intell. Syst. Appl. **13**(4), 18–28 (1998)
8. Huynh-Thu, Q., Ghanbari, M.: Scope of validity of PSNR in image/video quality assessment. Electron. Lett. **44**(13), 800–801 (2008)
9. Jiang, Y., Moseson, S., Saxena, A.: Efficient grasping from RGBD images: learning using a new rectangle representation. In: 2011 IEEE International Conference on Robotics and Automation, pp. 3304–3311. IEEE (2011)
10. Johnson, K.O., Phillips, J.R.: Tactile spatial resolution. I. Two-point discrimination, gap detection, grating resolution, and letter recognition. J. Neurophysiol. **46**(6), 1177–1192 (1981)
11. Lepora, N.F.: Soft biomimetic optical tactile sensing with the tactip: a review. IEEE Sens. J. **21**(19), 21131–21143 (2021)
12. Lepora, N.F., Ward-Cherrier, B.: Superresolution with an optical tactile sensor. In: International Conference on Intelligent Robots and Systems, pp. 2686–2691 (2015)
13. Lepora, N.F., Ward-Cherrier, B.: Tactile quality control with biomimetic active touch. IEEE Robot. Autom. Lett. **1**(2), 646–652 (2016)
14. Mahler, J., et al.: Dex-net 1.0: a cloud-based network of 3D objects for robust grasp planning using a multi-armed bandit model with correlated rewards. In: 2016 IEEE International Conference on Robotics and Automation (ICRA), pp. 1957–1964. IEEE (2016)
15. Okamoto, S., Nagano, H., Yamada, Y.: Psychophysical dimensions of tactile perception of textures. IEEE Trans. Haptics **6**(1), 81–93 (2012)
16. Olczak, D., Sukumar, V., Pruszynski, J.A.: Edge orientation perception during active touch. J. Neurophysiol. **120**(5), 2423–2429 (2018)
17. Roscow, E., Kent, C., Leonards, U., Lepora, N.F.: Discrimination-based perception for robot touch. In: Lepora, N.F.F., Mura, A., Mangan, M., Verschure, P.F.M.J.F.M.J., Desmulliez, M., Prescott, T.J.J. (eds.) Living Machines 2016. LNCS (LNAI), vol. 9793, pp. 498–502. Springer, Cham (2016). https://doi.org/10.1007/978-3-319-42417-0_53
18. Russakovsky, O., et al.: ImageNet large scale visual recognition challenge. Int. J. Comput. Vision **115**, 211–252 (2015)
19. Sathian, K., Zangaladze, A., Green, J., Vitek, J., DeLong, M.: Tactile spatial acuity and roughness discrimination: impairments due to aging and Parkinson's disease. Neurology **49**(1), 168–177 (1997)

20. Stassi, S., Cauda, V., Canavese, G., Pirri, C.F.: Flexible tactile sensing based on piezoresistive composites: a review. Sensors **14**(3), 5296–5332 (2014)
21. Van Boven, R.W., Johnson, K.O.: A psychophysical study of the mechanisms of sensory recovery following nerve injury in humans. Brain **117**(1), 149–167 (1994)
22. Wang, X., Yang, Y., Zhou, Z., Xiang, G., Liu, H.: HSVTac: a high-speed vision-based tactile sensor for exploring fingertip tactile sensitivity. IEEE Sens. (2023)
23. Wang, Z., Bovik, A.C., Sheikh, H.R., Simoncelli, E.P.: Image quality assessment: from error visibility to structural similarity. IEEE Trans. Image Process. **13**(4), 600–612 (2004)
24. Ward-Cherrier, B., et al.: The TacTip family: Soft optical tactile sensors with 3d-printed biomimetic morphologies. Soft Rob. **5**(2), 216–227 (2018)
25. Ward-Cherrier, B., Pestell, N., Lepora, N.F.: NeuroTac: a neuromorphic optical tactile sensor applied to texture recognition. In: 2020 IEEE International Conference on Robotics and Automation (ICRA), pp. 2654–2660. IEEE (2020)

Towards Intensifying Perceived Pressure in Midair Haptics: Comparing Perceived Pressure Intensity and Skin Displacement Between LM and AM Stimuli

Tao Morisaki$^{(\boxtimes)}$ (ID) and Yusuke Ujitoko (ID)

NTT Communication Science Laboratories, Nippon Telegraph and Telephone Corporation, Atsugi, Japan
{tao.morisaki,yusuke.ujitoko}@ntt.com

Abstract. Ultrasound Midair Haptics (UMH) can present various non-contact tactile patterns by focusing ultrasound on human skin. With UMH, a steady pressure sensation can be presented by periodically shifting a stimulus point (ultrasound focus) at several hertz. Such stimulus with a periodic focal shift is called Lateral Modulation (LM). The perceived intensity of this pressure sensation was several times stronger than the applied radiation force (e.g., 0.22 N for 27 mN of radiation force). Further intensifying the pressure sensation by LM expands the range of reproducible tactile sensations such as a hard object; however, a stimulus design guideline for the intensification has not been established because the perception mechanism of the LM-evoked pressure sensation is still unclear. Towards intensifying the pressure sensations in UMH, this study investigates the effects of the main frequency components of skin vibrations produced by LM and that of the amplitude on the perceived pressure intensity. We first confirmed that the perceived pressure intensity of LM 5 Hz was stronger than that of 5 Hz amplitude modulation (AM). AM is a simple vibration with a fixed stimulus position. We also measured the 5 Hz vibration amplitude of the skin during stimulation and confirmed no significant difference in the amplitude between LM and AM. The results showed that a 5 Hz skin vibration and the amplitude alone cannot explain the perceived intensity of the pressure sensation by LM. These results indicate that other factors in LM such as focal shifts would be necessary to present stronger pressure sensations.

Keywords: Pressure sensation · Skin Vibration · Ultrasound Midair Haptics

1 Introduction

Noncontact tactile displays by focusing ultrasound, known as Ultrasound Midair Haptics, are promising tools for haptics since they can present various tactile patterns without users needing to wear any devices [19]. Noncontact force, called

H. Kajimoto et al. (Eds.): EuroHaptics 2024, LNCS 14768, pp. 107–119, 2025.
https://doi.org/10.1007/978-3-031-70058-3_9

acoustic radiation force, is generated at an ultrasound focus [25], conveying a noncontact tactile stimulus onto the human skin [2, 8, 10]. An airborne ultrasound tactile display (AUTD), an array of independently controllable ultrasound transducers [8], is used for ultrasound focusing. A list of acronyms used in this paper is shown in Table 1. Many applications of noncontact tactile presentation using ultrasound have been proposed, such as a midair touchable 3D image [14, 15], and a midair gesture interface with haptic feedback [13, 26]. To make a focused ultrasound stimulus perceptible, modulation of radiation force distribution is necessary since the radiation force of focused ultrasound is too weak for humans to perceive in a typical AUTD setup [7]. The acoustic radiation force is in the several tens of mN even if a large aperture AUTD (1 m × 1 m) is used [20]. Such a weak constant force cannot be perceived (cannot continuously exceed human perception threshold) due to tactile receptor adaptation for stationary forces [23]. To enable humans to perceive an ultrasound tactile stimulus more intensely, Amplitude Modulation (AM) [7] and Lateral Modulation (LM) [21, 22] have been proposed. A schematic illustration of these modulations is shown in Fig. 1. In the case of AM, the radiation force at the ultrasound focus is periodically varied. In the case of LM, the focus is periodically moved by several millimeters. A line and circle whose representative length are several millimeters have been used for the focal trajectory in LM. When a large focal trajectory is used with continuous focal movement, the stimulus method is called Spatiotemporal Modulation (STM) [5, 6]. With the AM, LM, and STM, tactile stimuli are typically perceived as vibrations since the radiation force presented to each point on the skin constantly fluctuates.

Recently, Morisaki et al. experimentally demonstrated that an LM stimulus is perceived as a stationary pressure rather than a vibration when the focus shifts at a low frequency (5–15 Hz) [17, 18]. Although the evoked sensation was not perfectly static, the remaining nonstationary sensations (i.e., the focus movement sensation or vibration sensation) were significantly suppressed. Perception of pressure is important since when humans contact objects during daily life, they inevitably perceive pressure in addition to any vibration [1]. Morisaki et al. presented a 5 Hz linearly moving focus (i.e., a linear 5 Hz LM) to the palm of a hand. They showed that the presented LM stimulus was perceived as 0.21 N pressure sensation for the physically applied radiation force of 27 mN [17]. They also showed that a 5 Hz circular focus trajectory with a radius of 6 mm or less was perceived as a 0.24 N pressure sensation on a finger pad [18]. Although the level of pressure sensation currently achievable with LM is already acceptable for the tactile presentation of some virtual objects, these are still limited to soft objects, such as a gel [16]. Further intensification of the LM-evoked pressure sensations expands the range of reproducible tactile sensations such as harder objects or stronger impact sensations.

For stronger pressure sensation presentation with LM, it is necessary to optimize the stimulus pattern (e.g., the focal trajectory shape), based on an understanding of the human perceptual mechanism, since the maximum radiation force at a focus is limited by acoustic saturation [20]. However, the perceptual mechanism of the pressure sensation presented by LM is yet to be elucidated, and thus a stimulus design guideline for stronger pressure sensation presentation remains out of reach. The LM-evoked pressure sensation is only experimentally found, and it has not been investigated which factors in skin stimulus patterns (i.e., periodic focal shifts) caused by the LM are crucial for producing pressure sensation. Although the mechanism has not been explored, previous studies' results implied that one of the important factors for producing pressure sensations is a low-frequency (several hertz) skin vibration produced by low-frequency LM [12,17]. Morisaki et al. reported that the tactile sensation evoked by LM at 5 Hz approached a pure stationary pressure sensation when the pressure fluctuation pattern on the skin approached a sinusoidal wave from a square wave (i.e., the harmonic component was decreased) [17]. Konyo et al. confirmed that pressure sensation was evoked by presenting 5 Hz vibrations using a tactile vibrator [12]. Moreover, the used low-frequency bandwidth (several hertz) is matched the bandwidth in which SA-I (slowly adaptive type-I mechanoreceptors) mainly respond. It is known that pressure sensation is induced by the firing of SA-I [11]. Based on these previous implications, we first focused on the effect of low-frequency (several Hz) skin vibrations on the perceived intensity of pressure sensations by LM.

In this study, we investigated the contribution of the low-frequency component of skin vibration produced by LM and its amplitude to the perceived intensity of pressure sensations. These investigations contribute to establishing a design guideline for stronger presser sensation presentation. We conducted two series of experiments: a perceived intensity evaluation of LM and AM, and a physical measurement of skin vibration. In Experiment 1, to examine the effect of the low-frequency component, we compared the intensity of 5 Hz LM- and 5 Hz AM-evoked pressure sensations and found that the 5 Hz LM-evoked pressure sensations were significantly stronger. To examine whether this intensity difference is due to the skin vibration amplitude, in Experiment 2, we measured the actual amplitude of the produced 5 Hz skin vibration with a laser displacement sensor. Although previous studies measured skin displacement for ultrasound stimulus, skin vibrations produced by AM and LM stimuli of the same frequency were not compared [3,4]. The measurement results showed that the amplitude was not significantly different between 5 Hz AM and 5 Hz LM. Since the 5 Hz AM is a vibration with a fixed stimulus position, this difference in the pressure sensation intensity must be attributed to factors other than the 5 Hz component of the produced skin vibration. The other factors (e.g., repetitive focus shift) would be needed for strong presser sensation presentation by ultrasound.

2   Experiment 1: Comparing LM and AM in the Intensity of Evoked Pressure Sensation

In this experiment, we compared the intensity of the pressure sensations evoked by a 5 Hz LM and a 5 Hz AM to investigate whether the intensity of the sensations is attributable only to a 5 Hz skin vibration. The perceived intensity of pressure

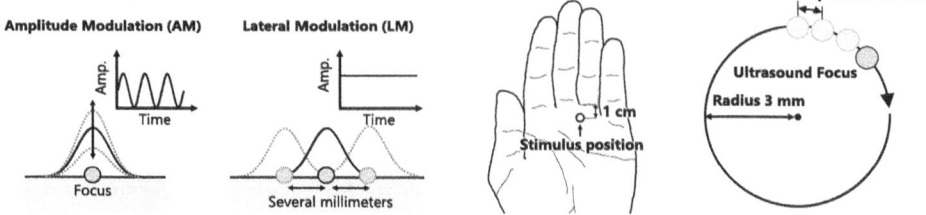

Fig. 1. Schematic of Amplitude Modulation (AM) and Lateral Modulation (LM). In the case of AM, the radiation force was modulated with a fixed stimulus position. In the case of LM, the stimulus position was repeatedly moved several millimeters.

Fig. 2. Stimulus position and schematic of used circular lateral modulation. In the case of the LM, a focus was moved along a 3 mm radius circle at 5 Hz. The movement step width was 0.2 mm.

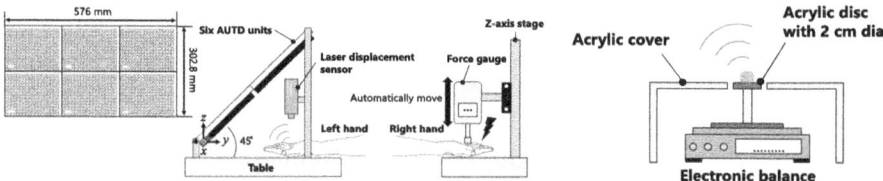

Fig. 3. Setup for evaluating pressure sensations evoked by LM and AM. An ultrasound tactile stimulus was presented to the left hand. An indentation stimulus created by a z-axis stage was presented to the right hand. Participants compared the perceived intensity of these stimuli.

Fig. 4. Setup for measuring radiation force at ultrasound focus. The radiation force was 1.92 mN for the maximum device output of the AUTDs.

sensations by ultrasound was quantified by comparing a physical indentation of the palm of a hand. Although the comparison between an LM stimulus and a palm indentation has been conducted previously [17, 18], a comparison with the intensity of AM-evoked pressure sensations has not been investigated.

In a previous study [17, 18], a palm indentation of over 2 s duration with a constant indentation depth (i.e., a continuous and physically static stimulus) causes adaptation of tactile receptors (i.e., a change in the perceived intensity over time) [9]. In our experiment, the indentation depth was continuously varied to suppress this adaptation effect for a more stable evaluation of the pressure sensation intensity (as shown in Fig. 5).

2.1 Stimulus

We presented 5 Hz LM and 5 Hz AM stimuli in this experiment. 10 Hz AM stimuli were also used because 10 Hz pressure fluctuations also occurred with 5 Hz LM stimuli [17]. In the case of 5 and 10 Hz AM, the radiation force at the focus varied in a pattern with 5 and 10 Hz sinusoidal waves, respectively. In the case of 5 Hz LM, the focal movement frequency was 5 Hz. The schematic of the LM applied in this experiment is shown in Fig. 2. The trajectory of the LM stimulus was a circle with a radius of 3 mm, and the spatial movement step width of the focus was 0.2 mm. The parameters of the LM stimulus were determined based on previous studies [17, 18].

2.2   Experimental Setup

Overview. A schematic illustration of the experimental setup and its coordinate system is shown in Fig. 3. The origin of the coordinate system is the lower left transducer of the AUTDs. The setup consisted of six AUTDs, a single-axis stage (ARS-7036-GM and QT-AMM3, CHUO PRECISION INDUSTRIAL Co., Ltd.), and a force gauge (IMADA ZTS-50N). The 249 ultrasound transducers driven at 40 kHz were arrayed in a single unit of AUTD. Each AUTD unit communicated by EtherCat was synchronously driven [20]. The AUTDs were used to present ultrasound tactile stimulus with the maximum power to the participant's left palm. As described in the later section, the radiation force at maximum power was 19.2 mN. The AUTDs were tilted 45° around the y-axis for the skin displacement measurement (as described in Experiment 2). The force gauge was mounted on a single-axis stage and automatically moved up and down to push the participant's right palm with its tip. A beveled plastic cylinder with a diameter of 6 mm was attached to the force gauge tip. The radius of this cylinder matched the radius of the focal trajectory of the LM stimulus. Participants compared the pressure sensations on their palms evoked by the ultrasound stimulus with that evoked by the force gauge stimulus and evaluated which force was greater. During the experiment, the participants listened to white noise through headphones to mask the noise created by the experimental equipment.

Measurement of Radiation Force at Focus. In this section, we measured the radiation force at the presented ultrasound focus in the experimental setup (shown in Fig. 3). The radiation force was 19.2 mN for the maximum output of the AUTDs.

Figure 4 shows the setup for measuring radiation force at the focus. The focus position was the same as in Experiment 1. A 2 cm diameter acrylic circle disc was placed in the focus position and an electronic balance measured the radiation force applied to the disc. The diameter was determined based on a previous study [17] so that the disc completely covers a created focus. An acrylic cover was also placed over the electronic balance to remove the side lobe of the focus. The side lobe is a radiation force distribution that appears extensively within a few millimeters of the focus position.

2.3   Procedure

Seven males (6 in their 20 s and 1 in their 30 s) and six females (5 in their 20 s and 1 in their 30 s) participated in this experiment. Ethical approval for this

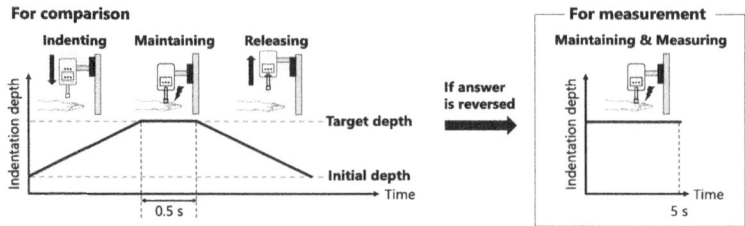

Fig. 5. Schematic of indentation depth control. For force comparison, the stage stopped 0.5 s after indenting the palm to the target depth and then returned to the original position. The stage movement speed was 3 mm/s. When participants' answer was reversed, the indentation force was measured. For the measurement, the stage indented the right palm with the same target depth again and kept it during 5 s.

study was obtained from the ethics committee at Nippon Telegraph and Telephone Corporation (Approval number: R05-018). The experiments were conducted according to principles that have their origin in the Helsinki Declaration. Written informed consent was obtained from all participants in this study. The experiment was conducted in November 2023.

In advance of the experiments, a marker indicating the stimulus position (1 cm below the base of the middle finger) was drawn on both palms of the participants' hands with a water-based pen. The schematic of the stimulus position is shown in Fig. 2. The participants first placed their left hand so that the position of the drawn marker matched the ultrasound stimulus position (284.3, 150, 0) mm. The right hand was also placed so that its marker was aligned directly under the tip of the force gauge. After hand placement, an ultrasound and an indentation stimulus were presented to the left and right hands, respectively.

The depth of the physical indentation made on the right palm was increased or decreased by 0.1 mm based on the interleaved staircase. The initial value was determined in advance for each participant so that the initial indentation force was 0 N for the ascending series and 1 N for the descending series. First, ultrasound and indentation stimuli were presented to a left and right hand, respectively. Participants compared the force perceived by their right and left hands and indicated which they considered the stronger one using a foot switch. An example of the indentation depth control for the comparison is depicted in Fig. 5. Figure 6 shows a measured one participant's temporal profile of the indentation force. The negative indentation force was unintended noise caused by the force gauge movement. The stage stopped at 0.5 s after indenting the palm to a predetermined depth, and then it returned to its original position. The stage movement speed (indentation speed) was 3 mm/s. When participants' answer was reversed (i.e., the stimulus that participants perceived more strongly switched), the change in the indentation depth was reversed on the next trial. In addition, the applied indentation force at the reverse timing was measured and saved. For the force measurement, the stage indented the right palm with the same depth again and kept the indentation depth during the 5-s measure-

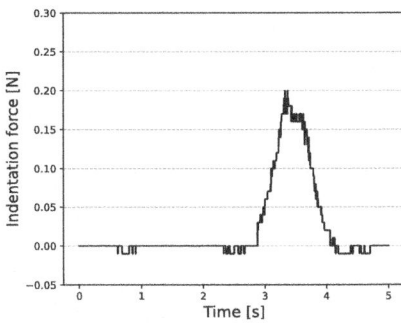
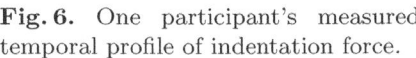
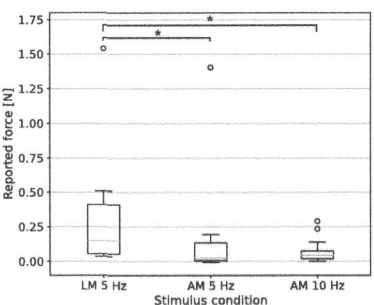

Fig. 6. One participant's measured temporal profile of indentation force.

Fig. 7. Evaluation result for presser sensation evoked by AM and LM stimuli (as performed in Experiment 1). A white circle indicated outliers.

ment time. The indentation depth control for the measurement is depicted in Fig. 5. When the number of reversals reached six, the ascending/descending series ended. The stimuli of the ascending and descending series were randomly mixed. This experiment was performed once for each stimulus type (LM 5 Hz, AM 5 Hz, and AM 10 Hz), and the order of presentation was randomized.

To accurately evaluate the intensity of the pressure sensation evoked by the LM and AM, the participants were instructed to focus only on the steady force component and ignore the vibratory component (vibration sensation and movement sensation of the stimulus position) when comparing the ultrasound and the indentation stimulus.

2.4 Result

A box-and-whisker plot of the evaluated perceived intensity of the pressure sensation is shown in Fig. 7. To calculate the perceived intensity, the indentation depth values at the points of the last three reversals obtained in the interleaved staircase were used [24]. Six indentation forces were obtained (from the ascending and descending series) per participant, and the average of these was used as the perceived intensity of the LM- or AM-evoked pressure sensation.

Among all stimulus conditions, the highest median value of the perceived intensity was 0.15 N (LM stimulus at 5 Hz). The lowest median value was 0.025 N (AM at 5 Hz).

We performed a statistical analysis to evaluate the differences in the perceived intensity among the stimulus conditions. First, a Shapiro-Wilk test showed that the data did not belong to a normal distribution in all conditions ($p < 0.005$). For this reason, a Wilcoxon signed-rank test with Bonferroni correction was used. The analysis results showed that the pressure sensation by 5 Hz LM was significantly stronger than that by 5 Hz AM ($p = 0.0007$ and $r = 0.88$) and 10 Hz AM ($p = 0.0007$ and $r = 0.88$). p and r represent the p-value and effect size,

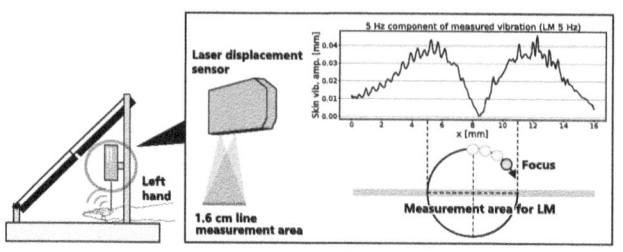
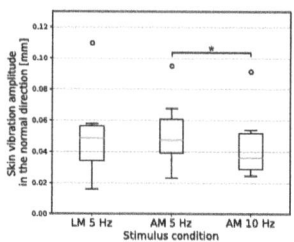

Fig. 8. Measurement setup for skin vibration. The laser displacement sensor can measure displacement on a 1.6 cm line at a time. The measured 1D spatial distribution of 5 Hz amplitude for 5 Hz LM is also shown.

Fig. 9. Measurement result of skin displacement (Experiment 2). A white circle indicates outliers.

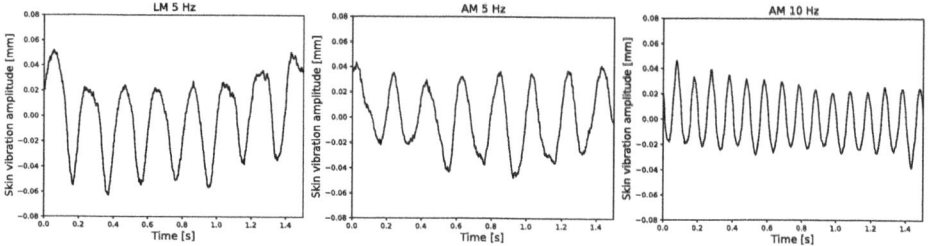

Fig. 10. Measured skin vibration waveform of one participant.

respectively. The effect size was calculated as $r = \frac{Z}{\sqrt{N}}$. Z and N are z-score and sample size, respectively. There was no significant difference between 5 Hz AM and 10 Hz AM ($p = 0.9 > 0.05$, $r = 0.28$).

These results indicate that the skin vibration at 5 Hz cannot alone explain the strong perceived intensity of the pressure sensation evoked by LM.

3   Experiment 2: Measurement of Skin Vibration Produced by LM and AM Stimuli

The results of Experiment 1 showed that the pressure sensations evoked by the LM stimuli were stronger than those by the AM stimuli. In Experiment 2, we measured the amplitude of the skin displacement caused by LM and AM stimuli to examine whether the strong pressure sensation evoked by LM was explained by the amplitude.

3.1   Experimental Setup and Procedure

Figure 8 shows a schematic illustration of the experimental setup. The setup consisted of six AUTDs and a laser displacement sensor (LJ-X8060, KEYENCE CORPORATION). The arrangement of the AUTDs was the same as in Experiment 1. This displacement sensor can measure displacement on a 1.6 cm line at a time. The skim vibration was measured. The displacement sensor was calibrated in advance so that the center of the line-shaped measurement range matched the presentation position of the AM stimulus (center of the focal trajectory in LM). A schematic illustration of the measurement range is shown in Fig. 8.

The experimental participants were the same as in Experiment 1. Participants placed their left hand so that the drawn marker (as described in Experiment 1) aligned with the stimulus position. Ultrasound stimuli (5 Hz LM and 5, 10 Hz AM) with the maximum power were presented by the AUTDs.

3.2 Result and Analysis

We performed a Fourier transform on the measured amplitude of the skin vibration, and the skin vibration amplitude at the modulation frequencies in the presented stimuli are shown in Fig. 9. The 5 Hz component was plotted for the 5 Hz LM and AM stimuli, and the 10 Hz component was plotted for the 10 Hz AM stimulus. The skin vibration amplitude was measured in the range of a 1.6 cm line, and only the maximum amplitude was plotted in Fig. 9. The one participant's 1D spatial distribution data for 5 Hz LM is shown in Fig. 8.

Among stimulus conditions, the highest median value of the amplitude was 0.049 mm (LM at 5 Hz). The lowest median value was 0.036 mm (AM at 10 Hz).

We performed a statistical analysis to evaluate the differences in skin amplitude between each stimulus condition. First, the Shapiro-Wilk test showed that the 5 Hz LM and 10 Hz AM data did not belong to a normal distribution ($p < 0.005$). We conducted a Wilcoxon signed rank test with Bonferroni correction. We set a significance level at 0.05. The analysis results showed there are no significant differences between pairs of LM-AM (5 Hz) ($p = 0.78$ and $r = 0.087$), and LM-AM(10 Hz) ($p = 0.57$ and $r = 0.53$). The results also showed that the skin amplitude for AM at 5 Hz was significantly larger than that for AM at 10 Hz ($p = 0.006$, $r = 0.72$).

4   Discussion

4.1   Effect of 5 Hz Skin Vibration and Its Amplitude on Pressure Sensation Intensity

This study investigated the effects of 5 Hz skin vibration and its amplitude on the perceived intensity of the pressure sensations evoked by LM stimuli at 5 Hz.

The experimental results showed that the 5 Hz skin vibration and its amplitude cannot alone explain why LM evokes strong pressure sensations of over 0.15 N for a 19.2 mN physically applied radiation force. As considered in previous studies [12, 17], if 5 Hz skin vibration produced by LM is the sole reason why LM at 5 Hz is perceived as a strong pressure sensation, the perceived intensity of pressure sensation must be equal between AM and LM when the 5 Hz amplitude of the produced skin vibration is the same. However, the results presented in Fig. 7 and Fig. 9 show that the pressure sensation evoked by LM was significantly higher than that by AM even though the 5 Hz amplitude of skin vibration was the same. Therefore, the strong perceived intensity of pressure sensation by LM at 5 Hz cannot be explained solely by the 5 Hz skin vibration. This suggests that other factors (e.g., a harmonic component of skin vibration or repetitive focus movement) will be essential for evoking strong pressure sensations by ultrasound.

In the future, we will further investigate the perception mechanism of the LM-evoked pressure sensation. First, since this paper used only single-point local stimulation, we will investigate the effect of the stimulation area on the pressure sensation intensity by manipulating the focus shape and the focal trajectory in LM. We will also investigate the effect of harmonic components in LM-produced skin displacement shown in Fig. 10, which may trigger rapidly adaptive mechanoreceptors.

4.2   Comparison with Previous Study

The result of Experiment 1 was consistent with previous works [17, 18, 21, 22]. First, the perceived pressure intensity by LM at 5 Hz was consistent between our results and previous studies. The previous study reported that the perceived pressure intensity was 0.21 N and 0.24 N for a 27 and 20 mN applied radiation force, respectively [17, 18]. Second, the comparison result between LM and AM in perceived intensity was also consistent with the previous study. Takahashi et al. reported that the perception threshold of the LM was lower than that of the AM in the 10–200 Hz range [21, 22]. They did not compare the pressure sensation intensity between LM and AM at 5 Hz; thus, our study is the first comparison between them.

4.3   Limitation and Future Work

The palm indentation stimulus conducted in Experiment 1 has two limitations. The first limitation is that stepwise control of the indentation force could not be conducted due to the hardware limitations of the experimental setup. In Experiment 1, we controlled the indentation depth (i.e., increased or decreased it) using a single-axis stage. Thus, the indentation force was not controlled directly and the step size of the indentation force control in the interleaved staircase was not constant. The second limitation is the difference in temporal stimulus profile between the ultrasound and the indentation stimulus. Although the physical intensity of the ultrasound stimulus was constant (continuously presented with the maximum power of the AUTDs), the indentation force was continuously varied to avoid adaptation of the mechanoreceptors (as shown in Fig. 5). This difference in the temporal property between the ultrasound stimulus and the indentation stimulus would affect the comparison.

In spite of these limitations, we conjecture that the evaluation of pressure sensation intensity was stably conducted and the measurements of intensity obtained in this study were appropriate since they were consistent with the results obtained in previous studies [17, 18] as described in the previous section.

In Experiment 2, we only compared the maximum skin vibration amplitude and ignored its spatial distribution. We will evaluate the effect of the spiritual distribution on the perceived pressure intensity.

5 Conclusion

This study verified that the perceived intensity of the pressure sensation evoked by a 5 Hz LM stimulus cannot be explained only by the produced low-frequency skin vibration and its amplitude. In the experiments, we compared the intensity of the pressure sensation evoked by an AM stimulus, in which the radiation force at the stimulus point fluctuated at 5 Hz, and evoked by an LM stimulus, in which a stimulus point moved at 5 Hz circularly. The results showed that the pressure sensation evoked by LM was stronger than that by AM, even though there was no difference in the 5 Hz amplitude of the skin vibration produced by each stimulus.

Table 1. List of acronyms used in this paper.

Acronym	
UMH	Ultrasound Midair Haptics
AUTD	Airborne Ultrasound Tactile Display
LM	Lateral Modulation
AM	Amplitude Modulation

Our experimental results indicate that factors other than low-frequency vibration on the skin are necessary for strong pressure sensation presentation using ultrasound. In the future, as an additional factor, we will examine the effect of stimulus area since LM stimulates a larger area than AM.

References

1. Bolanowski Jr., S.J., Gescheider, G.A., Verrillo, R.T., Checkosky, C.M.: Four channels mediate the mechanical aspects of touch. J. Acoust. Soc. Am. **84**(5), 1680–1694 (1988)
2. Carter, T., Seah, S.A., Long, B., Drinkwater, B., Subramanian, S.: UltraHaptics: multi-point mid-air haptic feedback for touch surfaces. In: Proceedings of the 26th annual ACM Symposium on User Interface Software and Technology, pp. 505–514. ACM (2013)
3. Chilles, J., Frier, W., Abdouni, A., Giordano, M., Georgiou, O.: Laser doppler vibrometry and fem simulations of ultrasonic mid-air haptics. In: Proceedings of 2019 IEEE World Haptics Conference (WHC), pp. 259–264. IEEE (2019)
4. Frier, W., Abdouni, A., Pittera, D., Georgiou, O., Malkin, R.: Simulating airborne ultrasound vibrations in human skin for haptic applications. IEEE Access **10**, 15443–15456 (2022)

5. Frier, W., et al.: Using spatiotemporal modulation to draw tactile patterns in mid-air. In: Prattichizzo, D., Shinoda, H., Tan, H.Z., Ruffaldi, E., Frisoli, A. (eds.) EuroHaptics 2018. LNCS, vol. 10893, pp. 270–281. Springer, Cham (2018). https://doi.org/10.1007/978-3-319-93445-7_24

6. Frier, W., Pittera, D., Ablart, D., Obrist, M., Subramanian, S.: Sampling strategy for ultrasonic mid-air haptics. In: Proceedings of the 2019 CHI Conference on Human Factors in Computing Systems, pp. 1–11 (2019)

7. Hasegawa, K., Shinoda, H.: Aerial vibrotactile display based on multiunit ultrasound phased array. IEEE Trans. Haptics **11**(3), 367–377 (2018)

8. Hoshi, T., Takahashi, M., Iwamoto, T., Shinoda, H.: Noncontact tactile display based on radiation pressure of airborne ultrasound. IEEE Trans. Haptics **3**(3), 155–165 (2010)

9. Iggo, A., Muir, A.R.: The structure and function of a slowly adapting touch corpuscle in hairy skin. J. Physiol. **200**(3), 763 (1969)

10. Iwamoto, T., Shinoda, H.: Ultrasound tactile display for stress field reproduction-examination of non-vibratory tactile apparent movement. In: Proceedings of First Joint Eurohaptics Conference and Symposium on Haptic Interfaces for Virtual Environment and Teleoperator Systems, pp. 220–228. IEEE (2005)

11. Kajimoto, H., Kawakami, N., Tachi, S.: Electro-tactile display with tactile primary color approach. In: Proceedings of 2004 IEEE/RSJ International Conference on Intelligent Robots and Systems. IEEE (2004)

12. Konyo, M., Tadokoro, S., Yoshida, A., Saiwaki, N.: A tactile synthesis method using multiple frequency vibrations for representing virtual touch. In: Proceedings of 2005 IEEE/RSJ International Conference on Intelligent Robots and Systems, pp. 3965–3971. IEEE (2005)

13. Korres, G., Chehabeddine, S., Eid, M.: Mid-air tactile feedback co-located with virtual touchscreen improves dual-task performance. IEEE Trans. Haptics **13**(4), 825–830 (2020)

14. Makino, Y., Furuyama, Y., Inoue, S., Shinoda, H.: HaptoClone (haptic-optical clone) for mutual tele-environment by real-time 3D image transfer with midair force feedback. In: Proceedings of the 2019 CHI Conference on Human Factors in Computing Systems, pp. 1980–1990 (2016)

15. Monnai, Y., Hasegawa, K., Fujiwara, M., Yoshino, K., Inoue, S., Shinoda, H.: HaptoMime: mid-air haptic interaction with a floating virtual screen. In: Proceedings of the 27th Annual ACM Symposium on User Interface Software and Technology, pp. 663–667 (2014)

16. Morisaki, T., Fujiwara, M., Makino, Y., Shinoda, H.: Midair haptic-optic display with multi-tactile texture based on presenting vibration and pressure sensation by ultrasound. In: Proceedings of SIGGRAPH Asia 2021 Emerging Technologies, pp. 1–2 (2021)

17. Morisaki, T., Fujiwara, M., Makino, Y., Shinoda, H.: Non-vibratory pressure sensation produced by ultrasound focus moving laterally and repetitively with fine spatial step width. IEEE Trans. Haptics **15**(2), 441–450 (2021)

18. Morisaki, T., Fujiwara, M., Makino, Y., Shinoda, H.: Noncontact haptic rendering of static contact with convex surface using circular movement of ultrasound focus on a finger pad. IEEE Trans. Haptics (2023)

19. Rakkolainen, I., Freeman, E., Sand, A., Raisamo, R., Brewster, S.: A survey of mid-air ultrasound haptics and its applications. IEEE Trans. Haptics **14**(1), 2–19 (2020)

20. Suzuki, S., Inoue, S., Fujiwara, M., Makino, Y., Shinoda, H.: AUTD3: scalable airborne ultrasound tactile display. IEEE Trans. Haptics **14**(4), 740–749 (2021)

21. Takahashi, R., Hasegawa, K., Shinoda, H.: Lateral modulation of midair ultrasound focus for intensified vibrotactile stimuli. In: Prattichizzo, D., Shinoda, H., Tan, H.Z., Ruffaldi, E., Frisoli, A. (eds.) EuroHaptics 2018. LNCS, vol. 10894, pp. 276–288. Springer, Cham (2018). https://doi.org/10.1007/978-3-319-93399-3_25
22. Takahashi, R., Hasegawa, K., Shinoda, H.: Tactile stimulation by repetitive lateral movement of midair ultrasound focus. IEEE Trans. Haptics **13**(2), 334–342 (2019)
23. Vallbo, A.B., Johansson, R.S., et al.: Properties of cutaneous mechanoreceptors in the human hand related to touch sensation. Hum. Neurobiol. **3**(1), 3–14 (1984)
24. Wojna, K., Georgiou, O., Beattie, D., Frier, W., Wright, M., Lutteroth, C.: An exploration of just noticeable differences in mid-air haptics. In: 2023 IEEE World Haptics Conference (WHC), pp. 410–416. IEEE (2023)
25. Yosioka, K., Kawasima, Y.: Acoustic radiation pressure on a compressible sphere. Acta Acust. Acust. **5**(3), 167–173 (1955)
26. Young, G., Milne, H., Griffiths, D., Padfield, E., Blenkinsopp, R., Georgiou, O.: Designing mid-air haptic gesture controlled user interfaces for cars. Proc. ACM Hum.-Comput. Interact. **4**(EICS), 1–23 (2020)

Virtual Hand Illusion Induced by Suction Pressure Stimulation to the Face

Takayuki Kameoka[1]([✉])(iD), Taku Hachisu[1](iD), and Hiroyuki Kajimoto[2](iD)

[1] Tsukuba University, Tsukuba, Iabaraki 305-8573, Japan
{kameoka,hachisu}@ah.iit.tsukuba.ac.jp
[2] The Universiry of Electro-Communications, Chofu, Tokyo 182-8585, Japan
kajimoto@kaji-lab.jp

Abstract. The body ownership toward avatars in virtual environments is a key factor for enhancing the quality of virtual reality experiences. Despite previous beliefs that spatiotemporal congruence of visuo-motor and visuo-tactile stimulation was essential for inducing body ownership, this paper shows that body ownership can occur with spatial incongruence of visuo-tactile stimulation, specifically when tactile stimuli are applied to the face. We developed a suction pressure stimulation display system integrated with a head-mounted display in an immersive virtual environment, where tactile feedback corresponded to the interaction between a virtual hand and virtual object. Three tactile stimulation conditions were compared: no tactile feedback, stimulation to the fingertips, and stimulation to the face. In an experiment involving ten participants, we observed that body ownership was induced with tactile feedback provided to the fingertips or face. These results indicate that body ownership can be induced even under a large spatial incongruence between visuo-tactile stimulation, and provide insights for designing tactile displays that present tactile stimuli to different body parts.

Keywords: Body ownership · Suction pressure stimulation · HMD

1 Introduction

With the increasing use of head-mounted display (HMD) with optical hand tracking systems, allowing for hands-free virtual reality (VR) experiences [7], various studies have proposed stimulating different body parts instead of hands to provide tactile feedback in virtual environments [19,20,23]. Furthermore, we have previously developed a tactile feedback system that substitutes hand stimulation with facial stimulation using suction pressure and evaluated its effect on enhancing the quality of VR experiences [12–15]. The face and hands exhibit equivalent sensitivity in terms of the two-point discrimination threshold. Additionally, they are situated proximally within the somatosensory cortex of the brain, suggesting their potential suitability as alternative sites for tactile stimulation [10,22].

Supported by JSPS KAKENHI, Grant Numbers JP20K20627 and JP20J23626.

H. Kajimoto et al. (Eds.): EuroHaptics 2024, LNCS 14768, pp. 120–132, 2025.
https://doi.org/10.1007/978-3-031-70058-3_10

However, these studies primarily focused on evaluating the quality of VR experiences and did not evaluate body ownership toward virtual avatars [19,20,23]. In VR experience, virtual avatars are used in various contexts [33] and can reduce racial biases and treat phobias, with more effective outcomes observed when body ownership toward the avatar is stronger [3,9]. For measuring body ownership induced by combined visuo-tactile stimuli, the Rubber Hand Illusion (RHI) offers key insights. In this illusion, participants perceive a rubber hand as their own when both it and their hidden hand receive synchronized tactile stimuli [4,5]. Particularly, the feeling felt toward an avatar's virtual hand in a VR environment, termed Virtual Hand Illusion (VHI) [5]. Previous research indicates that spatiotemporal congruence of visuo-motor and visuo-tactile stimuli is essential for inducing body ownership [16]. Therefore, while tactile stimulation to different body parts enhances the quality of the VR experience, it could decrease body ownership, thereby necessitating the measurement of both.

Furthermore, we use suction pressure stimulation as a tactile stimulus. Previously, measurements of body ownership often involved presenting incongruent tactile stimuli, such as vibration feedback in response to contact with virtual objects [16,24]. Our approach uses suction pressure stimulation capable of creating illusory pressure sensations [18], allowing for the presentation of a realistic sense of pressure corresponding to contact with virtual objects. Suction pressure stimulation devices can also be designed with minimal size, facilitating their integration into wearable devices [28].

This study aims to evaluate the quality of VR experiences and body ownership toward the virtual hand during suction pressure stimulation to the face. The experiment result showed that body ownership was induced even with suction pressure stimulation to the face, with no significant difference in body ownership intensity between stimulation to the fingertips and the face.

2   Related Work

To induce body ownership, spatiotemporal congruence of visuo-motor and visuo-tactile stimuli is important [4,26]. Arata et al. demonstrated that body ownership is not induced when the visuo-motor temporal delay in active movement exceeds 100 ms using a robotic arm [1]. Costantini et al. demonstrated that body ownership is induced with a visuo-tactile stimulation temporal delay of up to 300 ms [8]. Riemer et al. demonstrated that body ownership is not induced when presenting spatially incongruent visuo-tactile stimuli to the index and little fingers [25]. Thus, the induction of body ownership requires a multisensory spatiotemporal congruence.

However, there is flexibility in inducing body ownership, and seemingly conflicting reports have been made. Ramentol et al. showed that body ownership can be induced by synchronizing the movement of a fake arm with the participant's active motions, thus preserving visuo-motor congruence, even without tactile stimuli [6]. Kokkinara et al. have shown that body ownership can be induced when either visuo-motor or visuo-tactile stimulus congruence is achieved and that body ownership is more strongly induced when both visuo-motor and visuo-tactile congruences stimulate synchronously [17]. An important study focusing on the congruence of stimulation sites is by Scandola et al. They experimented on participants with paralysis, tetraplegics, and also healthy participants, synchronously stimulating the rubber hand and the participants' cheeks. In this experiment, only tetraplegia participants reported experiencing body ownership toward the rubber hand [27]. The variation of body ownership intensities under various conditions can be attributed to the different intensities of stimuli presented in the experiments [26]. For example, Heed et al. used the grasping object task as tactile stimulation [11] and Wen et al. used the moving hand task to a target as motor stimulation [31]. Using strong stimuli can enhance the induction of body ownership. Therefore, there is a possibility that body ownership can still be induced even when tactile stimuli are presented to areas that do not coincide with the visual stimulus, particularly when participants can actively move the virtual hand.

To assess body ownership, questionnaires based on subjective evaluation are widely used [4,24,26]. These questionnaires typically consist of items assessing body ownership, control questions, and disownership, answered using either the Visual Analog Scale or the Likert scale. Alternative methods include assessing proprioceptive drift by prompting participants to report the perceived position of their actual arm [29], and measuring skin conductance response [2,30]. Considering the comparable results between the questionnaire-based approach and objective measurements, along with their ease of administration, we opted to employ this method for this paper in this study.

3   Experiment

In this experiment, we investigate whether people feel body ownership to the virtual hand when suction pressure stimuli are presented to the face in response to the virtual hand's contact with the virtual object.

3.1   System

The experimental system consists of a host computer, suction pressure control unit, suction ports, HMD (Meta Quest 2, Meta Inc.), noise-canceling headphones (WH-1000XM5, SONY Inc.), and numeric keypad (Fig. 1). Pink noise was played through noise-canceling headphones to mask auditory cues.

Suction Pressure Stimuli System. The suction pressure stimulus presentation device consists of a suction pressure control mechanism and suction port.

The suction pressure control system is shown in Fig. 2. The suction port connects to the air suction pump (SC3701PML, SHENZHEN SKOOCOM ELECTRONIC) via a 3-way solenoid valve (SC415GF, SHENZHEN SKOOCOM ELECTRONIC) and to the external air via a 2-way solenoid valve (SC0526GF, SHENZHEN SKOOCOM ELECTRONIC). By switching these solenoid valves,

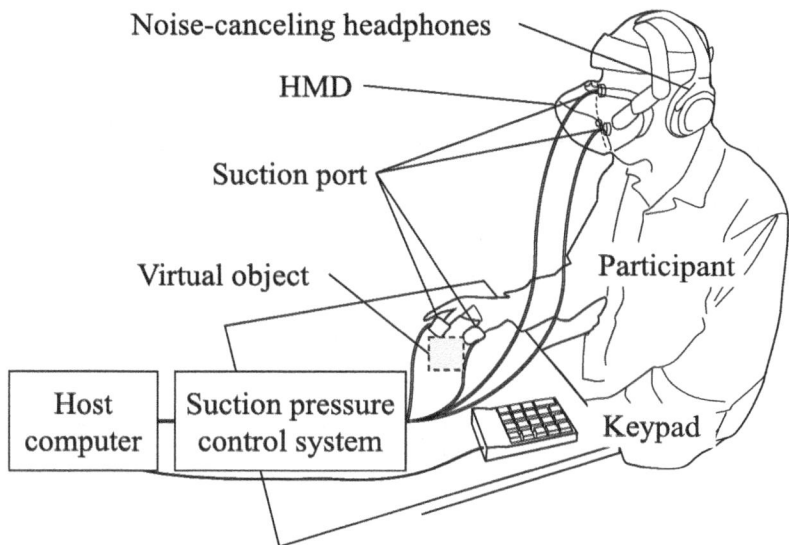

Fig. 1. Overview of the experimental system, including host computer, suction pressure control system, suction port, HMD, noise-canceling headphones, and numeric keypad.

the system toggles between suction, exhaust, and closure. A microcontroller (ESP32-DecKitC, Espressif Systems Ltd.) controls the two solenoid valves with field transistors based on the air pressure sensor (MIS-2503-015V, Metrodyne Microsystem Corp.) value. The control loop operates at a frequency of 1 kHz.

Figure 3 shows the detailed suction port size and the layout of the suction holes. Two types of suction ports were created, for fingertips and for the face. The suction ports were made with a flexible material (Elasticresin50A, Formlab., Inc.) on a photo-fabricated 3D printer (Form3, Formlab., Inc.). For the fingertips, a 16 × 10 mm rectangle with 15 suction holes of 2 mm diameter was placed at a center distance of 3 mm. For the face, 37 suction holes of 2 mm diameter were placed with 2.5 mm distance between centers in a hexagonal close-packed structure on a hexagon circumscribed circle of 21 mm diameter. Four suction ports were attached to the right index finger, right thumb, right cheek, and right forehead. Suction pressure stimuli were presented to the index finger or forehead for the right index finger contact of the virtual hand and to the thumb or cheek for the thumb contact of the virtual hand.

VR Environment. The host computer simulated the VR environment and communicated with the microcontroller. The VR environment was developed using the game engine (Unity, Unity Inc.), with visual stimuli presented to participants through the HMD. Within this environment, a white virtual cube (5 × 5 × 5 cm), was positioned on a virtual desk. Participants could interact with this cube using the virtual hand's right index finger and thumb, as illustrated in Fig. 4. Meta Quest2's hand tracking system was used to track participants'

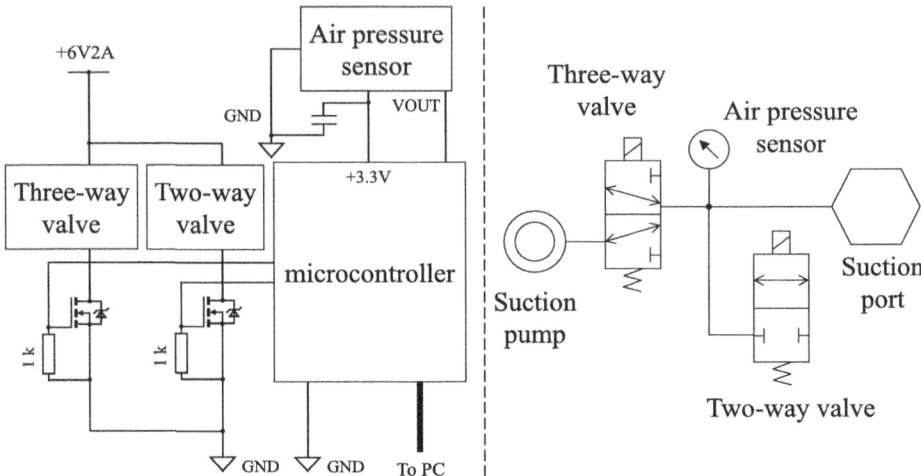

Fig. 2. Suction pressure control system: The air pressure is controlled by the micro-controller through the air pressure sensor, the three-way solenoid valve, the two-way solenoid valve, and the field-effect transistors. Right is electrical circuit diagram, left is pneumatic circuit diagram.

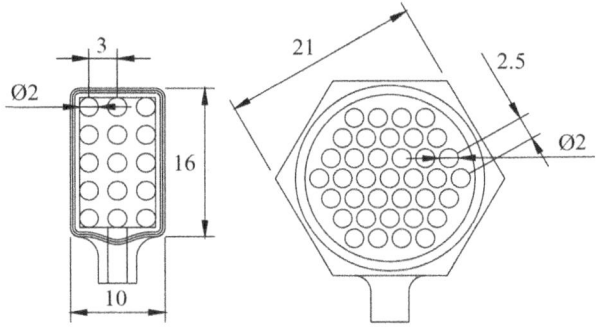

Fig. 3. Suction parts. Sketch of suction parts. Left: for finger, right: for cheek.

active hand movements on the virtual hand. A graphical user interface (GUI) was displayed, enabling participants to input their body ownership with a numeric keypad.

3.2   Design

We designed a questionnaire based on the one used in previous work [24] (Table 1). The questionnaire consisted of several questions for each item (body ownership, control questions, and sense of disownership) and answers were measured using the Visual Analog Scale (VAS, $-100 \sim +100$). In preliminary experiments, Question 6 from the reference questionnaire, which asked about tactile

Fig. 4. Experiment overview in VR environment: A participant controlled the virtual hand to touch the virtual object by moving the right hand.

stimulation from non-fingertip body parts, was excluded as it was not suitable as a control question for the conditions of this experiment. In addition, to evaluate the quality of the VR experience, participants rated the realism of the virtual objects and the overall quality of the VR experience on a 7-point Likert scale (1: bad, 7: good).

To investigate whether body ownership of the virtual hand is induced by tactile stimuli to the face, we introduced three conditions: the *Finger* condition stimulating the fingertips (index finger and thumb), the *Face* condition stimulating the face (forehead and cheeks), and the *NoHaptics* condition with no tactile stimuli. During the tactile conditions (*Finger* and *Face* conditions), suction pressure stimuli were applied upon contact between the virtual hand's fingers and the virtual object. Before the experiment, we asked participants to adjust the intensity of suction pressure stimuli to the fingers and face subjectively equal to the tactile intensity perceived from the visual of touching the virtual object.

The following hypotheses were formulated for this experiment

H0: Body ownership is induced in all conditions.
H1: Body ownership is rated higher in *Finger* than *NoHaptics*.
H2: Body ownership is rated higher in *Finger* than *Face*.
H3: Realism of virtual object is rated higher in *Face* than *NoHaptics*.
H4: Quality of experience is rated higher in *Face* than *NoHaptics*.

For H0, we hypothesized that body ownership is induced in all conditions. First of all, the *NoHaptics* condition, based on previous studies indicating the induced body ownership under conditions of visuo-motor congruence without tactile stimulation [6,24]. Next, for the *Face* condition, compared to prior studies that induced RHI with tactile stimulation to the face [27], we expected actively

Table 1. Questionnaire for body ownership, disownership, and quality of VR experience

Category	Question item
Body Ownership	Q1. I felt as if I was looking at my own hand
	Q2. I felt as if the Virtual Hand was part of my body
	Q3. It felt as if the touch I experienced was directly caused by the cube that was touching the virtual hand
Control	Q4. It felt as if I had more than one right hand
	Q5. I felt as if my real hand was turning virtual
Disownership	Q6. It seemed as if my hand had disappeared
	Q7. It seemed as if I could not really tell where my hand was
	Q8. It seemed as if I was unable to move my hand
Quality of VR experience	Q9. The realism of the cube (1: bad, 7: good)
	Q10. The overall quality of experience (1: bad, 7: good)

moving a virtual hand would induce body ownership in this experiment, in alignment with findings from previous research. For the *Finger* condition, as it represents a typical condition for eliciting VHI, the emergence of body ownership was expected.

For H1, we hypothesized that the *Finger* condition, which induces the traditional virtual hand illusion, would receive higher body ownership ratings than the *NoHaptics* condition without tactile stimuli.

For H2, the *Finger* condition is most expected to induce body stronger body ownership because the visual stimuli representing fingertip contact and the tactile stimuli presented to the fingertips are spatially congruent. Therefore, we hypothesized that the participant would rate the *Finger* condition higher for body ownership than the *Face* condition.

For H3 and H4, by adding tactile stimuli to the visual stimuli provided by the HMD, we hypothesized that the *Face* condition would enhance the realism of virtual object and the overall quality of the experience compared to the *NoHaptics* condition. For H3 and H4, we hypothesized that by integrating tactile stimuli with the visual stimuli presented through the HMD, the *Face* condition would enhance the realism of the virtual objects and improve the overall quality of the experience compared to the *NoHaptics* condition.

3.3   Procedure

Ten participants (eight males and two females) with an average age of 23.7 years (SD 2.41) voluntarily participated in the experiment. All were right-handed. At the outset, participants took a seat at a desk and were fitted with an HMD, suction ports on their index finger, thumb, cheek, and forehead, and noise-canceling headphones. Before the experiment, participants were instructed to adjust the suction air pressure applied to their fingers and face. The target air pressure value fell within the pressure sensation range, equivalent to the intensity experienced when touching the virtual cube. Air pressure adjustments were made using GUI via the numeric keypad with the left hand. During the main task, one of the three conditions (*NoHaptics*, *Finger*, and *Face*) was randomly presented by the system, and participants were given one minute to freely interact with the virtual cube using their right hand. Possible interactions include touching, pinching, lifting, and throwing. The virtual cube remained constantly visible in front of the participants. If it moved for away, it was automatically returned to its initial position. Participants responded to questions regarding their body ownership and their assessment of the overall experimental quality via a GUI, utilizing a numeric keypad. Each participant completed one trial for each condition, totaling three trials per participant. After all trials were finished, participants were invited to share their overall impressions and feedback on the experiment.

3.4 Analysis

Given the non-parametric nature of the data, the Wilcoxon signed-rank test was employed for the analysis. This test was used to determine if the VAS scores for Q1 to Q3 in each condition were significantly greater than zero, thereby investigating the induction of body ownership under each condition. A one-sided test ('greater') was employed.

Subsequently, we transformed the responses from VAS and Likert scale using the aligned rank transformation (ART) for non-parametric factorial analysis [32]. An analysis of variance (ANOVA) was performed with the within-participant factors being the different stimulus conditions (*NoHaptics*, *Finger*, and *Face*). The significance level was set at $\alpha = 0.05$. If significant differences were detected, multiple comparisons with Bonferroni's correction were applied.

4 Result

Figure 5 presents the body ownership results as measured by VAS. The x are mean values, the boxes are the interquartile ranges (IQRs), and the whiskers extend to 1.5 × the IQRs. To confirm whether body ownership is induced under each stimulus condition, the *NoHaptics* condition showed no significant difference from score zero across all items: Q1 ($p = 0.28$), Q2 ($p = 0.50$), and Q3 ($p = 0.98$). In contrast, under the *Finger* condition, significant body ownership was evident with Q1 ($p = 0.006$), Q2 ($p = 0.026$), and Q3 ($p < 0.001$). Similarly, in the *Face* condition, significant body ownership was observed with Q1 ($p < 0.001$), Q2 ($p = 0.031$), and Q3 ($p = 0.005$).

To confirm if there are differences in the intensity of body ownership induction among each stimulus condition. For Q1, there was a significant difference in the stimulus condition ($p = 0.009$), with post hoc tests indicating *Face* > *NoHaptics* conditions ($p = 0.009$). For Q2, there was a significant difference in the stimulus condition ($p = 0.012$), with post hoc tests indicating *Finger* >

NoHaptics conditions ($p = 0.012$). Finally, for Q3, there was a significant difference in the stimulus condition ($p < 0.001$), with post hoc tests indicating *Face*
$>$ *NoHaptics* ($p < 0.001$), and *Finger* $>$ *NoHaptics* conditions ($p < 0.001$).

The results of Q9 and Q10 are shown in Fig. 6. The x are mean values, the
boxes are the interquartile ranges (IQRs), and the whiskers extend to 1.5 × the
IQRs. For Q9, there was a significant difference in the stimulus condition ($p =$
0.001), with post hoc tests indicating *Finger* ¿ *NoHaptics* conditions ($p = 0.001$)
and *Face* ¿ *NoHaptics* conditions ($p < 0.001$). For Q10, there was a significant
difference in the stimulus condition ($p = 0.001$), with post hoc tests indicating
Finger ¿ *NoHaptics* conditions ($p < 0.001$) and *Face* ¿ *NoHaptics* conditions
($p < 0.001$).

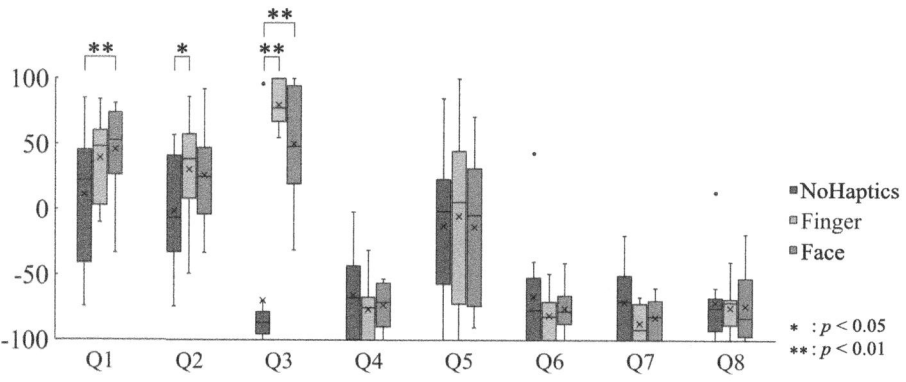

Fig. 5. Response values for body ownership in Experiment are plotted based on raw
measurements, with significant differences between conditions identified using ART-
ANOVA marked accordingly.

5   Discussion

There were significant differences between the zero score and *Finger*, *Face* conditions in Q1 to Q3, which evaluate body ownership. This result partially supports
H0. The reason no significant difference was observed in the *NoHaptics* condition could be attributed to the presence of tactile stimulation in the comparison
conditions. Despite potentially induced body ownership [6], participants might
have assigned lower scores in the *NoHaptics* condition compared to the tactile
stimulation conditions. To avoid this, conducting a between participants experiment or administering the *NoHaptics* condition as the baseline condition first
might have been necessary.

There were significant differences between the *Finger* and *NoHaptics* conditions in Q2 and Q3, which evaluate body ownership. This result partially
supports H1, suggesting that the *Finger* condition facilitated body ownership.

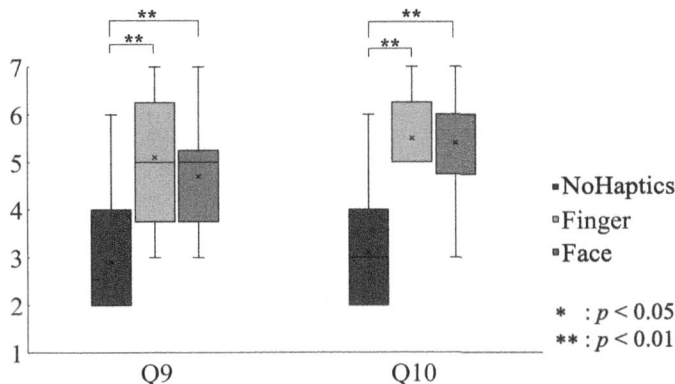

Fig. 6. Response values for realism of virtual object (Q9) and quality of VR experience (Q10) are plotted based on raw measurements, with marks significant differences between conditions identified using ART-ANOVA marked accordingly.

There were no significant differences between the *Face* and *Finger* conditions in Q1 to Q3, which evaluate the body ownership. This result partially supports H2. A potential explanation is that the temporal synchronization of tactile stimuli plays a crucial role in enhancing body ownership in actively moving the virtual hand, whereas the spatial congruence of tactile stimulation might be less.

In comparisons between the *Face* and *NoHaptics* conditions, ratings for the *Face* condition were significantly higher for Q9 and Q10, both of which assess realism and the quality of the VR experience. These findings lend support to both H3 and H4. Drawing parallels to a previous experiment documented in [15], where the sensation of contact between the virtual object and the virtual hand was achieved using vibration to the palm and suction pressure on the face. The face's suction pressure stimulation received higher ratings for realism and VR experience quality. The current results align with those observations.

In summary, these results demonstrate that spatial congruency between visual and tactile sensation is not important for the occurrence of body ownership for an actively movable virtual hand, and that the virtual hand illusion is induced by suction stimuli. Additionally, this experiment focused primarily on body ownership, therefore, it did not distinguish between body ownership and agency [5]. A dedicated experimental paradigm is required to investigate the mechanisms of body ownership, agency, and embodiment [21].

6 Conclusion

In this study, we demonstrated the induction of the virtual hand illusion using suction pressure stimulation on the fingertips, face, and no stimulation. Participants actively interacted with a virtual object in a VR environment, receiving tactile feedback at specified sites. The results revealed that body ownership was induced both when tactile feedback was presented to the fingertips and face. From these findings, it can be inferred that for perceiving body ownership in a virtual hand spatial inconsistencies in stimulus may not significantly influence the perception. Moreover, subjective evaluations showed that providing tactile stimuli to the face significantly enhanced the virtual object's realism and the VR experience's quality compared to no stimulation.

Future research will focus on modifying the experimental paradigm and investigating the extent to which spatial incongruities between visual and tactile stimuli are permissible.

References

1. Arata, J., Hattori, M., Ichikawa, S., Sakaguchi, M.: Robotically enhanced rubber hand illusion. IEEE Trans. Haptics **7**(4), 526–532 (2014). https://doi.org/10.1109/TOH.2014.2304722
2. Armel, K.C., Ramachandran, V.S.: Projecting sensations to external objects: evidence from skin conductance response. Proc. Ro.l Soc. London Ser.: B Biol. Sci. **270**(1523), 1499–1506 (2003). https://doi.org/10.1098/rspb.2003.2364
3. Banakou, D., Hanumanthu, P.D., Slater, M.: Virtual embodiment of white people in a black virtual body leads to a sustained reduction in their implicit racial bias. Front. Hum. Neurosci. **10**(NOV2016), 226766 (2016). https://doi.org/10.3389/FNHUM.2016.00601/BIBTEX
4. Botvinick, M., Cohen, J.: Rubber hands 'feel' touch that eyes see. Nature **391**(6669), 756 (1998). https://doi.org/10.1038/35784
5. Braun, N., et al.: The senses of agency and ownership: a review. Front. Psychol. **9**(APR), 334248 (2018). https://doi.org/10.3389/FPSYG.2018.00535/BIBTEX
6. Brugada-Ramentol, V., Clemens, I., de Polavieja, G.G.: Active control as evidence in favor of sense of ownership in the moving Virtual Hand Illusion. Conscious. Cogn. **71**, 123–135 (2019). https://doi.org/10.1016/J.CONCOG.2019.04.003
7. Buckingham, G.: Hand tracking for immersive virtual reality: opportunities and challenges. Front. Virtual Reality **2**, 728461 (2021). https://doi.org/10.3389/FRVIR.2021.728461/BIBTEX
8. Costantini, M., Robinson, J., Migliorati, D., Donno, B., Ferri, F., Northoff, G.: Temporal limits on rubber hand illusion reflect individuals' temporal resolution in multisensory perception. Cognition **157**, 39–48 (2016). https://doi.org/10.1016/J.COGNITION.2016.08.010
9. Freeman, D., et al.: Automated psychological therapy using immersive virtual reality for treatment of fear of heights: a single-blind, parallel-group, randomised controlled trial. Lancet Psychiatry **5**(8), 625–632 (2018). https://doi.org/10.1016/S2215-0366(18)30226-8
10. Gordon, E.M., et al.: A somato-cognitive action network alternates with effector regions in motor cortex. Nature **617**(7960), 351–359 (2023). https://doi.org/10.1038/s41586-023-05964-2
11. Heed, T., et al.: Visual information and rubber hand embodiment differentially affect reach-to-grasp actions. Acta Psychol. **138**(1), 263–271 (2011). https://doi.org/10.1016/J.ACTPSY.2011.07.003
12. Kameoka, T., Kon, Y., Nakamura, T., Kajimoto, H.: Haptopus: haptic VR experience using suction mechanism embedded in head-mounted display. In: UIST 2018

Adjunct - Adjunct Publication of the 31st Annual ACM Symposium on User Interface Software and Technology (2018). https://doi.org/10.1145/3266037.3271634

13. Kameoka, T., Kajimoto, H.: Tactile transfer of finger information through suction tactile sensation in HMDs. In: 2021 IEEE World Haptics Conference, WHC 2021, pp. 949–954 (2021). https://doi.org/10.1109/WHC49131.2021.9517176

14. Kameoka, T., Kajimoto, H.: Design of suction-type tactile presentation mechanism to be embedded in HMD. Front. Virtual Reality 75 (2022). https://doi.org/10.3389/FRVIR.2022.894873

15. Kameoka, T., Kon, Y., Nakamura, T., Kajimoto, H.: Haptopus: transferring the touch sense of the hand to the face using suction mechanism embedded in HMD. In: Proceedings of the Symposium on Spatial User Interaction, pp. 11–15. ACM, New York (2018). https://doi.org/10.1145/3267782.3267789

16. Kilteni, K., Maselli, A., Kording, K.P., Slater, M.: Over my fake body: body ownership illusions for studying the multisensory basis of own-body perception. Front. Hum. Neurosci. **9**(MAR), 119452 (2015). https://doi.org/10.3389/FNHUM.2015.00141/BIBTEX

17. Kokkinara, E., Slater, M.: Measuring the effects through time of the influence of visuomotor and visuotactile synchronous stimulation on a virtual body ownership illusion. **43**(1), 43–58 (2019). https://doi.org/10.1068/P7545

18. Makino, Y., Asamura, N., Shinoda, H.: Multi primitive tactile display based on suction pressure control. In: 2004 Proceedings of the 12th International Symposium on Haptic Interfaces for Virtual Environment and Teleoperator Systems, HAPTICS 2004, pp. 90–96. IEEE (2004). https://doi.org/10.1109/HAPTIC.2004.1287182

19. Moriyama, T., Kajimoto, H.: Wearable haptic device presenting sensations of fingertips to the forearm. IEEE Trans. Haptics **15**(1), 91–96 (2022). https://doi.org/10.1109/TOH.2022.3143663

20. Moriyama, T., Takahashi, A., Asazu, H., Kajimoto, H.: Simple is vest: high-density tactile vest that realizes tactile transfer of fingers. In: SIGGRAPH Asia 2019 Emerging Technologies, pp. 42–43. ACM, New York (2019). https://doi.org/10.1145/3355049.3360532

21. Mottelson, A., Muresan, A., Hornbæk, K., Makransky, G.: A systematic review and meta-analysis of the effectiveness of body ownership illusions in virtual reality. ACM Trans. Comput.-Hum. Interact. **30**(5), 76 (2023). https://doi.org/10.1145/3590767

22. Penfield, W., Rasmussen, T.: The cerebral cortex of man: a clinical study of localization of function. J. Am. Med. Assoc. **144**(16), 1412–1412 (1950). https://doi.org/10.1001/JAMA.1950.02920160086033, https://jamanetwork.com/journals/jama/fullarticle/307907

23. Pezent, E., Macklin, A., Yau, J.M., Colonnese, N., O'Malley, M.K.: Multisensory pseudo-haptics for rendering manual interactions with virtual objects. Adv. Intell. Syst. **5**(5), 2200303 (2023). https://doi.org/10.1002/aisy.202200303

24. Pyasik, M., Tieri, G., Pia, L.: Visual appearance of the virtual hand affects embodiment in the virtual hand illusion. Sci. Rep. **10**(1), 1–11 (2020). https://doi.org/10.1038/s41598-020-62394-0

25. Riemer, M., et al.: The rubber hand illusion depends on a congruent mapping between real and artificial fingers. Acta Psychol. **152**, 34–41 (2014). https://doi.org/10.1016/J.ACTPSY.2014.07.012

26. Riemer, M., Trojan, J., Beauchamp, M., Fuchs, X.: The rubber hand universe: on the impact of methodological differences in the rubber hand illusion. Neurosci. Biobehav. Rev. **104**, 268–280 (2019). https://doi.org/10.1016/j.neubiorev.2019.07.008, https://linkinghub.elsevier.com/retrieve/pii/S014976341930051X

27. Scandola, M., Tidoni, E., Avesani, R., Brunelli, G., Aglioti, S.M., Moro, V.: Rubber hand illusion induced by touching the face ipsilaterally to a deprived hand: evidence for plastic "somatotopic" remapping in tetraplegics. Front. Hum. Neurosci. 8(JUNE), 87142 (2014). https://doi.org/10.3389/FNHUM.2014.00404/BIBTEX
28. Shtarbanov, A.: FlowIO development platform - the pneumatic raspberry Pifor Sof robotics. In: Conference on Human Factors in Computing Systems - Proceedings (2021). https://doi.org/10.1145/3411763.3451513
29. Tsakiris, M., Haggard, P.: The rubber hand illusion revisited: visuotactile integration and self-attribution. J. Exp. Psychol. Hum. Percept. Perform. 31(1), 80–91 (2005). https://doi.org/10.1037/0096-1523.31.1.80, https://pubmed.ncbi.nlm.nih.gov/15709864/
30. Tsuji, T., et al.: Analysis of electromyography and skin conductance response during rubber hand illusion. In: Proceedings of IEEE Workshop on Advanced Robotics and its Social Impacts, ARSO, pp. 88–93 (2013). https://doi.org/10.1109/ARSO.2013.6705511
31. Wen, W., et al.: Goal-directed movement enhances body representation updating. Front. Hum. Neurosci. 10, 210789 (2016). https://doi.org/10.3389/FNHUM.2016.00329/BIBTEX
32. Wobbrock, J.O., Findlater, L., Gergle, D., Higgins, J.J.: The aligned rank transform for nonparametric factorial analyses using only ANOVA procedures. In: Conference on Human Factors in Computing Systems - Proceedings, pp. 143–146 (2011). https://doi.org/10.1145/1978942.1978963
33. Zhang, G., Cao, J., Liu, D., Qi, J.: Popularity of the metaverse: embodied social presence theory perspective. Front. Psychol. 13, 997751 (2022). https://doi.org/10.3389/FPSYG.2022.997751/BIBTEX

Task-Adapted Single-Finger Explorations of Complex Objects

Lisa Pui Yee Lin[1]([✉]), Alina Böhm[2], Boris Belousov[2,3], Alap Kshirsagar[2],
Tim Schneider[2], Jan Peters[2,3,4], Katja Doerschner[1], and Knut Drewing[1]

[1] Department of General Psychology, Justus-Liebig University Gießen, Gießen, Germany
Pui.Lin@psychol.uni-giessen.de
[2] Intelligent Autonomous Systems Group, Computer Science Department, Darmstadt,
Darmstadt, TU, Germany
[3] Research Department SAIROL, German Research Center for AI (DFKI), Darmstadt, Germany
[4] Hessian Centre for Artificial Intelligence, Darmstadt, Germany

Abstract. The perception of material/object properties plays a fundamental role
in our daily lives. Previous research has shown that individuals use distinct and
consistent patterns of hand movements, known as exploratory procedures (EPs),
to extract perceptual information relevant to specific material/object properties.
Here, we investigated the variation in EP usage across different tasks involving
objects that varied in task-relevant properties (shape or deformability) as well
as in task-irrelevant properties (deformability or texture). Participants explored 1
reference object and 2 test objects with a single finger before selecting the test
object that was most similar to the reference. We recorded their finger movements
during explorations, and these movements were then categorised into different
EPs. Our results show strong task-dependent usage of EPs, even when exploration
was confined to a single finger. Furthermore, within a given task, EPs varied
as a function of material/object properties unrelated to the primary task. These
variations suggest that individuals flexibly adapt their exploration strategies to
obtain consistent and relevant information.

Keywords: Exploratory procedures · haptic perception · active touch

1 Introduction

The perception of haptics properties of objects plays a fundamental role in our daily
lives, whether it's stoking a cat's fur, gauging the weight of a rock, or pressing one's
palm on a chair to evaluate its sturdiness. Different exploratory movement patterns are
intentionally employed during active touch to perceive different dimensions [1]. The
selection of exploratory procedures (EP) is closely linked to the material/object properties and the perceiver's exploration objectives [2, 3]. Previous research has demonstrated
that individuals habitually use distinct EPs, to extract perceptual information related to
the target haptic properties [1, 4, 5]. For example, to perceive compliance, people tend
to indent or apply pressure to the object; for texture perception, they engage in lateral

© The Author(s), under exclusive license to Springer Nature Switzerland AG 2025
H. Kajimoto et al. (Eds.): EuroHaptics 2024, LNCS 14768, pp. 133–146, 2025.
https://doi.org/10.1007/978-3-031-70058-3_11

motion, repetitively rubbing their fingers across the surface to discern its texture. In contrast, when judging the shape of an object, they use contour following, with one or more of their fingers or the entire hand tracing along the contour of the object. Hence, the selection of EP is strongly determined by the target haptic properties one aims to discern.

However, objects often possess many haptic properties beyond the one targeted, which raises the question of whether the properties of an object beyond the primary focus of exploration would influence the selection of EPs. For example, when the primary objective is to explore the shape of an object, would other properties, like its texture or deformability, influence the perceiver's selection of EP when exploring the object? There is limited research on how task-irrelevant object properties modulate exploration behaviours. Klatzky et al. [5] conducted an experiment in which multidimensional objects varying in shape, size, hardness and texture were used, and participants had to sort objects along a designated dimension (e.g., texture). Their results demonstrated that while exploring objects with variations in multiple haptic properties, the utilization of EPs varied according to the specific perceptual task. However, while they confirmed that the task strongly influenced specific EP frequency, they did not investigate whether EP usage also systematically varied with object properties unrelated to the primary aim of exploration.

Therefore, our current study examines how varying object properties unrelated to the primary aim of exploration may affect EP patterns during haptic exploration. However, in contrast to [5], in our experiment, we constrained the exploration to the use of the index finger. While effective executions of many EPs may usually involve coordinating both hands or using the whole hand, several studies have also shown that a single finger usage is adequate for haptic exploration in various tasks and contexts [e.g., 6–8]. It is likely, however, that the single finger restriction may – as a function of task and/or object property - alter EP frequency and/or EP character, a possibility that we will consider below.

Taken together, the goals of our study are to examine the variation in EP usage for two different tasks performed with the same set of objects, that either varied in task-relevant (first goal) or in task-irrelevant properties (second goal). By examining variations in EP usage in the presence of task-relevant and task-irrelevant properties, our study adds to the growing body of literature in haptics research, offering insights into the adaptive nature of haptic exploration strategies, in particular under the constraints of single-finger explorations, which holds potential for real-world application such as robotic explorations and teleoperation. For the first goal, we used a set of objects that varied in both shape and deformability, and we had participants explore and evaluate the similarity of objects based on either shape or deformability. For the second goal, we used data from the set of objects that varied in both shape and deformability and further data from another set of rigid objects that varied in shape and texture. We compared how participants explore and assess the similarity of objects in the two sets based on the shapes.

2 Methods

2.1 Participants

10 participants (5 females, $M_{age} = 28$yr, $SD_{age} = 4.59$yr, Range $= 21–34$) were recruited from Giessen University. All participants but one were right-handed and had no history of motor or cutaneous impairments. All participants had a 2-point discrimination threshold of <4 mm on their right-hand index fingertips. All participants provided informed consent and received compensation of ($8€$/h) for their participation. The video recordings of 2 participants were excluded from the video analyses due to recording errors or incomplete data sets. This study was approved by the local ethics committee at Giessen University and conducted in accordance with the declaration of Helsinki (2013).

2.2 Aparatus

Participants sat at a table opposite the experimenter. A monitor and keyboard were placed on the experimenter's left to run the experiment and collect participants' responses. The experiment was programmed using Psychopy (version 2022.2.4). During the experiment, participants' hand movements were recorded with a Sony Digital 4K Video Camera (recording 28-bit videos with a resolution of 1920 × 1080 pixels); the camera was placed on a tripod on the left of the table. Each of these stimuli was placed in a 3D-printed tray (65 mm × 65 mm) mounted on the table, and a thin layer of silicone was applied to the interior of the tray to minimise potential displacement during explorations.

2.3 Stimuli

We organised our stimuli into two primary categories: smooth-deformable shapes and textured-rigid shapes. Within each category, we created two subsets labelled as set A and set B, resulting in 4 subsets in total: smooth-deformable set A, smooth-deformable set B, textured-rigid set A, and textured-rigid set B. Each subset consisted of five objects.

   The smooth-deformable shapes were cast in 3D-printed moulds. To achieve varying levels of deformability in the stimuli, a two-component silicone rubber solution (Alpa Sil EH A & B) was mixed with different amounts of silicone oil. There were five levels of deformability (Most soft - d1: 1.13 mm/N; d2: 1.02 mm/N; d3: 0.79 mm/N; d4: 0.68 mm/N; Least soft - d5: 0.44 mm/N). On the other hand, the textured-rigid shapes were 3D-printed plastic objects covered with different textured fabrics (i.e., t1: corduroy, t2: tweed, t3: velvet, t4: jersey cotton, t5: burlap – note: we randomly assigned a number to each texture, and it bears no relation to nature of the texture itself).

   Across all sets (set As and set Bs), the shapes were defined by varying numbers and prominence of concavities and convexities, and these remained consistent across all sets (labelled S1 to S5). However, within each set, we varied the texture and deformability levels to create different texture × shape and deformability × shape combinations across set A and B. (see Fig. 1).

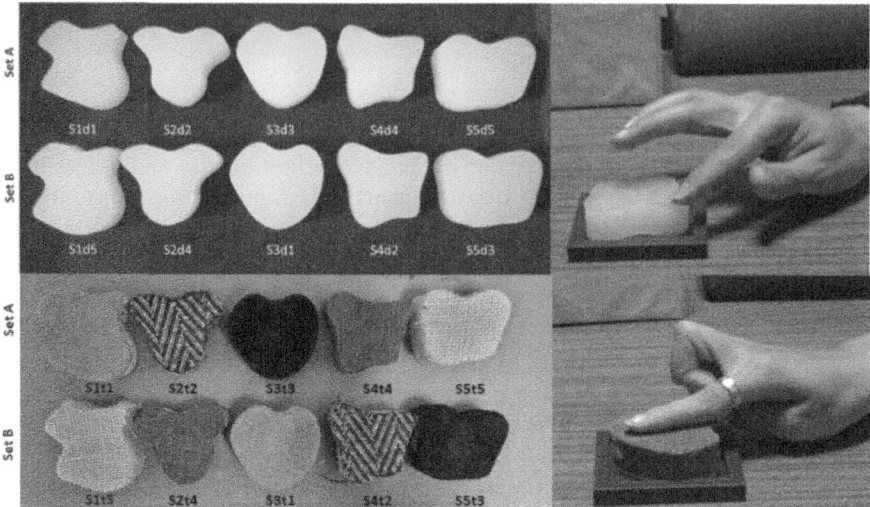

Fig. 1. Depiction of the experimental set-up, as well as the smooth-deformable shapes (top left panel) and textured-rigid shapes (bottom left panel) used in the experiment.

2.4 Design

Over two consecutive days, participants engaged in two sessions, each lasting approximately 1.5–2 h. In one session, they made judgments on deformability or shape (using the smooth-deformable shapes –deformability/shape condition), and in the other session, they made judgments on texture or shape (using the textured-rigid shapes – texture/shape condition).

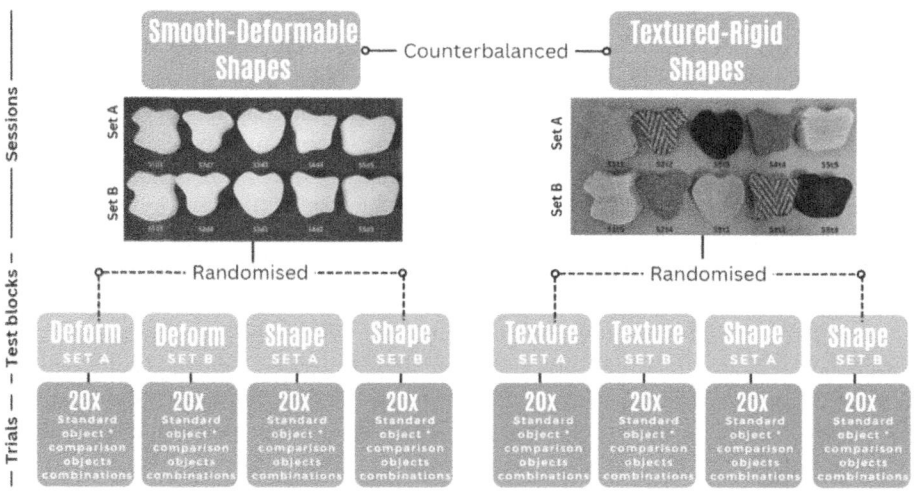

Fig. 2. The flow chart of the experimental conditions.

Within each session, participants underwent a practice block followed by four test blocks. In the practice block, participants were given ten practice trials featuring random standard object comparison object combinations. In the deformability/shape session, these four test blocks comprised two deformability blocks and two shape blocks. Conversely, in the texture/shape session, participants completed two texture blocks and two shape blocks.

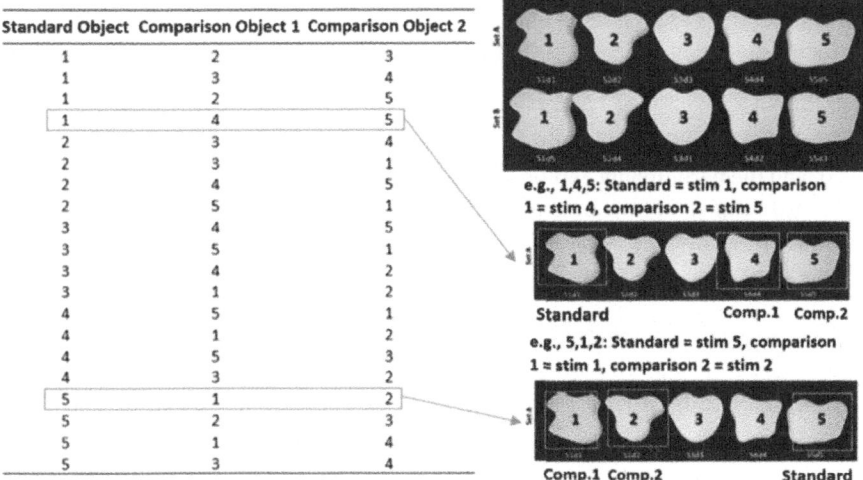

Standard Object	Comparison Object 1	Comparison Object 2
1	2	3
1	3	4
1	2	5
1	4	5
2	3	4
2	3	1
2	4	5
2	5	1
3	4	5
3	5	1
3	4	2
3	1	2
4	5	1
4	1	2
4	5	3
4	3	2
5	1	2
5	2	3
5	1	4
5	3	4

e.g., 1,4,5: Standard = stim 1, comparison 1 = stim 4, comparison 2 = stim 5

Standard   Comp.1 Comp.2

e.g., 5,1,2: Standard = stim 5, comparison 1 = stim 1, comparison 2 = stim 2

Comp.1 Comp.2   Standard

Fig. 3. Depiction of 20 standard object × comparison objects combinations used in the current experiment, along with an explanation of the numerical references within our list and the corresponding object in our stimuli.

Each test block exclusively utilised stimuli from either set A or B, with the order of these test blocks randomised across participants. Within each test block, participants completed 20 trials, each corresponding to a distinct standard object comparison-object combination, resulting in 80 trials completed in total for each session (See Fig. 2).

Each set of stimuli (e.g., smooth-deformable set A) comprised 5 objects, allowing us to generate 20 distinct standard object comparison-object combinations. To ensure comprehensive coverage, each object served as the standard four times, while the remaining 4 objects were presented as comparison objects twice (see Fig. 3). In each trial, one object was designated as the standard, accompanied by two comparison objects.

2.5 Procedure

After providing their informed consent, participants were asked to sit facing the table, they were blindfolded, and we assessed the 2-point discrimination thresholds. Afterwards, they were given instructions for the experimental task. Participants were told to perform a match-to-sample task: they were presented with a standard object, and two comparison objects, which they would explore for one of the two object properties (deformability or shape; texture or shape), and their task was to select the comparison object that best matches the standard in terms of the specified dimension. It should be noted that while we have collected the data and video recordings for texture judgements, our focus in this study was mainly on the deformability and shape data of textured and deformable objects.

Each trial began with participants placing their right hand on the table, palm facing upwards, and began their exploration on the experimenter's signal. Using their right hand index finger, participants explored the stimulus at their own pace, concluding by placing their hand on the table to signal completion. Afterwards, the experimenter presented two comparison objects sequentially. Once the participant had explored all three objects, they made their judgements by saying 'first' or 'second', referring to the number of the best matching comparison objects. Participants were reminded that there is no right or wrong answer, and there would be no identical match to the standard object, and they would have to select the best matching one. Simultaneous comparison of the stimuli was not allowed, and the participants explored each object only once. Participants were not given explicit instructions on how to explore the objects to avoid biases in their explorations; they were only instructed to explore the object with a single finger. There was no time limit for exploration or response.

3 Data Analysis

3.1 Similarity Judgements

Using participants' perceptual judgments, we calculated Cronbach's alpha coefficients between participants for every standard object comparison object combination to assess the consistency of their perceptual judgments across each dimension - deformability, smooth-deformable shape, and textured-rigid shape. Additionally, we analysed how frequently objects were rated as being most similar to a given standard object within each dimension. We did this by examining the instances in which an object was chosen as more similar when compared to another, counting the number of such occurrences, and dividing it by the total number of comparisons.

3.2 Finger Movements

We categorised participants' finger movements into nine different EPs and computed the relative frequency of the EPs. We conducted two separate MANOVAs to examine whether EP usage varied as a function of 1) the object dimension of the task (shape vs. deformability judgements) and 2) the material properties of objects (smooth-deformable shapes vs. textured-rigid shapes). Additionally, we used video recordings of finger movements to quantify the duration of exploration per trial. We compared the mean exploration time as a function of object dimension and material properties using paired t-tests.

4 Results

4.1 Similarity Judgements

In the texture/shape condition, participants made judgements about the similarity of objects based on texture and shape, whereas in the deformability/shape condition, they

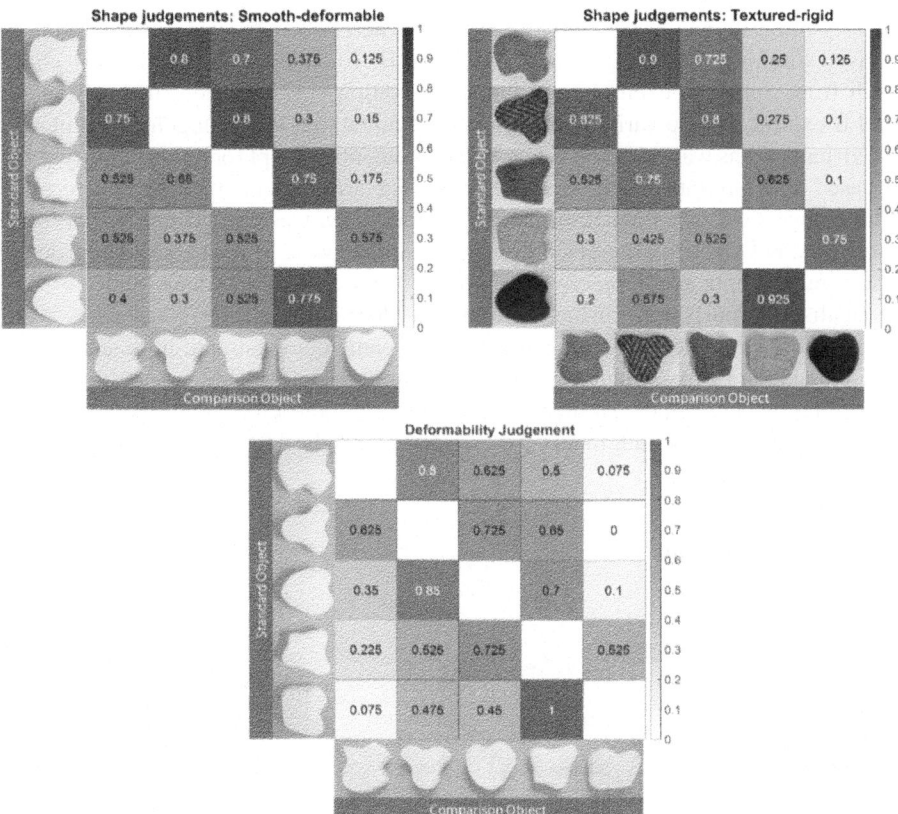

Fig. 4. Top panel: Perceived similarity in smooth-deformable (left) or textured-rigid (right) shape-judgments. Objects are arranged from those with a higher number and more prominent convexity/concavity on the left to those with fewer on the right. Bottom panel: Perceived similarity in deformability judgements, the objects are arranged from most soft to least soft, progressing from left to right. These similarity matrices illustrate how often an object was rated as most similar to a given standard object, with light colours indicating lower values and darker colours indicating higher values.

made judgements about the similarity of objects based on deformability or shape. The Cronbach's alpha values for smooth-deformable shape, textured-rigid shape and deformability were 0.476, 0.758 and 0.895, respectively. Deformability judgements exhibited a high level of interobserver consistency, while the interobserver consistency for textured-rigid shapes was slightly lower; it still had an overall good level of consistency. In contrast, similarity judgments on smooth-deformable shapes exhibited a lower level of consistency, suggesting substantial variability among participants in their perceptual assessments of shape in this condition.

Next, we assessed how often an object was rated to be most similar to a given standard object by examining the instances in which an object was chosen as more similar when compared to another. In Fig. 4, we show a visualisation of the perceived similarity in participants' perceptual judgements across dimensions, incorporating the perceived similarity ratings from both Set A and B. To enhance clarity, we have chosen stimuli from Set A as exemplars in our visualisation since the deformability in Sets A and B exhibit inverse variations. It should be noted that the arrangement of items in these visualisations was based on subjective judgements, except for deformability, which was based on compliance level measurements. The arrangement of shapes, determined by the number and prominence of convexity/concavity, was carried out by two of the authors (LL, K.Doerschner) and a naive participant unaware of the experiment's details. These arrangements are subjective judgments from these individuals and are solely for visualisation purposes. Furthermore, despite low interobserver consistency within the smooth-deformable shape judgement, participants demonstrated high similarity in shape judgements across material properties (smooth-deformable vs. textured-rigid). The correlation between the similarity matrices of smooth-deformable shape judgments and textured-rigid shape judgments (i.e., top left and right figure) was notably strong: r $= 0.924, p < .001$.

4.2 Finger Movements

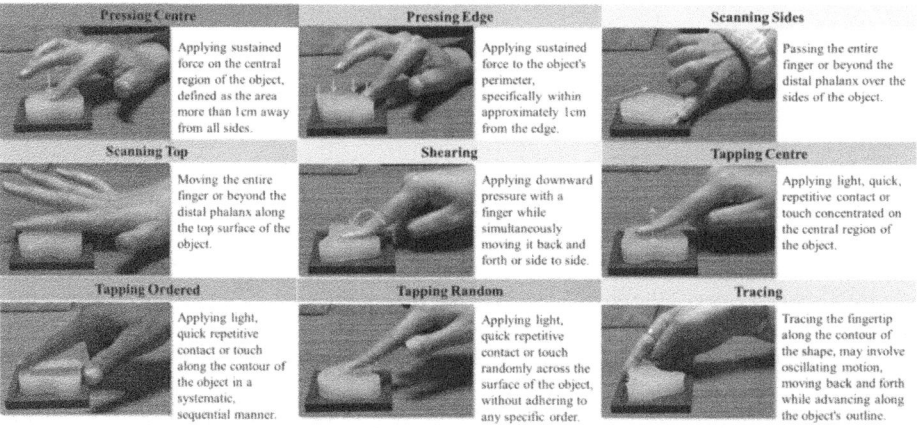

Fig. 5. Illustration and description of the nine EPs proposed in this experiment. Yellow arrows depict the direction and trajectories of movement.

Prior to video coding, 80 video snippets of the shape and deformability conditions were randomly selected from all participants' video recordings. The authors and two raters watched these videos together and discussed the observed events. Through discussion and further refinement, we identified nine specific finger movements that were frequently observed among participants during shape and deformability judgements. Examples of each of these EPs are shown in Fig. 5.

EP Coding and Relative EP Frequencies. For the deformable data set (smooth-deformable shape/deformability judgements), one rater coded the entire video dataset, while an additional rater independently coded the same 50% of videos to assess interrater reliability. For textured-rigid data set (textured-rigid shape judgements), one rater coded the entire video dataset, and additional two raters coded the same 25% of videos to assess interrater reliability, and once again, interrater reliability was assessed. In general, the interrater reliability was high (Cronbach's alpha = .992-.994), suggesting that all raters have coded the videos and EP in a consistent manner.

Using the nine EPs proposed above, participants' finger movements during exploration were coded. In each trial, the occurrence of EPs was coded by annotating each EP's start and end points. Subsequently, we computed the relative frequency of EPs by dividing the duration of a specific EP by the total duration of all EPs in that trial so that the sum of relative EP frequencies for all EPs executed in each trial equals 1. The relative EP frequencies from the coding of the videos were averaged across raters, and these relative EP frequencies were used in all subsequent analyses.

4.3 EP Patterns During Smooth-Deformable and Deformability Judgements

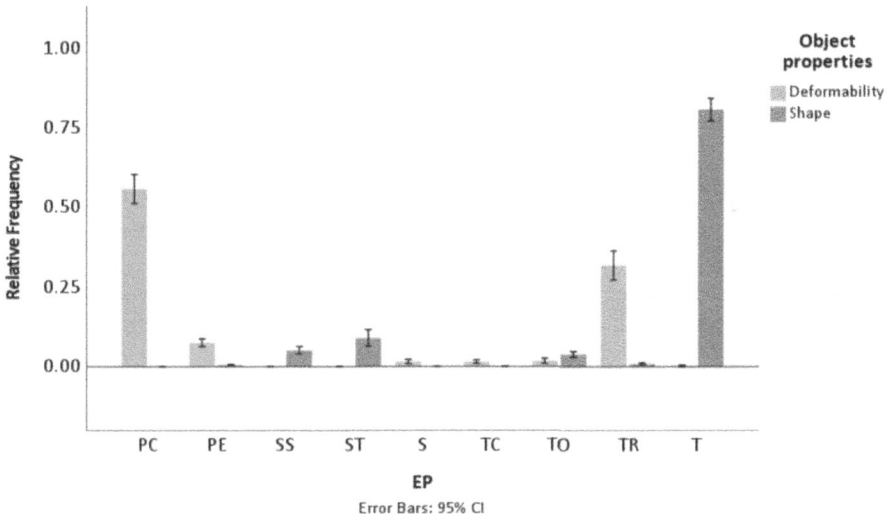

Fig. 6. Relative EP frequencies plotted as a function of object properties (deformability vs. shape). **PC** = pressing centre, **PE** = pressing edge, **SS** = scanning sides, **ST** = scanning top, **S** = shearing, **TC** = tapping centre, **TO** = tapping ordered, **TR** = tapping random, **T** = tracing.

Using these relative EP frequencies (40 trials per participant per condition), we conducted a MANOVA to investigate the effects of object properties on EP patterns, where the frequencies of the nine EP were the dependent variables and the object property (deformability/shape) was the independent variable. We found a significant effect of dimension on EP pattern, $F(9,630)$, $p < .001$, Wilk's $\Lambda = 0.013$, $\eta_p^2 = .987$, which suggests that the frequencies of EP differed depending on the object dimension of the task (i.e., deformability vs shape judgements). Subsequent Univariate ANOVAs indicated that the EP frequencies differed significantly across object properties for all EPs (See Table 1). Overall, during shape judgments, tracing emerged as the predominant EP, followed by scanning and ordered tapping movements. Conversely, pressing and random tapping movements were the most frequently observed EPs during deformability judgments (See Fig. 6).

4.4   The Influence of Material Properties on EP Patterns During Shape Judgements

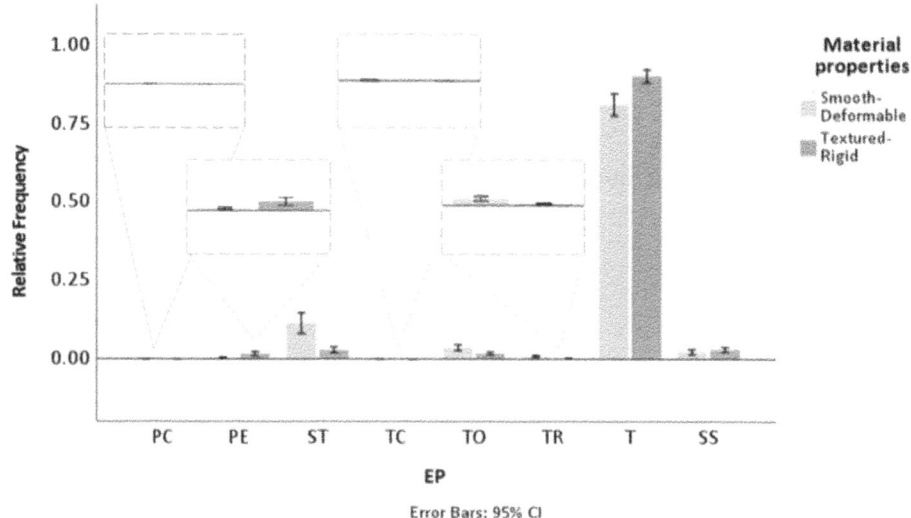

Fig. 7. Relative EP frequencies plotted as a function of material properties (smooth -deformable vs. textured-rigid). **PC** = pressing centre, **PE** = pressing edge, **SS** = scanning sides, **ST** = scanning top,**TC** = tapping centre, **TO** = tapping ordered, **TR** = tapping random, **T** = tracing. The EP shearing was omitted from the plot as it was not used.

We conducted a MANOVA to investigate the effects of material properties on EP patterns during shape judgements, where the frequencies of the nine EP were the dependent variables and the material property (Smooth-deformable/Textured-rigid) was the independent variable. Analysis revealed a significant effect of material properties on EP patterns, $F(7,632)$, $p < .001$, Wilk's $\Lambda = 0.910$, $\eta_p^2 = .090$, indicating that EP frequencies differed based on the material properties of stimuli. Univariate ANOVAs indicated significant differences in EP frequencies across material properties for all EPs except for Scanning sides and Tapping center (see Table 1). However, the EP shearing was not used during shape perception.

Our results suggest that material properties have some influence on the EPs used during shape perception. Although tracing was the most commonly used exploratory procedure across smooth-deformable and textured-rigid shapes, our participants also utilised additional exploratory procedures such as scanning top and tapping ordered more often when exploring smooth-deformable shapes (see Fig. 7).

Table 1. Univariate analysis of variances across object properties and material properties.

Object Properties	Deformability vs. Shape judgments (Dfs: 1,638 for all dependent variables)			Material Properties	Textured-Rigid vs. Smooth-Deformable Shape judgements (Dfs: 1,638 for all dependent variables)		
Dependent Variable	F value	p value	η_p^2	**Dependent Variable**	F value	p value	η_p^2
Pressing centre	556.955	<.001	0.466	Pressing centre	4.802	.029	.007
Pressing edge	127.328	<.001	0.166	Pressing edge	9.822	.002	.015
Scanning sides	62.589	<.001	0.089	Scanning sides	2.413	.121	.004
Scanning top	45.037	<.001	0.066	Scanning top	20.691	<.001	.031
Shearing	17.188	<.001	0.026	Shearing	.	.	.
Tapping centre	24.931	<.001	0.038	Tapping centre	3.391	.066	.005
Tapping ordered	11.000	<.001	0.017	Tapping ordered	15.023	<.001	.023
Tapping random	176.529	<.001	0.217	Tapping random	18.435	<.001	.028
Tracing	1979.163	<.001	0.756	Tracing	21.208	<.001	.032

4.5 Exploration Time as a Function of Object Properties and Material Properties

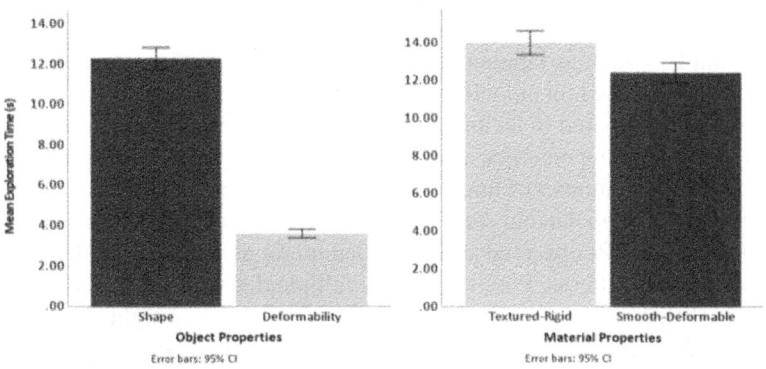

Fig. 8. The mean exploration time plotted as a function of object dimension (left) and as a function of material properties (right).

We investigated if participants' exploration times varied across the object object properties and material properties. Paired t-test revealed significant differences in exploration time based on object properties, with participants spending more time exploring before making shape judgments ($M = 12.28$, $SD = 8.715$) compared to deformability judgments ($M = 3.614$, $SD = 3.342$, $t(959) = 33.847$, $p < .001$). Additionally, exploration time also varied based on material properties, with participants spending more time exploring textured-rigid shapes ($M = 13.94$, $SD = 10.129$) compared to smooth-deformable shapes ($M = 12.398$, $SD = 8.556$, $t(959) = 5.498$, $p < .001$), see Fig. 8.

5 Discussion

Previous research has demonstrated that observers are apt at identifying the most efficient EP for perceiving the target haptic properties, and they strategically adapt their movement parameters based on task or stimulus properties to optimise performance. In this experiment, we explored variations in EP usage during different tasks performed with the same set of objects varying in two dimensions in single-finger exploration. Additionally, we investigated whether EP usage differs with respect to object properties unrelated to the aim of exploration.

In line with previous studies, our results demonstrated a substantial influence of the perceptual task on the selection of EPs [1, 4, 5]. When executing different perceptual tasks using the same sets of objects varying in dimensions, participants exhibited varying EP usage based on the specific task. Furthermore, we found that even under the constraint of single-finger exploration, participants consistently use task-specific EPs. For shape judgements, participants predominately used tracing, whereas, for deformability judgements, they relied on pressing to assess deformability [9, 10]. These findings suggest that, in the current context, the tasks and targeted haptic properties strongly shape the use of EPs, and the effectiveness of these EPs remains relatively unaffected by the constraint of restricted single-finger exploration.

Additionally, within a given task, we observed slight variations in the use of EPs as a function of material properties. When exploring textured-rigid shapes, tracing clearly dominated the EP patterns, whereas when exploring smooth-deformable shapes, participants not only traced along the contour of the shape but also often supplemented it with additional EPs such as tapping and scanning. Although these supplementary EPs were not exclusively used for smooth-deformable shape judgments, they were more frequently observed compared to textured-rigid shape judgments. This finding aligns with existing literature, tracing emerged as the predominant EP for acquiring shape information, regardless of the number of fingers used for exploration [e.g. 1, 9]. However, given the limited participant pool in our study and the constraints of single-finger explorations, it is possible that the EPs observed and described might not encompass the full possible range of exploration patterns. Possibly, using multiple digits or whole-hand exploration may result in different patterns of results, and would provide insights into how exploration constraints impact EP usage, particularly in relation to object properties unrelated to the primary aim of exploration. This question would be very interesting to explore in the future.

Nevertheless, we speculate that the more frequent usage of additional EPs during deformable shape judgements likely served as a strategy to obtain shape information while compensating for single-finger explorations and the deformable nature of the objects. This is supported by the similarity in participants' shape judgements across conditions, even with variations in EP usage during the exploration of textured-rigid and smooth-deformable shapes. It is plausible that the observed EPs provide comparable perceptual information, thus contributing to participants' consistent shape judgements. The observed variations in EP usage suggested that individuals strategically modify their exploration strategies when encountering object properties unrelated to the primary task to enhance the acquisition of consistent and relevant information for the primary task. These findings also have potential applications in other domains, such as in the development of robotic exploration strategies. It offers potential for transferring human single-finger exploration strategies to single-digit robotic explorations across diverse contexts.

Interestingly, despite the increased use of additional EPs when exploring smooth-deformable shapes, participants spent less time exploring them overall compared to textured-rigid shapes. Hence, we wonder whether the perceived pleasantness of the stimuli, potentially influenced by factors such as the stickiness of silicone material, might have influenced exploration behaviour. It is possible that participants utilised additional EPs during the exploration of smooth-deformable shapes not only to compensate for the deformable nature of the objects but also to quickly obtain sufficient perceptual information in a shorter amount of time to avoid prolonged discomfort. Although we did not directly measure perceived pleasantness in our experiment, it raises the intriguing possibility that EP usage may also vary based on hedonistic goals unrelated to the primary haptic task.

Regarding perceptual judgements, we observed low interobserver consistency in participants' similarity judgements about smooth-deformable shapes, suggesting substantial variability among participants in their perceptual assessments of deformable shapes. This lower agreement among participants possibly stems from the challenge of forming stable shape representations due to the deformable nature of the stimuli, thus introducing more variability in their perception of shape similarity. Yet, interestingly, participants had similar shape judgements across textured-rigid and smooth-deformable conditions, implying the ability to perceive shape similarity irrespective of material properties. Plausibly, participants may have focused on certain underlying structural information, for instance, shape or geometric features, such as the magnitude of curvature of the number of convex/concave elements in their judgements. This hints at the possibility that participants may have either ignored or attenuated local tactile information unrelated to the primary task, e.g. deformability, and compensated for it by extracting shape information through other available cues such as kinematic or proprioceptive cues [e.g. 11]. Furthermore, there were no identical shapes in the current experiment, which raises the question of whether participants prioritised certain shape locations as more informative in their similarity assessment. Future research using similar stimuli can explore whether individuals exhibit a preference for certain shape locations or find certain areas more informative than others. This would allow us to identify what kind of shape cues contribute to perceiving one shape as more similar to a given reference shape than others and provide insights into factors that influence our shape similarity perception.

Taken together, the current study examined the variation in EP usage based on different tasks performed with the same set of objects varying in two dimensions, and we explored whether EP usage differs based on object properties unrelated to the aim of exploration. Our findings revealed a robust influence of perceptual tasks on the selection of EPs, even when exploration was confined to a single finger. Additionally, we demonstrated that perceivers could strategically adapt their use of EPs to obtain information pertinent to the primary task, even in the presence of additional properties that are unrelated to the primary task.

Acknowledgments. L.L., K.D., A.K. and J.P. were supported by the Hessisches Ministerium für Wissenschaft und Kunst (HMWK; project 'The Adaptive Mind'), K.D. and K.D. were supported by the Deutsche Forschungsgemeinschaft (DFG, German Research Foundation) – project number 222641018 – SFB/TRR 135, A5 & B8. The authors would also like to thank Viktoria Neuwirt for data collection and data coding. Manuela Kußler and Sara Vitagliano for data coding.

Disclosure of Interests. The authors have no competing interests to declare that are relevant to the content of this article.

References

1. Lederman, S.J., Klatzky, R.L.: Hand movements: a window into haptic object recognition. Cogn. Psychol. **19**(3), 342–368 (1987)
2. Cavdan, M., Doerschner, K., Drewing, K.: Task and material properties interactively affect softness explorations along different dimensions. IEEE Trans. Haptics **14**(3), 603–614 (2021)
3. Dövencioğlu, D.N., Üstün, F.S., Doerschner, K., Drewing, K.: Hand explorations are determined by the characteristics of the perceptual space of real-world materials from silk to sand. Sci. Rep. **12**(1), 14785 (2022)
4. Lederman, S.J., Klatzky, R.L.: Extracting object properties through haptic exploration. Acta Physiol (Oxf.) **84**(1), 29–40 (1993)
5. Klatzky, R.L., Lederman, S.J., Reed, C.: There's more to touch than meets the eye: the salience of object attributes for haptics with and without vision. J. Exp. Psychol. Gen. **116**(4), 356 (1987)
6. Norman, J.F., Adkins, O.C., Dowell, C.J., Hoyng, S.C., Gilliam, A.N., Pedersen, L.E.: Aging and haptic-visual solid shape matching. Perception **46**(8), 976–986 (2017)
7. Klatzky, R.L., Loomis, J.M., Lederman, S.J., Wake, H., Fujita, N.: Haptic identification of objects and their depictions. Percept. Psychophys. **54**, 170–178 (1993)
8. Zoeller, A.C., Drewing, K.: A systematic comparison of perceptual performance in softness discrimination with different fingers. Atten. Percept. Psychophys. **82**, 3696–3709 (2020)
9. Withagen, A., Kappers, A.M., Vervloed, M.P., Knoors, H., Verhoeven, L.: The use of exploratory procedures by blind and sighted adults and children. Atten. Percept. Psychophys. **75**, 1451–1464 (2013)
10. Mizrachi, N., Nelinger, G., Ahissar, E., Arieli, A.: Idiosyncratic selection of active touch for shape perception. Sci. Rep. **12**(1), 2922 (2022)
11. Hayward, V.: Haptic shape cues, invariants, priors and interface design. In: Grunwald, M. (eds.) Human Haptic Perception: Basics and Applications, pp. 381–392. Birkhäuser, Basel (2008). https://doi.org/10.1007/978-3-7643-7612-3_31

Exploring Frequency Modulation in Decoding Edge Perception Through Touch

Mounia Ziat$^{(\boxtimes)}$ ⓘ, Iliyas Tursynbek, Thu Pham, and Allison Ling

Bentley University, Waltham, MA 02452, USA
{mziat,itursynbek}@bentley.edu

Abstract. This research explores how different vibration frequencies influence our ability to perceive and identify edges through touch. The study used a PinArray device to simulate various edge shapes, testing if participants could recognize these shapes based on pairing two frequencies (LF-HF). The findings showed that certain frequency combinations in low and medium ranges were particularly effective in conveying step edge and ridge shapes. This has important implications for enhancing tactile technology in fields like robotics and assistive devices by improving the realism of tactile virtual shapes.

Keywords: Frequency Modulation · Edge Detection · Mechanoreceptors

1 Introduction

The quest to understand and replicate human tactile perception has been an evolving field of research, intertwining physiological, psychophysical, and engineering insights. Our work, building on the foundational concepts introduced by other researchers, seeks to address the complexities inherent in tactile perception and haptic technology, and more specifically, edge perception.

In our ongoing research, we extend this exploration by questioning some of the long-standing assumptions in the field, particularly those pertaining to mechanoreceptor specificity in tactile perception. Pioneering studies, focused on the correlation between mechanoreceptor activation and tactile sensations, often confined to threshold-level stimuli [2,3,9–12]. However, recent physiological evidence advocates for a more integrative approach, challenging the traditional frequency-specific models [6,13,22,23]. We propose that tactile perception, especially in complex tasks like edge detection, necessitates the cooperation of multiple receptor types, aligning with emerging trends in sensory research, emphasizing pattern coding over receptor-specific responses [1,18,22].

2 Related Work

It has been established that changes in the shape of dermal papillae significantly influence our ability to discern edges [4,15]. Whenever a finger touches an edge,

H. Kajimoto et al. (Eds.): EuroHaptics 2024, LNCS 14768, pp. 147–158, 2025.
https://doi.org/10.1007/978-3-031-70058-3_12

it exerts pressure on the skin, enabling the brain to interpret these nerve signals as edge characteristics [7]. In particular, when a finger lies on a surface featuring a groove (see Fig. 1), the skin deforms, causing the dermal papillae to move due to a shift in the gradient of the papillary ridges. Since SAII and SAI receptors are situated in the dermal papillae, they too are subject to this deformation. The Merkel receptors, which are believed to be vital for perceiving stationary edges, and the Meissner corpuscles, found in the papillary ridges (fingerprints), sense the motion of the papillae caused by sideways stress, and relay this data to the brain where it is processed as information about the edge [4]. Kuroki et al. mention that the side-to-side stress corresponds directly to the gradient alteration [15]. Additionally, the papillae are thought to enhance the skin's deformation as a signal booster, transmitting detailed information about the object's shape [14].

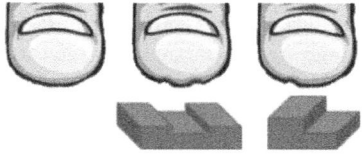

Fig. 1. Skin deformation when a finger is resting on two edges or one edge.

Various research efforts have focused on replicating edges in tactile displays through specific skin distortions [16,20]. However, such studies, aimed at understanding the dynamics of simulated edges in haptic displays, have been limited to threshold detection levels. To truly replicate the sensation of a synthetic edge as if it were real, a comprehensive approach is needed. Recent research indicates that FAI and FAII receptors might also be involved in edge detection due to their capacity to decode frequencies across a wide range. In this context, both Ruffini receptors, reacting to constant pressure, and Pacinian receptors, sensitive to motion, play a part in recognizing objects, including edges and corners [17,21,25].

Lim et al. conducted a comprehensive analysis using a vibrotactile display to assess participants' ability to discern shapes by modulating vibration frequency without changing amplitude [16]. The findings revealed that vibration frequency can effectively be employed to sense shapes with different heights even under a fixed amplitude stimulus. This insight into the vibrotactile threshold and frequency modulation's impact on tactile perception has significant implications for advancing haptic technology, particularly in enhancing the realism and fidelity of tactile displays.

In this paper, our focus was on understanding the impact of frequency modulation on the perception of edges. Specifically, we altered the frequency while keeping the amplitude constant to induce a perception of varying amplitude.

The main goal of this study was to determine if a specific shape could be created using a frequency-pair with minimal amplitude changes. To explore how participants perceive differences in frequency as a tactile representation of a surface's protrusion, we conducted two experiments:

Experiment 1 - Combining Different Frequencies: This setup (illustrated in Fig. 2b) involved vibrating pins at varied frequencies (F1 and F2) to form the shapes shown in Fig. 2a. This method was based on the approach by Lim et al. 2006 [16] and expanded the frequency range tested from 3 Hz to 250 Hz to include a range of receptors.

Fig. 2. a) Shapes used from left to right: S1 (ascending step edge), S2 (descending step edge), S3 (groove), S4 (ridge); b) Example of 2 × 2 frequencies for a descending edge; c) A 2 × 2 frequency shift paradigm.

Experiment 2 - Gradual Shift in Frequencies: In this case (depicted in Fig. 2c), two frequencies were quickly shifted from one side to the other. Initially, this may create a blurred sensation, but it eventually gives the impression of an edge emerging from a flat surface. In a preliminary test [8], aimed at confirming the accuracy of frequency shifts, participants successfully identified edges with shift durations of 500 ms or 1100 ms. Shorter intervals, however, led to more errors. Interestingly, even when a low frequency was used to represent a non-edge condition, participants still perceived an edge, suggesting the complexity of creating a tactile edge illusion. This indicates the need for further experimentation and validation of different conditions to accurately simulate haptic perceptions.

3   Experiment 1

3.1   Participants

Before the experiment, we conducted a power analysis to determine the optimal number of participants. Using G*Power, we aimed to detect a large effect size (f = 0.25) in a repeated measures ANOVA (within factor), with a significance level α of 0.05 and a desired power of 0.90 with 7 repetitions. The recommended total sample size was determined to be 22. This study involved 23 individuals (13 female and 10 male participants), with an average age of 21.8 years (SD = 5.8). None of the participants had any reported issues with hand cutaneous or kinesthetic functions. Before starting the experiment, they completed an electronic consent and the Edinburgh Handedness Questionnaire [19], a 10-item tool used to identify their dominant hand. Results showed 20 right-handed and 3 left-handed participants. All participants received compensation for their time. The experiment's methods received approval from Bentley University Institutional Review Board (IRB).

3.2 Apparatus

Stimuli were generated using the PinArray, a custom vibrotactile device. The PinArray device consisted of twelve (3×4) flat-topped pins shown in Fig. 3b, each with a diameter of 1.5 mm and a center-to-center distance of 2.5 mm between them. Each pin was vertically actuated by an independent DC voice coil linear actuator (NCM02-05-005-4JBL by H2W Technologies, Inc., CA, USA) attached underneath. These voice coil actuators were controlled by a Python 2.7 script via an Ethernet connection and a custom controller box (Sigma Design, Camas, WA, USA). The primary function of the pins was to deliver vibrotactile stimulation to the fingerpad resting on top.

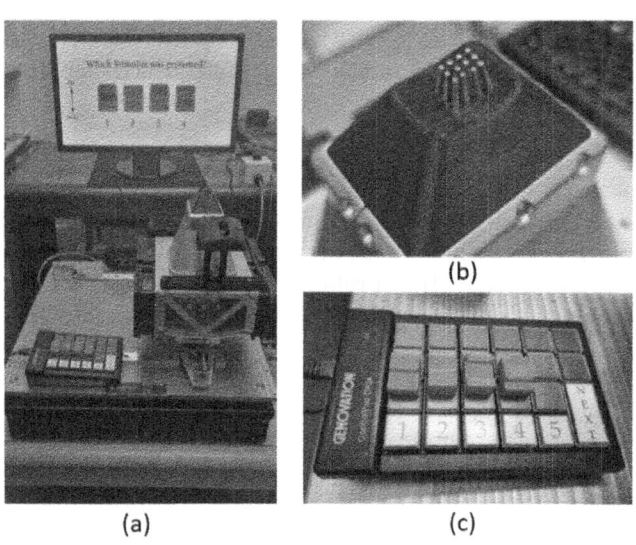

(a)   (b)   (c)

Fig. 3. a) A depiction of the experimental setup. b) Twelve vibrotactile pins. c) Response keypad.

The frequency, amplitude, and duration of each pin's activity could be set independently, enabling the creation of relatively complex patterns of vibrotactile stimulation across the fingertip. Prior to the experiments, we conducted a comprehensive evaluation of each pin in the PinArray, testing across frequency ranges of 3–350 Hz and amplitudes. Each assessment, lasting about 10 s, involved analyzing hall sensor data for amplitude range and frequency stability. This rigorous calibration ensures that the device performs consistently across all experimental conditions, providing reliable data for our study. Hence, a constant ratio of 1.5 between frequency and amplitude was selected. Table 1 displays the corresponding amplitude for each frequency.

Table 1. Frequencies and corresponding amplitudes.

Frequency	Amplitude	Frequency	Amplitude	Frequency	Amplitude
50 Hz	0.030 mm	60 Hz	0.025 mm	75 Hz	0.020 mm
100 Hz	0.015 mm	110 Hz	0.014 mm	125 Hz	0.012 mm
150 Hz	0.010 mm	200 Hz	0.008 mm	225 Hz	0.007 mm
240 Hz	0.006 mm	250 Hz	0.006 mm		

3.3 Stimuli

Vibrotactile stimuli were generated via vertical oscillatory motion in each pin. The four edge shape configurations (S1 to S4), shown in Fig. 2, were created by pairing two frequencies: F1 and F2. This pairing is illustrated in Fig. 4, where red dots represent pins operating at the higher frequency (HF) and blue dots represent pins at the lower frequency (LF).

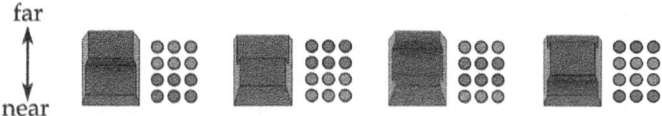

Fig. 4. Four edge shape configurations: S1 - ascending step edge (LF to HF), S2 - descending step edge (HF to LF), S3 - groove (HF-LF-HF), and S4 - ridge (LF-HF-LF).

As shown in Table 2, the tested frequency pairs (F1-F2) fell into three categories: Low Range (LF), Medium Range (MF), and High Range (HF). The differences between the frequencies in each F1-F2 pair were set at 10, 25, 50, and 100 Hz. This configuration resulted in 48 unique test conditions (4 shapes × 4 frequency pairs × 3 frequency ranges).

Table 2. Frequency pairs (FP) used in Experiment 1.

	FP1 (10 Hz ≠)	FP2 (25 Hz ≠)	FP3 (50 Hz ≠)	FP4 (100 Hz ≠)
LF	50–60 Hz	50–75 Hz	50–100 Hz	50–150 Hz
MF	100–110 Hz	100–125 Hz	100–150 Hz	100–200 Hz
HF	240–250 Hz	225–250 Hz	200–250 Hz	150–250 Hz

3.4   Procedure

Participants were seated in front of a table on which the experimental computer and the connected PinArray device were placed, as illustrated in Fig. 3. The position of the pin tactile display was adjusted to accommodate the participant's dominant hand: to the right for right-handed participants and to the left for left-handed participants. Upon signing the consent form, each participant was instructed to gently place the index fingertip of their dominant hand on the PinArray display, where the pins were located. A handrest was positioned at the front of the stimulator surface for their comfort (see Fig. 3a).

The experiment consisted of three stages: familiarization, training, and testing. In the familiarization stage, participants were introduced to the vibrotactile stimuli through a sequence of 12 test conditions. These conditions were randomly selected to ensure exposure to each shape (S1 to S4) across all frequency ranges (LF, MF, HF). During this stage, participants were simply required to feel the stimuli without providing any response. Each stimulus was paired with a visual representation of its corresponding shape on the computer screen. No information regarding the frequency range of the stimuli was disclosed to the participants. During the training stage, participants were exposed to the four shapes randomly selected and were asked to select the corresponding key on the keypad in front of them, as shown in Fig. 3c. They then received immediate feedback indicating whether their response was correct or incorrect.

Transitioning to the testing stage, participants underwent a total of 336 trials, with the 48 test conditions repeated 7 times in random order. During this stage, they wore noise-cancelling headphones playing pink noise to block any potential auditory cues from the PinArray device. The primary task for participants was to identify the shape presented, as shown in Fig. 5. Following the delivery of each stimulus for 500 ms, they had a 5-second window to respond. If a response was not provided within this time frame, the missed trial was randomly reintroduced later in the experiment. The delivery of test trials was fully automated, eliminating the need for participants to manually proceed to the next trial. The experiment lasted approximately 45 min, during which two 2-min breaks were given. At the end of the experiment, participants were invited to assess the difficulty level on a 5-point Likert scale, ranging from 1 (very easy) to 5 (very hard). Additionally, they were encouraged to share any feedback or comments regarding their overall experience.

3.5   Results

In the three-way repeated measures ANOVA, the main effects and interactions of shape, range, and pair were examined. The main effect of shape [$F(3, 66) = 8.62$, $p = .002$, $\eta_p^2 = .28$], range [$F(2, 44) = 13.74$, $p < .001$, $\eta_p^2 = .38$], and pair [$F(3, 66) = 30.21$, $p < .001$, $\eta_p^2 = .58$] were highly significant. Significant interaction effects between shape and range [$F(6, 132) = 5.88$, $p < .001$, $\eta_p^2 = .21$], range and pair [$F(6, 132) = 16.13$, $p < .001$, $\eta_p^2 = .42$], and between shape, range, and pair [$F(18, 396) = 3.16$, $p < .001$, $\eta_p^2 = .13$] were also found. To analyze

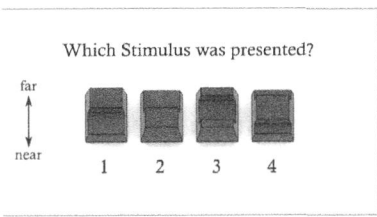

Fig. 5. The response screen displays the shapes presented to participants at the end of each trial.

these interaction effects, we conducted simple pairwise comparisons. The post-hoc analysis revealed significant mean differences across various combinations of shapes (S1 to S4), frequency ranges (LF, MF, HF), and pairs (FP1 to FP4), indicating distinct patterns of interaction between these factors. Notably, in the low frequency range, S1 with FP1 and FP2 was significantly different (p < .001) from all the remaining shapes. Furthermore, performance was near chance level (.25), as confirmed by a one-sample t-test (p > .05), for S2, S3, and S4 when paired with FP3 and FP4 (Fig. 6). In the medium frequency range, performance was significantly above chance level (p > .05) for all conditions. More specifically, S1, S2, and S4 were most significantly recognized when paired with FP2 and FP4 (p < .001). In the high frequency range, except for FP3, performance was significantly different from the chance level (.25) for most conditions. However, there was a notable decrease in performance compared to the low and medium frequency (LF and MF) conditions. Additionally, no significant differences were observed between shapes for each pair condition. These findings indicate that frequency effects are especially pronounced in perceiving step edges, with these effects varying across different pairings.

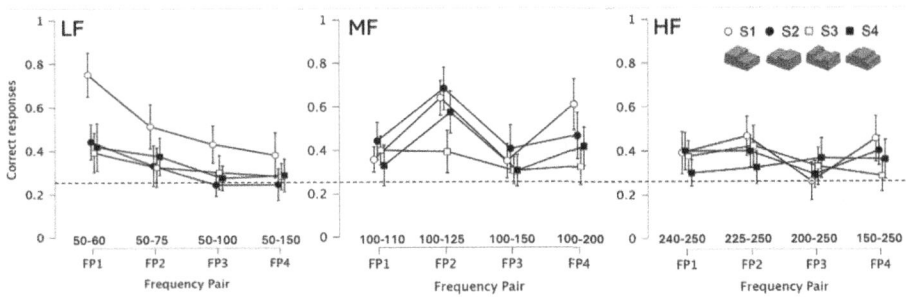

Fig. 6. Percentage of correct responses per shape (S1, S2, S3, S4) and frequency pairs (FP1, FP2, FP3, FP4) for low, medium, and high ranges (LF, MF, and HF). The error bars represent 95% confidence intervals (CIs). The dashed line represent chance level (25%).

4 Experiment 2

4.1 Participants

Twenty-two participants (14 female, 8 male, Mean age = 23.1, SD = 4.6) were involved in this study. Nineteen participants were right-handed, while the remaining three were left-handed. None of the participants had reported issues related to hand cutaneous or kinesthetic functions. Each participant received compensation for their time. The experimental methods were approved by the Institutional Review Board (IRB) at Bentley University.

4.2 Apparatus and Stimuli

In the second experiment, the primary modification involved dynamically shifting the vibration frequencies across the PinArray to simulate edge detection. This was achieved by gradually transitioning the frequencies across the pins: half of the pins initially set to vibrate at frequency F1 switched to F2, and vice versa, over a period of 500 ms. This change was symmetrically mirrored on the opposite side, creating a dynamic frequency shift. The entire sequence lasted for 2 s (Fig. 7). This approach, modeled after previous implementations with the device [8], allows us to examine how abrupt changes in frequency influence tactile perception and edge discernment in real-time.

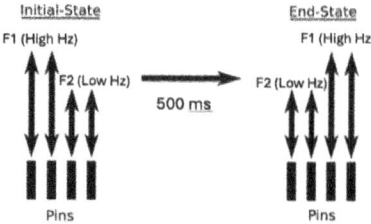

Fig. 7. The spatial location of the frequencies within a pair switched over time.

Due to the low performance observed with high-frequency (HF) pairs in Experiment 1, specifically in participants' ability to accurately identify shapes, only low-frequency (LF) and mid-frequency (MF) ranges were tested in this subsequent experiment. As in the first experiment, frequency pairs were chosen to maintain differences of 10, 25, 50, and 100 Hz between F1 and F2. Combining these frequency differences with four shape configurations, the experiment yielded 32 test conditions (8 frequency pairs × 4 shape configurations) (Table 3).

4.3 Procedure

Experiment 2 followed the same protocol as Experiment 1, with minor modifications. During the familiarization stage, eight stimuli were presented to the

Table 3. Frequency pairs used in experiment 2.

	FP1 (10 Hz $\neq$)	FP2 (25 Hz $\neq$)	FP3 (50 Hz $\neq$)	FP4 (100 Hz $\neq$)
LF	50–60 Hz	50–75 Hz	50–100 Hz	50–150 Hz
MF	100–110 Hz	100–125 Hz	100–150 Hz	100–200 Hz

participants. In the testing stage, the experiment featured 32 test conditions, each randomly repeated seven times, totaling 224 trials. Participants were asked to respond after each stimulus. The entire duration of Experiment 2 was around 30 min.

4.4   Results

The three-way repeated measures ANOVA with the factors as shape, pair, and range showed significant effect of shape [F(3, 63) = 5.79, p < .001, $\eta_p^2 = .22$], range [F(1, 21) = 6.20, p < .05, $\eta_p^2 = .23$], and pair [F(3, 63) = 2.98, p < .001, $\eta_p^2 = .12$]. Significant interaction effects between the shape and range factors [F(3, 63) = 14.09, p < .001, $\eta_p^2 = .4$], and between shape and pair [F(9, 189) = 3.54, p < .001, $\eta_p^2 = .14$] were also found (Fig. 8).

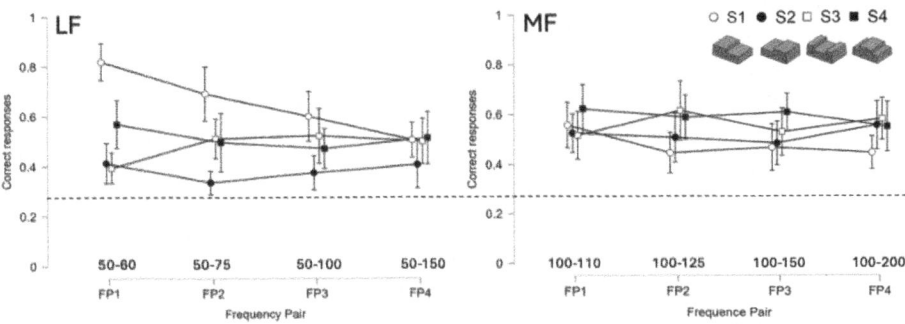

Fig. 8. Percentage of correct responses per shape (S1, S2, S3, S4) and frequency pairs (FP1, FP2, FP3, FP4) for Low and Medium ranges (LF and MF). The error bars represent 95% confidence intervals (CIs). The dashed line represent chance level (25%).

Post-hoc analysis of simple main effects revealed that for the low-frequency range, S1 and S4 were significantly different from S2 and S3 when using FP1. S1 was also significantly different from S2 using FP2 and FP3. Across all frequency pairs in the medium range, there were non-significant differences between the shapes. However, with the understanding that recognition is still above chance in all these conditions (p < .001), it indicates that participants were effectively identifying shapes, albeit without significant variation in performance between different shapes. Comparing the results of both experiments, independent t-tests showed that performance increased significantly in Experiment 2 for S3 and S4, suggesting that shifts in frequency improve the perception of two-edge shapes such as grooves and ridges.

5   Discussion and Conclusion

The results from Experiments 1 and 2 contribute to a growing body of research related to edge detection [4, 7, 20]. Our study aims to clarify how edge recognition via vibrotactile feedback is influenced by variations in frequency ranges and specific frequency pairings, while maintaining constant amplitude changes. A key finding from Experiment 1 was the differentiation in shape recognition across frequency ranges. Specifically, S1 was distinctly recognized when paired with FP1 (50–60 Hz) and FP2 (50–75 Hz) in the LF range. This unique recognition of S1 in the LF range may be linked to increased spatial acuity in the proximal third of the distal phalanx of the fingerpad [5]. Research by Ziat and colleagues [26] has demonstrated that tactile perception is enhanced when stimuli move distally, transitioning from areas of higher receptor density to lower density. Conversely, perception tends to diminish when stimuli remain in the same proximal region, which is inherently less sensitive. The perceptual distinction for S1, an ascending step edge, could be attributed to this phenomenon, as the higher frequency in the pair consistently engages the receptor-dense area on the fingertip. The medium frequency range expanded the perceptible spectrum, with S1, S2, and S4 being distinctly recognized with FP2 (100–125 Hz) and FP4 (100–200 Hz), suggesting optimal edge perception within this range. However, high frequencies were associated with a general decline in perceptual accuracy. Notably, S3 was consistently the most challenging for participants to recognize across all frequency ranges. Its difficulty in recognition suggests complexities or ambiguities in how frequency modulation simulates grooves and other tactile features. This challenge may be due to various factors, including the inherent difficulty in conveying certain shapes or textures through frequency changes alone.

Experiment 2 expanded these findings by introducing dynamic frequency shifts. In the low-frequency range, there was significant differentiation in recognizing S1 from S2 and S3 across multiple frequency pairs (FP1, FP2, and FP3). Additionally, S4 was distinguishable from S2 and S3 with FP1. In contrast, the medium frequency range showed consistent performance across different shapes, suggesting perceptual stability and adaptability. This was evidenced by uniformly and significantly above-chance performance across these shapes, where no single shape was notably more perceived than another. The performance for S3 and S4, representing a groove and a ridge, respectively, improved significantly from Experiment 1 in terms of accuracy and perceptual clarity. In tactile edge detection, discerning both grooves and ridges is crucial. A groove is a linear indentation, felt as a drop in the surface followed by a rise, allowing for the detection of its edges, depth, and width. Conversely, a ridge is a raised band on a surface, perceived through an elevation in the surface followed by a descent back to the original level. Detecting these opposing features-the inward dip of a groove and the outward rise of a ridge-is essential in applications such as robotics telemanipulation and assistive tactile displays, providing crucial spatial information. The consistent recognition of various shapes in the medium frequency range (100–200 Hz) aligns with pattern coding, suggesting that the human tactile system might be proficient at processing complex patterns of stimulation, rather than relying solely on specific receptor activation. This could indicate a more holistic approach to tactile perception, where integrating multiple sensory inputs is key. The distinct performances in different frequency ranges and with specific frequency pairings in Experiment 1, along with the dynamic frequency temporal shift in Experiment 2, demonstrate the complexity of edge perception.

The next stage of this research is to explore the role of action (passive, marginally active, and active) on tactile edge perception. The goal is to determine whether the direction of movement influences performance. This will enable us to investigate the significant role that proprioception plays in everyday tasks, such as typing on a keyboard without looking at our hands. When the proprioceptive sense is diminished or lost, perceptual thresholds for angle discrimination increase [15,24,25], suggesting that angle discrimination is an integrative process involving both the cutaneous and proprioceptive senses. Moreover, further exploration into how multiple sensory inputs such as visual-tactile or auditory-tactile are integrated and processed by the central nervous system could offer deeper insights into the nature of perceiving an edge in a multimodal setting.

Acknowledgments. This work is funded by the National Science Foundation (NSF) - grant number 2216763.

References

1. Birznieks, I., McIntyre, S., Nilsson, H.M., Nagi, S.S., Macefield, V.G., Mahns, D.A., Vickery, R.M.: Tactile sensory channels over-ruled by frequency decoding system that utilizes spike pattern regardless of receptor type. Elife **8**, e46510 (2019)
2. Bolanowski, S., Jr., Zwislocki, J.J.: Intensity and frequency characteristics of pacinian corpuscles. I. Action potentials. J. Neurophys. **51**(4), 793–811 (1984)
3. Bolanowski, S.J., Jr., Gescheider, G.A., Verrillo, R.T., Checkosky, C.M.: Four channels mediate the mechanical aspects of touch. J. Acoust. Soc. Am. **84**(5), 1680–1694 (1988)
4. Chorley, C., Melhuish, C., Pipe, T., Rossiter, J.: Tactile edge detection. In: SENSORS, 2010 IEEE, pp. 2593–2598. IEEE (2010)
5. Craig, J.C.: Grating orientation as a measure of tactile spatial acuity. Somatosens. Motor Res. **16**(3), 197–206 (1999)
6. Enander, J.M., Jörntell, H.: Somatosensory cortical neurons decode tactile input patterns and location from both dominant and non-dominant digits. Cell Rep. **26**(13), 3551–3560 (2019)
7. Gerling, G.J., Thomas, G.W.: The effect of fingertip microstructures on tactile edge perception. In: WorldHaptics 2005, pp. 63–72. IEEE (2005)
8. de Grosbois, J., Di Luca, M., King, R., Ziat, M.: The predictive perception of dynamic vibrotactile stimuli applied to the fingertip. In: HS2020, pp. 848–853 (2020)

9. Johansson, R.S.: Tactile sensibility in the human hand: receptive field characteristics of mechanoreceptive units in the glabrous skin area. J. Physiol. **281**(1), 101–125 (1978)

10. Johansson, R.S., Vallbo, Å.B.: Tactile sensory coding in the glabrous skin of the human hand. Trends Neurosci. **6**, 27–32 (1983)

11. Johnson, K.O., Lamb, G.D.: Neural mechanisms of spatial tactile discrimination: neural patterns evoked by braille-like dot patterns in the monkey. J. Physiol. **310**(1), 117–144 (1981)

12. Johnson, K.O., Phillips, J.R.: Tactile spatial resolution. I. Two-point discrimination, gap detection, grating resolution, and letter recognition. J. Neurophysiol. **46**(6), 1177–1192 (1981)

13. Jörntell, H., Bengtsson, F., Geborek, P., Spanne, A., Terekhov, A.V., Hayward, V.: Segregation of tactile input features in neurons of the cuneate nucleus. Neuron **83**(6), 1444–1452 (2014)

14. Kikuuwe, R., et al.: The tactile contact lens. In: Sensors, pp. 535–538. IEEE (2004)

15. Kuroki, S., Kajimoto, H., Nii, H., Kawakami, N., Tachi, S.: Proposal of the stretch detection hypothesis of the meissner corpuscle. In: EH2008, pp. 245–254 (2008)

16. Lim, S.C., Kim, S.C., Kyung, K.U., Kwon, D.S.: Quantitative analysis of vibrotactile threshold and the effect of vibration frequency difference on tactile perception. In: 2006 SICE-ICASE International Joint Conference, pp. 1927–1932. IEEE (2006)

17. Macefield, V.G.: Physiological characteristics of low-threshold mechanoreceptors in joints, muscle and skin in human subjects. Clin. Exp. Pharmacol. Physiol. **32**(1–2), 135–144 (2005)

18. Muniak, M.A., Ray, S., Hsiao, S.S., Dammann, J.F., Bensmaia, S.J.: The neural coding of stimulus intensity: linking the population response of mechanoreceptive afferents with psychophysical behavior. J. Neurosc. **27**(43), 11687–11699 (2007)

19. Oldfield, R.C.: The assessment and analysis of handedness: the edinburgh inventory. Neuropsychologia **9**(1), 97–113 (1971)

20. Park, J., Doxon, A.J., Provancher, W.R., Johnson, D.E., Tan, H.Z.: Haptic edge sharpness perception with a contact location display. IEEE ToH **5**(4), 323 (2012)

21. Park, J., Kim, M., Lee, Y., Lee, H.S., Ko, H.: Fingertip skin-inspired microstructured ferroelectric skins discriminate static/dynamic pressure and temperature stimuli. Sci. Adv. **1**(9), e1500661 (2015)

22. Saal, H.P., Bensmaia, S.J.: Touch is a team effort: interplay of submodalities in cutaneous sensibility. Trends Neurosci. **37**(12), 689–697 (2014)

23. Spanne, A., Jörntell, H.: Questioning the role of sparse coding in the brain. Trends Neurosci. **38**(7), 417–427 (2015)

24. Voisin, J., Benoit, G., Chapman, C.E.: Haptic discrimination of object shape in humans: two-dimensional angle discrimination. Exper. Brain Res. **145**, 239–250 (2002)

25. Westling, G., Johansson, R.S.: Responses in glabrous skin mechanoreceptors during precision grip in humans. Exp. Brain Res. **66**, 128–140 (1987)

26. Ziat, M., Hayward, V., Chapman, C.E., Ernst, M.O., Lenay, C.: Tactile suppression of displacement. Exp. Brain Res. **206**, 299–310 (2010)

The Role of Implicit Prior Information in Haptic Perception of Softness

Didem Katircilar$^{(\boxtimes)}$ and Knut Drewing

Justus Liebig University, 35390 Giessen, Germany
didem.katircilar@psychol.uni-giessen.de

Abstract. People regularly use active touch to perform daily life tasks. Imagine choosing a comfortable pillow and how you would explore its softness. It is known that people tune their exploratory behavior to get the most relevant information. In the exploration process, also prior information is used, which is available before we touch an object. For softness perception, object indentation plays a crucial role; indentation forces were higher, when people implicitly expected to explore harder as compared to softer objects. This force-tuning improved perception, and was observed when trials of the same softness level (hard or soft) were presented in longer blocks. However, it was not reported for predictable patterns in that hard and soft stimuli alternate in every or every other trial. Here, we investigated when and how implicit prior information about the softness level becomes accessible for successful force-tuning in softness discrimination. Participants were presented with hard and soft stimulus pairs in sequences of the length of 2, 4 or 6 trials. In predictable conditions, same-length sequences of hard and soft trials alternated constantly. In unpredictable conditions, we presented sequences of lengths 2, 4 and 6 randomly. We analyzed initial peak indentation forces. Participants applied higher forces to harder stimuli in the predictable condition in longer sequences (4 and 6) as compared to the unpredictable condition and shorter sequences of 2. We interpret the findings in terms of an anticipatory and incremental mechanism of force-tuning, which needs to be triggered by an initial predictable stimulus.

Keywords: softness perception · prediction · implicit mechanisms · prior information

1 Introduction

Humans interact with different objects and assess their properties while performing everyday tasks. Imagine choosing a comfortable pillow or buying an avocado in grocery shopping. The sense of touch provides highly relevant information here. It is known that people adjust their hand movements according to the object property that they explore to extract the most relevant information–such as indenting or squeezing regarding softness [1]. The interaction with the object typically starts before the first contact with the object. Prior information about the object property such as our previous experiences, information from other sensory modalities (i.e. vision) or what we have been told about

© The Author(s), under exclusive license to Springer Nature Switzerland AG 2025
H. Kajimoto et al. (Eds.): EuroHaptics 2024, LNCS 14768, pp. 159–170, 2025.
https://doi.org/10.1007/978-3-031-70058-3_13

the object can guide the exploratory movement and also contribute to perception. For instance, people can use visual prior information to adjust their exploratory direction and improve perception in texture perception [2] or they use prior information about the compliance of upcoming stimuli and tune their forces accordingly to increase their precision in softness perception [3]. Here, we study mid-term mechanisms of the usage of implicit prior information in force-tuning mechanisms for softness perception. In the following, we will (1) explain the basic dissociation between implicit vs. explicit information, report what is known about prior information in softness perception in general (2) and with respect to the mechanisms of its usage (3), and detail our present research question (4).

Prior information can be acquired and utilized implicitly or explicitly. Implicit learning corresponds to information acquired without any conscious attempt to learn and when there is no clear indication of what has been learned. In contrast, explicit learning implies acquiring the knowledge that is available to consciousness and can be reported verbally [4, 5]. One example from literature is implicit sequence learning: Repeated sequences of stimuli presented in predictable patterns can be learned [6, 7] and markers of learning can be seen in participants' improved performance as compared to the performances in randomized presentation of the stimuli as unpredictable patterns. For example, in serial reaction time tasks, participants are asked to respond as quickly as possible to a series of stimuli, such as different letters on a screen, with stimulus-specific responses (e.g., pressing different buttons). If there is a constant, repeated pattern in the series, participants responded faster as compared to a random presentation [8]. They were however not aware of their learning. Also, more complex motor behaviors than button presses can benefit from implicit sequence learning. Baird and Stewart used a motor task that required three-dimensional whole-arm reach movements to a series of target locations [9]. If target locations followed a repeated predictable pattern, participants were able to complete this complex task faster than with a random series.

How prior information affects haptic exploration and perception has been studied, in particular, for the guidance of exploratory movements in softness perception. Softness is one of the main dimensions in haptic perception [10], and it is a multidimensional percept [11, 12]. One important perceptual correlate of perceived softness is compliance, that is the deformability of an object surface under the application of physical force. It is known that people choose specific hand movements based on the haptic property that they aim to explore [1], that they finely adjust movement parameters to improve perception and that prior information can guide this fine tuning [3, 13]. For compliance-related softness exploration, people often choose indentation movements and they tune their peak indentation forces. Peak forces appear to have a crucial impact on softness perception [13]. In [3], Kaim and Drewing showed that fine tuning of peak forces can take place with the very first indentation if participants can predict the softness level of the upcoming stimuli. They observed that participants applied from the beginning of each trial more force to harder stimuli versus to softer stimuli when trials with the same softness level (harder or softer) were presented in a block, and this behavior improved perceptual performance. Such force tuning was not observed when harder and softer trials were presented to the participants in a randomized order and thus the softness level was unpredictable. Initial peak forces are a reliable indicator for the use of prior information since sensory information from the present trial is not available yet. The findings, thus, suggested that prior information plays a crucial role for force tuning in softness perception. In [14], the effect of different types of prior information was systematically investigated: Participants either explored only hard or only soft stimulus pairs in a blocked fashion, or they received semantic information (verbal label "soft" or "hard") about the compliance of the upcoming stimulus pair just before the exploration started, or a video with an indentation of a probe into a rendered stimulus (deep or shallow indentation). The former condition was considered implicit, whereas the latter two rather presented explicit prior information. In the explicit conditions harder or softer trials were presented to participants in random order. The expected force tuning (higher initial force to harder stimuli) was only found when stimuli were presented to participants in blocks of only harder or softer trials, i.e. the presumably implicit condition. In a second experiment, they investigated if explicit information interferes with implicit force tuning mechanisms [14]. One group of participants received only implicit prior information from blocking while another group additionally received explicit verbal information about the softness of the upcoming block of trials (e.g., "The next stimuli will be harder"). Force tuning was observed when participants received only implicit information, but it vanished with extra explicit information. Taken together, these studies showed that force tuning mechanisms in softness perception require prior information to be implicit, whereas explicit information interferes with the feedforward mechanism of force tuning.

To gain deeper insight into how people use implicit prior information in softness perception, Drewing and Zoeller studied force adaptation when prior information on stimulus levels could be in principle predicted from implicit sequence learning [15]. They conducted a study in which hard or soft trials were presented in 3 different predictable, but implicit patterns to the participants: 1) longer blocks in which only hard or soft stimuli were presented, 2) short pattern in which hard and soft trials were presented alternately, and 3) long pattern in which always two hard and two soft trials alternated. Indeed, participants were not aware of any of these patterns. Still, participants showed successful force tuning in the blocked condition, i.e. applied higher force to harder versus softer stimuli. In contrast, in short and long pattern conditions, the result was inverse and participants applied slightly higher forces to the softer stimuli than to the harder stimuli. The authors concluded that also in these patterns implicit mechanisms were used for force control, but that under the given conditions–i.e. the relatively short same-softness sequences—their effect was not sufficient for successful force tuning. Due to the inverse effect, the authors also discussed whether force tuning might alternatively be based on reactive trial-by-trial mechanisms rather than on anticipatory mechanisms that use prior information, because, in the alternating patterns, hard and soft trials consistently followed each other in short time. So, forces for one softness-level could have been reactive adaptations to the other level and hence inverse. We know from the literature on object lifting that anticipatory mechanisms can help people to adjust their grip force to predictable changes of object weight [16], but that people are also able to adjust their grip force to unexpected weights in a few trials by reactive trial-to-trial mechanisms [17]. However, post-hoc analyses in [15] rather tended to show that reactive trial-to-trial mechanisms were not responsible for the inverse effect.

That is, overall the above study [15] suggests that prior information from implicit sequence learning influences force control based on anticipatory mechanisms. However, it is not clear why the prior information in the given patterns was not sufficient for successful force tuning. Here, we aim to unravel how mid-term mechanisms achieve that prior information in softness perception becomes accessible for successful force-tuning, in order to better understand the processes by which humans use implicit prior information and optimize their exploratory movements. This is not only of relevance in itself, but may also contribute to improving sensing capabilities in technical systems such as robots. In particular, we studied how long same-softness sequences need to be in order to allow for successful force tuning, and we experimentally tested the assumption that anticipatory mechanisms rather than reactive ones are involved in the force tuning. We presented hard and soft stimuli in predictable and unpredictable patterns. In predictable patterns, hard and soft trials were alternating in same-softness sequences of 2, 4 or 6 trials. Thus, participants were able to learn the patterns and predict the softness level of the upcoming stimulus pair. In unpredictable patterns, we presented sequences of 2, 4 and 6 same-softness trials in random order. Hence, presentation of hard and soft trials did not follow a constant pattern and participants were not able to implicitly learn and predict the softness level of the upcoming stimulus pair. Participants did not receive any information about patterns in the experiment. We expected that when stimuli were presented in a predictable pattern, participants successfully tune their force (higher force to a harder stimulus) compared to the unpredictable pattern thanks to implicit predictive mechanisms. We also expected that participants tune their force only in the longer predictable sequences based on the previous findings.

2   Method

2.1   Participants

A total of 24 participants (19 females, aged between 19 and 35 with mean age of 24.5, SD = 4.04) took part in the study. All of the participants were right-handed and they did not report any cutaneous and motor impairments in the past. Their two-point discrimination threshold was equal to or smaller than 4 on the index finger of their dominant hand. We calculated the required sample size using G*Power [18] for a repeated measure ANOVA, a factor with 2 levels, a power of 80%, alpha level of 5% and a medium-to-large effect size f = 0.325 (from data in [15]) resulting in a sample size of 21. All participants were naïve to the purpose of the study and gave written informed consent before the experiment. They received 8€ per hour for participating. Methods and procedures of the experiment were approved by the Local Ethics Committee of Giessen (LEK FB06), and were carried out in accordance with the Declaration of Helsinki (2013) except for preregistration.

2.2   Setup and Stimuli

Participants performed the experiment on a custom-made visuo-haptic workbench (Fig. 1a) which includes a 24" 3D computer screen with 120 Hz and 1600 × 900 pixel, a PHANToM 1.5 A haptic force feedback device with 1000 Hz temporal and 0.03 spatial resolution, and a force sensor with 682 Hz and 0.05 N resolution to collect normal force data. Participants sat in front of the experimental set up and connected the index finger of their dominant hand to the PHANToM arm via a spherical magnet adapter which sticks to the finger nail and enables them to move their finger in the 38 × 27 × 20 cm³ workspace. Participants were able to explore stimuli with the bare finger since the adapter was only fixed to the finger nail and did not cover the finger pad. A chin rest stabilized their head and participants wore stereo glasses (NVidia 3D Vision 2) to see a 3D visual scene through a mirror. 3D stimuli in the visual scene were aligned with the haptic stimuli to allow for an exploration that is as natural as possible while eliminating direct visual feedback from the haptic stimuli. Haptic stimulus pairs were placed on a plate above the force sensor next to each other, with a distance of approximately 11 cm from center to center. The finger's position was presented as a green dot (8 mm) which however disappeared as soon as the force applied to the stimulus was higher than 0.01 N to prevent any visual feedback about deformation. All participants wore ear plugs to eliminate any potentially confounding noise.

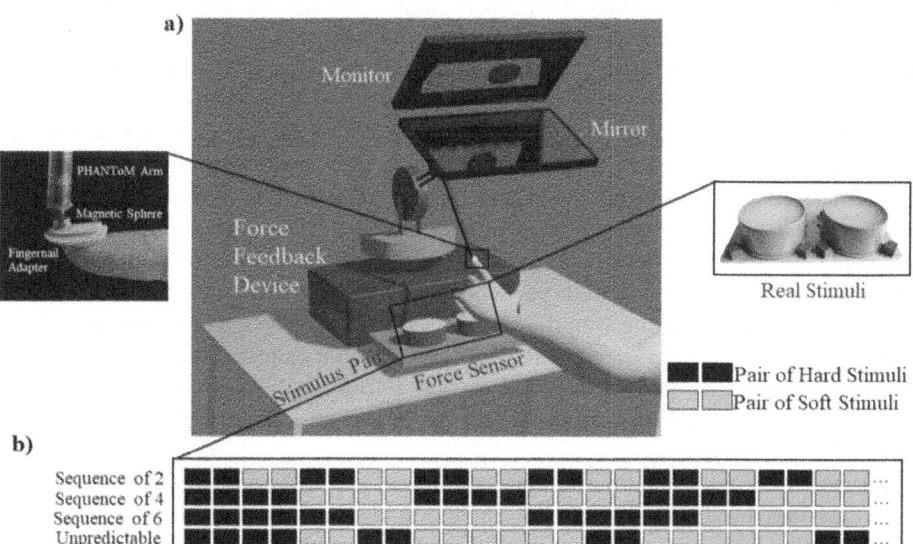

Fig. 1. **a)** Illustration of custom made visuo-haptic setup. **b)** Depiction of finger connection of Phantom to finger nail. **c)** Picture of real stimuli. **d)** Visualization of stimulus presentation in Predictable Condition with Sequence Length 2, 4, 6 and Unpredictable Condition (random presentation of Sequence Length 2, 4 and 6).

We used six cylindric silicone rubber stimuli (height: 38 mm, diameter: 75 mm, see [3] for details of production). Three of the six stimuli were classified as hard, and the other three were classified as soft. We used one standard stimulus (hard: 159.62 kPa, soft: 50.12 kPa) and two comparison stimuli (hard: 174.95 kPa and 132.00 kPa, soft: 60.33 kPa and 42.34 kPa) from each category. The stimuli from both groups were selected based on their elasticity measured as Young's modulus and based on their perceptual distinctness. We followed the standard method to measure the elasticity of the stimuli introduced by [19]. For that purpose, we obtained standard samples from each stimulus by pouring, during stimulus creation, the mixed solutions also into small cylinders with a defined height of 10 mm and a diameter of 10 mm. The standard samples were positioned on a force sensor. Using the PHANToM we pressed an aluminium plate with diameter of 24 mm on the sample–increasing the force by 0.005 N every 3 ms until a minimum force of 1 N and a minimum displacement of 1 mm were detected. We transformed the force and displacement data into stress-strain data and fitted a linear regression line to determine Young's modulus (in MATLAB R2022a). We selected stimuli so that the differences between the two comparisons per category and their standard is as similar as possible regarding their Young's moduli and that each comparison was correctly discriminated from its standard in about 80% of trials (73%–87% in a pilot study with 4 participants).

2.3 Design and Procedure

The experiment comprised three within-participant variables: Predictability (Predictable or Unpredictable), Compliance category (Hard and Soft) and Length of Sequences (2, 4 and 6), (Fig. 1b): Stimulus pairs of each softness-level were presented in sequences of 2, 4 or 6 trials including pairs of the same softness level (either hard or soft). Hard and soft sequences always alternated. In the predictable condition sequences of the same length were blocked, so that participants would be able to learn patterns implicitly and predict the softness level of upcoming stimuli. Each predictable block included 48 trials, i.e. 24, 12 or 6 sequences (of lengths 2, 4, or 6, respectively). In the unpredictable condition, sequences of different lengths appeared in random order, and the unpredictable block included 144 trials (again, 48 trials per sequence length). In this block, participants were not able to predict upcoming stimuli since the sequence length of a hard or soft sequence differed due to randomization and did not follow a regular pattern. We balanced the order of the 4 block types with a Latin-square design. Further, half of the participants started the experiment with a soft sequence while the other half started with a hard sequence pair.

In each trial the standard stimulus and one of the two comparison stimuli of the same category (hard or soft) were presented. Standard and comparison were randomly assigned to the left or right position. In the beginning of each trial, a visual representation of the left stimulus indicated the participant to start exploration. After they had indented the left stimulus two times, a visual representation of the right stimulus appeared. After two indentations of the right stimulus, the participant was asked to choose which of the stimuli they had perceived to be softer using a virtual button.

Participants performed 8 test trials before the main experiment to familiarize themselves with the task. In the test trials, hard or soft pairs were presented to the participants in a random order. After they completed the experiment, participants were given a questionnaire to measure their explicit knowledge about presentation patterns in the experiment (blocks of 2-4-6 same category trials). The questionnaire included 8 different presentation patterns, including the 3 actual patterns in the experiment (alternating sequences of 2 soft - 2 hard, 4 soft - 4 hard, 6 soft - 6 hard) and 5 distractor patterns (alternating sequences of 1 soft – 1 hard, 3 soft – 3 hard, full soft, full hard, 3 hard – 1 soft). They were first asked if they had realized any presentation pattern and if yes, they were required to choose which of the patterns had been included.

2.4 Data Analysis

First of all, we analyzed the discrimination performance in the experiment by calculating the percentage of correct responses. In order to warrant that participants had focused on the task, we excluded participants with bad discrimination performance. We eliminated data from 3 participants (64%, 66%, 73%) setting a criterion of 75% correct, on average.

Mainly, we analysed initial peak forces during the exploration of each stimulus pair, indicating the use of prior information [3]. In the first step, we smoothed the measured normal force values with a moving-averaging window with a kernel of 45 ms [15]. Afterwards, we captured the force maxima by detecting turning points where the first derivative of force shifted from positive to negative over time. We filtered out force maxima below 5 N (N) and defined that the time interval between two valid peak forces has to exceed 180 ms. With these restrictions we aimed to minimize artifacts in the data stemming from local maxima, small finger tremor movements or movement rests occurring after valid indentations. In case there was more than one local maximum force within the specified time frame, the highest maximum was considered as peak force. The initial peak force was the very first peak force in a trial, the maximum force captured at least 180 ms after the initial peak force was assigned as second peak force. We applied standardized data filtering procedures consistent with those used in previous research [15] to ensure comparability. In addition to this, inspection of individuals force profiles suggested by all appearances that 180 ms is an appropriate border to dissociate separately planned indentation movements from each other—in line with previous reports suggesting that a single indentation lasts around 200–230 ms [20, 21]. As for the force criterion of 5 N, we also checked for a lower value of 2N, which however did not change our main conclusions.

We performed a repeated-measures ANOVA to compare initial peak forces with Predictability (Predictable and Unpredictable), Length of Sequence (2, 4 and 6) and Compliance category (Hard or Soft) being within-participant variables. p-values and degrees of freedom were corrected according to Greenhouse-Geiser [22] in case the sphericity assumption was violated. Afterwards, we compared initial peak forces between the predictable and unpredictable conditions for each compliance and length of sequence condition with planned paired t-tests. We expected participants to apply higher forces to hard stimuli in the predictable condition versus the unpredictable condition when they use prior information.

Furthermore, we averaged the initial peak forces corresponding to the same trial position in a sequence of certain length (2, 4, or 6) to see how initial peak forces develop across trials. On these data we conducted per length of sequence a within-participants linear contrast over trial positions separately for Hard and Soft softness levels and for Predictable and Unpredictable conditions to check for systematic developments of applied peak force. We expect that forces increase for hard stimuli in the predictable condition, which would show force-tuning, but we do not expect any other developments.

3 Results

The repeated-measure ANOVA on initial peak forces revealed a significant main effect of the Length of Sequences, $F_{2,40} = 3.770$, $p = .032$, $\eta_p^2 = 0.159$ and a significant interaction effect between Predictability and the Length of Sequences, $F_{2,40} = 3.542$, $p = .038$, $\eta_p^2 = 0.150$. These results support our hypothesis that participants adjusted their initial peak forces dependent on the length of the sequence when the compliance of the upcoming stimulus pair was predictable (see in Fig. 2.). Other effects were not significant (main effects of Predictability, $F_{1,20} = 3.542$, $p = .331$, and Softness Level, $F_{1,20} = 3.542$, $p = .091$; interactions Predictability X Level, $F_{1,20} = 2.168$, $p = .156$, Softness Level X Length of Sequences, $F_{1,20} = .204$, $p = .816$, Predictability X Length of Sequences X Softness Level also failed to reach significancy, $F_{1,40} = 1.014$, $p = .372$.

The planned paired t-tests of initial peak forces confirmed the expected results in detail: A) participants applied higher forces to hard stimuli in the predictable condition than in the unpredictable condition for patterns with sequence length 4, $t(20) = 1.996$, $p = .030$, and for patterns with sequence length 6, $t(20) = 1.948$, $p = .033$. B) Also as expected, we did not find significant predictability effects on force for hard stimuli in patterns of sequence length 2, nor for soft stimuli for any sequence length (all $p > .91$).

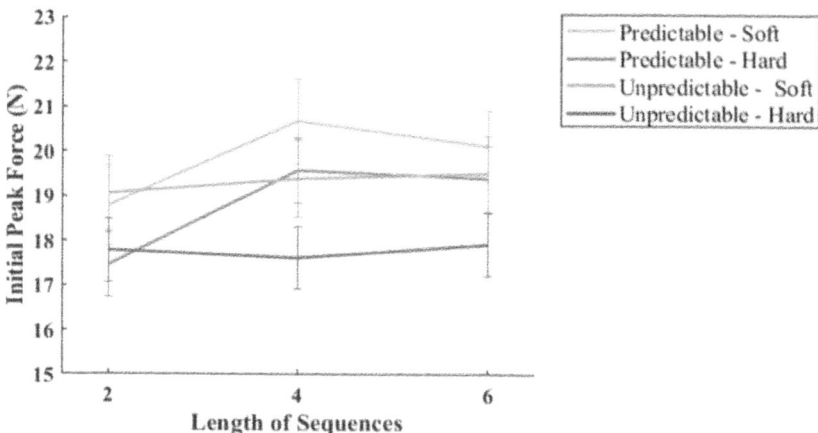

Fig. 2. Initial peak forces as a function of Length of Sequence and Compliance Category. The error bars show the standard error of the mean (SEM). Note that SEMs are based on and represent dispersion between participants and thus provide an additional aspect of the data that is not informative for the significance of within participant analyses.

The averaged initial peak forces as a function of trial position in the sequences of different lengths are depicted in Fig. 3. Within-participant linear contrasts per predictability and softness level revealed the following insights about how initial peak force develop across trials:

A) For Sequence Length 2, there is a linear contrast of increasing forces from the first to the second trial for hard stimuli in the predictable condition as expected, $F_{1,20} =$

4.741, $p = .021$ (one-tailed), $\eta_p^2 = 0.192$, but also for soft stimuli in the unpredictable condition, $F_{1,20} = 12.269$, $p = .002$, $\eta_p^2 = 0.380$ (Fig. 3a) (all other $ps > .209$).

B) For Sequence Length 4, we also found the expected significant linear contrast showing increasing forces for hard stimuli in the predictable condition $F_{1,20} = 11.692$, $p = .002$ (one-tailed), $\eta_p^2 = 0.369$, and unexpectedly for soft stimuli in the unpredictable condition $F_{1,20} = 7.771$, $p = .011$, $\eta_p^2 = 0.280$ (Fig. 3b) (all other $ps > .061$).

C) For Sequence Length 6, we again observed the expected contrast for hard stimuli in the predictable condition, $p = .035$ (one-tailed) (Fig. 3c) (all other $ps > .071$).

Lastly, we analyzed the participants' response to the survey. 56% of the participants did not realize the patterns, further 20% responded there were patterns, but marked the same number of wrong and correct patterns in the questionnaire. Also, none of the remaining participants accurately exclusively marked all the correct patterns. So, overall while a minority of participants might have had limited recognition, most did not.

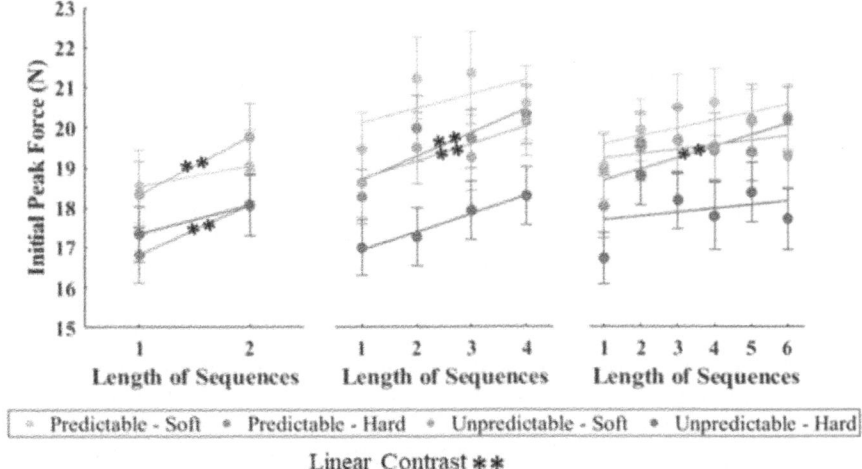

Fig. 3. Initial peak forces as a function of Trial Position for each Length of Sequence: **a)** Sequence Length of 2. **b)** Sequence Length of 4. **c)** Sequence Length of 6. Lines and the significance level are included to graphically represent linear contrasts (lines are actually regression lines, but only serve representational purposes here). The error bars indicate the standard error of the mean.

4 Discussion

In the present study, we investigated how and where prior information on softness level (hard or soft) becomes accessible for successful force tuning in softness perception. Previous studies had indicated that participants apply higher force to harder stimuli versus softer stimuli (=force tuning) when stimulus pairs were presented in longer sequences with trials of the same softness level [3, 14]. An inverse effect was observed when soft and hard pairs alternate in every, or every other trial [15]. To provide insight into the mechanisms of force tuning and to understand how and where the prior information becomes accessible, we varied the length of sequences from the same softness level to be 2, 4 or 6 trials and presented them in predictable and unpredictable conditions. We found out that participants successfully adjusted their force by applying higher force to harder stimuli when the longer sequences followed a predictable pattern compared to the unpredictable pattern. This confirms that anticipatory mechanisms are involved in force-tuning in softness perception. According to the questionnaire, most participants were not aware or did not recognize the presentation patterns, suggesting that we studied implicit mechanisms here.

Note that, in [15], authors compared the initial peak forces applied to the hard and soft stimuli across softness levels. In the present study, we compared hard and soft stimuli only with themselves between predictable and unpredictable conditions to investigate anticipatory mechanisms by studying hard and soft stimulus independent from each other. Overall, our results are in line with the previous findings: If participants anticipated that they will explore harder stimuli in the predictable condition, they exerted a higher peak force in the very initial contact with the stimuli compared to unpredictable conditions. We found this effect between predictable and unpredictable conditions only if the same-softness sequences were a length of 4 or 6 trials. We already knew that longer blocks result in successful force tuning. The present results now specify that force tuning is already successful when same-softness sequences are just 4 trials long, whereas a length of 2 is insufficient as already observed by [15].

The analyses of the development of initial peak forces across trial positions showed that participants apply increasing force to the hard stimuli in the predictable condition during each same-softness sequence of lengths of 2, 4 and 6. We did not observe this linear contrast for hard stimuli in unpredictable conditions, nor for soft stimuli in the predictable conditions. These findings are well in line with our conclusions from previous studies. The increasing forces over trials for hard stimuli fit with the expectation that participants apply higher forces when they expect to explore hard stimuli. The lack of this linear trend in the unpredictable conditions and in the predictable soft conditions is in line with a crucial role for anticipatory mechanisms: Only if harder stimuli can be really predicted, participants tune their forces by increasing them. Moreover, we know from previous study [15] and from our results that sequence length of 2 is not long enough for successful force-tuning, which we also did not observe in the overall analysis. However, the linear contrast observed for sequence length 2 promotes a view that participants still use implicit learning cues to trigger the process to optimize their sensory intake. Taken together, the analysis of peak forces over position, and the overall findings on the effect of predictability suggest an anticipatory mechanism of force tuning, which however needs to be triggered by an initial stimulus of the predictable stimulus level. Only, then forces will be slowly tuned over about 2–3 trials. This may be because the participants were learning whether the sequences were predicably longer or not, rather than learning exactly how long the sequences were. As a result, based on predictability of some sequence, they might undertake the effort to fine-tune their forces.

Note that we also observed unexpected trends: Participants applied increasing forces over trial position to the soft stimuli in the unpredictable condition with sequence lengths of 2 and 4. One idea could be that this again results from optimizing of sensory intake but this time because of uncertainty: participants might develop a strategy to apply a high force to the following stimulus, because a hard stimulus might appear in the next trial. However, this is highly speculative and we did not observe such a strategy for unpredictable sequences of length 6. Note besides that 80% of our participants were females. However, it seems very unlikely that our results are considerably biased by this unbalanced gender distribution, because psychological studies including a study on haptic perception [23] hardly report gender differences, least of all ones of considerable magnitude.

Considering all these findings, our present knowledge suggests that usage of implicit prior information and anticipatory mechanisms play a crucial role in successful force-tuning for softness perception as long as presentation conditions are stable enough which is sequence length of 4 in our study. This finding enhances the understanding of exploratory mechanisms guided by prior information in haptic perception. In previous studies, it already became evident that the type [14] or the quality of prior information [2] affects the exploration in haptic perception. The present study makes a valuable supplement to previous studies in terms of understanding mid-term mechanisms by that prior information becomes accessible for exploratory control. In particularly, we suggest that there is an anticipatory and incremental mechanism of force tuning based on prior information, which however needs to be triggered by an initial stimulus corresponding to the predicted stimulus conditions. These insights could have practical implications to improve haptic guidance systems such as robotic arms and artificial robotic skin. Robots can use the prior tactile information while learning new objects [24]. A comprehensive understanding of mid-term mechanisms regarding usage prior information in human haptic perception is essential in advancing the development of these systems, also in terms of implementing optimized learning and exploration process. Future studies are needed to explore whether similar mechanisms apply to other types of prior information. Last but not least, it may also be of future interest to investigate the roles of memory and learning over longer periods of prior information use.

Acknowledgments. This research was supported by the DFG (SFB/TRR135/1-3, A05, project 222641018). The authors would like to thank Aleksandra Mijailovic for the data collection and Marai Soehngen and Marlene Sophia Oster for stimulus preparation.

References

1. Lederman, S.J., Klatzky, R.L.: Hand movements: a window into haptic object recognition. Cogn. Psychol. **19**, 342–368 (1987)
2. Jeschke, M., Zöller, A.C., Drewing, K.: Influence of prior visual information on exploratory movement direction in texture perception. Haptics: Sci. Technol. Appl. 30–38 (2022)
3. Kaim, L., Drewing, K.: Exploratory strategies in haptic softness discrimination are tuned to achieve high levels of task performance. IEEE Trans. Haptics **4**, 242–252 (2011)
4. Reber, A.S.: Implicit learning and tacit knowledge. J. Exp. Psychol. Gen. **118**, 219–235 (1993)
5. Berry, D.C., Broadbent, D.E.: On the relationship between task performance and associated verbalizable knowledge. Q. J. Exp. Psychol. Sect. A **36**, 209–231 (1984)
6. Dienes, Z., Perner, J.: A theory of implicit and explicit knowledge. Behav. Brain Sci. **22**, 735–808 (1999)

7. Sun, R., Slusarz, P., Terry, C.: The interaction of the explicit and the implicit in skill learning: a dual-process approach. Psychol. Rev. **112**, 159–192 (2005)
8. Nissen, M.J., Bullemer, P.: Attentional requirements of learning: evidence from performance measures. Cogn. Psychol. **19**, 1–32 (1987)
9. Baird, J., Stewart, J.C.: Sequence-specific implicit motor learning using whole-arm three-dimensional reach movements. Exp. Brain Res. **236**, 59–67 (2017)
10. Bergmann Tiest, W.M., Kappers, A.M.L.: Analysis of haptic perception of materials by multidimensional scaling and physical measurements of roughness and compressibility. Acta Psychologica **121**, 1–20 (2006)
11. Dövencioglu, D., Doerschner, K., Drewing, K.: Aspects of material softness in active touch, Trieste, 2018
12. Cavdan, M., Doerschner, K., Drewing, K.: The many dimensions underlying perceived softness: how exploratory procedures are influenced by material and the perceptual task. In: 2019 IEEE World Haptics Conference (WHC) (2019)
13. Srinivasan, M.A., LaMotte, R.H.: Tactual discrimination of softness. J. Neurophysiol. **73**, 88–101 (1995)
14. Zoeller, A.C., Lezkan, A., Paulun, V.C., Fleming, R.W., Drewing, K.: Integration of prior knowledge during haptic exploration depends on information type. J. Vis. **19**, 20 (2019)
15. Drewing, K., Zoeller, A.C.: Influence of presentation order on force control in softness exploration. In: 2021 IEEE World Haptics Conference (WHC) (2021)
16. Mawase, F., Karniel, A.: Evidence for predictive control in lifting series of virtual objects. Exp. Brain Res. **203**, 447–452 (2010)
17. Wolpert, D.M., Flanagan, J.R.: Motor prediction. Current Biol. **11** (2001)
18. Faul, F., Erdfelder, E., Buchner, A., Lang, A.-G.: Statistical Power analyses using G*Power 3.1: Tests for correlation and regression analyses. Behav. Res. Methods **41**, 1149–1160 (2009)
19. Gerling, G.J., Hauser, S.C., Soltis, B.R., Bowen, A.K., Fanta, K.D., Wang, Y.: A standard methodology to characterize the intrinsic material properties of compliant test stimuli. IEEE Trans. Haptics **11**, 498–508 (2018)
20. Metzger, A., Lezkan, A., Drewing, K.: Integration of serial sensory information in haptic perception of softness. J. Exp. Psychol. Hum. Percept. Perform. **44**, 551–565 (2018)
21. Lezkan, A., Metzger, A., Drewing, K.: Active haptic exploration of softness: indentation force is systematically related to prediction, sensation and motivation. Front. Integr. Neurosci. **12** (2018)
22. Greenhouse, S.W., Geisser, S.: On methods in the analysis of profile data. Psychometrika **24**, 95–112 (1959)
23. Drewing, K.: Perceptuo-affective organization of touched materials in younger and older adults. PLoS ONE **19**, e0296633 (2024)
24. Feng, D., Kaboli, M., Cheng, G.: Active prior tactile knowledge transfer for learning tactual properties of new objects. Sensors **18**, 634 (2018)

Discovering the Causal Structure of Haptic Material Perception

Jaime Maldonado[1]([✉])[ID], Christoph Zetzsche[1][ID], and Vanessa Didelez[2][ID]

[1] Cognitive Neuroinformatics, University of Bremen, Bremen, Germany
`jmaldonado@uni-bremen.de`
[2] Department of Biometry and Data Management, Leibniz Institute for Prevention Research and Epidemiology - BIPS, Bremen, Germany

Abstract. The sensory signals that occur when we touch or interact with objects carry the information necessary to perceive and reason about object properties. Research on material perception has provided evidence that humans can categorize materials and assess their similarity based solely on haptic information. This evidence is based on the performance on classification tasks and correlation analyses, which, by definition, provide no information on the causes of the observed behavior. This paper explores the use of causal discovery methods to analyze human haptic perception of material categories. Causal discovery algorithms analyze statistical patterns in the data, such as conditional (in)dependence relationships, and then determine causal relationships between the variables that are compatible with these patterns. The result is a set of causal graphs with nodes representing the variables and directed edges representing empirically plausible causal relationships. In this paper, a causal discovery algorithm is used to analyze material category ratings and vibratory signals from haptic exploration. The goal is to understand the underlying cause-effect structure linking material samples, vibration signals, and category similarity ratings. The identified causal structure indicates that the information represented by the slope of the vibratory signal plays a key role in rating a material's similarity to different categories, but in parts, it is only an indirect cause. The practical use of causal discovery methods for analyzing haptic perception data is demonstrated.

Keywords: haptic perception · material category · material similarity · causality · causal discovery

1 Introduction

Humans can perceive the physical properties of objects (material and structural) by touching their surfaces thanks to the mechanical and thermal receptors located in the hand and fingertips [10]. Haptic information is necessary to perceive object properties and reason about them, for example, to make judgments about material classes or properties such as friction or elasticity [3,10]. Research

H. Kajimoto et al. (Eds.): EuroHaptics 2024, LNCS 14768, pp. 171–184, 2025.
https://doi.org/10.1007/978-3-031-70058-3_14

in material classification has shown that humans can categorize materials based solely on the haptic information generated during unconstrained exploratory movements in experimental settings where participants could neither see nor hear the interaction with the material samples [3,21]. The analysis of classification performance has shown that, although humans achieve better-than-chance performance, misclassification occurs often [3,21]. Additionally, correlation analyses have shown that perceived material categories are associated with each other [3] and that vibratory signals obtained from the classified materials are associated with haptic properties [21].

Correlations have limited explanatory power as they provide no information about the cause(s) of the observed associations between variables. Furthermore, it has been argued that results obtained in classification tasks do not provide explanations of the cognitive phenomena behind human performance [6]. These limitations can be addressed by applying causal analysis methods to identify cause-effect relationships in the data generation process of a given system. Among the different causal analysis methods, causal discovery algorithms identify the underlying cause-effect relations in data by analyzing its statistical properties [6,7]. When applied to analyze human behavior, the learned causal relations can provide insight to understand the cognitive mechanisms that generate the observed data [6,11].

In this paper, we explore the use of causal discovery methods to analyze human haptic perception. We apply a causal discovery algorithm to analyze the material category ratings and vibratory signals from haptic exploration provided in the *ViPer* database [21]. In particular, we are interested in the cause-effect structure that links the material samples, vibration signals, and category similarity ratings. We demonstrate the practical use of causal discovery methods for analyzing haptic perception and category similarity. To the best of our knowledge, no other study has applied causal discovery methods in the context of haptic perception of material categories.

We learn a causal structure that provides insight into the cognitive mechanisms underlying the perception of material similarity based on haptic exploration. Our results indicate that the information contained in a feature of the vibratory signal represents the information used by participants to rate a material's similarity to different categories. This feature is a direct cause for some categories and an indirect cause for others.

2   Related Work and Background

This section summarizes the sensory mechanisms that enable humans to perceive object properties from the vibrations elicited while performing exploratory movements. Subsequently, we review the features computed from vibration signals acquired during exploratory movements that represent surface properties relevant to the analysis of human perception. Finally, we provide a short introduction to causal analysis and discovery methods. We introduce the formalism used throughout the paper to represent causal models and highlight how causal discovery algorithms retrieve cause-effect relations from data that cannot be identified using associative (e.g., correlation) analyses.

2.1 Perception of Material Category Based on Haptic Exploration

When touching an object, its surface characteristics are perceived by the skin deformation patterns in the fingertips [5,10,22]. While coarse features can be perceived just by touching, movement is necessary to perceive fine textural features [8,9]. Running the fingers over a surface elicits skin vibrations, which reflect the microstructure of the surface [8,22]. These vibrations convey the information necessary to perceive fine textural properties such as roughness and to assess (dis)similarity between materials [4,5,8,22]. Information about a texture's fine properties is encoded in the responses of the rapidly adapting tactile nerve fibers of type I and II, associated with Meissner and Pacinian corpuscles, respectively, which are highly sensitive to skin vibrations [5,10,22]. It has been observed that the ability to discriminate different textures and perceive their characteristics is independent of the speed of the exploratory movement [5,10]. Evidence from the analysis of neural activity in primates indicates that perceptual constancy across different speeds is achieved by a systematic dilation/contraction of the spiking patterns of afferent responses [22].

Human material classification based on haptic exploration has been investigated using experimental setups where participants could freely explore different materials but could neither see nor hear the interaction [3,21]. In a setup where participants could explore materials from seven material classes (plastic, paper, fabric, fur and leather, stone, metal, and wood) with their bare hands, it was observed that materials with similar surface properties were confused most often and that, on average, 66% of the stimuli were consistently assigned to their material class [3]. The authors conclude that haptic information alone does not allow perfect material recognition.

The perception of material category based on indirect haptic perception has been investigated using a setup where participants explored materials using a hand-held tool [21]. The tool recorded the vibration signals during the haptic exploration. After the exploration, participants were asked to rate the sample's similarity to each of the following categories: wood, plastic, fabric, paper, metal, stone, and animal. A classifier of the perceived material category was trained using the vibration signals. The slope of the signals' frequency spectrum was used as a classification feature, achieving an accuracy of 38.27% (empirical chance level = 16.67%). Based on these results, the authors indicate that perception is partly based on the signal's slope, where materials showing similar slopes can explain human misclassifications. It is important to note that when a tool (e.g., a knife) is used to explore an object, haptic perception is regarded as remote or indirect [10].

2.2   Features from Vibration Signals

Acquiring the signals that result during the material exploration enables the joint analysis of human subjective perception and the signal features (e.g., [21]). In experimental settings, measurement instruments are equipped with acceleration sensors to acquire vibration signals during haptic exploration (e.g., [18,19,21]). The vibration signal is typically summarized with hand-crafted features representing surface properties (e.g., hardness or roughness) [19,20]. In contrast to the features corresponding to specific material properties, it has been proposed to use the slope of the vibration signal's frequency spectrum [21]. Based on a simulation analysis, it has been determined that the slope feature correlates with the activity of the rapidly adapting fibers of type I and II [21], which sense temporal changes in skin deformation for vibration detection and fine texture perception [10]. Additionally, it has been shown that the slope feature varies systematically between material categories and correlates with the perceived properties of the material (roughness, hardness, elasticity, and friction) [21].

2.3   Causal Models and Causal Discovery

Causal analysis methods enable the evaluation and modeling of a data-generating process in terms of cause-effect relationships. Associations, i.e., dependencies, between variables in a data-generating process might have different causal explanations. Considering two associated variables, X and Y, the following scenarios are possible: 1) X causes Y, 2) Y causes X, or 3) there is a third variable that causes both X and Y. An associative analysis describes and quantifies the extent to which variables are associated, irrespective of any existing or absent cause-effect relations. The methods specifically developed for causal analysis make use of the fact that the absence of a marginal or conditional dependence indicates (under weak assumptions) the absence of indirect or direct causal relations and thus can at least partially reconstruct the underlying cause-effect structure of a data-generation process [7,13] .

Causal models can be expressed as Directed Acyclic Graphs (DAG) [15], where nodes represent the variables and directed edges (i.e., arrows) their direct causal relations. The nodes can represent categorical, ordinal, or continuous variables. The DAG formalism allows the inclusion of unmeasured variables (also graphically represented). The DAG arrows represent direct causation between variables. For example, $X \rightarrow Y$ indicates that any manipulation of the value of X will have a probabilistic effect on the value of Y when all the other variables are held fixed [7,15]. $X \rightarrow Y$ also indicates that inducing changes in the value of Y (e.g., by forcing Y to take a specific value) will not affect X. The structure of the DAG implies conditional independence statements among the variables, known as the causal Markov assumption [7,15]. This assumption states that a variable is statistically independent of its non-effects, conditional on its direct causes. Consequently, conditional independence tests can be applied to verify or validate causal relations against data [2,7].

Given a system with two variables of interest X and Y, it would be possible to determine the underlying cause-effect relationship by conducting an experiment in which X and Y are manipulated while leaving all the other variables unaltered: while manipulating the cause variable yields changes in the effect, manipulating the effect would leave the cause unchanged. Whenever experimental data are not available and only observational data are provided (i.e., coming from an unmanipulated data-generation process), it is impossible to determine whether $X \rightarrow Y$ or $X \leftarrow Y$ from the purely statistical perspective. Causal discovery algorithms, also known as structure learning algorithms, aim to determine the causal structure that could have produced the given observational data, typically by a systematic analysis of many possible causal structures and testing probabilistic independence and dependence between the variables [6,7,13]. The result is a set of causal graphs with nodes representing the variables and directed edges representing the causal relationships consistent with the statistical properties of the data. The algorithms can also report un- or bidirected edges, which indicate undecidable directionality. Many algorithms have been developed during the last decades for different types of variables (discrete, continuous, or mixed data). While some algorithms assume that all the relevant variables are available in the data, others allow the possibility of unobserved(latent) variables. For recent reviews on causal discovery, see [7,13,23].

3   Materials and Methods

3.1   ViPer Database

The ViPer database[1] contains vibratory signals and perceptual judgments of material category similarity from haptic exploration [21]. Eleven naive participants explored 81 material samples (14 × 14 cm) belonging to seven material categories: wood, plastic, fabric, paper, metal, stone, and animal. There are twelve different material samples within each category, except for the metal category, which has only nine samples [21]. The animal category includes fur and leather samples. Participants explored the material samples with a custom pen equipped with a steel tip and an accelerometer to record the vibrations during the exploration movement. Each participant explored each material sample once, resulting in 891 exploration trials.

Participants were instructed to explore each material sample freely for 14 s. During the material exploration, participants wore earplugs and headphones, and their hands were occluded, so they could not see or hear the interaction with the material sample [21]. For each trial, a 1D vibratory signal of 10 s, computed from the 3D accelerometer signal, is provided. The vibratory signals were recorded with a sampling rate of 3200 Hz [21].

At the end of each exploration trial, participants were requested to rate the similarity of the material sample to each of the seven material categories [21].

---

[1] The database is publicly available at https://github.com/matteo-toscani-24-01-1985/ViPer, access 10.11.2023.

The database contains the similarity ratings as continuous values ranging from 0 (very different) to 10 (very similar). Further details about the data acquisition setup are also provided in [12].

3.2 Discovery Variables

The variables processed with the causal discovery algorithm represent the factors involved in rating the similarity of a material exemplar to different categories. The variables are obtained from the ViPer database described in Sect. 3.1. The first variable is the *category*, so to say the ground truth, which describes the material category of the sample explored in each trial. It comprises 7 levels: wood, plastic, fabric, paper, metal, stone, and animal.

The second variable is the *slope* feature [21], which aims to represent the vibrations elicited during the exploration of the material sample, which carry information about its textural properties. The use of this feature to represent vibration signals has been motivated by the observation that the spectral power P of the signal is related to the temporal frequency f following a power law $P = 1/f^s$ [21]. Following the procedure described in [21], computing the natural logarithm to both sides of the equation yields $ln(P) = -s \cdot ln(f)$. Thus, the parameter s, i.e., the slope, can be retrieved by fitting a line in the transformed space. It has been shown that the *slope* feature provides a concise measure of the vibratory signals elicited during the haptic exploration of materials and is correlated with the activity of the skin afferents involved in texture perception and human perceptual category judgments [21].

Finally, the scores of the perceived similarity provided by the subjects after exploring each material sample are included in the following variables: *wood*, *plastic*, *fabric*, *paper*, *metal*, *stone*, and *animal*. These variables contain continuous values ranging from 0 (very different) to 10 (very similar).

Figure 1 illustrates a hypothetical DAG containing the variables selected for causal discovery. Subjects were presented with material samples from different categories. The actual category of each sample was unknown to the subjects. The samples were perceived through the vibration of the pen. It is assumed that materials from different categories cause distinctive vibration patterns, which are represented by the *slope* feature. This is expressed in the edge *category* $\rightarrow$ *slope*, which indicates that manipulating the material's *category* yields changes in the *slope* of the vibration signal. The vibration signal, represented by the *slope* feature, provided sensory evidence of the sample's properties. The graph represents the assumption that this evidence is directly used by the subjects in their subjective rating of the sample's similarity to the seven categories. This is expressed in the edges from *slope* to the *wood*, *plastic*, *fabric*, *paper*, *metal*, *stone*, and *animal* nodes, which contain ratings of the perceived similarity. These edges indicate that any change in the *slope* produces changes in the similarity ratings. The DAG expresses two key assumptions: that the information participants use is entirely captured by the *slope*, i.e., there is no further information on the category used to come to the perceptions (no edges from *category* to any of the seven similarity scores); and that the dependence between the material ratings is entirely captured by the slope (no latent variables explaining correlations between the ratings). This hypothetical DAG can be contrasted against the discovered causal structure described in Sect. 4.

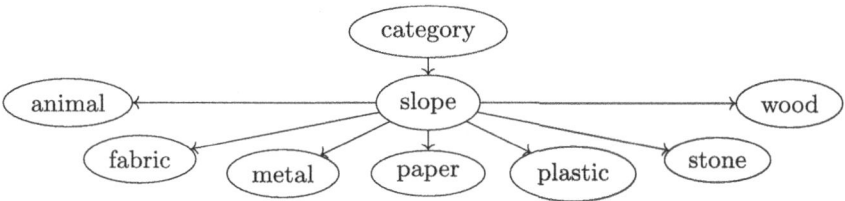

Fig. 1. Hypothetical DAG of the data-generating process of material similarity rating. The node *category* describes the category of the explored material. The node *slope* corresponds to the slope of the vibration signal. Finally, the nodes *wood, plastic, fabric, paper, metal, stone,* and *animal* contain the scores of the perceived similarity of the explored material to each category.

3.3   Causal Discovery Algorithm

We analyze the selected variables with the Greedy Fast Causal Inference (GFCI) algorithm [14]. This algorithm is a modification of the Fast Causal Inference algorithm (FCI), a well-established discovery method [7,11,13], which improves accuracy on small sample sizes. The GFCI algorithm allows for discovering causal relations in datasets containing mixed continuous and discrete data and discovering latent (also termed *unmeasured*) confounders[2]. In a nutshell, the GFCI starts with an undirected graph where all the variables are fully connected. This initial graph is pruned by performing a sequence of statistical tests to remove the edges that connect conditionally independent variables, and the remaining edges are subsequently oriented based on a set of rules (for a detailed description of the algorithm, see [14]). Table 1 summarizes the output edge types of the GFCI algorithm and the interpretation of the relationship between the variables they represent.

We use the implementation of the GFCI algorithm available in Tetrad[3] (version 7.6.1-0), a software toolbox for causal discovery [17]. In order to obtain reliable results, we chose the algorithm parameters based on configurations reported in benchmarking studies of causal discovery on mixed continuous and categorical data [1,16]. Furthermore, we validate the stability and reliability of the causal relationships inferred from the data by performing a bootstrapping analysis [7].

---

[2] Given two variables X and Y, an *unmeasured confounder* is an unobserved variable U that causes both X and Y, that is, $U \rightarrow X$ and $U \rightarrow Y$.

[3] Available at: https://www.ccd.pitt.edu/tools/ , access: 29.11.23.

Table 1. Graph edge types discovered by the GFCI algorithm.

Edge type	Present relationship	Absent relationship
$A \rightarrow B$	A is a cause of B. Also, there may be an unmeasured confounder of A and B.	B is not a cause of A.
$A \leftrightarrow B$	There is an unmeasured confounder of A and B.	A is not a cause of B. B is not a cause of A.
$A \circ\!\!\rightarrow B$	Either A is a cause of B (i.e, $A \rightarrow B$) or there is an unmeasured confounder of A and B (i.e, $A \leftrightarrow B$) or both.	B is not a cause of A.
$A \circ\!\!-\!\!\circ B$	Exactly one of the following holds: 1. *A is a cause of B* 2. *B is a cause of A* 3. *there is an unmeasured confounder of A and B* 4. both 1 and 2 5. both 3 and 3	
$A \xrightarrow{NLC} B$	A is a cause of B and there is no latent confounder. Also, A may not be a direct cause of B.	B is not a cause of A.
$A \xrightarrow{DD} B$	A is definitely a direct cause of B and there is no latent confounder.	B is not a cause of A.

We use the Degenerate Gaussian Likelihood Ratio Test (DG-LRT) [1], which has shown good performance on causal discovery with datasets containing both categorical and continuous variables and small sample sizes [1]. The decision threshold α for the DG-LRT was set to 0.01. This parameter indicates the value at which the test results are regarded as dependent. The *penalty discount* parameter controls for false positives and negative edges [16]. We set this parameter 4, as used in benchmarking studies [1,16], where it has shown good discovery performance compared to lower and higher values. To strengthen the confidence in the discovery results, internally we also assessed the results obtained when the penalty discount is set to 3 and 5 and obtained similar results. Therefore, we report the results with the penalty discount set to 4. We also observed that setting this parameter to values ≤ 2 and ≥ 6 led to qualitatively different results, yielding DAGs with more and fewer edges, respectively, which could be due to false positive and false negative edge detections, as reported in benchmarking studies [16].

The Tetrad implementation of the GFCI algorithm allows the incorporation of background knowledge about the precedence of the variables (e.g., X occurs before Y; therefore, Y cannot cause X). Precedence is set by grouping variables into tiers, which specify the temporal order in which the variables occur. We allocate the variables in three tiers. The first tier contains the *category*, the second contains the *slope*, and the third contains the seven similarity scores.

In order to validate the causal relationships inferred from the data we performed a bootstrapping analysis. Bootstrapping provides an estimate of the stability and reliability of the causal relationships inferred from data. If the results vary widely over the different bootstrap samples, the output of the algorithm is regarded as unstable [7]. We run the discovery algorithm on 500 bootstraps with 80% of the records in each bootstrap. As a result, we obtain 500 different structures. As a summary of the bootstrapping results, we report the frequency of the edge types between variables. For ease of interpretation and comparison, the frequency of edge type is reported as a proportion of the number of bootstraps. The edge-type frequencies obtained from bootstrapping indicate whether the discovered causal relationships are stable across different samples of the data [7]. Given the different edge types (including "no edge"), we interpret an edge frequency larger than 0.5 as stable. Finally, we present a DAG which includes the stable edges, termed *discovered DAG*.

4   Results

In this section, we report the frequency of the edge types discovered by the GFCI algorithm. The frequency of the edge type is indicated in parentheses. The edge *category* $\circ\!\!\rightarrow$ *slope* (1.00) was discovered in all the bootstraps (see Fig. 2(a)). This edge indicates that the data support the assumption that material samples from different categories have an effect on the slope of the vibratory signal.

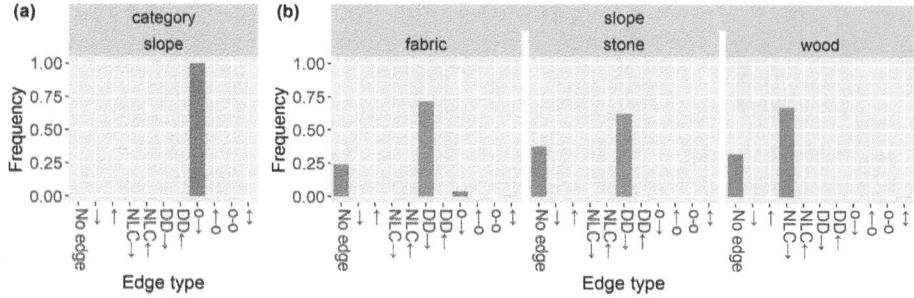

Fig. 2. (a) Edge types between the sample *category* and vibratory signal's *slope*. (b) Edge types between the vibratory signal's *slope* and the category similarity scores.

Figure 2(b) shows the edge types discovered between the vibratory signal *slope* and the category similarity scores. In all the bootstraps, no edges were discovered between the *slope* and the *animal*, *metal*, *paper*, and *plastic* similarity scores; thus, they are not shown in the figure for simplicity. The edges *slope DD*→ *fabric* (0.72) and *slope DD*→ *stone* (0.62) provide evidence that the information represented with the slope of the vibratory signal is the direct cause of the perceived similarity. The edge *slope NLC*→ *wood* (0.67) indicates no latent confounder, and that *slope* causes the perceived similarity to *wood*.

Figure 3 shows the edge types discovered among the category similarity scores. For simplicity, the figure only shows the variable pairs with edge types other than *no edge*. The edge *animal DD→ wood* (0.67) indicates that the perceived similarity to *animal* is a direct cause of the similarity to *wood*. The edges *fabric NLC→ animal* (0.72) and *fabric NLC→ metal* (0.95) indicate that the perceived similarity to *fabric* causes the *animal* and *metal* similarity scores and that there are no latent confounders. The edge frequencies for the pair *fabric − plastic* lay below 0.5 and show partly contradicting causal relations. Therefore, the evidence from the data about the *fabric − plastic* relation is considered as unclear. The edge *metal DD→ animal* (0.54) indicates that the perceived similarity to *metal* is a direct cause of *animal*. The edge frequencies for the pair *stone − fabric* lie below 0.5; thus, their relation is considered as unclear. Finally, the edge *stone DD→ metal* (0.81) indicates that the perceived similarity to *stone* is a direct cause of *metal*. In all the bootstraps, no edges between *paper* and the other variables were discovered.

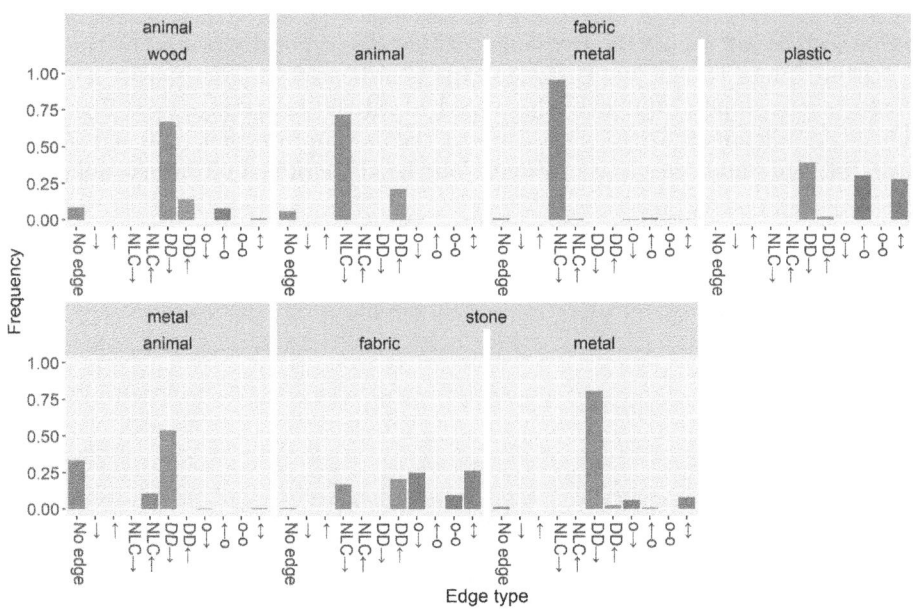

Fig. 3. Edge types between category similarity scores.

The causal discovery results are summarized in Fig. 4. The results indicate that, in general, the information represented in the *slope* variable causes the perceived similarity to different material categories. This causal relationship might be direct (e.g., *slope* to *fabric*) or indirect (e.g., *slope* to *metal* via *fabric* and *stone*). It is important to note that no edges were discovered between the *category* and the category similarity scores in all the bootstraps. This suggests that the variable *slope* mediates all the information between the material sample and the similarity scores.

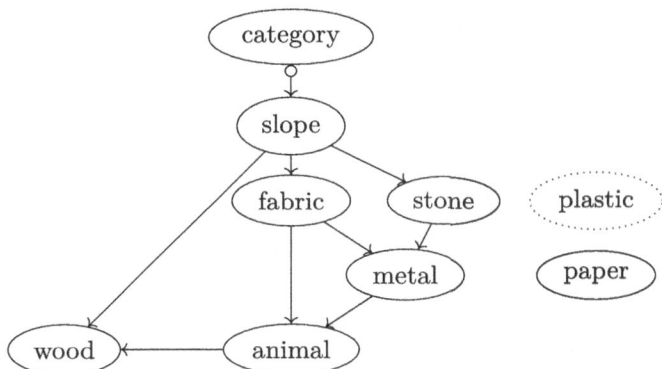

Fig. 4. Discovered DAG. The dotted node *plastic* indicates that the discovery results are unclear about its relation with other variables.

5   Conclusions

In this paper, we applied the GFCI causal discovery algorithm to a mixture of categorical and continuous variables representing the process of rating material similarity based on haptic exploration. The results indicate that *category* causes *slope*; that is, changing the category of the material sample causes a probabilistic change in the value of the vibratory signal's slope. Furthermore, the results indicate that the information represented by the slope causes the perceived material similarity. However, there may be only an indirect cause-effect relationship for some categories and none for *paper* and *plastic*. These results are consistent with previous work in which the slope has been used to classify the actual and the perceived material category [21], where it has been argued that differences in the slope can explain perception and human (mis)classifications. The discovery results differ from the hypothesized data generation process (Fig. 1) in that some similarity scores are only indirect effects of *slope*. Furthermore, the results of the causal relations between *plastic* and the other variables are unclear. This indicates that the data are too ambiguous to distinguish any edge type.

Regarding the interpretation of the discovered causal relations, it is important to recall that causal discovery methods can not demonstrate causality. Rather, the algorithms determine potentially underlying causal relations that correspond to the statistical properties of the data, specifically, the conditional (in)dependencies [7,11,13]. The learned causal structure can provide insights into the data-generating process. For example, the discovered edges *fabric* → *animal* ← *metal* indicate that the perceived similarity to *fabric* and *metal* causes the similarity assigned to *animal*. These causal relations suggest a structure compatible with the observed data of the internal computations performed by the subjects to complete the perceptual task. If we take the learned causal structure at face value, it suggests that humans decide (mainly using the slope) first on the similarities to *fabric* and *stone*, and these judgments may then suggest or rule out *metal*, *animal*, and finally, *wood*. Latent variables (so other information used for the assessments not captured by slope) can, however, not be ruled out.

In addition, it has to be noted that any causal structure may be compatible with different cognitive models. For example, the causal structure may reflect a sequential process in which the similarities are determined in a particular temporal order. The similarities to *fabric* and *stone* may be determined first, then the similarity to *metal*, and so forth. However, the similarities may also be computed in a network-like fashion, where the network activations converge towards an equilibrium state, and the causal relations reflect, e.g., a dominant influence of the *fabric* and *stone* units on the *metal* unit.

In general, the learned causal relations can be regarded as hypotheses that can be further tested and validated, for example, by conducting further experiments [7,11]. For example, for relations like *stone* → *metal* → *animal* (see Fig. 4) may suggest the temporal precedence of similarity judgments between categories. Thus, further experiments could be conducted to investigate the order in which subjects submit their answers or the effect of the categories' ordering in the response display[4]. The variables used for causal discovery can be further examined. We used the slope of the vibration signal as a proxy of the representation of the signal used by participants to rate the similarity of the sample to a category. Further work can be used to assess other signal features, like those proposed in [19], and determine their possible causal relation to the perceived similarity. The results of causal discovery depend on the characteristics of the algorithm and, crucially, its parameters [7,11]. For a given algorithm, using different statistical conditional independence tests might yield different results. We chose the algorithm based on the characteristics of the data (891 samples of mixed categorical and continuous variables) and its parameters based on benchmarking results reported in the literature [1,14]. We followed the approach of reporting stable results over a range of settings [11]. The graph reported in Fig. 4 is based on edge types discovered with a frequency larger than 0.5 over 500 bootstraps, aiming to provide a summary of the stable discovered edges, thus informing about the confidence in the causal structure [7].

The conditions of haptic perception experiments differ from real-world situations where individuals use visual, haptic, and acoustic cues to assess material properties. It could be assumed that humans can categorize materials based on an internal haptic representation available to cope with the task. Such a unimodal representation could be constructed by enabling a learning phase where participants could establish how materials from different categories feel. How-

---

[4] No details about the interface provided to the subjects to give the ratings are provided with the ViPer database.

ever, uni-modal material categorization is likely to result from the interplay between haptic information and high-level cognitive mechanisms (e.g., memory or heuristics) in the lack of feedback or a learning phase. Baumgartner et al. [3] discuss the potential role of heuristics or lifelong associations (e.g., we always experience that metal feels cold) in participants judging category membership based on uni-modal perception. The discovered causal structure provides insight into the human reasoning process behind haptic-based material perception, presenting causal relations that can be tested and validated in further experiments.

Acknowledgments. The research reported in this paper has been supported by the German Research Foundation DFG, as part of Collaborative Research Center (Sonderforschungsbereich) 1320 Project-ID 329551904 "EASE - Everyday Activity Science and Engineering", University of Bremen (http://www.ease-crc.org/). The research was conducted in subproject H01 - Sensorimotor and Causal Human Activity Models for Cognitive Architectures.

Disclosure of Interests. The authors have no competing interests to declare that are relevant to the content of this article.

References

1. Andrews, B., Ramsey, J., Cooper, G.F.: Learning high-dimensional directed acyclic graphs with mixed data-types. In: Proceedings of Machine Learning Research. Proceedings of Machine Learning Research, vol. 104, pp. 4–21. PMLR (2019), https://proceedings.mlr.press/v104/andrews19a.html
2. Ankan, A., Wortel, I.M.N., Textor, J.: Testing graphical causal models using the r package "dagitty". Curr. Protoc. **1**(2), e45 (2021). https://doi.org/10.1002/cpz1.45
3. Baumgartner, E., Wiebel, C.B., Gegenfurtner, K.R.: Visual and haptic representations of material properties. Multisens. Res. **26**(5), 429–455 (2013). https://doi.org/10.1163/22134808-00002429
4. Bensmaia, S., Hollins, M.: Pacinian representations of fine surface texture. Percept. Psychophys. **67**(5), 842–854 (2005). https://doi.org/10.3758/bf03193537
5. Bensmaia, S.J., Hollins, M.: The vibrations of texture. Somatosens. Motor Res. **20**(1), 33–43 (2003). https://doi.org/10.1080/0899022031000083825
6. Danks, D., Davis, I.: Causal inference in cognitive neuroscience. WIREs Cogn. Sci. **14**(5), e1650 (2023). https://doi.org/10.1002/wcs.1650
7. Glymour, C., Zhang, K., Spirtes, P.: Review of causal discovery methods based on graphical models. Front. Genet. **10**, 524 (2019). https://doi.org/10.3389/fgene.2019.00524
8. Greenspon, C.M., McLellan, K.R., Lieber, J.D., Bensmaia, S.J.: Effect of scanning speed on texture-elicited vibrations. J. R. Soc. Interface **17**(167), 20190892 (2020). https://doi.org/10.1098/rsif.2019.0892
9. Hollins, M., Bensmaia, S.J., Washburn, S.: Vibrotactile adaptation impairs discrimination of fine, but not coarse, textures. Somatosens. Motor Res. **18**(4), 253–262 (2001). https://doi.org/10.1080/01421590120089640
10. Lederman, S.J., Klatzky, R.L.: Haptic perception: a tutorial. Attent. Percept. Psychophys. **71**(7), 1439–1459 (2009). https://doi.org/10.3758/app.71.7.1439

11. Malinsky, D., Danks, D.: Causal discovery algorithms: a practical guide. Philos. Compass **13**(1), e12470 (2017). https://doi.org/10.1111/phc3.12470

12. Metzger, A., Toscani, M.: Unsupervised learning of haptic material properties. eLife **11**, e64876 (2022). https://doi.org/10.7554/elife.64876

13. Nogueira, A.R., Pugnana, A., Ruggieri, S., Pedreschi, D., Gama, J.: Methods and tools for causal discovery and causal inference. WIREs Data Min. Knowl. Discov. **12**(2), e1449 (2022). https://doi.org/10.1002/widm.1449

14. Ogarrio, J.M., Spirtes, P., Ramsey, J.: A hybrid causal search algorithm for latent variable models. In: Antonucci, A., Corani, G., Campos, C.P. (eds.) Proceedings of the Eighth International Conference on Probabilistic Graphical Models. Proceedings of Machine Learning Research, vol. 52, pp. 368–379. PMLR, Lugano, Switzerland (2016). https://proceedings.mlr.press/v52/ogarrio16.html

15. Pearl, J.: Causal Diagrams and the Identification of Causal Effects, chap. 3, pp. 65–106. Cambridge University Press (2009). https://doi.org/10.1017/cbo9780511803161.005

16. Ramsey, J., Glymour, M., Sanchez-Romero, R., Glymour, C.: A million variables and more: the fast greedy equivalence search algorithm for learning high-dimensional graphical causal models, with an application to functional magnetic resonance images. Int. J. Data Sci. Anal. **3**(2), 121–129 (2016). https://doi.org/10.1007/s41060-016-0032-z

17. Ramsey, J.D., et al.: Tetrad-a toolbox for causal discovery. In: 8th International Workshop on Climate Informatics (2018)

18. Strese, M., Boeck, Y., Steinbach, E.: Content-based surface material retrieval. In: 2017 IEEE World Haptics Conference (WHC). IEEE (2017). https://doi.org/10.1109/whc.2017.7989927

19. Strese, M., Brudermueller, L., Kirsch, J., Steinbach, E.: Haptic material analysis and classification inspired by human exploratory procedures. IEEE Trans. Haptics **13**(2), 404–424 (2020). https://doi.org/10.1109/toh.2019.2952118

20. Strese, M., Schuwerk, C., Iepure, A., Steinbach, E.: Multimodal feature-based surface material classification. IEEE Trans. Haptics **10**(2), 226–239 (2017). https://doi.org/10.1109/toh.2016.2625787

21. Toscani, M., Metzger, A.: A database of vibratory signals from free haptic exploration of natural material textures and perceptual judgments (ViPer): analysis of spectral statistics. In: Haptics: Science, Technology, Applications, pp. 319–327. Springer International Publishing (2022). https://doi.org/10.1007/978-3-031-06249-0_36

22. Weber, A.I., Saal, H.P., Lieber, J.D., Cheng, J.W., Manfredi, L.R., Dammann, J.F., Bensmaia, S.J.: Spatial and temporal codes mediate the tactile perception of natural textures. Proc. Natl. Acad. Sci. **110**(42), 17107–17112 (2013). https://doi.org/10.1073/pnas.1305509110

23. Zanga, A., Ozkirimli, E., Stella, F.: A survey on causal discovery: theory and practice. Int. J. Approx. Reason. **151**, 101–129 (2022). https://doi.org/10.1016/j.ijar.2022.09.004

The Visual and Haptic Contributions to Hand and Foot Representation

Lara A. Coelho[1]([✉]) [iD], Anna Vitale[1] [iD], Carolina Tammurello[1], Claudio Campus[1] [iD], Claudia L. R. Gonzalez[2] [iD], and Monica Gori[1] [iD]

[1] Unit for Visually Impaired People, Italian Institute of Technology, Genova, Italy
`lara.coelho@iit.it`
[2] Brain in Action Laboratory, Department of Kinesiology, University of Lethbridge, Lethbridge, AB, Canada

Abstract. Research has shown systematic distortions in hand representation. It has been argued that these distortions reflect the somatosensory homunculus, and it is through vision that the distortions are corrected. However, in those previous studies the haptic and visual tasks used were considerably different from one another. Therefore, in the current study, we devised a task that had identical requirements for each sensory condition. To be specific, the participants haptically and visually deciphered if various sized gloves and shoes were bigger than their hands and feet. We found that participants overestimated their hand size significantly more in the haptic condition, while the feet were estimated similarly between conditions. Moreover, hand distortions in the haptic condition were significantly larger than feet distortions. Lastly, we also found sex differences, as females overestimated both hand and feet size significantly more than males. Taken together, our results support the suggestion that distortions in haptic body representation tasks are more somatotopic.

Keywords: Body representation · vision · haptics · distortions · somatosensory

1 Introduction

The internal representation of the size, shape, and state of the body are characteristics of body representation. While body representations are perceptual in nature, they also provide us with the necessary tools to interact with our environments. In fact, it has been argued that to metrically guide our actions, our brains must rely on a mental image of our body [1]. Intriguingly, these internal representations do not necessarily reflect anatomical body size. For example, it has long been established that the somatosensory homunculus (S1) is a distorted representation of the human body in the cortex [2]. In this representation, body parts with fewer tactile receptive fields are represented as smaller (e.g., back), and those with larger amounts are represented as big (e.g., hands, mouths).

The formation and updating of body representations occur through the available sensory modalities. For example, when typing on a keyboard, we may visually see the position of our hands, feel the keys with our fingertips, and hear each key being pressed.

H. Kajimoto et al. (Eds.): EuroHaptics 2024, LNCS 14768, pp. 185–193, 2025.
https://doi.org/10.1007/978-3-031-70058-3_15

Therefore, a complete body representation is a by-product of multisensory integration. However, some research has proposed that vision is necessary to build an accurate representation of the body [3]. These studies have argued that the haptic representation of our bodies retains characteristics from the somatosensory homunculus, and it is through vision that these distortions are reduced [3]. In one such study, participants completed a localization and a template matching task [4]. For the localization task, the participants hand was hidden beneath a tabletop, and the participant had to estimate where various landmarks on the hand were located. This task has been argued to be primarily haptic in nature [5, 6]. The template matching task, however, is visual. In this task, participants must identify from an array of hand photographs, which best matches their own physical dimensions. Surprisingly, participants showed highly distorted hand maps in the localization task but were accurate at identifying the dimension of their own hand in the template matching task. This dissociation may be due to differences in sensory modality required for the two tasks [7]. Alternatively, it is also possible that the considerably different demands between the localization and template matching tasks, resulted in different accuracies in hand representation.

The purpose of the present study was to compare the visual and haptic representation of the hands and feet, using tasks with identical requirements. We chose to focus on the hand because the distortions in body representations have been repeatedly reported in the hands [3–7], making it the ideal target for our study. In addition, as we know that the hand is overestimated in S1, if the homunculus is influencing haptic body representations, we expect the hands to be overestimated if measured haptically. With regards to the feet, there are not that many studies, and to our knowledge none of the foot representation studies have investigated how vision and haptics contribute to the representation of this body part. By investigating both hand and foot representation, we increase the novelty of the current study. In addition, the feet would make an ideal control for the hands because they are physically similar but have a smaller cortical representation. Therefore, if haptic representations retain characteristics of the somatosensory homunculus, participants should overestimate hand more than foot size in this condition. Furthermore, if vision reduces distortions driven by the somatosensory homunculus, participants' estimates should be less distorted in the visual condition.

2 Methods

32 participants (18 females) were told to visually (visual condition) and haptically (haptic condition) judge if the size of a glove or shoe, was larger than their hands or feet (see Fig. 1). Each target item was placed in front of the participant on a tabletop. Gloves ranged in length from 17 cm–23.5 cm (XS-XXL), and shoes ranged 23.7 cm–28.3 cm (size 35–46). For a description of object size, see Table 1. All participants were presented with the same six sizes of each clothing item. To keep the number of stimuli identical between the gloves and shoes, we either presented the participant with the odd or even shoes (this was counterbalanced across participants). So, each participant saw/felt 6 of the 12 pairs of shoes. There were separate blocks for the two clothing items, during which the experimenter presented a different sized object on every trial in a randomized fashion. Within each block, the participant saw/felt each item six times. There was a

total of 36 trials for each item, meaning every participant completed a total of 72 trials per condition (144 trials total). In the vision condition, the participant placed their hands on their lap below the tabletop (no vision of their hands) and had to visually judge if each clothing item was larger than their target body part (e.g., hand). If they believed the object was larger than their hand, they were told to respond "yes", otherwise they were instructed to say "no" (two-alternative forced choice task). Participants were given a maximum of 10 s to explore each item. The instructions were identical in the haptic task, except participants were blindfolded, thus having to use haptics. Condition, block, and trial order were all randomized between participants.

Table 1. The sizes of each of the presented gloves or shoes. Each participant was presented with the same 6 pairs of gloves and shoes. For the shoes, we selected either the odd or even pairs (counterbalanced across participants), to keep the number of stimuli identical for both the hands and the feet.

Object	Size	CM
Gloves	XS	17
	S	18
	L	19
	XL	21
	XXL	21.5
	XXXL	23.5
Shoes	35	23.7
	36	24
	37	24.4
	38	25
	39	25.3
	40	25.8
	41	26.3
	42	27
	43	27.5
	44	28
	45	28.3
	46	29

2.1 Analysis

We calculated perceived body part size by plotting the proportion of trials that the participant said "yes" (i.e., the item was bigger than their body part) against the size (cm)

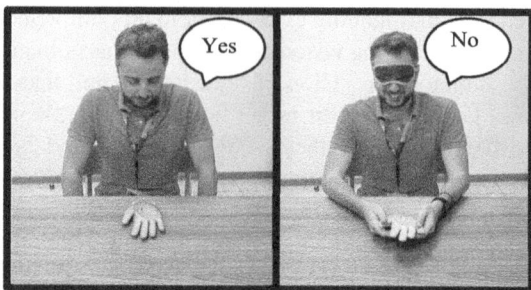

Fig. 1. Participants in both conditions responded if the glove was bigger than their hand. We presented various sizes of gloves (XS-XXL) to the participants in a randomized fashion. In the vision condition, they could only see the object (not touch), and in the haptics condition the participant was blindfolded and therefore could only haptically explore the glove.

of each of the target items. This was then fit with cumulative Gaussian functions. The point at which the participant was 50% likely to say yes/no defines the point of subjective equality (PSE), see Fig. 2 frame A. We took the PSE as the measure of perceived body size. We also calculated Just Noticeable Differences (JNDs) as a measure of precision. As with similar studies [8], we defined the JND as the standard deviation of the Gaussian fit. We calculated PSE biases by taking the PSE and subtracting the participants actual body part size. If there was an overestimation the PSE bias would be positive and vice versa if participants underestimated.

We conducted a series of pair-samples t-tests to determine if the PSE bias (cm) was distorted compared to the participants physical body part size. This analysis was similar to those from previous studies on hand representation [e.g., 6, 9]. These results are listed in Table 2.

To determine if PSE distortions differed between body parts, conditions, and sexes we conducted a mixed-design repeated measures ANOVA. We used the normalized value of PSE bias, meaning all variables are expressed as percent of physical body size. In addition, to control for the differences in physical hand/foot size (i.e., extremely big or extremely small), we reran these analyses with participants divided into two groups: above mean hand/foot size, and below mean hand/foot size. To ensure that any difference between body part, condition, and sex was not due to the level of precision in these variables, we repeated the analyses with the JND's as the dependent variables. For all results, means and standard errors are reported. Multiple comparisons were corrected with Bonferroni.

Table 2. The results of the series of pair-samples *t*-tests: Female participants significantly over-estimated hand and foot size in both conditions (*p*'s < .01). Male participants, however, were accurate at estimating hand/foot size in the visual condition, but significantly overestimated the body parts in the haptic condition.

Sex	Condition	Body part		Mean	SE	df	t	p
Females	Vision	Hands	PSE	19.41	.30	17	5.88	<.01
			Physical	17.56	.29			
		Feet	PSE	25.30	.30	17	7.85	<.01
			Physical	23.09	.25			
	Haptics	Hands	PSE	20.81	.36	17	8.86	<.01
			Physical	17.56	.29			
		Feet	PSE	25.83	.27	17	7.86	<.01
			Physical	23.09	.25			
Males	Vision	Hands	PSE	19.75	.36	13	1.64	.50
			Physical	19.08	.31			
		Feet	PSE	26.31	.35	13	2.54	.10
			Physical	25.34	.31			
	Haptics	Hands	PSE	21.04	.43	13	4.41	<.01
			Physical	19.08	.31			
		Feet	PSE	26.40	.34	13	3.21	.03
			Physical	25.34	.31			

3   Results

3.1   Differences Between Body Part, Condition, and Sex

PSE: There was a main effect of body part [$f(1, 30) = 6.0, p = .02, n^2 = .04$], in which participants overestimated their hand size significantly more than their feet. There was a main effect of condition [$f(1, 30) = 29.2, p < .01, n^2 = .06$], which was due to participants overestimating body part size more in the haptic condition compared to in the vision condition. There was also a body part x condition interaction [$f(1, 30) = 22.8, p < .01, n^2 = .04$] which better explains the results, see Fig. 2 frames B and C. Specifically, participants overestimated hand size in the haptic condition (14.68 ± 2.34) significantly more than in the vision condition (7.31 ± 2.07, t(30) $= -7.21, p < .01$). The same was not true for the feet (vision $= 7.32 \pm 1.33$, haptics $= 8.14 \pm 1.47, p = 1$). Lastly, there was also a main effect of sex [$f(1, 30) = 13.5, p < .02, n2 = .18$], where females ($13.1 \pm 1.76$) overestimated body parts significantly more than males (5.6 ± 1.84). No other interactions were significant.

JNDs: There was a main effect of body part [$f(1, 30) = 8.9, p < .01, n^2 = .06$], where estimates of the hand ($4.3 \pm .57$) were more variable compared to the feet ($3.08 \pm .4$).

There was also a main effect of condition [$f(1, 30) = 12.11, p < .01, n^2 = .13$], in which the participants estimates in the haptic condition (4.57 ± .55) were more variable than in the visual condition (2.8 ± .42). There were no other main effects or interactions.

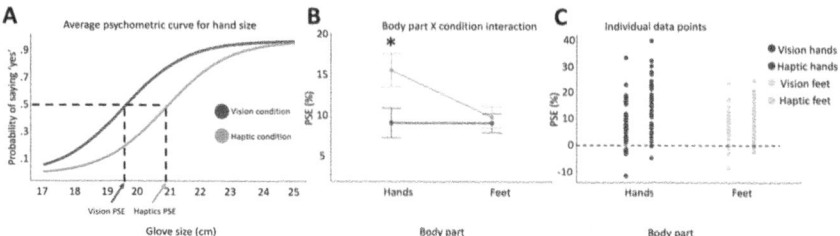

Fig. 2. Frame A: The average psychometric curve for participants during the hand estimation (i.e., gloves) in the vision (blue) and haptic (green) conditions. Expressed as the probability of saying "yes". Here, .5 is the value at which participants were equally likely to say "yes" or "no" (PSE). Participants (n = 32) had larger PSEs for the haptic condition. Frame B: The interaction between body parts and condition. Participants overestimated (%) the size of their hands significantly more in the haptic compared to the vision condition. For the feet, estimates were the same between conditions. Frame C: The individual data points for hand and feet size estimates in the vision (blue) and haptic (green) conditions. The dashed black line indicates 0% PSE distortion, in other words an accurate estimate. There were larger PSE% in the haptic hand condition.

When we divided participants into two groups, above mean hand/foot size, and below mean hand/foot size, we found the exact same results. Furthermore, there were no differences between groups, indicating no influence of physical body size on the results.

4   Discussion

The current study finds that hand and foot representations are distorted, particularly when measured under haptic guidance. One proposal is that vision reduces somatotopic distortions in body representation tasks [3]. The results from this experiment support this claim, as we find overestimation of hand size in the haptic condition, while its visual representation was relatively accurate (particularly for males). In addition, there were no differences between the visual and haptic representations of the feet. All together, these results support that under haptics, the representation of the body retains some characteristics from the somatotopic representation. A recent study found that in a mental rotation task, when visual experience was degraded by 60%, the resulting body representation became more somatotopic [10]. Our results support that vision and haptics rely on different representations of the body, one being based on the distorted somatotopic representation, and the other on a relatively accurate spatial visual map [10].

The present study builds off previous reports which have found haptically distorted hand representations, while more accurate representations when visually driven [3–5]. Those previous works used different methodologies for the haptic and visual representations, making it difficult to directly compare the accuracy of the two modalities.

In the current study, we used an ecologically valid method that is easy to implement. This allowed us to directly compare haptic and visual representations. Without using an identical paradigm for haptic and visual body representation tasks, it is impossible to disentangle the different demands of the tasks from the sensory influence. Our results confirm that the differences found in the localization and template matching tasks in previous work are driven by different sensory modalities. In addition, using this paradigm, future research can include special populations like those with no visual experience (e.g., blind) or younger children, who might find difficult to complete other body representation tasks due to the methodological demands (e.g., the localization task).

The reverse distortion hypothesis posits that areas with less sensory receptors should be visually perceived as larger, to counteract the haptic pattern of representation [11]. Our results do not support this hypothesis, as we found that visually, the hands and feet were perceived similarly. This suggests that there must be other compensatory mechanisms occurring that reduce the somatotopic distortions. One such proposal is that that accurate information about the tactile size and shape of the hand is elucidated through object interaction [12]. This speculation warrants future research.

The finding that females overestimated body part size more than the males replicates a study on hand representation [9]. In that study the authors proposed that an overestimation of hand size in healthy females may be a contributing factor as to why females are more likely to suffer from body representation disorders that are characterized by an overestimation of body size. The results from the current study suggest that both in the visual and haptic domains, females overestimate hand and foot size in comparison to males. To our knowledge this is the first study that has reported sex differences in foot representation. Future studies should investigate if these differences are exaggerated in body part areas that are more sensitive to weight gain/loss (e.g., stomach). Overall, our results support that body representation overestimation in females is not just a feature of clinical body representation disorders, but rather of the population as a whole.

The results from our analysis of the JNDs indicate that increased variability does not drive the significant interaction between body part and condition along with the sex differences found in the PSEs. We did find that participants were more precise overall in the visual condition, probably due to experience with visually selecting clothing items. In addition, judging the feet (overall) was less variable than judging hand size. It is possible that because our hands change size by spreading the fingers in various positions during the day, we have a fluctuating image of hand size. Future research needs to investigate this possibility.

As we previously mentioned we chose to investigate hand representation because it is the body part where haptic distortions have been most identified. In addition, in the cortex this representation is enlarged. We chose to compare the hands to the feet for two reasons 1) anatomically they are similar, but the feet have smaller cortical representation, and 2) few studies have investigated the accuracy of foot representations both visually and haptically. It would be possible and important to explore other areas of the body such as the forearm using a similar paradigm. The forearms have one of the smallest somatotopic representations [11], so we would expect that this would follow an opposite pattern of the hand, i.e., be haptically underestimated.

Cumulatively, the results from this study suggest that when vision is restricted the resulting representation is more somatotopic, retaining some characteristics from the homunculus. One way to further this finding is to test long-term visually deprived individuals to investigate if the perception of their bodies more closely aligns to a somatosensory representation. Research is ongoing to investigate this possibility. In the future, researchers and engineers should consider the enlarged haptic representation of the hand when designing handheld haptic devices.

There are a few limitations with the present study. First, as participants had different sized hands and feet, but were presented with the same stimuli, each participant had a different number of gloves/shoes that were bigger or smaller than their own body part. To control for this, we first normalized all data when we compared between participants. In addition, we ran secondary analyses where we split our participants into those who had above mean hand/foot size, and those who had below mean hand/foot size. We found identical results for both groups, i.e., the sex differences remained and there was a body part x condition interaction in which participants overestimated their hand size more in the haptic condition. In addition, there were no differences between the groups. Therefore, we do not believe that this limitation influenced the results. Future research could measure hand and foot size prior to the experiment and include only a couple smaller and larger options for each participant. One additional limitation is that in the visual condition we asked participants to place their hands on their lap, underneath the table, so that they could not use vision of their hands to guide their judgments. However, it is possible that in this position, the participants could have used tactile feedback to guide their estimates in the visual condition. This setup has been used in previous studies to evaluate haptic hand representation (for example see [6]). In addition, it is likely that within a few seconds of having their hands placed still on their lap sensory adaptation occurred. Sensory adaptation is the phenomenon that after a stimulus is presented for a given amount of time (usually <14 s [13]), neurons stop firing in response to said stimulus. Therefore, in the present study it is unlikely that participants relied on haptic feedback to base their estimates in the visual condition. However, it is possible that it played a small role. In the future, researchers could use numbing cream (e.g. EMLA cream) to fully investigate the role of haptics in visual body representations.

Acknowledgments. This study was supported by the MYSpace project awarded to Monica Gori. This project is funded by a European Research Council (ERC) Horizon 2020 research and innovation grant (No 948349).

Disclosure of Interests. The authors have no competing interests to declare that are relevant to the content of this article.

References

1. Gadsby, S.: Anorexia nervosa and body representation. Doctoral dissertation, Macquarie University (2022)
2. Penfield, W., Rasmussen, T.: The cerebral cortex of man; a clinical study of localization of function (1950)

3. Longo, M.R.: Types of body representation. In: Perceptual and emotional embodiment, pp. 125–142. Routledge (2015)
4. Longo, M.R., Haggard, P.: An implicit body representation underlying human position sense. Proc. Natl. Acad. Sci. **107**(26), 11727–11732 (2010)
5. Longo, M.R.: The effects of immediate vision on implicit hand maps. Exp. Brain Res. **232**, 1241–1247 (2014)
6. Coelho, L.A., Lee, R., Gonzalez, C.L.: The distorted hand: systematic but 'independent' distortions in both explicit and implicit hand representations in young female adults. Exp. Brain Res. **241**(1), 175–186 (2023)
7. Longo, M.R., Haggard, P.: Implicit body representations and the conscious body image. Acta Physiol (Oxf.) **141**(2), 164–168 (2012)
8. Cuturi, L.F., Gori, M.: Biases in the visual and haptic subjective vertical reveal the role of proprioceptive/vestibular priors in child development. Front. Neurol. **9**, 1151 (2019)
9. Coelho, L.A., Gonzalez, C.L.: Chubby hands or little fingers: sex differences in hand representation. Psychol. Res. **83**, 1375–1382 (2019)
10. Giovaola, Y., Rojo Martinez, V., Ionta, S.: Degraded vision affects mental representations of the body. Vis. Cogn. **30**(10), 686–695 (2022)
11. Linkenauger, S.A., et al.: The perceptual homunculus: the perception of the relative proportions of the human body. J. Exp. Psychol. Gen. **144**(1), 103 (2015)
12. Tamè, L., Limbu, S., Harlow, R., Parikh, M., Longo, M.R.: Size constancy mechanisms: empirical evidence from touch. Vision **6**(3), 40 (2022)
13. Wark, B., Lundstrom, B.N., Fairhall, A.: Sensory adaptation. Curr. Opin. Neurobiol. **17**(4), 423–429 (2007)

How Visualizing Touch Can Transform Perceptions of Intensity, Realism, and Emotion?

Xin Zhu[1]([✉])[iD], Zhenghui Su[1], Jonathan Gratch[2][iD], and Heather Culbertson[1][iD]

[1] Department of Computer Science, University of Southern California,
Los Angeles, USA
{xinzhu,suzhengh,hculbert}@ict.usc.edu
[2] Institute for Creative Technologies, University of Southern California,
Los Angeles, USA
gratch@ict.usc.edu

Abstract. Social touch is a common method of communication between individuals, but touch cues alone provide only a glimpse of the entire interaction. Visual and auditory cues are also present in these interactions, and increase the expressiveness and recognition of the conveyed information. However, most mediated touch interactions have focused on providing only haptic cues to the user. Our research addresses this gap by adding visual cues to a mediated social touch interaction through an array of LEDs attached to a wearable device. This device consists of an array of voice-coil actuators that present normal force to the user's forearm to recreate the sensation of social touch gestures. We conducted a human subject study (N = 20) to determine the relative importance of the touch and visual cues. Our results demonstrate that visual cues, particularly color and pattern, significantly enhance perceived realism, as well as alter perceived touch intensity, valence, and dominance of the mediated social touch. These results illustrate the importance of closely integrating multisensory cues to create more expressive and realistic virtual interactions.

Keywords: Haptic System Design · User Interface · Tactile Mapping · Affective Social Touch · Multimodal Interaction

1 Motivation

As we explore the complexities of human-computer interaction, it becomes clear that our experiences are influenced not only by basic functionality, but also by a blend of multisensory cues. Traditional interfaces often rely on visual and auditory feedback, overlooking one of the most primal human senses: touch. Our

This research was supported by the National Science Foundation under Grant No. 2047867.

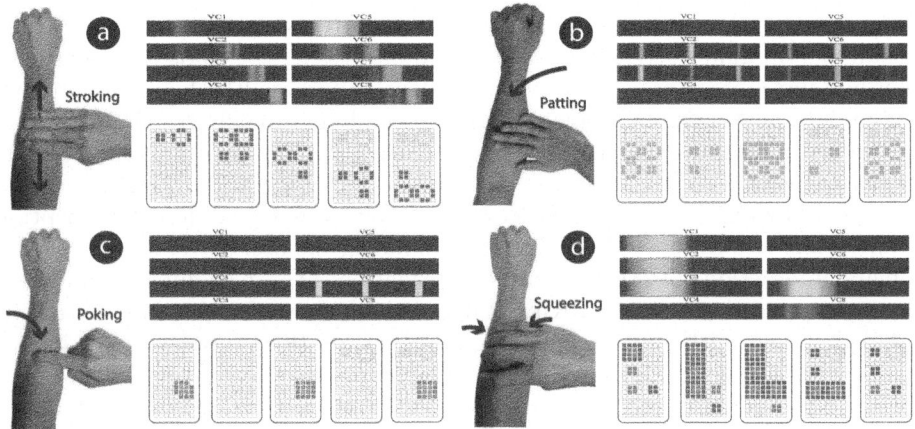

Fig. 1. Haptic and visual cues for social touch gestures a) Stroking gesture, red concentric circles visual cue. b) Patting gesture, orange concentric circles visual cue. c) Poking gesture, green expansion visual cue. d) Squeezing gesture, blue expansion visual cue.

sense of touch significantly influences how we feel and act toward others, as it intertwines with the expression of emotions through changes in speech, facial expressions, posture, and physiological processes [7]. Haptics reintroduces this tactile dimension into interactions, fostering a richer and more immersive experience [14]. Recent studies in haptics have concentrated on integrating affective touch to create diverse tactile experiences and to elicit and convey emotions [9,19]. Research has shown that mediated social touch can convey emotions with similar effectiveness as direct human-human touch [15]. Similarly, Hertenstein et al. showed that participants can communicate emotions through touch with an accuracy similar to facial expressions [8]. Gallace et al. suggests that combining social touch with congruent visual information could increase the effectiveness and accuracy of conveying emotions [6].

However, our social interactions are complex, and touch plays only part of the emotion equation. Visual cues also play important roles in emotion perception and conveyance. Since Goethe first associated specific color groups with emotional reactions such as warmth and excitement [20], our understanding of the interplay between color and emotion has deepened. Studies reveal that bright colors elicit a mainly positive emotion association and dark colors elicit the opposite, such as increased red conveying anger or embarrassment, whereas increased blue or green tint conveying illness [4]. Motion of the visual feedback also carries emotional weight, being intricately tied to the principles of animation that have long governed our visual media [18]. The manner in which the visual feedback moves can influence a user's emotional response [17]. Gentle, flowing motions - reminiscent of the animation principle 'slow in and slow out' - might convey calmness. In contrast, abrupt or erratic movements, which can be likened to the 'exaggeration' principle, might signify alarm or urgency. These principles extend to everything from animation on a screen to interaction designs such as human mental models [12]. Recognizing and harnessing the emotive potential of color and motion patterns can elevate the depth and breadth of communication, making interactions more nuanced, intuitive, and emotionally resonant.

Understanding the connections of color, motion, and emotion in user interaction, there is a clear advantage in multisensory integration. This approach, fundamental to multimodal interfaces, effectively combines different perception channels like vision, touch, and hearing, enhancing the user's overall comprehension and interaction with the presented information [13]. An example of multisensory perception is the "Bouba-Kiki" Effect, where people tend to associate soft, rounded shapes with soft, rounded-sounding words (like "bouba") and jagged shapes with sharper, more angular-sounding words (like "kiki") [11]. This effect was further studied to show that visual imagery plays a role in crossmodal integration [5]. The use of multisensory stimuli usually involves specific tasks where adding more sensory cues would be beneficial to the goal [10]. Akshita et al's investigations into combinations of visual stimuli and grounded haptic feedback suggested that haptic stimulus affects the arousal of the visual stimulus [1].

Research has shown that emotion perception across visual, touch, and auditory modalities share processing channels in the brain [16]. Despite the complementary benefit of multisensory integration, the best way to effectively represent affective concepts like valence and arousal through a multimodal interaction remains unclear [19]. Addressing the intimate nature of touch, this paper proposes integrating touch sensations with visual cues, particularly in visualizing social touch gestures. In the sections that follow, we present our haptic wearable system that embodies this fusion of haptic and light animation (Fig. 2). We also present a human subject study to evaluate the effect that visual feedback has on user's perception of a mediated social touch's intensity, realism, and emotion.

2   Experimental Setup

2.1   Hardware Setup

To display mediated social touch gestures, we designed our device as a fabric sleeve with an array of voice coil actuators that display normal indentation to the forearm. Prior work has shown that this method of haptic feedback is effective at providing both pleasant and realistic social touch cues [3,15]. The voice coil actuators (Tectonic Elements TEAX19C01-8) have 2 cm $\times$ 2 cm contact area and are arranged in a 2×4 array to create a total skin contact area of 8×16 cm^2 for the actuation layer. To create the visualization layer on the top of the sleeve, we added an LED matrix panel (BTF-LIGHTING WS2812B-8x8) containing 128 individually addressable digital pixels in an 8×16 array, shown in Fig. 2.

2.2   Touch Gesture Rendering and Visual Mapping

We use actuation methods developed in prior work [21] to create the social touch experience with the hardware setup. A set of four data-driven social touch

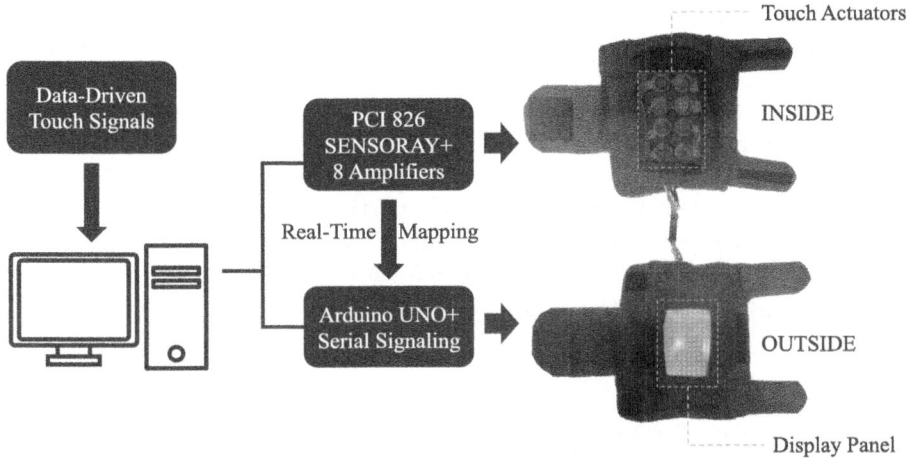

Fig. 2. Real-time LED Illumination System for Touch Gesture Generation.

gestures (stroking, patting, poking and squeezing) were collected using a 2 × 4 array of force-sensing resistors at 1000 Hz. These four gestures were selected based on their inherent capacity to communicate a diverse range of both positive and negative emotions and information [8]. The recorded spatio-temporal force data for each gesture was low-pass filtered and normalized. Depending on the types of gestures, algorithmic methods were applied to the processing progress to convert the measured force data to actuation signals used to control the motion of the voice coil actuators. Full details on the data mapping algorithms are in [21]. Each stimuli is 5 s to 20 s long depending on the gesture type. Using this data-driven rendering method, users can feel a range of social touch gesture patterns through the actuation sleeve.

Our system is designed to showcase a diverse combination of tactile and visual cues, encompassing four distinct social touch gestures (stroking, patting, poking, squeezing), four illuminating colors (red, yellow, green, white), and three unique lighting patterns. Integrating the "Bouba-Kiki" [11] effect, we designed three haptic-visual mapping algorithms (direct mapping, expansion, and concentric circles) that vary in both shape and motion. These algorithms were tailored to transform haptic signals into discernible patterns on an LED panel, aiming to encapsulate the essence of the social touch gesture's motion and intensity, shown in Fig. 1. The full implementation of the visual mapping algorithms can be found in Appendix A.

2.3 User Study

We conducted three experiments, which were each designed to evaluate a distinct component of the user's perceptual experience in the presence of our integrated visual feedback.

A total of 20 volunteers (aged 20–58, 8 female, 12 male, with no prior haptics experience) participated in the study. The study was approved by the University of Southern California's Institutional Review Board under Protocol UP-19-00712, and all participants gave informed consent. Each full study of three experiments took participants approximately 40 min to complete. The order of the experiments was randomized, and participants were given a 5-min break between each experiment. During the study, participants wore the sleeve on their non-dominant arm and answered questionnaires using a mouse with their dominant hand, as shown in Fig. 3(a). Participants were instructed to keep their visual attention on the device and wore headphones playing white noise.

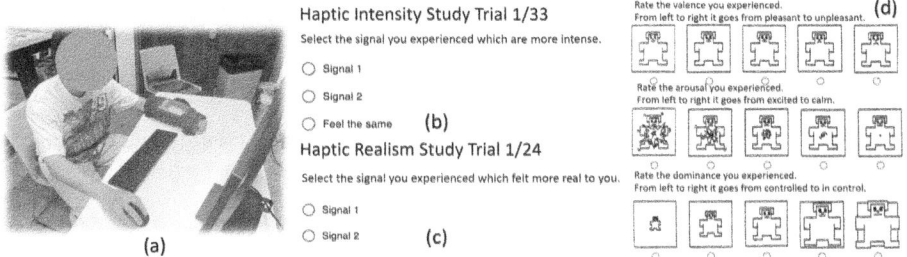

Fig. 3. a) Experimental setup b) GUI for rating perceived intensity. c) GUI for rating perceived realism. d) Self-Assessment Manikins scales used to evaluate valence, arousal, and dominance (reprinted from [2]).

In Experiment 1, we hypothesize that **specific lighting colors** can significantly modulate users' perceptions of the **intensity** of social touch gestures. We selected two lighting colors (Green and Orange), and a third case where the visual cues were disabled (OFF). The touch signal was provided at two intensity levels, low and high, with the low-intensity (Low) set at 60% of the high-intensity (High) level by applying a scaling factor of 0.6 to the actuation signals, which will give a difference of 1 mm for normal indentation to the skin. The gestures assessed in this experiment were stroking, poking, and squeezing. In Experiment 1-A, shown in Table 1, we evaluated the effectiveness of lighting cues in improving users' ability to distinguish different intensity of interactions by comparing by asking users to compare high-intensity with low-intensity touch signals under different lighting conditions. For each gesture, three lighting conditions were provided, resulting in 9 trials (3 gestures × 3 lighting conditions). Note that although the visual mapping pattern was the same between the two interactions, the intensity (brightness) of the visual cues did differ between the high and low intensity conditions. In Experiment 1-B, we compared how different colors affect users' perception of the intensity of the touch signals when played at the same level of intensity, shown in Table 1. There were 24 trials in this experiment, 8 for each gesture (4 lighting conditions × 2 intensity levels).

In Experiment 2, we hypothesize that different **lighting patterns** will influence users' perception of **realism** in relation to specific social touch gestures.

Table 1. Comparison Variables in Experiment 1. Note that in Experiment 1-A the lighting brightness is proportional to the intensity of signal.

Gesture	Experiment1-A		Experiment1-B	
	Variable$_1$	Variable$_2$	Variable$_1$	Variable$_2$
Poking Squeezing Stroking	Green High	Green Low	Green High	Orange High
	Orange High	Orange Low	Green High	OFF High
	OFF High	OFF Low	Green Low	Orange Low
			Green Low	OFF Low
			Orange High	OFF High
			Orange Low	OFF Low
			OFF Low	OFF Low
			OFF High	OFF High

We selected three distinct lighting patterns (direct mapping (DP), expansion (EX), concentric circles (CC)) (shown in Fig. 1) and also presented a condition with visual cues disabled (OFF). The gestures evaluated in this experiment were stroking, patting, poking, and squeezing. Participants were presented with a set of two combinations of haptic visual cues presented one after the other and were asked to indicate which one felt more realistic using the scales shown in Fig. 3(c). The comparisons were structured to evenly span across the combinations of lighting patterns and gestures. As a result, each gesture had 6 unique lighting pattern comparisons for each of the 4 gestures, resulting in a total of 24 trials, shown in Table 2.

Table 2. Comparison Variables in Experiment 2. Different Lighting Patterns were compared to understand human perception of the realism of the mediated touch.

Gesture	Experiment2	
	Variable$_1$	Variable$_2$
Stroking Poking Squeezing Patting	DP	EX
	DP	CC
	DP	OFF
	EX	CC
	EX	OFF
	CC	OFF

In Experiment 3, we hypothesize that the interplay between **lighting color and pattern** plays a significant role in shaping users' **perceived emotions** during the interaction, measured through ratings of valence, arousal, and dominance. For this study, we selected four lighting colors (red, yellow, green, white), three lighting patterns (expansion, concentric circles, and OFF). The gestures evaluated in this experiment were: stroking, poking, and squeezing. Unlike previous experiments, this experiment only examines one signal in each trial. There were 27 trials in this experiment, 9 trials for each gesture (4 lighting colors × 2 lighting patterns + OFF lighting condition). After each trial, participants rated their perceived valence, arousal, and dominance of the interaction using the Self-Assessment Manikin (SAM) rating scale [2], shown in Fig. 3(d).

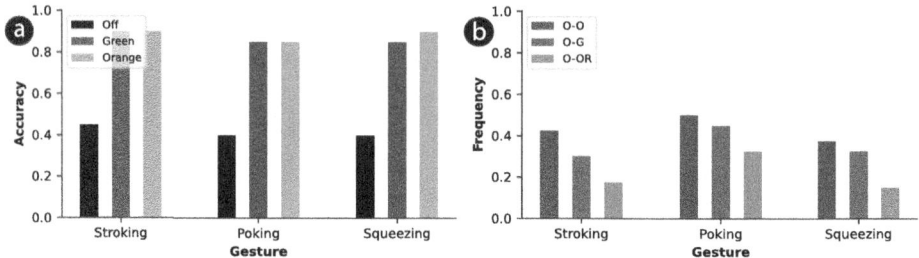

Fig. 4. Experiment 1: a) Accuracy in identifying the stronger haptic signal under different lighting colors and touch gestures. b) Frequency of perceiving touch signals to have the same intensity: O-O (both visual cues off), O-G (one off, one green), and O-OR (one off, one orange).

3   Results

We first evaluated the effectiveness of providing any visual cue on participants' ability to correctly identify the more intense haptic signal. We gathered information on participants' feedback of interactions with identical visual mapping patterns, but varying haptic intensities in Experiment 1-A (Fig. 4(a)). We then conducted a Chi-Square test to assess the associations between the presence of visual cues and the participants' capacity to distinguish touch signal intensities. This evaluation showed that the presence of a visual cue made participants significantly more accurate at detecting the stronger haptic signal (for stroking, $\chi^2(1) = 12.1, p < 0.01$; for poking, $\chi^2(1) = 10.804, p < 0.01$; for squeezing, $\chi^2(1) = 12.568, p < 0.01$).

Next, we evaluate how visual cues could influence individuals' perception of the intensity of a touch signal, by providing identical haptic signals, but varying visual colors in Experiment 1-B. We calculated the frequency with which participants perceived the touch signals as having the same intensity (Fig. 4(b)). We then conducted a Chi-Square test to examine the relationship between the lighting color and participants' perceptions of touch signal intensity (Table 3), which revealed significant associations between the orange lighting color and participants' perceptions of signal intensity for stroking and squeezing. Following these observations, we utilized the Binomial test to determine whether the presence of a specific color (orange) influenced participants' perception of an increased feeling of signal intensity. The results show that participants were significantly more inclined to perceive an increased intensity in conditions with orange lighting cues for both stroking ($p < 0.01$) and squeezing ($p < 0.01$).

Table 3. Experiment 1: Chi-Square test results assessing the relationship between lighting color and participants' perceptions of touch signal intensity across different gestures. $^*p < 0.05$.

Gesture	Variable$_1$		Variable$_2$		χ^2	df	p-value
	Signal$_1$ Color	Signal$_2$ Color	Signal$_1$ Color	Signal$_2$ Color			
Stroking	Off	Off	Off	Orange	4.821	1	.028*
Stroking	Off	Off	Off	Green	0.865	1	.352
Poking	Off	Off	Off	Orange	1.857	1	.173
Poking	Off	Off	Off	Green	0.050	1	.823
Squeezing	Off	Off	Off	Orange	4.132	1	.042*
Squeezing	Off	Off	Off	Green	0.054	1	.815

To evaluate how different lighting patterns affect the perceived realism of mediated social touch gestures, we measured how frequently each combination of lighting and touch was selected as the most realistic in Experiment 2. We conducted a Chi-Square test comparing this frequency of selection to chance in order to determine how much a specific lighting pattern increases the frequency of the interaction. Result shows that interactions with the direct mapping (DP) cues were significantly more realistic than interactions without visual cues (OFF) for stroking ($\chi^2(1) = 7.2, p < 0.01$), patting ($\chi^2(1) = 5.0, p < 0.05$), and poking ($\chi^2(1) = 5.0, p < 0.05$). Similarly, interactions with the expansion mapping (EX) cue were significantly more realistic than interactions without visual cues (OFF) for patting ($\chi^2(1) = 7.2, p < 0.01$) and poking ($\chi^2(1) = 12.8, p < 0.01$). Furthermore, interactions with the concentric circles (CC) mapping were more realistic than interactions without visual cues for patting ($\chi^2(1) = 7.2, p < 0.01$) and squeezing ($\chi^2(1) = 5.0, p < 0.05$). The results of this analysis are shown in Fig. 5.

To assess the interplay between lighting color and pattern in shaping how users perceive the conveyed emotions of the interaction, we evaluated how the valence, arousal, and dominance ratings change as a function of the haptic-

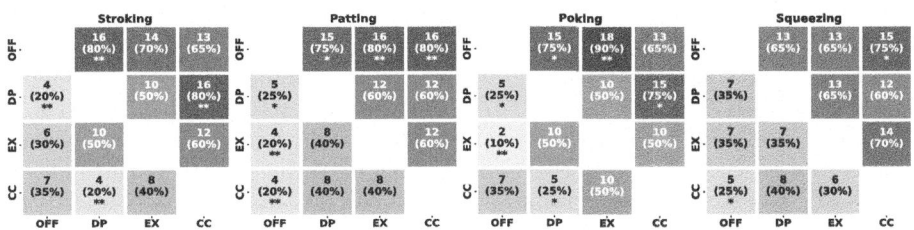

Fig. 5. Experiment 2: Perceived realism over different mapping patterns, frequency of selected lighting patterns across different gestures. The displayed numbers and percentages represent the frequency with which each lighting pattern (shown on the x-axis) was chosen in each pairwise comparison. $^*p < 0.05$, $^{**}p < 0.01$.

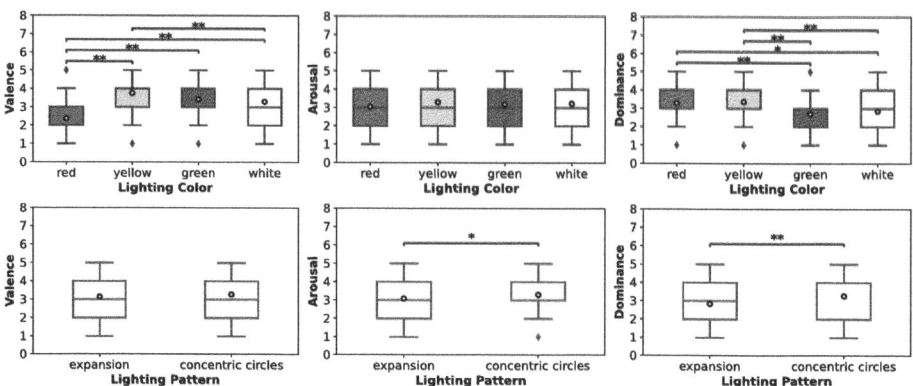

Fig. 6. Experiment 3: Ratings of valence, arousal, and dominance across different lighting colors and patterns. $^*p < 0.05$, $^{**}p < 0.01$.

visual mapping condition (Experiment 3). We conducted a set of three two-way ANOVAs individually on each of these three scales with lighting color and lighting pattern as factors. We first concentrated on the interaction effects between the two factors. If no significant interaction effect emerged, the main effects were highlighted; but if there was a significant interaction, we reported the simple main effects. Beyond the main effects, we also employed a Tukey post-hoc test to pinpoint significant differences between each lighting color and pattern through multiple pairwise comparisons (Fig. 6).

Valence. The analysis shows that there was a main effect of lighting color on associated valence ($F(3, 472) = 40.4, p < 0.01, \eta^2 = 0.199$). No significant effect of lighting pattern on valence was observed ($F(1, 472) = 2.16, p = 0.142$). There was no interaction effect between lighting color and lighting pattern ($F(3, 472) = 2.18, p = 0.089$). Post-hoc pairwise comparisons highlighted discernible variations in perceived valence across the different lighting colors. Specifically, the color red was associated with a markedly lower valence ($p < 0.01$) relative to the other colors (yellow, green, and white). Conversely, the color yellow exhibited significantly higher valence associations ($p < 0.01$) when compared to the colors green and white.

Arousal. There was a main effect of lighting pattern on associated arousal ($F(1, 472) = 4.31, p < 0.05, \eta^2 = 0.009$), but no significant effect of lighting color ($F(3, 472) = 1.04, p = 0.374$) and also no interaction ($F(3, 472) = 0.214, p = 0.886$). Post-hoc pairwise comparisons indicated marked distinctions in perceived arousal between the two lighting patterns. Notably, the lighting pattern of concentric circles exhibited significantly elevated arousal associations ($p < 0.05$) in comparison to the expansion pattern.

Dominance. There were significant main effects of both lighting color ($F(3, 472) = 9.98, p < 0.01, \eta^2 = 0.057$) and lighting pattern ($F(1, 472) = 18.2, p < 0.01, \eta^2 = 0.035$) on associated dominance, but no statistically significant interaction ($F(3, 472) = 1.82, p = 0.142$). Post-hoc pairwise comparisons revealed that participants significantly associated dominance more with the colors red and yellow than with the other colors (green and white), $p < 0.05$. Additionally, the lighting pattern of concentric circles was associated with a notably higher level of dominance ($p < 0.01$) compared to the expansion pattern.

4 Discussion

The findings from our experiments demonstrate that the integration of lighting with mediated touch significantly influences perceived intensity, realism, and emotions. Experiment 1 revealed that people discern stronger sensations more effectively when assisted by lighting, as visual cues provide an intuitive grasp of touch intensity. The relatively low accuracy in conditions without lighting (40%) suggests a difficulty in distinguishing variations in touch intensity, especially when these variations are subtle (e.g., 60% compared to 100% intensity). This underscores the crucial role of haptic-visual mapping, particularly when differences in touch intensity are minimal. Additionally, lighting color plays a vital role in modulating the perceived intensity of haptic signals. Specific visualization patterns and colors can either enhance or moderate the sensation. For example, under identical touch intensities and lighting patterns, subjects often perceived orange visualizations as stronger than those in green or without any light. However, it was also observed that for certain gestures, like poking, visual cues did not alter perceived intensity across different lighting colors. Moreover, green lighting did not impact the perceived intensity of any gestures. These intriguing findings suggest a complex interaction between tactile and visual modalities. Incongruent pattern pairings may lead to cognitive dissonance, whereas synergistic combinations can intensify the overall experience.

We observed marked disparities in perceived realism across various lighting patterns and gestures. Generally, the presence of lighting enhances the realism of mediated touch gestures, but the visualization patterns have different effects on increasing the realism level. For instance, stroking and poking were perceived as more realistic with Concentric Circle (CC) patterns than with Direct Mapping (DP). This might be because the CC pattern aligns with the Bouba-kiki effect, which suggests that humans associate rounded shapes with softer, more gentle sensations, enhancing the perception of realism in tactile interactions. These findings, reinforced by the frequency data presented in Fig. 5, highlight the effectiveness of visual cues in augmenting the realism of artificial touch sensations. Notably, while significant variances were noted, it remains to be determined which specific lighting pattern is most effective in optimizing perceived realism when visual cues are employed. This inquiry opens avenues for further research, potentially guiding the development of more sophisticated and user-responsive haptic methods.

Drawing upon the color emotion theory, which shows that each color triggers distinct psychological reactions, we find that integrating haptic and visual illumination is pivotal in emotion regulation. The color red evokes negative valence during haptic experiences, whereas yellow predominantly elicits positive valence or optimism. Moreover, the Concentric Circle (CC) pattern distinctly conveys a more pronounced sense of dominance relative to other lighting patterns, possibly due to its spreading shape and human motion perception preference nature. Interestingly, we did not observe a significant effect on the perceived arousal, which might be further explored with other visualization factors like brightness. However, our exploration of visualization techniques for affective touch is constrained by the choices of color and mapping algorithms. A more meticulous examination of tactile-visual integration in the future will enrich the narrative of mediated social touch.

5   Conclusion

This paper explored the effect of visual cues on individuals' perception of mediated social touch. Specifically, we aimed to understand how specific lighting colors and visual-haptic mapping patterns modulate users' perceptions of the intensity, realism, and emotion of social touch gestures. Our results show that visual cues can affect the perceived valence and dominance of the interactions. Additionally, the incorporation of certain visual patterns enhances the realism of the touch sensation and alters the perception of mediated touch strength or intensity. This work paves the way for touch devices with visual illumination, enabling more nuanced and realistic emotional communication through wearable technology. In future work, we envision such integration being used in various applications, from providing intuitive robot task execution to enhancing safety in human-computer interaction. Yet, there remains gaps that needs to be further addressed, such as effective and granular mapping methods to capture the smallest of feedback changes, as well as accurate representation for a emotional perception. As understanding of human perception advances, the integration of tactile feedback with visual representation will be key to unlocking more sophisticated and intuitive human-computer interaction.

A   Haptic–Visual Mapping Algorithms

Algorithm 1: Direct Mapping

Input: LED Matrix M of size $N \times N$
Input: Intensity factor f where $0 \leq f \leq 0.9$
Input: Base color (R, G, B)

1 **for** $i = 1$ *to* N **do**
2   **for** $j = 1$ *to* N **do**
3          $r \leftarrow \text{round}(f \times R)$;
4          $g \leftarrow \text{round}(f \times G)$;
5          $b \leftarrow \text{round}(f \times B)$;
6   setPixelColor(i,j,r,g,b);

Algorithm 2: Centered Square Expansion Mapping

Input: LED Matrix M of size $N \times N$
Input: Intensity factor f where $0 \leq f \leq 0.9$
Input: Base color (R, G, B)

1 $c \leftarrow \frac{N}{2}$;
2 $x \leftarrow \lceil f \times c \rceil$;
3 **for** $i = 1$ *to* N **do**
4   **for** $j = 1$ *to* N **do**
5   **if** $c - x < i \leq c + x$ *and* $c - x < j \leq c + x$ **then**
6              $r \leftarrow \text{round}(f \times R)$;
7              $g \leftarrow \text{round}(f \times G)$;
8              $b \leftarrow \text{round}(f \times B)$;
9   setPixelColor(i,j,r,g,b);
10   **else**
11   setPixelColor(i,j,0,0,0);

Algorithm 3: Concentric Circle Mapping

Input: LED Matrix M of size $N \times N$
Input: Intensity factor f where $0 \leq f \leq 0.9$
Input: Base color (R, G, B)

1 $c \leftarrow \frac{N}{2}$;
2 **if** $f \approx 0$ *or* $f \approx 1$ **then**
3      $x \leftarrow \lceil f \times c \rceil$;
4   **for** $i = 1$ *to* N **do**
5   **for** $j = 1$ *to* N **do**
6   **if** $i = c - x + 1$ *or* $i = c + x$ *or* $j = c - x + 1$ *or* $j = c + x$ **then**
7                  $r \leftarrow \text{round}(f \times R)$;
8                  $g \leftarrow \text{round}(f \times G)$;
9                  $b \leftarrow \text{round}(f \times B)$;
10   setPixelColor(i,j,r,g,b);
11   **else**
12   setPixelColor(i,j,0,0,0);

13 **else**
14     $r \leftarrow f \times \frac{c - 0.5}{\cos(\arctan(\frac{1}{3}))}$;
15   **for** $i = 1$ *to* N **do**
16   **for** $j = 1$ *to* N **do**
17             $d \leftarrow \sqrt{(i - c - 0.5)^2 + (j - c - 0.5)^2}$;
18   **if** $d \approx r$ **then**
19                 $r \leftarrow \text{round}(f \times R)$;
20                 $g \leftarrow \text{round}(f \times G)$;
21                 $b \leftarrow \text{round}(f \times B)$;
22   setPixelColor(i,j,r,g,b);
23   **else**
24   setPixelColor(i,j,0,0,0);

References

1. Akshita, Alagarai Sampath, H., Indurkhya, B., Lee, E., Bae, Y.: Towards multimodal affective feedback: interaction between visual and haptic modalities. In: Proceedings of ACM Conference on Human Factors in Computing Systems, pp. 2043–2052 (2015)
2. Bradley, M.M., Lang, P.J.: Measuring emotion: the self-assessment manikin and the semantic differential. J. Behav. Therapy Exp. Psych. **25**(1), 49–59 (1994)
3. Culbertson, H., Nunez, C.M., Israr, A., Lau, F., Abnousi, F., Okamura, A.M.: A social haptic device to create continuous lateral motion using sequential normal indentation. In: Proceedings of IEEE Haptics Symposium, pp. 32–39 (2018)
4. Elliot, A.J.: Color and psychological functioning: a review of theoretical and empirical work. Front. Psychol. **6**, 368 (2015)
5. Fryer, L., Freeman, J., Pring, L.: Touching words is not enough: How visual experience influences haptic-auditory associations in the " bouba-kiki ' 'effect. Cognition **132**(2), 164–173 (2014)
6. Gallace, A., Spence, C.: The science of interpersonal touch: an overview. Neurosci. Biobehav. Rev. **34**(2), 246–259 (2010)
7. Gratch, J., Marsella, S., Petta, P.: Modeling the cognitive antecedents and consequences of emotion. Cogn. Syst. Res. **10**(1), 1–5 (2009)
8. Hertenstein, M.J., Holmes, R., McCullough, M., Keltner, D.: The communication of emotion via touch. Emotion **9**(4), 566 (2009)
9. Huisman, G.: Social touch technology: a survey of haptic technology for social touch. IEEE Trans. Haptics **10**(3), 391–408 (2017)
10. Jalapati, P., Sweidan, S., Zhu, X., Culbertson, H.: Vocalization for emotional communication in crossmodal affective display. In: International Conference on Affective Computing and Intelligent Interaction Workshops and Demos, pp. 1–8. IEEE (2023)
11. Köhler, W.: Gestalt psychology: an introduction to new concepts in modern psychology, vol. 18. WW Norton & Company (1970)
12. Nikolaidis, S., Shah, J.: Human-robot teaming using shared mental models. In: Proceedings of ACM/IEEE Conference on Human-Robot Interaction (2012)
13. Obrenovic, Z., Starcevic, D.: Modeling multimodal human-computer interaction. Computer **37**(9), 65–72 (2004)
14. Raisamo, R., Salminen, K., Rantala, J., Farooq, A., Ziat, M.: Interpersonal haptic communication: review and directions for the future. Int. J. Hum Comput. Stud. **166**, 102881 (2022)
15. Salvato, M., Williams, S.R., Nunez, C.M., Zhu, X., Israr, A., Lau, F., Klumb, K., Abnousi, F., Okamura, A.M., Culbertson, H.: Data-driven sparse skin stimulation can convey social touch information to humans. IEEE Trans. Haptics **15**(2), 392–404 (2021)
16. Schirmer, A., Adolphs, R.: Emotion perception from face, voice, and touch: comparisons and convergence. Trends Cogn. Sci. **21**(3), 216–228 (2017)
17. Schulz, T., Torresen, J., Herstad, J.: Animation techniques in human-robot interaction user studies: a systematic literature review. ACM Trans. Hum. Robot Interact. (THRI) **8**(2), 1–22 (2019)
18. Thomas, F., Johnston, O.: The Illusion of Life: Disney Animation. Disney Editions, New York (1995)
19. Van Erp, J.B., Toet, A.: Social touch in human-computer interaction. Front. Digit. Hum. **2**, 2 (2015)

20. Von Goethe, J.W.: Theory of Colours. MIT Press, Cambridge (1970)
21. Zhu, X., Feng, T., Culbertson, H.: Understanding the effect of speed on human emotion perception in mediated social touch using voice coil actuators. Front. Comput. Sci. **4**, 826637 (2022)

Technology and Systems

Asymmetric Hit-stop for Multi-user Virtual Reality Applications: Reducing Discomfort with the Movement of Others by Making Hit-stop Invisible

Shinnosuke Noguchi[1]([✉]), Keigo Matsumoto[1] [iD], Yuki Ban[2] [iD], and Takuji Narumi[1] [iD]

[1] The University of Tokyo, 7-3-1 Hongo, Bunkyo-ku, Tokyo 113-0033, Japan
{noguchi-shinnosuke,matsumoto,narumi}@cyber.t.u-tokyo.ac.jp
[2] The University of Tokyo, 5-1-5 Kashiwanoha, Kashiwa-shi, Chiba 277-8561, Japan
ban@edu.k.u-tokyo.ac.jp

Abstract. A hit-stop is an expression technique primarily used in fighting games that emphasizes the response to the impact felt by a player when the target's movement is momentarily delayed at the moment of impact. It has been reported that the hit-stop technique can make users feel pseudo-impact forces when applied to virtual reality (VR). However, existing research has focused on a single user, and the effects of hit-stops when used by multiple users in the same virtual environment in a competitive game-like situation have not been verified. There is concern that users may feel discomfort, and the quality of the VR experience may be degraded by observing the difficult-to-predict movements of other players due to hit-stop. Therefore, this study examines the degree to which hit-stops in a multi-user VR cause discomfort to users and investigates the extent to which such discomfort can be mitigated by presenting hit-stops asymmetrically. We conducted an experiment in which two players played a tennis rally in a virtual environment to see whether they felt discomfort in the movements of the other player and themselves under four conditions: a combination of showing the player's hit-stop compensated movement (symmetric condition) or their real movements without hit-stop correction (asymmetric condition) to their opponent, and showing the opponent's hit-stop compensated movement or the opponent's real movement without hit-stop correction to the player. The results showed that the asymmetric presentation of hit-stop can reduce discomfort while making the participant feel a pseudo-impact force.

Keywords: Hit-stop · Pseudo-haptics · Discomfort · Multi-user virtual reality

1 Introduction

A hit-stop is an expressive technique that emphasizes the response to an attack by stopping or slowing the animation at the moment of impact. This technique is mainly used in fighting games, such as Street Fighter[1] and Super Smash Bros[2]. When applied to virtual environments (VEs), it evokes a pseudo-impact force on the user [1]. Moreover, it has been reported that the hit-stop improves the enjoyment and presence of the virtual reality (VR) experience when combined with controller vibration. Thus, a hit-stop presentation in VE not only appears as a fresh experience to the user but also improves the sense of immersion in VR through an inexpensive and simple system.

While a hit-stop can easily present a sense of impact force, the effectiveness of applying hit-stop to a multi-user VE has not yet been clarified. In multi-user VR, users observe the detailed movements of avatars, which symbolize their real bodies, and communicate in a realistic manner. In such a situation, we must be careful because sudden changes in the avatar behavior that do not occur in reality may confuse the observers. Freiwald et al. [2] showed that in competitive multi-user VR, a non-continuous movement method using teleportation makes it more difficult for the observer to keep track of the instantaneous movement of the user than a continuous movement method using a joystick, and significantly reduces the observer's sense of fairness and co-presence in the game. Based on this result, users who observe hit-stop caused by others in the virtual space may feel uncomfortable with their unpredictable movements, which may degrade the quality of the VR experience. For example, in video games, such as Super Smash Bros., there is no sense of discomfort when a player observes his or her opponent's hit-stop. However, because VR immerses users in a virtual space, it is easier for them to perceive the events that occur in that world as closer to reality than in a game. Therefore, it is anticipated that they will feel unnatural about the hit-stop behavior, which is far from reality.

For this reason, this study proposes a hit-stop asymmetric presentation method to reduce the discomfort caused by hit-stops in multi-user VR. When the conventional hit-stop presentation method is applied to multi-user VR, the user observes the other player's movement with hit-stop, which is expected to be confusing because the movement of the avatar changes rapidly. Therefore this method converts the opponent's movement into a natural movement by making the hit-stop invisible to the user (Fig. 1). Our experiments investigated the effectiveness of the proposed method. The hit-stop asymmetric presentation method is expected to contribute to multi-user VR, such as competitive VR games, metaverse VR, and rehabilitation tools in virtual shared spaces.

2   Proposed Method

2.1   Previous Work

Pseudo-haptics is a method of creating a pseudo-tactile sensation without the need for a physical device [3]. This is created by making the user observe the displacement between the real and virtual bodies. Using pseudo-haptics, it has

---

[1] https://www.streetfighter.com/6/ja-jp.
[2] https://www.smashbros.com/ja_JP/.

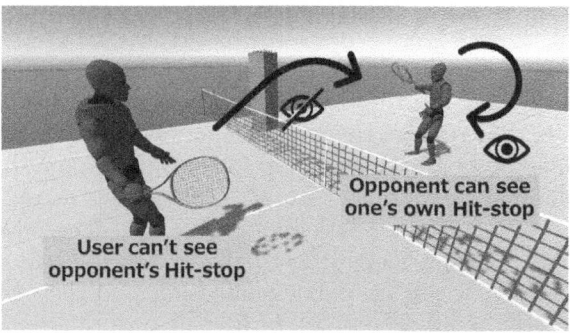

Fig. 1. Hit-stop Asymmetric Presentation Method

been shown that force perception can be presented by making the user wear an HMD and displaying a virtual hand in a different position from the actual hand [4]. There are numerous examples of pseudo-haptic applications, including frictional forces with input devices [5] and spring compliance [6,7].

Ban et al. [1] found that a hit-stop in a VE causes pseudo-haptics. They applied a hit-stop to a virtual tennis environment and showed that it evoked a pseudo-impact sensation in the user. In the case of hit-stopping, after the collision between the racket and ball was detected, the movement of the racket was delayed compared to the movement of the controller, as shown by the solid and dotted red lines in Fig. 2 (solid line: stop-motion; red dotted line: slow-motion). The racket was then rapidly forwarded from the end of the hit-stop to catch up with the controller. The experimental results showed that the longer the hit-stop duration, up to 200 ms, the stronger the sense of the impact force felt by the user. The combination of vibration and slow motion enhanced the sense of agency, thereby improving the sense of impact force, presence, and enjoyment of the VR experience.

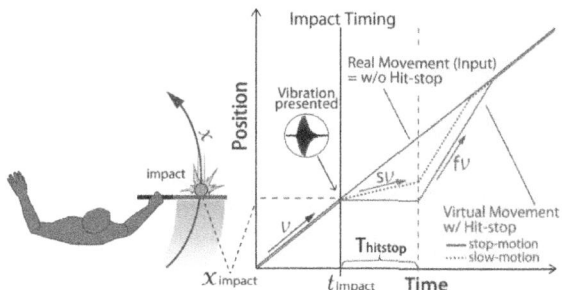

Fig. 2. Relationship between the racket and the actual body movement (cited from [1]).

2.2   Asymmetric Presentation of Hit-Stop in Multi-user VR

A hit-stop has the property of stopping the whole- body motion of the subject who causes it, and then returning to the normal state with a rapid motion. Therefore, in a VE that reflects the smooth motion of a real body, such a sudden change in whole-body motion is expected to confuse a hit-stop observer. We conducted a preliminary experiment on six individuals to confirm this expectation. In this experiment, the participants were asked to play a tennis rally with an NPC in VR, and hit-stops were applied when the ball returned. In the post-experiment interview, all six participants indicated they felt more uncomfortable with the NPC's hit-stop than with their own. This discomfort would interfere with the user's immersion in the VE and detract from the enjoyment of the VR experience. In addition, there is a concern that it may harm the benefits of hit-stops in VR, such as enjoyment and presence. Based on this discussion, a hit-stop presentation method in which observers feel natural in multi-user VR is required.

In this study, we propose a novel hit-stop asymmetric presentation method in which hit-stop applied to one's own movement is presented as they are, while hit-stop applied to the opponent's movement is not presented, but instead, the same movement as in reality is presented (Fig. 1). For example, when a hit-stop is applied to the opponent, they will see their virtual avatar, racket, and the ball move later than they should. On the other hand, to the user, the opponent's avatar, racket, and the ball will appear to be in motion without the hit-stop applied. This proposed method allows the opponent to observe hit-stopping oneself and the user to continue playing without feeling that the opponent is hit-stopping at all.

3   Experiment

Participants. The total 24 participants in the experiment (ten of whom were female) were between 20 and 27 years old (M = 23.0, SD = 1.60). Two participants took part in each experiment, totaling 12 pairs (six MM, four FF, and two FM). The participants were recruited through social networking services. Of the 24 participants, 22 were right-handed and 2 were left-handed. None of the participants had any prior knowledge of the experiment. After the experiment, the participants were rewarded with an Amazon gift card worth 2,000 JPY.

Task. In this experiment, two participants wore HMDs and played a rally in a VE in which they hit a tennis ball back and forth with a racket for as long as possible. When the trigger button on the right controller was pressed, the tennis ball was ejected from the ball ejector located in front of the left side for each participant (Fig. 3). The participants started the rally by hitting the ejected ball over the net. This sequence of events is referred to as a "serve" in this experiment. To avoid bias because only one participant continued to serve, we conducted a change of serve in which the other participant served after a certain period of rally time.

In the experiment, the direction of the ball hit back by the racket was fixed because we focused on investigating the discomfort caused by the hit-stop. We judged that if the ball's direction were not fixed, the rally would not continue, and participants' attention would be diverted to other factors besides the hit-stop. In addition, the range within which the racket could detect collisions with the ball was set larger than what appeared to make it easier for the racket to return the ball.

Fig. 3. Scene of the experiment. The ball is ejected from the green box on the left toward the participant. The blue box on the right ejects the ball toward the other participant.

Study Design. The experiment was conducted using a within-participant design with five conditions: four conditions with two factors (one's own symmetry and the other's symmetry, Figs. 4 and 5) and one baseline condition.

		The Other's Symmetry	
		Symmetric	Asymmetric
One's Own Symmetry	Sym-metric	Sym-Sym presentation	Sym-Asym presentation
	Asym-metric	Asym-Sym presentation	Asym-Asym presentation

Fig. 4. Four conditions with two factors

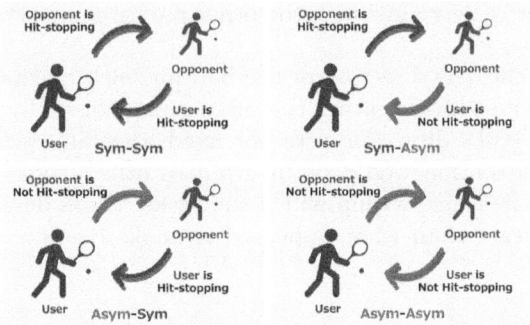

Fig. 5. Conceptual diagrams of the four conditions

Here, one's own symmetry refers to whether the hit-stop was presented symmetrically or asymmetrically to the first participant, the survey respondent him/herself, and the other's symmetry refers to whether the hit-stop was presented symmetrically or asymmetrically to the second participant, the opponent of the survey respondent. A method in which the other person's hit-stop is shown is called a symmetric presentation of the hit-stop (the hit-stop is presented equally to the two participants), and a method in which the other person's hit-stop is not shown is called an asymmetric presentation (the hit-stop is presented differently to the two participants). Let the two participants be Players A and B. The five conditions from the perspective of Player A are as follows:

– *Baseline*: Not applied hit-stop to both players.
– *Symmetric-symmetric presentation*: Applied a hit-stop to both players. Symmetric presentations to A and B. (Player A can see Player B's hit-stop, and B can see A's hit-stop.)
– *Symmetric-asymmetric presentation*: Applied a hit-stop to both players. Symmetric presentation to A and asymmetric presentation to B. (Player A can see Player B's hit-stop, but B cannot see A's hit-stop.)
– *Asymmetric-symmetric presentation*: Applied a hit-stop to both players. Asymmetric presentation to A and symmetric presentation to B. (Player A cannot see Player B's hit-stop, but B can see A's hit-stop.)
– *Asymmetric-asymmetric presentation*: Applied a hit-stop to both players. Asymmetric presentation to A and B. (Player A cannot see Player B's hit-stop, and B cannot see A's hit-stop.)

Two points should be noted in this regard. First, the two participants always witnessed their own hit-stops for the four conditions other than the baseline condition. Therefore, the only thing that changed depending on the condition was the visibility of the other's hit-stop from the participant's point of view. Second, as the two participants experimented simultaneously, their own experiment conditions were related to each other's conditions. For example, if Player A conducts the experiment under symmetric-asymmetric conditions, then Player B will be in asymmetric-symmetric conditions.

The rallies were conducted five times, with one trial per condition. The order of the trials was randomized to suppress the influence of trial order.

As a common setting to all five conditions, the right controller vibrated for 100 ms when the ball impacted. For the four conditions, except for the baseline condition, we adopted the stop-motion condition from a previous study [1]. Specifically, at the time of the collision, all the movements of the ball, racket, and avatar were stopped for 200 ms, and the avatar in the VE was set to catch up with the real body at three times the normal speed.

Experimental System. Each participant wore Meta Quest 3. Meta Quest 3 was wired to a PC via Quest Link for real-time control. The two PCs used in this experiment were Katana 15 B13V, which is powered by a 13th-generation Intel Core processor and NVIDIA GeForce RTX 4070.

The VR tennis environment used in this experiment was implemented using Unity 2021[3]. The VE was built using Oculus Integration[4], a plugin widely used in VR research. This made it possible to acquire the positions and rotations of the HMD and controllers in real time using Unity. The Unity plugin Final IK[5] was used to track the avatar to the real body. In addition, a plugin called Mirror[6] was used to synchronize the position and rotation information of the objects among multiple PCs connected by a LAN cable.

The tennis ball in Unity had a diameter of 68 mm and a weight of 58 g, similar to a real hardball. The ball was ejected from the ball ejector at approximately 10 m/s, and the distance between the ejector and the player was set to approximately 9 m.

Procedure. All the participants received a detailed explanation of the experiment and signed an informed consent form. After answering the pre-questionnaire and simulator sickness questionnaire (SSQ) [8], the participants repeated the rally task and answered the questionnaire five times. The following procedure was used for each trial.

1. Wear the HMD, practice rally, and adjust the standing position under the no hit-stop (baseline) condition.
2. One of the participants serves and plays a rally for 2 min.
3. Change the serve, and the other participant serves and rallies for 2 min.
4. Remove the HMD and answer the questionnaire to evaluate the tennis experience, the virtual embodiment questionnaire (VEQ) [9], and the igroup presence questionnaire (IPQ) [10].

---

[3] https://unity.com/.

[4] https://assetstore.unity.com/packages/tools/integration/oculus-integration-deprecated-82022.

[5] https://assetstore.unity.com/packages/tools/animation/final-ik-14290.

[6] https://assetstore.unity.com/packages/tools/network/mirror-129321.

In the questionnaire to evaluate their tennis experience, the respondents answered questions with a numerical value ranging from 0 (completely disagree) to 100 (completely agree). We used these values in our analysis as a measure of discomfort, impact force sense, and enjoyment. The questions were as follows.

- I felt "my" play was natural. (Discomfort with oneself)
- I felt "the other's" play was natural. (Discomfort with the other person)
- I felt the impact force on "my" racket. (Impact force sense on oneself)
- I felt the impact force on "the other's" racket. (Impact force sense on the other person)
- I felt the VR experience was enjoyable. (Enjoyment)

After completing all the trials, the participants answered the SSQ again and then answered the verbal interview.

The duration of the experiments ranged from 60 to 90 min. The total VR experience time was approximately 40–50 min, and the maximum continuous HMD wearing time was less than 15 min.

4   Results

When analyzing the data, analysis of variance (ANOVA) with an aligned rank transform (ART) was performed for each indicator (discomfort, impact force sense, etc.), with two factors: one's own symmetry and the other's symmetry, for the four conditions, excluding the baseline condition. In addition, multiple comparisons using the Wilcoxon signed-rank test were repeated four times between the baseline group and the other four conditions, and corrected using Holm's method. These comparisons were conducted to confirm that the proposed method does not decrease enjoyment and improves one's own sense of the impact force, as in a previous study [1]. In this section, feelings toward the participant's own body will be expressed using "oneself," such as "discomfort with oneself," and toward the opponent's body using "the other person," such as "impact force sense on the other person."

4.1   Discomfort

Discomfort was calculated from the questionnaire responses regarding naturalness as follows (Fig. 6):

$$(Discomfort) = 100 - (Naturalness) \tag{1}$$

Discomfort with Oneself. First, the results of two-way ANOVA for discomfort with oneself confirmed that the main effect was significant for the other's symmetry ($F(1, 23) = 7.65$, $p = .011$, $\eta_p^2 = 0.25$), but not for one's own symmetry ($F(1, 23) = 1.14$, $p = .297$, $\eta_p^2 = 0.05$) and interaction ($F(1, 23) = 0.208$, $p = .653$, $\eta_p^2 = 0.009$) were not significant.

In addition, the results of multiple comparison tests show significant differences between the baseline group in all four conditions (Table 1).

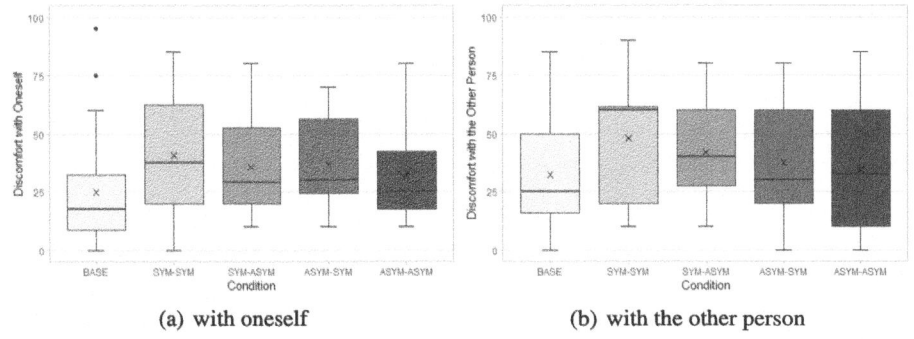

(a) with oneself   (b) with the other person

Fig. 6. Discomfort (a) with oneself and (b) with the other person

Table 1. Value of discomfort with oneself and comparison between baseline and the other four conditions

Condition	Mean	SD	Effect Size r	p	Corrected p
Baseline	24.8	25.2	–	–	–
Sym-Sym	40.8	25.6	0.435	0.002	0.010
Sym-Asym	35.5	20.2	0.354	0.014	0.028
Asym-Sym	36.8	18.8	0.395	0.006	0.019
Asym-Asym	32.3	23.5	0.291	0.044	0.044

Discomfort with the Other Person. Next, the results of ANOVA for discomfort with the other confirmed that the main effect was significant for one's own symmetry (F (1, 23) = 6.63, p = .017, η_p^2 = 0.22), but not for the main effect of the other's symmetry (F (1, 23) = 1.06, p = .314, η_p^2 = 0.04) and the interaction effect(F (1, 23) = 0.005, p = .944, η_p^2 = 0.0002).

In addition, multiple comparisons with the baseline condition detected a significant difference between the symmetry-symmetry conditions. However, no significant differences were found for the other three conditions (Table 2).

Table 2. Value of discomfort with the other person and comparison between baseline and the other four conditions.

Condition	Mean	SD	Effect Size r	p	Corrected p
Baseline	32.4	23.9	–	–	–
Sym-Sym	47.9	24.2	0.609	0.00002	0.0001
Sym-Asym	41.9	19.8	0.256	0.076	0.229
Asym-Sym	37.5	24.8	0.191	0.185	0.370
Asym-Asym	34.4	27.7	0.088	0.543	0.543

4.2   Impact Force Sense

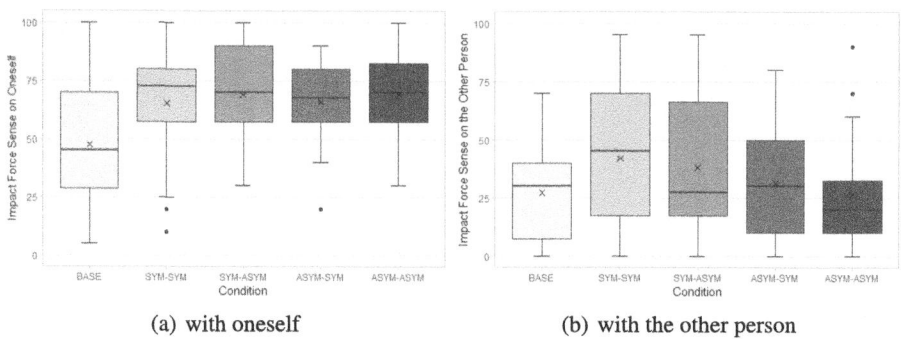

(a) with oneself   (b) with the other person

Fig. 7. Impact Force Sense (a) with oneself and (b) with the other person

Impact Force Sense on Oneself. The ANOVA results for the sense of impact force on oneself were not significant for either the main effect of one's own symmetry, the main effect of the other's symmetry, or the interaction effect.

In the multiple comparison tests, significant differences were detected for all four conditions (Table 3).

Table 3. Value of impact force sense on oneself and comparison between baseline and the other four conditions.

Condition	Mean	SD	Effect Size r	p	Corrected p
Baseline	47.5	27.5	–	–	–
Sym–Sym	65.4	25.2	0.359	0.013	0.013
Sym–Asym	69.0	21.0	0.463	0.001	0.005
Asym–Sym	65.8	19.0	0.374	0.010	0.019
Asym–Asym	69.1	21.4	0.450	0.002	0.005

Impact Force Sense on the Other Person. The results of ANOVA for a sense of impact force on the other confirmed that the main effect was significant for one's own symmetry (F (1, 23) = 6.08, p = .022, η_p^2 = 0.21), but the main effect for the other's symmetry (F (1, 23) = 2.06, p = .165, η_p^2 = 0.08) and the interaction effect (F (1, 23) = 0.033, p = .857, η_p^2 = 0.0014) were not significant.

The the multiple comparison tests detected significant differences for the symmetric-symmetric condition, but not for the other three conditions (Table 4 and Fig. 7).

Table 4. Value of impact force sense on the other person and comparison between baseline and the other four conditions.

Condition	Mean	SD	Effect Size r	p	Corrected p
Baseline	27.2	21.1	–	–	–
Sym-Sym	42.1	29.3	0.439	0.002	0.009
Sym-Asym	38.2	30.3	0.249	0.084	0.253
Asym-Sym	31.2	25.2	0.131	0.363	0.726
Asym-Asym	26.7	22.8	0.054	0.710	0.710

5   Other Indicators

No significant differences were detected in enjoyment in the questionnaire to evaluate tennis experience, Ownership, and Change-two indicators calculated from the VEQ- and G1, SP, INV, and REAL-four indicators calculated from the IPQ.

In contrast, no significant main or interaction effects were found for Agency in the ART ANOVA, but significant differences were found in all four conditions in multiple comparisons with the baseline condition (Fig. 8).

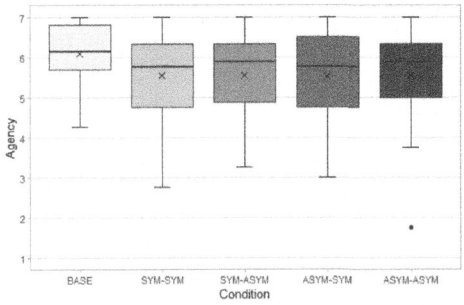

Fig. 8. Agency calculated from VEQ

6   Discussion

First, the mean value of discomfort toward the other person in the asymmetric presentation was significantly lower than that in the symmetric presentation of the participant's own body. More specifically, applying the hit-stop asymmetric presentation method to multi-user VR makes the hit-stops of others invisible to the user, thereby reducing the sense of discomfort.

Next, we showed that an asymmetric presentation to another person could reduce the user's sense of discomfort with his or her own movement. Note that the user did not know in advance whether the other player would observe the user's hit-stop. Thus, the user was influenced in some way to perceive his/her movement as natural or unnatural without knowing the symmetry of the other. It is thought that the discomfort caused by seeing the opponent's unnatural movement due to the hit-stop affected the observer's movement, which in turn caused the user who observed it to feel discomfort in the opponent's movement, which propagated.

In the interview, some participants answered that they felt discomfort with the other player's movement because it was different from the tennis movement in the real world. Thus, it is expected that the cause of the discomfort of hit-stop lies in its disconnection from the real world. In addition, many respondents answered that they felt the hit-stop applied to themselves and those of their opponents presented a problem in the game system, such as lag, communication errors, and screen freezes. This is expected to be caused by the combination of the hit-stop duration of 200 ms and the stop-motion condition, and can be moderated by reducing the duration or by adopting the slow-motion condition. Additionally, several participants responded that the ball returned by the other player moved too quickly after the tennis ball hit-stopped, suggesting that it was difficult for observers to predict the effect of the hit-stop caused by the other player.

The results of multiple comparisons of the impact force sensation on one's own body show that the impact force sensation was stronger with hit-stops than without, which corresponds to the results of Ban et al. [1].

The results also show that the impact force felt against the other's body was enhanced when his/her hit-stop was observable by the user, suggesting that the impact force felt by the hit-stop can be recalled not only for oneself but also for one's counterpart. However, four participants answered during the interview that they did not feel the impact force against the other person, indicating that there were individual differences in whether they felt the impact force against the other person's hit-stop.

Additionally, multiple comparison tests revealed that the sense of agency was reduced in four conditions compared with the baseline condition. These results are consistent with the experimental results of Ban et al. [1], who found that the sense of agency decreased more significantly when the hit-stop duration was 200 ms than when it was 0 ms under the stop-motion condition. It can be said that it is necessary to select an appropriate hit-stop time that does not lower the sense of agency when applying the proposed method to multi-user VR.

7   Conclusion

In this study, we proposed a novel pseudo-haptic presentation method for multi-user VR called asymmetric hit-stop. Our user study revealed that asymmetric hit-stop can evoke a pseudo-impact force on the player while reducing the discomfort caused to the opponent. Furthermore, the results suggest a phenomenon known as "discomfort propagation," in which a player's sense of discomfort is transmitted to the other player by the appearance of discomfort in his or her movements. The hit-stop presentation method in multi-user VR is expected to be applied in various fields in the future, such as competitive VR games, metaverse VR, and rehabilitation tools in virtual shared spaces.

Acknowledgments. This work was partially supported by a JSPS Grant-in-Aid for Scientific Research (B) (JP22H03628) and Grant-in-Aid for Early-Career Scientists (22K17929). We would like to thank Editage (https://www.editage.jp) for English language editing.

Disclosure of Interests. The authors have no competing interests to declare that are relevant to the content of this article.

References

1. Ban, Y., Ujitoko, Y.: Hit-stop in VR: combination of pseudo-haptics and vibration enhances impact sensation. In: 2021 IEEE World Haptics Conference (WHC), pp. 991–996 (2021)
2. Freiwald, J.P., Schenke, J., Lehmann-Willenbrock, N., Steinicke, F.: Effects of avatar appearance and locomotion on co-presence in virtual reality collaborations. In: Proceedings of Mensch und Computer, pp. 393–401 (2021)
3. Lécuyer, A.: Simulating haptic feedback using vision: a survey of research and applications of pseudo-haptic feedback. Presence Teleoper. Virtual Environ. **18**(1), 39–53 (2009)
4. Pusch, A., Martin, O., Coquillart, S.: Hemp-hand-displacement-based pseudo-haptics: a study of a force field application. In: 2008 IEEE Symposium on 3D User Interfaces, pp. 59–66 (2008)
5. Ujitoko, Y., Ban, Y., Hirota, K.: Presenting static friction sensation at stick-slip transition using pseudo-haptic effect. In: 2019 IEEE World Haptics Conference (WHC), pp. 181–186 (2019)
6. Lécuyer, A., Coquillart, S., Kheddar, A., Richard, P., Coiffet, P.: Pseudo-haptic feedback: Can isometric input devices simulate force feedback?. In: Proceedings IEEE Virtual Reality 2000 (Cat. No. 00CB37048), pp. 83–90 (2000)
7. Lécuyer, A., Burkhardt, J. M., Coquillart, S., Coiffet, P.: "Boundary of illusion": an experiment of sensory integration with a pseudo-haptic system. In: Proceedings IEEE Virtual Reality 2001, pp. 115–122 (2001)
8. Kennedy, R.S., Lane, N.E., Berbaum, K.S., Lilienthal, M.G.: Simulator sickness questionnaire: an enhanced method for quantifying simulator sickness. Int. J. Aviat. Psychol. **3**(3), 203–220 (1993)
9. Roth, D., Latoschik, M.E.: Construction of the virtual embodiment questionnaire (VEQ). IEEE Trans. Vis. Comput. Graph. **26**(12), 3546–3556 (2020)
10. Schubert, T., Friedmann, F., Regenbrecht, H.: The experience of presence: factor analytic insights. Presence Teleoper. Virtual Environ. **10**(3), 266–281 (2001)

Design of Haptic Rendering Techniques for Navigating with a Multi-actuator Vibrotactile Handle

Pierre-Antoine Cabaret[1]([✉]), Claudio Pacchierotti[2], Marie Babel[1],
and Maud Marchal[1,3]

[1] Univ Rennes, INSA Rennes, IRISA, Inria, CNRS, Rennes, France
{Pierre-Antoine.Cabaret,Marie.Babel,Maud.Marchal}@irisa.fr
[2] CNRS, Univ Rennes, IRISA, Inria, Rennes, France
Claudio.Pacchierotti@irisa.fr
[3] Institut Universitaire de France (IUF), Paris, France

Abstract. This paper presents the design and experimental evaluation of haptic rendering techniques for navigating using localized vibrotactile stimuli provided by a custom multi-actuator haptic handle. We present two haptic rendering schemes which are then used in combination with three navigation strategies to guide users along a path. We evaluate these techniques in a user study where 18 participants walk in a 8×8 m room, following haptic cues displayed by the handle. Results show that participants are able to navigate along the path successfully, with success rates ranging from 80% to 100% across conditions.

Keywords: Navigation · Handle · Multi-actuator · Vibration

1 Introduction

Safe and efficient navigation in complex or unfamiliar environments poses a daily challenge for many individuals. Nowadays, smartphone applications serve as the predominant mean for wayfinding, using GPS to display location data on screens. However, these solutions demand visual or auditory attention and may not suit individuals with visual impairments or those using mobility aids. In addressing these limitations, haptic feedback has emerged as a promising method for effectively delivering navigational cues [3,11], catering to users both with and without disabilities. Some implementations integrate haptic cues into existing assistive devices [24], while others exist as standalone systems [23]. Importantly, utilizing haptic cues avoids overloading the auditory or visual sensory channels, allowing them to be used for alternate tasks or, in some cases, for users with diminished sensory capabilities [18,19]. One of the challenges of using haptic feedback for navigation is finding a rendering scheme able to provide necessary information while staying easily understandable. Localized sensations appear to be a good fit for providing rich directional information, but they need to be coupled with an intuitive and informative navigation paradigm.

© The Author(s), under exclusive license to Springer Nature Switzerland AG 2025
H. Kajimoto et al. (Eds.): EuroHaptics 2024, LNCS 14768, pp. 224–236, 2025.
https://doi.org/10.1007/978-3-031-70058-3_18

In this paper, we evaluate a set of haptic rendering techniques for navigation, combining three navigation strategies and two rendering schemes using the same custom multi-actuator haptic handle , which can provide localized vibrotactile stimuli at four distinct locations within the users' hand [4]. It also has the advantage of not being as cumbersome or challenging to equip as some wearable devices, which are often not suitable for users with physical disabilities. The three navigation strategies offer guidance by indicating a direction to follow, the deviation from the path, or the direction to return. The two rendering schemes offer different level of granularity of the navigation information, utilizing the localized feedback as a distinctive directional cue relative to the user orientation.

The main contributions of the paper can be outlined as follows:

- Design of haptic rendering techniques for navigation tailored for a custom multi-actuator haptic handle, combining three navigation strategies with two rendering schemes;
- User study evaluation involving 18 participants to assess their effectiveness in following predetermined paths within a 8×8 m room using the provided haptic navigation information.

2   Related Work

Today, GPS-based systems employing screens, audio guidance, or a blend of both are the predominant navigation devices aiding users in unfamiliar settings. Effective navigation requires instructions that users can quickly and easily interpret. To address this, haptic feedback has been proposed as an alternative means of guiding users. Indeed, as the haptic modality remains usually unengaged, providing haptic navigation information avoids interference with already overstimulated visual or auditory cues. Moreover, for individuals with disabilities or impairments, auditory or visual information might not be a suitable choice, in which case haptic feedback can be a more compatible option.

Different types of haptic cues have been employed to provide navigation information, such as skin stretch [5,13], temperature [16], pressure [17], kinesthetic feedback [7], with the most popular being vibrations [1,2,6,12]. Interfaces can be standalone devices [3,12,23,26] or integrated into a mobility assistance device, such as white canes [9,16,21,25] or walkers [24]. Kappers et al. [11] have recently presented a survey of hand-held haptic devices in walking applications, highlighting the popularity and effectiveness of vibrotactile solutions. Interfaces with a single vibrotactile actuator can be used to provide simple proximity information [25] or more complex patterns, e.g., by using tactons [8] to encode the desired information. For example, PocketNavigator [20] uses different vibration patterns to communicate the angle between the user and the next waypoint, increasing the length of the vibration signal with the angle. NaviRadar [22] evaluates different combination of a radar sweep metaphor with intensity, duration, or rhythm to give a similar directional information.

Interfaces with multiple actuators often use the location of the vibration across the hand [16,23,26] or other body parts, such as the waist [10] or the arms [2,6], to encode spatial information. For example, HALO [25] links the actuation of each motor to a proximity sensor. In a similar way, haptic belts and wristbands use the location of the haptic stimuli to communicate a direction around the considered user limb [6,15]. The haptic cricket [23] uses its three actuators differently: the one at the front of the device has its intensity linked to the proximity to the target, while the two on the sides encode right and left deviation from the target, respectively. Lacôte et al. [14] employ a custom 5-actuator haptic handle, using the apparent haptic motion illusion to provide directional cues towards the target.

Overall, the effectiveness of vibrotactile feedback for navigation application has been proven in diverse scenarios , with one or multiple actuators. However, the use of multi-actuator in hand-held devices is less common, mainly due to the fact that providing multiple vibrations in a small area is usually not easy to understand and differentiate. Existing hand-held multi-actuator devices have rarely been evaluated in navigation application. For these reasons, in this paper, we aim to design a set of haptic-based navigation strategies by providing rich and intuitive vibration information on the hand. These strategies apply previous navigation techniques to our multi-actuator handle, capitalizing on localized feedback to improve navigation efficiency.

3 Design

3.1 Haptic Hand-Held Vibrotactile Interface

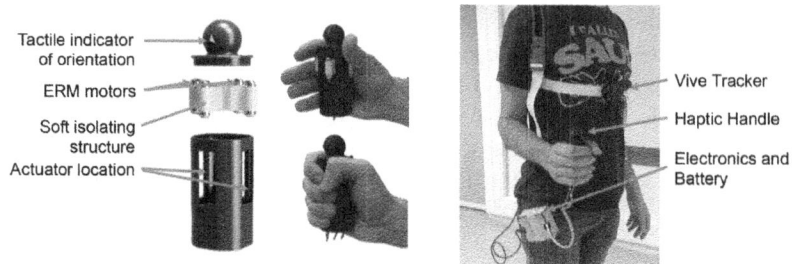

Fig. 1. Our custom haptic handle provides localized vibrotactile feedback within the user's hand thanks to four ERM actuators isolated with a soft 3D-printed structure (left). We evaluated three navigation strategies with two haptic rendering schemes, having participants follow three invisible target paths using the haptic navigation information provided by the haptic handle (right).

In our study, we utilized a custom 3D-printed hand-held haptic interface, shown in Fig. 1. It is designed to offer easily comprehensible directional haptic cues. It achieves this objective through localized multi-actuator feedback in an anthropomorphic design : the handle is topped by a spherical part, resembling a head, with a tactile indent on one side, resembling a nose. Doing this, haptic cues from the handle can be easily mapped to a direction relative to the user. Within the handle, a soft component made of a deformable plastic (TPU) houses four 7-mm-diameter Eccentric Rotating Mass (ERM) motors, facilitating local- ized vibrotactile feedback. This design element was engineered to minimize the propagation of the vibrations across the handle, so as to help the user better identifying the stimuli source. This design was evaluated in more details in [4]. All electronic components were integrated with an ESP32 micro-controller and battery, that users could easily carry on their belt. Navigation commands were computed within Unity and transmitted wirelessly to the handle via ROS2.

3.2 Guidance Feedback

We designed three navigation strategies and two haptic rendering schemes for our haptic handle to guide users along reference paths. Each strategy starts by computing an angle θ, defined as the angle between the user and a navigation target on the path to follow, dependant on both the user orientation and position, as detailed in Fig. 2. θ is then mapped to a directional stimuli to be displayed by the haptic handle depending on the active rendering scheme, as described in Fig. 3.

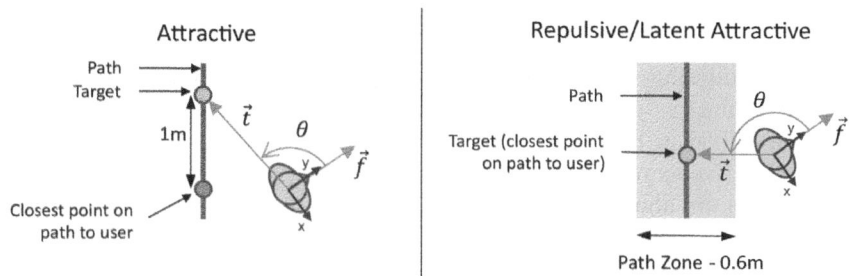

Fig. 2. Each navigation strategy (Attractive A, Repulsive R, Latent Attractive LA) computes a signed angle θ between the user heading direction $\vec{f}$ and a vector pointing toward a target on path $\vec{t}$. The Attractive strategy sets the target as the point on the path one meter ahead from the closest point on path to the user. The Repulsive and Latent strategies set the target to the closest point on path to the user.

The three navigation strategies provide information regarding the position of the user with respect to the target path. The Attractive strategy (A) pro- vides navigation information through 0.2-s-long vibratory cues conveyed regu- larly every second. The Repulsive strategy (R) provides navigation information only when the user is more than 30 cm away from the path (i.e., out of the 60-cm- wide path zone or neighbourhood seen in Fig. 2, corresponding to the width of a side step on each side of the path), continuously conveying vibratory cues oppo- site to the desired motion direction. The Latent Attractive strategy (LA) also

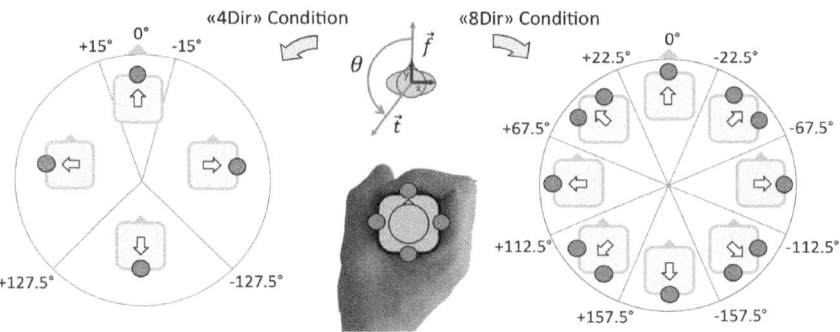

Fig. 3. Each navigation strategy can map θ into vibratory stimuli with two rendering schemes, using four (4Dir) or eight (8Dir) directions. In 4Dir, θ is mapped into the four directions left, right, front, back, using one motor at a time; 8Dir considers four additional directions, front-left, front-right, back-left, back-right, which simultaneously use two actuators. The location and interpretation of each stimulus are linked, e.g., the "left" stimulus is provided on the left side of the handle. Depending on the strategy used to guide the user, these stimuli are played as 0.2-s-burst every second (A) or continuously (R, LA).

provides navigation information only when the user is more than 30 cm away from the path, but it continuously convey vibratory cues *towards* the desired motion direction.

The two rendering schemes provide the directional information with different levels of granularity: 4Dir provides four directional cues while 8Dir provides eight, as described in Fig. 3.

Representative examples of the considered combination of strategies and rendering schemes are shown in Fig. 4.

4   User Study

We recruited 18 participants (14M, 4F, aged 20-47: mean = 27.28, SD = 6.25) to take part in the study, all of whom gave their written informed consent. The study has been approved by Inria's ethics committee (COERLE, No. 2023-49). Before the start of the experiment, participants were introduced to the haptic handle and familiarized with the different haptic cues.

4.1   Experimental Setup

We conducted the experiment in a 8×8 m room (see Fig. 1). A Vive tracking system tracked the user pose. Participants wore a harness on which the Vive tracker was positioned and held the handle in their dominant hand, with the control box attached to their belt. A noise-cancelling headset was used to hide potential audio cues from the vibrations in the handle. We considered three target paths, shown in Fig. 5, which were not visible to the participants. These

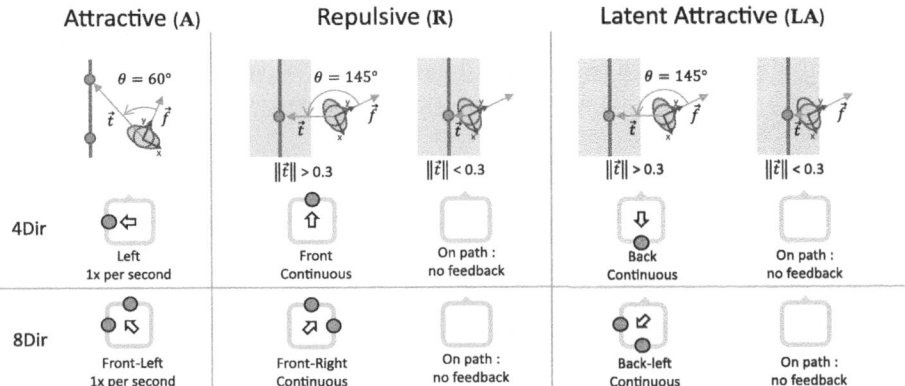

Fig. 4. Navigation strategies (Attractive A, Repulsive R, Latent Attractive LA) are combined with the two rendering schemes (4Dir, 8Dir). The orange circles (bottom) represent the actuators which are activated on the handle, depending on the position and orientation of the user relative to the path (top).

paths were designed to have similar lengths (13.5 ± 1, each with a different combination of angles between path segments (90° for P1, 45° for P1 and both 90° and 45° for P1), which could highlights differences between 4Dir and 8Dir rendering schemes. Two possible starting points on opposite sides of the room were used during the experiment to minimize learning effects (i.e., the target path could be rotated 180 degrees around the center of the room, with no changes in the overall trajectory participants would follow).

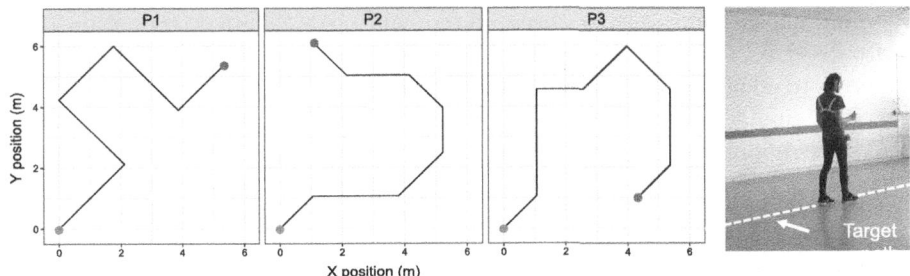

Fig. 5. Three paths P1, P2, and P3 were used throughout the experiment. The starting point of each path is displayed in green and the end point in red (left). During trials, participants were guided along one of these paths without seeing it (right). The paths were regularly rotated by 180°, so as to minimize learning effects without changing their shape.

4.2   Experimental Design

We evaluated the ability of the proposed haptic navigation strategies and rendering schemes to guide participants along target paths. We considered the following experimental variables:

- Navigation strategy: A, R, and LA, as presented in Figs 3 and 2;
- Rendering scheme (information granularity): 4Dir and 8Dir, as shown in Fig.3;
- Target paths: P1, P2, P3, as shown in Fig. 5.

The experiment was made of three blocks, one for each navigation strategy. Within each block, the considered navigation strategy was used with both rendering schemes on the three paths, one after the other (i.e., 6 trials per block). Strategies and rendering schemes presentation order was counter-balanced across participants, while path order was randomized. The starting point changed every three trials, i.e., when the rendering scheme changed.

At the beginning of each block, participants received explanations on the navigation strategy that would be used in the block. They were asked, for each trial, to follow the handle instructions to the best of their ability. Trials started with participants standing at the starting point and facing toward the opposite corner of the room. Trial stopped when reaching the end of the path (0.5 m or less from the end point) or after four minutes, whichever came first.

4.3   Collected Data

We collected position and orientation of the participant, duration of the trial, and information displayed by the handle. After each trial, participants were asked to evaluate their perceived success on a scale from 1 to 7. Participants also judged each rendering scheme, rating three statements on a 7-item Likert scale (1 = Totally disagree, 7 = Totally agree): Vibration were tiring (*Tiring*); Vibrations were easy to locate (*Locate*); Vibrations were difficult to interpret (*VibInterpret*). Participants also rated each navigation strategy through ten statements: I navigated confidently (*NavConfidence*); The navigation strategy was easy to use (*Use*); The guidance instructions were hard to interpret (*NavInterpret*); I could use it without preliminary instructions (*Instructions*); I learned to use it quickly (*LearnSpeed*); It is easy to learn to use it (*Learn*); It is fun to use (*Fun*); It is pleasant to use (*Pleasant*); The task is mentally demanding (*MantalDemand*); The task is physically demanding (*PhysicalDemand*). We also collected open comments throughout the experiment. At the end, participants were asked for their preferred strategy.

4.4   Results

We considered successful trials as trials where (i) participants reached the end of the path (i.e., 0.5 m or less from the end point) in less than four minutes and (ii) they did not walk from one section of the path to another, i.e., did not "cut" the path. Participants were successful in 90% of trials (see Table 1 for results per strategy and rendering scheme; see Fig. 6 for trajectories representative examples). We used a Generalised Linear Mixed Model to analyze the participants success. Navigation strategy, rendering scheme, and path were considered as the independent variables, while participants as a random effect. We observed a significant effect of the navigation strategy only ($\chi^2(2$, N $= 324) = 15.01$, p < 0.001). Post-hoc analysis using Tukey's test highlighted a greater success for A vs. R ($Z = 2.74$, p < 0.05) and A vs. LA ($Z = 3.61$, p < 0.01).

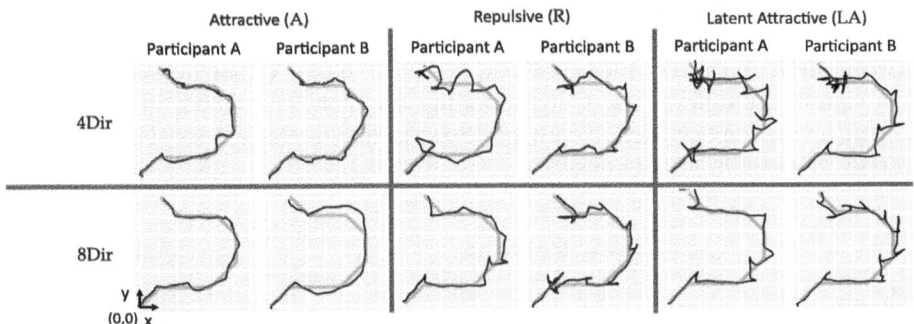

Fig. 6. Examples of trajectories of two participants on P2 (red: target path, black: participant trajectory). Trajectories tend to be smoother with the A strategy.

Table 1. Mean success rates for each combination of strategy and rendering scheme.

	Attractive	Repulsive	Latent Attractive
4Dir	100%	93%	83%
8Dir	98%	87%	80%

For the rest of this section, we only take into account successful trials. Path efficiency ratio, measured as the ratio between the distance walked by participants and the length of the path, is minimal for A (Mean $= 1.35$, SD $= 0.33$), and greater for LA (Mean $= 2.87$, SD $= 1.77$) than for R (Mean $= 2.32$, SD $= 1.09$). Wilcoxon signed-rank test show a significant difference between all pairs of navigation strategies (p < 0.01). While this difference is significant, it must be taken into account that the R and LA strategies provided feedback only when deviating more than 0.3 m from the path, so a higher error is expected. This effect can be observed when looking at the user's distance the path, which is lower for A than for the other strategies (A: mean $= 0.27$, median $= 0.19$; R: mean $= 0.35$, median $= 0.31$; LA: mean $= 0.36$, median $= 0.31$).

We also analyze the self-evaluated success from participants using a Cumulative link mixed-effects model. Results show a significant effect of the navigation strategy ($\chi^2(2$, N $= 291) = 10.06$, p < 0.01) and path ($\chi^2(2$, N$=291) = 7.15$, p < 0.05). As expected, post-hoc Tukey's test showed higher evaluated success for A vs. R (Z $= 2.38$, p < 0.05) and A vs. LA (Z$=2.95$, p < 0.01). P1 also showed significantly better results than P2 (p < 0.05) for this metrics.

Regarding the subjective questionnaire (see Sect. 4.3), Wilcoxon signed-rank tests showed significant differences for *Use* between A and LA (mean: 5.5 vs. 4.0 , p < 0.05), *Instructions* between A and LA (mean: 5.1 vs. 2.8 , p < 0.01) and A and R (mean: 5.1 vs. 3.0 , p < 0.01), *Learn* between A and R (mean: 5.9 vs. 4.6 , p < 0.05), and for *Fun* between A and LA (mean: 6.0 vs. 4.6 , p < 0.05). Significant differences were also observed for *VibInterpret* between A and LA (mean: 2.7 vs. 3.7 , p < 0.01) and A and R (mean: 2.7 vs. 3.8 , p < 0.01), as well as for *VibLocate* between 4Dir and 8Dir (mean: 5.8 vs. 5.1 , p < 0.01). Other questions did not show significant differences. Corresponding boxplots are shown in Fig. 7. At the end of the experiment, 11 out of 18 participants chose A as their preferred strategy, 3 chose R, and 4 chose LA.

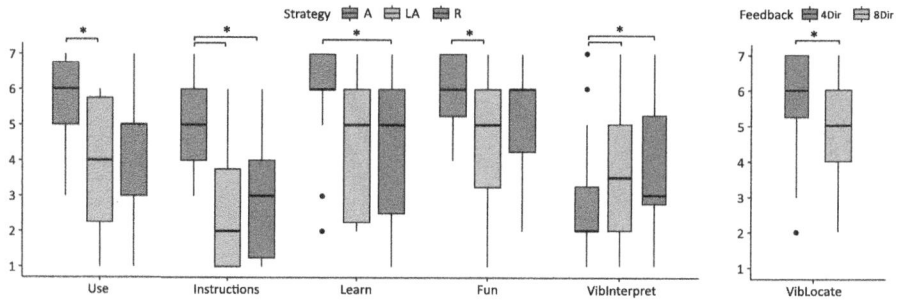

Fig. 7. Participants answers to items of the questionnaire over 7-item Likert scales.

5   Discussion

Our experiment evaluated the effectiveness of the proposed haptic navigation strategies and rendering schemes in guiding participants along target paths. Results showed that users were able to follow the proposed guidance successfully most of the time. Participants were able to use all three strategies successfully and quickly adapted to their use, confirming the intuitive nature of multi-actuator feedback in such a scenario. Overall, results show an advantage of A: trajectories carried out when being provided with this navigation strategy are smoother than the others (see, e.g., Fig. 6) and participants walked lower distances overall. This result is probably linked to the fact that the A strategy (i) is always active, continuously providing information, and that (ii) it conveys an intuitive information, i.e., the direction to go, similarly to turn-by-turn GPS systems.

However, the other navigation strategies still showed promising results. Indeed, several participants reported that evaluating their success in the attractive strategy A was harder, as they did not know how far they were from the target path (this strategy continuously provide feedback). These comments suggest that a promising approach could be to devise an adaptive navigation strategy, providing either information about the direction to follow or the deviation from the path, according to the user's performance or position: the A strategy could be used when users are close to the path, while LA or R could be used when deviation is too large.

After A, the R strategy seems to perform (slightly) better than LA. Comments from participants suggest that the information R provides is intuitive: vibrations can be interpreted as obstacles on which the user "bounces" and changes direction towards the correct path. Indeed, many participants described their experience with R and LA as "bumping into virtual walls", and compared the use of A to that of "a compass", which we found interesting. These two interpretations of the navigation strategies can be observed on the recorded trajectories. We highlight examples of these behaviors in Fig. 8. In failed trials, participants sometimes walked in the reverse direction of the path, which could be prevented by modifying the navigation strategy. In cases where users 'bounced' around the path (e.g., with R), some tended to turn around too much, thus walking from one side of the path to the other without making progress.

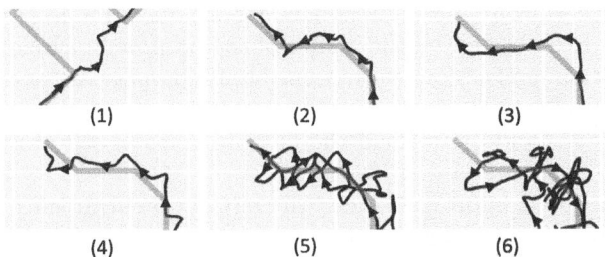

Fig. 8. Representative examples of participants behavior during navigation (red: target path, black: participant trajectory, arrows indicate the direction of motion). (1) Cut (unsuccessful trial): the participant goes too far from the path, directly reaching the other end and failing the trial; (2) Compass: a behavior often observed in the A strategy, the participant reorients at regular intervals using the handle as a compass; (3) Overshoot: the participant goes too far at turns, a result of either a fast walking speed or a long response time to navigation instructions; (4–5) Bouncing: a behavior often observed in the R and LA strategies, were participants tend to bounce from one side of the path to the other; (6) Never reaching the end (unsuccessful trial): the participant turned around before reaching the end of the path.

There does not seem to be a clear preference between 4Dir and 8Dir, with some participants finding the perception of 8Dir difficult and other finding it intuitive and richer. If interpreted correctly, 8Dir provides richer feedback, which allows users to change their trajectory more smoothly, as can be observed on Fig. 6. This result validates the use of both stimuli granularity for navigation, according to user's preference and expertise. Indeed, the personalization of the rendering stimuli is a challenging area of research that is promising in haptics [27].

Having vibrations displayed once every second for A was considered too much for some participants, while others found this frequency appropriate. Continuous vibration used in R and LA also received mixed feelings, showing again the importance of customization of the stimulation. The same applies for the intensity of the vibration, that some participants found appropriate and others too strong. Given the task and path chosen for the experiment, the frequency at which successive indications are provided seems appropriate: a longer delay between stimulation would have taken participants farther from the target path before the next instruction. A promising strategy in this respect is to consider a predictive guidance strategy, in which the instruction frequency depends on the (local) path complexity and speed of the user.

Finally, most participants reported that they enjoyed using the handle during the navigation task, with some of them finding R or LA strategies more "game-like", as if they were exploring a maze.

6   Conclusion

This paper introduces three navigation strategies for a multi-actuator haptic handle, considering two rendering schemes. We conducted a user study enrolling 18 participants. Results showed the ability of the proposed techniques to provide effective navigation instructions within our haptic handle. All navigation strategies were able to effectively guide participants along the target path. However, the "Attractive" strategy performed the best in terms of self-evaluated success, path efficiency ratio, and ease of learning.

Future work could improve the current navigation strategies by recalculating the path dynamically based on the user position (like in cars turn-by-turn navigation systems), considering the personalization and customization of the haptic stimuli, and combining instructions about the direction of the path with warnings about large deviations (e.g., combining the Attractive and Repulsive strategies presented here). The frequency of navigation instructions could also be improved, by adapting it depending on the local path shape and user movements. Although this study was carried out with able-bodied pedestrians, future work will focus on extending our haptic handle and navigation strategies to work with mobility assistance device, and testing with patients showing mobility and/or cognitive impairments.

Acknowledgements. This project has received funding from Inria – Défi "DOR-NELL".

Disclosure of Interests.   The authors have no competing interests to declare.

References

1. Bhatlawande, S., Mahadevappa, M., Mukherjee, J., Biswas, M., Das, D., Gupta, S.: Design, development, and clinical evaluation of the electronic mobility cane for vision rehabilitation. IEEE Trans. Neural Syst. Rehab. Eng. **22**(6), 1148–1159 (2014)
2. Bimbo, J., Pacchierotti, C., Aggravi, M., Tsagarakis, N., Prattichizzo, D.: Teleoperation in cluttered environments using wearable haptic feedback. In: Proceedings of IEEE/RSJ International Conference on Intelligent Robots and Systems (IROS), pp. 3401–3408 (2017)
3. Bouzbib, E., Kuang, L., Robuffo Giordano, P., Lécuyer, A., Pacchierotti, C.: Survey of Wearable Haptic Technologies for Navigation Guidance (2024). https://inria.hal. science/hal-04356277, pre-print
4. Cabaret, P.A., Bout, A., Manzano, M., Guégan, S., Pacchierotti, C., Babel, M., Marchal, M.: Multi-actuator haptic handle using soft material for vibration isolation. In: International Conference on Human Haptic Sensing and Touch Enabled Computer Applications. Springer (2024)
5. Chinello, F., Pacchierotti, C., Bimbo, J., Tsagarakis, N.G., Prattichizzo, D.: Design and evaluation of a wearable skin stretch device for haptic guidance. IEEE Robot. Autom. Lett. **3**(1), 524–531 (2017)
6. Devigne, L., Aggravi, M., Bivaud, M., Balix, N., Teodorescu, C.S., Carlson, T., Spreters, T., Pacchierotti, C., Babel, M.: Power wheelchair navigation assistance using wearable vibrotactile haptics. IEEE Trans. Haptics **13**(1), 52–58 (2020)
7. Devigne, L., Pasteau, F., Babel, M., Narayanan, V.K., Guegan, S., Gallien, P.: Design of a Haptic Guidance Solution for Assisted Power Wheelchair Navigation. IEEE SMC (2018)
8. Elvitigala, D.S., Matthies, D.J.C., Dissanayaka, V., Weerasinghe, C., Nanayakkara, S.: 2bit-tactilehand: Evaluating tactons for on-body vibrotactile displays on the hand and wrist. In: Proceedings of Augmented Humans Conference, pp. 1–8 (2019)
9. Gallo, S., Chapuis, D., Santos-Carreras, L., Kim, Y., Retornaz, P., Bleuler, H., Gassert, R.: Augmented white cane with multimodal haptic feedback. In: Proceedings of IEEE RAS & EMBS International Conference on Biomedical Robotics and Biomechatronics, pp. 149–155 (2010)
10. Heuten, W., Henze, N., Boll, S., Pielot, M.: Tactile wayfinder: a non-visual support system for wayfinding. In: Proceedings of Nordic Conference on Human-Computer Interaction, pp. 172–181 (2008)
11. Kappers, A.M., Oen, M.F.S., Junggeburth, T.J., Plaisier, M.A.: Hand-held haptic navigation devices for actual walking. IEEE Trans. Haptics **15**, 1–12 (2022)
12. Kawaguchi, H., Nojima, T.: STRAVIGATION: a Vibrotactile Mobile Navigation for Exploration-Like Sightseeing. In: Proceedings of International Conference on Advances in Computer Entertainment, pp. 517–520 (2012)
13. Kuang, L., Aggravi, M., Giordano, P., Pacchierotti, C., Robuffo Giordano, P.: Wearable Cutaneous Device for Applying Position/Location Haptic Feedback in Navigation Applications, pp. 1–6 (2022)
14. Lacôte, I., Pacchierotti, C., Babel, M., Gueorguiev, D., Marchal, M.: Investigating the haptic perception of directional information within a handle. IEEE Trans Haptics **16**, 680 (2023)
15. Monica, R., Aleotti, J.: Improving virtual reality navigation tasks using a haptic vest and upper body tracking. Displays **78**, 102417 (2023)

16. Nasser, A., Keng, K.N., Zhu, K.: ThermalCane: exploring thermotactile directional cues on cane-grip for non-visual navigation. In: Proceedings of International Conference on ACM SIGACCESS (2020)

17. Obermoser, S., Klammer, D., Sigmund, G., Sianov, A., Kim, Y.: A pin display delivering distance information in electronic travel aids. In: Proceedings of IEEE RAS & EMBS International Conference on Biomedical Robotics and Biomechatronics, pp. 236–241 (2018)

18. Pacchierotti, C., Prattichizzo, D.: Cutaneous/tactile haptic feedback in robotic teleoperation: motivation, survey, and perspectives. IEEE Trans. Robot. **40**, 978 (2023)

19. Pacchierotti, C., Sinclair, S., Solazzi, M., Frisoli, A., Hayward, V., Prattichizzo, D.: Wearable haptic systems for the fingertip and the hand: taxonomy, review, and perspectives. IEEE Trans. Haptics **10**(4), 580–600 (2017)

20. Pielot, M., Poppinga, B., Boll, S.: PocketNavigator: vibro-tactile waypoint navigation for everyday mobile devices. In: Proceedings International Conference on Human Computer Interaction with Mobile Devices and Services, pp. 423–426 (2010)

21. Pyun, R., Kim, Y., Wespe, P., Gassert, R., Schneller, S.: Advanced Augmented White Cane with obstacle height and distance feedback. In: IEEE International Conference on Rehabilitation Robotics, pp. 1–6 (2013)

22. Rümelin, S., Rukzio, E., Hardy, R.: NaviRadar: A tactile information display for pedestrian navigation. In: Proceedings of ACM UIST, pp. 293–302 (2011)

23. Spiers, A.J., Dollar, A.M.: Outdoor pedestrian navigation assistance with a shape-changing haptic interface and comparison with a vibrotactile device. In: Proceedings of IEEE Haptics Symposium, pp. 34–40 (2016)

24. Wachaja, A., Agarwal, P., Zink, M., Adame, M.R., Möller, K., Burgard, W.: Navigating blind people with walking impairments using a smart walker. Auton. Robot. **41**(3), 555–573 (2017)

25. Wang, Y., Kuchenbecker, K.J.: HALO: haptic alerts for low-hanging obstacles in white cane navigation. In: Proceedings of IEEE Haptics Symposium, pp. 527–532 (2012)

26. Yang, G.H., Jin, M.S., Jin, Y., Kang, S.: T-mobile: Vibrotactile display pad with spatial and directional information for hand-held device. In: Proceedings of IEEE International Conference on Intelligent Robots and Systems (IROS), pp. 5245–5250 (2010)

27. Young, E.M., Gueorguiev, D., Kuchenbecker, K.J., Pacchierotti, C.: Compensating for fingertip size to render tactile cues more accurately. IEEE Trans. Haptics **13**(1), 144–151 (2020)

Evaluating Tactile Interactions with Fine Textures Obtained with Femtosecond Laser Surface Texturing

G. Schuhler[1], H. Zahouani[1], J. Faucheu[3], Y. Di Maio[4], R. Vargiolu[1], and M. W. Rutland[1,2(✉)]

[1] Ecole Centrale de Lyon, CNRS, ENTPE, LTDS, UMR5513, 69130 Ecully, France
mark@kth.se
[2] Chemistry Department, KTH Royal Institute of Technology, Stockholm, Sweden
[3] LGF, UMR 5307 CNRS, Mines Saint-Etienne, Université de Lyon, Centre SMS, 42023 Saint-Etienne, France
[4] Manutech-USD, 20 rue du professeur Benoît Lauras, 42000 Saint-Etienne, France

Abstract. Tactile perception deteriorates with age, resulting in a negative impact on life quality. A clinical assessment of this decline could help to reduce its effects. Such a clinical apparatus for fine texture does not yet exist. Femtosecond laser surface texturing (LST) is capable of manufacturing fine textures on materials that are sufficiently robust for clinical requirements. This paper starts by addressing how LST can be used to manufacture surfaces for tactile tests, i.e. of sufficient dimensions to permit interrogation, and with a minimum quantity of uncontrolled surface features. Vibrotactile interrogation tests on textured surfaces demonstrate that the surface textures have controllable tactile signature and thus underline the suitability of the process for generating fine textures for tactile perception assessment.

Keywords: Tactile perception · fine textures · femtosecond laser surface texturing

1 Introduction

Tactile perception is one of the channels through which human beings can sense their environment. In the same way as sight or hearing, it deteriorates with age, resulting in a negative impact on life quality. This decline is due to a deterioration of the somatosensory nervous system and the biophysical properties of the skin [1–4]. The use of topical formulations can reduce or slow down this decline [5–7].

A clinical assessment of the tactile perception could help to propose the most suitable solutions to reduce or slow down its decline. Tactile thresholds are known to increase with age [1]. Therefore, tactile perception can be assessed with comparison tests between properly textured surfaces [6, 8, 9]. For textures with surface features below 100 μm (described as fine [10]), vibrations and friction are the main stimuli for tactile perception

[8]. Thus, prospective tests need to be active touch perception tests, which systematically vary in these physical dimensions.

At this texture scale, such a test is neither available nor designed for public clinical tests. Indeed, such an apparatus should withstand recurrent mechanical and chemical stresses in the long term as it would be subjected to daily finger rubbing sessions and mandatory cleaning procedures. Polymeric surfaces are widely used for tactile perception or aesthetic studies [8, 11–17], even as coatings on another material [18]. There are indeed numerous ways to generate a surface texture on them, such as UV surface wrinkling [8, 11], 3D-printing [11, 17], moulding [8, 12–15], sandpaper grinding [18] or laser surface texturing [16]. In their study, Skedung et al. used polymeric surfaces with fine wavy textures (270 nm–90 μm wavelengths) they obtained with surface wrinkling [8]. However, they show in the supplementary information of this study a degradation of the surface after multiple tests. Among the robust enough materials – both chemically and mechanically – are stainless steels, ceramics, and glass. Besides the required resistance to repeated use and disinfection in a clinical context, the selected materials need to be suitable for mass production. Moreover, to avoid uncontrolled tactile effects, their surface need to be smooth compared to the texture dimensions. Therefore, this study focuses on relatively unexpensive materials with easily controlled surface finishes or low initial roughness: stainless steel and glass.

In their review on micro-fabrication techniques on stainless steel sheets for skin friction, Zhang et al. concluded that LST is the most suitable process for this application [19]. It has been used to manufacture steel [9, 13, 20–23] or acrylic [16] surfaces for tactile perception studies but not with the same features dimensions as the ones addressed in this study.

The present work deals with fine grooved textures (30 μm and 45 μm wavelengths) on stainless steel and glass. The sense of touch was reported to be able to detect periodic surface variations with amplitudes down to 13 nm [8]. Therefore, in order to manufacture reliable tactile testing surfaces, the laser texturing process parameters must limit uncontrolled surface features such as redeposition or laser-induced periodic surface structures (or ripples) [24–27].

Laser impact duration has a direct effect on metal heating, as the time required to transfer laser energy from the electrons to the lattice is in the order of picoseconds [28, 29]. Ruf and Dausinger stated that at the high intensity level of a femto-second laser pulse the impact ionization results in an electron gas [28]. Thus, the dielectric and semi-conductors have a "metallic" reaction to ultrashort laser pulse impacts.

Di Maio reported that light polarization has an impact on absorption. Polarization perpendicular to the machining direction leads to more regular ablation [30]. In the present study, two polarization directions – parallel and perpendicular to the textured lines – were tested with no significant effect on the surfaces (for steel, compare Fig. 6 set "2s" and Fig. 4 with equivalent parameters. For glass, compare Fig. 6 set "2g" and Fig. 5 with equivalent parameters. See also the remark in Sect. 3.1, "Process and texture quality"). Apart from the results based on Fig. 1, Fig. 4, and Fig. 5, in which the polarization was parallel to the textured lines, the light was polarized perpendicular to the textured lines.

Five parameters are studied because of their impact on the pattern geometry and on the process:

- *Interspot spacing* (IsSp) – the distance between two laser impacts in the direction of the textured lines.
- *Line spacing (LSp)* – the space between the centres of two lines.
- *Number of passes* (NPas) – the number of times the laser passes in each line.
- *Pulse power* (P) – the mean laser power over one repetition period. It is dependent on the repetition frequency.
- *Repetition frequency* (F) – the frequency at which the laser hits the surface.

The fluence is defined as the energy delivered by a laser pulse per illuminated surface unit. It is thus impacted by P, as a higher P results in a higher pulse energy (at fixed F). Di Maio reported that a too high fluence leads to melting, cracks and spalling. A relatively lower fluence can generate ripples [30]. The textured surface also depends on how a similar quantity of laser impacts is brought to the surface. At fixed F and fluence, Di Maio compared 3 cases with different IsSp, so that the same amount of impacts was delivered in 1, 5 or 10 passes [30]. The higher the number of passes and IsSp were, the better the textured line quality was. The important overlapping between impacts led to potential melting of the textured area.

This study presents the results of a systematic parameters study. The process and parameters are then validated with early tactile tests.

2 Material and Methods

2.1 Materials

The first parameter tests – referred to as "set 1" – were performed on:

- AISI 316L stainless steel plates purchased and machined by Slic Valfor (Coreme group). Samples were polished with grit 1200 abrasive paper followed by 3 μm abrasive paste. The measured Sa is 11 nm;
- Microscope glass slides (Fisherbrand™ - product 12373118, 76 mm × 26 mm × 0.8–1.0 mm). The measured Sa is 0.13 nm.

  For all subsequent tests – referred to as "set 2":

- AISI 316L samples are supplied by Outokumpu Germany, in an annealed state. The sample dimensions are 80mm × 40 mm × 1.23 mm for the chosen surface finish (Outokumpu process 2R – 2R2). The measured Sa is 46 nm;
- Float glass samples have been selected (SCHOTT D263® T eco). The samples were cut from wider plates, so their dimensions vary from sample to sample. Their thickness is 1mm. The measured Sa is also 0.13 nm.

In the following sections, AISI 316L will be referred to as "316L" and the float glass as "glass".

2.2 Texturing Parameters

The surfaces were textured at Manutech-USD, with a femtosecond laser. A UV light source (343 nm) has been selected for its high energy and its small spot size (~7.5 µm). The laser impact duration is set to the minimal possible value (350 fs) to avoid metal heating [28, 29]. In this study, the five parameters listed in Sect. 1 were studied by manufacturing – both on 316L and on glass surfaces – matrices of 2 mm × 2 mm patterned squares with different parameters. These more than 200 4 mm² squares were systematically characterized with the equipment detailed in Sect. 2.3. This paper only presents a reduced selection to illustrate the process and the results.

For parametric studies, the default parameters are the following:

- Subsection 3.1, "Geometry", Fig. 2 and Fig. 3: 1st 316L, 5 µm IsSP, 20 µm LSp, 4 passes, 1.24 W P, 200 kHz F. Modified parameters are specified on the figures.
- Subsection 3.1, "Process and texture quality", Fig. 4 and Fig. 5: 2nd 316L/glass, 4 passes, 30 µm LSp.

The sets of parameters presented in Sect. 3.1 Fig. 6 are:

- *Set "1s"*: 1st 316L, 5 µm IsSP, 20 µm LSp, 4 passes, 1.24 W P, 200 kHz F;
- *Set "2s"*: 2nd 316L, 3 µm IsSP, 22.5 µm LSp, 4 passes, 0.14 W P, 100 kHz F;
- *Set "1g"*: 1st glass, 5 µm IsSP, 15 µm LSp, 4 passes, 0.28 W P, 200 kHz F;
- *Set "2g"*: 2nd glass, 3 µm IsSP, 15 µm LSp, 4 passes, 0.14 W P, 100 kHz F.

2mm × 2mm squares are suited for a parametric study, but not for tactile tests. Surfaces of 30 mm × 60 mm have been defined as wide and long enough to conduct tactile tests. However, due to the optical apparatus and the desired laser spot size, the maximum lens field without loss of focus is a 30 mm × 30 mm square. Indeed the authors confirmed that the limit for the maximum lens field is of the order of 30 × 30 mm: larger patterned surfaces had optical aberrations, and average IsSP and depth changed for areas beyond the 30 mm × 30 mm square limit (Fig. 1). Thus the patterns were constrained to be within this size limit.

The possibility of stitching several surfaces by moving the stage has been studied. Two approaches are quickly compared: stitching of 2 maximal lens field surfaces (30 mm × 30 mm) and stitching of 8 smaller surfaces (30 mm × 7.5 mm).

2.3 Patterned Surface Characterization

The textured surfaces were systematically characterized. The surface micron scale was observed with a Tescan Mira SEM. The surface topography was measured with a Bruker GT-K1 interferometer. The surface topography of the non-textured glass plates was measured with an AFM Park NX10. All the topography files were post-treated using a tilt removal, a standard deviation statistic filter (size 2 points) and a data restore legacy algorithm (5 iterations).

2.4 Patterned Surface Tactile Evaluation

Touchy Finger ® is a device that has been developed in LTDS [31, 32]. It can be equipped on a finger to measure the vibration acceleration and the applied force of this finger sliding on a surface. The vibration measurements were conducted by four users:

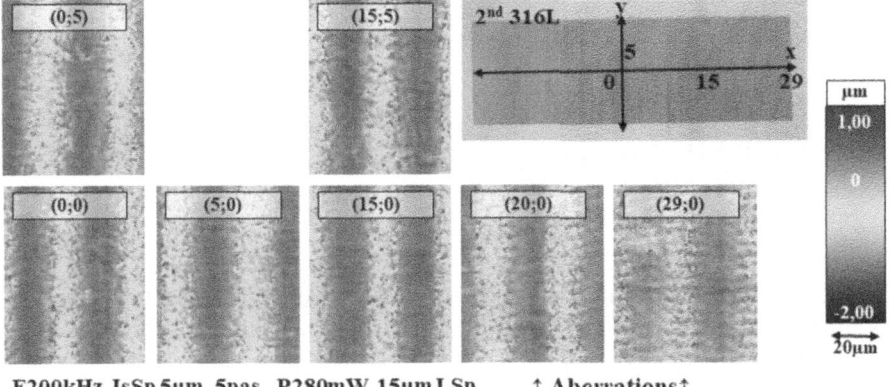

F200kHz, IsSp 5µm, 5pas., P280mW, 15µm LSp   ↑ Aberrations↑

Fig. 1. The texture beyond a 30 mm × 30 mm square area limit displays aberrations. Interferometry topology pictures of a 60 mm × 30 mm patterned surface (post-treated).

- *T1*: male, 32y old, right handed;
- *T2*: female, 27y old, left handed;
- *T3*: male, 55y old, right handed;
- *T4*: male, 35y old, right handed.

The four users were selected for their familiarity with the Touchy Finger ® technology which precludes any artifact due to a lack of training.

2 non-textured surfaces and 4 surfaces textured with the parameters from sets "2s" and "2g" (Fig. 6) were tested:

- *No texture (316L and glass)* – referred to as "s NT" and "g NT".
- *Line spacing 30 µm, 2 stitched 30 mm × 30 mm squares (316L)* – referred to as "s 30 µm 1st".
- *Line spacing 45 µm, 2 stitched 30 mm × 30 mm squares (316L and glass)* – referred to as "s 45 µm 1st" and "g 45 µm 1st".
- *Line spacing 45 µm, 8 stitched 7,5mm x 30mm rectangles (316L)* – referred to as "s 45 µm 7st".

For each surface, the users had to stroke the surface 3 times longitudinally from top to bottom with their dominant index finger. The sliding direction has been decided, based on the supplementary information of Skedung et al.'s study [8]. Additionally, due to the design of Touchy Finger ® [31, 32], the users can freely adapt the position and orientation of their finger in contact with the scanned surface. Since, it was not imposed by the instructions, the users involuntarily chose the most suitable area on their finger according to their habits of touch without the device on. After the measurement, each user was informed about the existence and the origin of the stitches, as well as their locations on the surfaces. They were asked to touch the surfaces with opened eyes and test if they could sense the stitches.

The identification of the start and finish of the sliding events is performed manually. The small number of participants greatly facilitated this process. A fast Fourier transform with a Blackman windowing function was applied on each sliding sequence acceleration signal [33]. The noise was removed from the obtained spectra, using the spectra of measured noise (equipped Touchy Finger ®, without any contact). For each user, the arithmetic mean of the three denoised spectra was then calculated.

3   Results and Discussion

3.1   Texturing Parameters

Geometry
The effect of line spacing on the patterned surface is as expected. The higher the LSp, the more spaced the lines. However, if the LSp is smaller than the spot diameter, the lines lose their definition (Fig. 2a). In some cases, a secondary pattern can result from these close lines (Fig. 3a).

The number of passes directly impacts the groove depth (Fig. 2b). For the NPas displayed in Fig. 2 b, depth is proportional to the NPas: for 2, 4, 10, 20 and 40 passes with the parameters used in Fig. 2 b depths are respectively 1.3 μm, 1.9 μm, 5.2 μm, 9.4 μm and 20.0 μm. Also, at each pass, a part of the molten matter is "splashed" outside the impact area. This effect is easily noticed on 316L (Fig. 2 b and Fig. 6 set "1s"), but also occurs on glass (Fig. 6 set "1g") though to a lesser extent. This splashing combined with matter redeposition leads to the presence of particles on top of the grooves. A higher NPas results in a higher quantity of these particles (Fig. 2b). This effect could be used to generate similar surfaces with different groove tops. However, the amount of redeposited material has to be controlled. The first step consisted in reducing it to the maximum to avoid uncontrolled disturbance in the tactile experience.

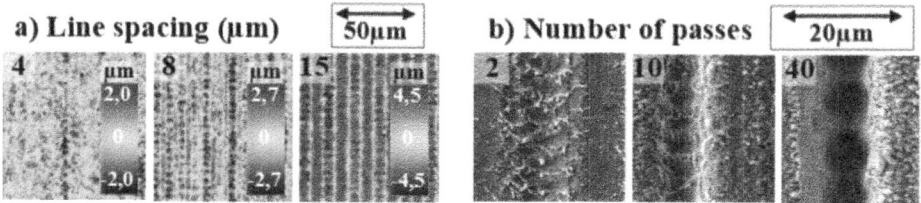

Fig. 2. Effect of geometry parameters – 316L – a) Interferometer topology pictures of patterns with different LSp, b) SEM pictures of patterns with different number of passes (SE probe)

The interspot spacing also has a direct effect on pattern geometry. Figure 2b of Fig. 6 (sets "1s" and "1g") shows clear spaces between impacted spots that are spaced distant from one IsSp (5 μm). This effect is reduced for smaller IsSp values (Fig. 4, Fig. 5 and Fig. 6 sets "2s" and "2g"), as the overlapping area between 2 impacts is increased. In Fig. 3b, the IsSp is increased to a point at which the pattern orientation is modified.

These three geometry parameters are expected to be combinable in order to obtain different wave shapes, such as the one simulated in Fig. 3c. To be able to obtain reliable pattern or pattern combinations, the process parameters must be selected to reduce any unwanted surface features such as redeposition or splashing effect.

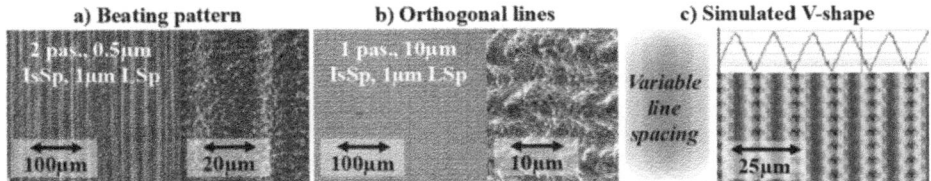

Fig. 3. Alternatives surfaces – a) "beating pattern" (SEM picture, SE probe), b) Orthogonal lines (SEM picture, SE probe), c) Simulated topography on a pattern with variable LSp

Process and Texture Quality

In this study, surface quality is defined as:

- Homogeneity of the surfaces features: line spacing, grooves bottoms, tops, and edges;
- No surface damage that could result in a shortened life of the samples, such as the cracks displayed in Fig. 6 set "1g".

With the set of parameters "1s" (Fig. 6), grooves are heterogeneously deep due to a too large IsSp. This affects the overall shape of the groove borders as well. Moreover, each impact is surrounded by a splashing effect and some redeposition. These results are also observed at lower power in Fig. 4 for IsSp at 3–5 μm and (F; P) values of (50 kHz; 70 mW) or below. Moreover, for lower IsSp values, ridges are generated at the centre of the grooves. This observation is important for the selection of a suitable set of parameters because the central ridges could act as starting points for corrosion.

The glass that was textured with the set "1g" (Fig. 6) displays the same features with a cleaner surface and a reduced splashing effect. In addition, cracks are observed in the direction perpendicular to the grooves. Figure 5 shows that cracks are generated for relatively high values of F and P. Moreover, redeposition is more important in these cases. As with steel, at IsSp 3–5 μm, the impact shape affects the overall shape of the groove borders.

As seen in Sect. 1, cracks and melting can be explained by too many impacts per second and per line or by a high fluence value [30]. This is confirmed by Fig. 4 and Fig. 5. The number of impacts per line and per second is impacted by the IsSp and the repetition frequency. Since a too large IsSp leads to noticeable spaces between impacts (Fig. 4 and Fig. 5), a decrease of both the IsSp and F was chosen. In order to reduce the number of impacts per second, while reducing the space between each impact, the frequency value was defined so that the frequency reduction was more important than the IsSp reduction.

The decisive factor to decide between the different possible (IsSp, F) pairs was the resulting process duration. Indeed, with the initial set "1s" and the (5 μm; 200 kHz) pair, the time to manufacture a 30mm x 60mm is around 12min. This time rises to 10h for the same set of parameters and a (1 μm; 20 kHz) pair. A compromise between the surface state and the process duration led to selecting the (3 μm; 100 kHz) pair – 40min for the discussed surface – in sets "2s" and "2g" (Fig. 6). With F set at 100 kHz, reduction of P was also considered. However with power values below 0.14 W, the ablation did not occur. Therefore P was set at 0.14 W for both 316L and glass.

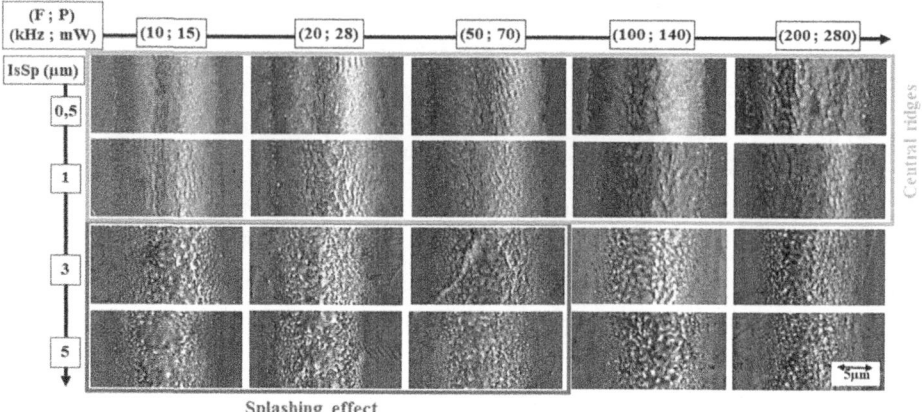

Fig. 4. Systematic study of F, P and IsSp on 316L: SEM pictures (SE probe). For lower IsSp, central ridges are generated. For higher IsSp and lower (F; P) values, a splashing effect is observed around the impacts.

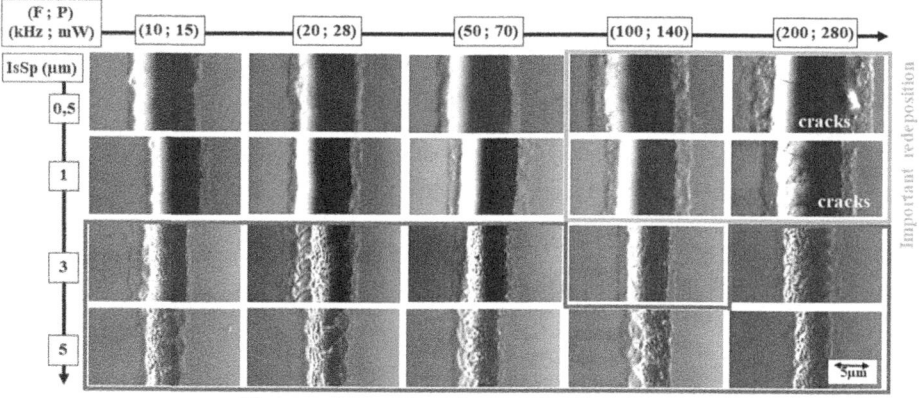

Fig. 5. Systematic study of F, P and IsSp on glass: SEM pictures (BSE probe). For lower IsSp and higher (F; P) cracks are generated and redeposition is relatively more important. For higher IsSp, the spaces between impacts are noticeable.

The grooves obtained with set 2 are wider than the ones from set 1 Fig. 6. Although, they appear more homogeneous on their edges, tops and bottoms:

- The bottom of the grooves on 316L displays some ripples in the same direction as the grooves;
- On glass, the redeposited matter seems to have recombined into smaller units.

The edges on glass are sharper than on 316L. This goes hand in hand with the grooves being deeper on glass than on 316L (with equivalent sets of parameters), as seen on the topology pictures in Fig. 6. Finally, the splashing effect – as well as the cracking of glass – did not occur and the material redeposition was reduced.

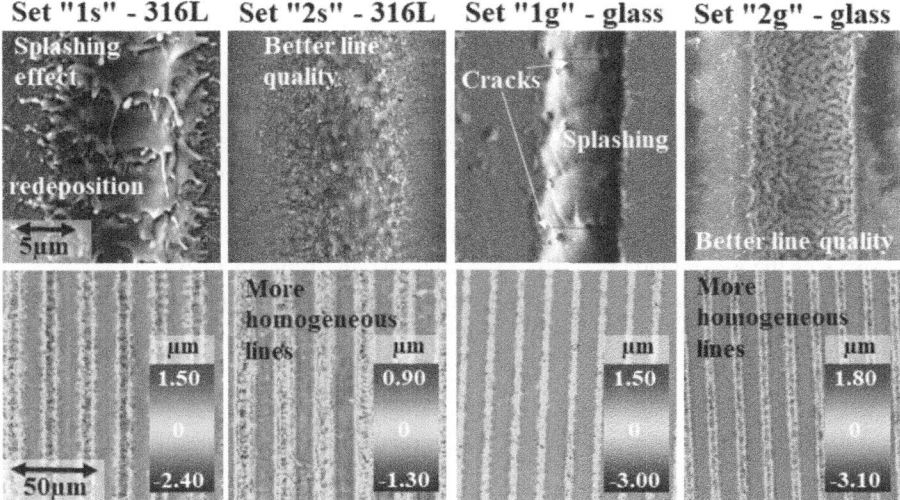

Fig. 6. Comparison of 2 sets of parameters – 316L and glass surfaces: SEM pictures (SE probe, excepted for glass set 1: BSE probe) and topology pictures

Remarks:

- the darker areas outside the grooves in Fig. 6 set "2g", are due to surface charging with the SEM electron beam in high vacuum. It offered a better picture definition and contrast than in low vacuum, but a more rapid build-up of charge on the glass surface.
- We observe one difference between the two polarization directions on the steel, namely the orientation of the laser-induced periodic surface structures (Fig. 4, Fig. 6 set "2s") [24–27]. They are oriented in the direction perpendicular to the polarization.

3.2 Long Surfaces for Tactile Tests

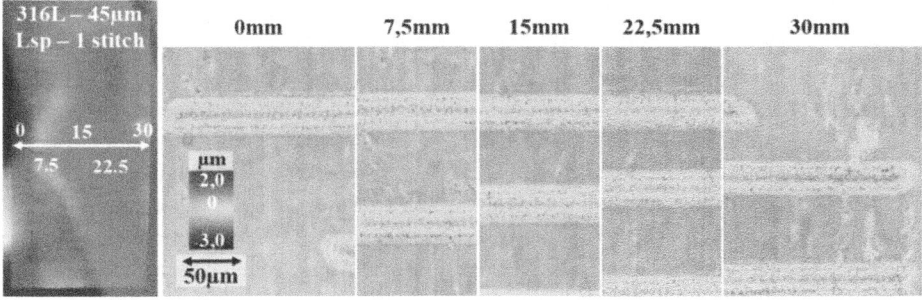

Fig. 7. Interferometry topology pictures of a stitch between two 30 mm × 30 mm squares textured with 45 μm-spaced lines

The process parameters have been selected in order to manufacture the most suitable surfaces for tactile tests. However, the use of long surfaces (to allow the finger to be moved over a sufficient distance to achieve a sufficient tactile interrogation) raises other parameters that need to be addressed. Figure 7 shows an example of interface between two stitched squares on the s 45 µm 1st. surface. An angle between the last line of the first square and the first line of the second square is observed. This angle stems from the shape default induced by the optical apparatus of the laser. The laser reference point is located at the centre of a square. The further a textured line is from this reference point, the more important the defaults get. In the case of two 30 mm × 30 mm squares (Fig. 7), the angle at the interface is 0.115°. For the 316L 45 µm 7st. and its eight 30 mm × 7.5 mm surfaces the mean angle is 0.020° ± 0.002° (mean of the 7 values).

The choice between the 2 stitching methods appears to be unresolvable in tactile perception. Indeed, none of the 4 users could feel any stitch. This observation will be the subject of more extensive psychophysics texts in the future.

3.3 Vibration Analysis

For each sliding sequence, the steps were similar (Fig. 8). The user placed their finger on the surface while starting to slide. Then the user slid their finger against the surface and removed it. These steps were repeated 3 times. The loading/beginning of sliding event, as well as the end of sliding/unloading event last a few tenths of second each. In the case of this study, the sliding step lasted between 0.6s and 3.2s depending on each user (Table 1). The sliding surfaces were 6cm long for the textured samples, and 7-8cm long for the non-textured samples. To simulate unconstrained natural conditions, the users did not start their sliding at the same distance from the edge of a surface. Therefore, the finger-tip of the users did not necessarily slide along the entire length. This observation led to consider possible sliding distance ranges (Fig. 8). With the measured sliding times (Table 1) – deduced from the acceleration and applied load curves – it is thus possible to determine possible sliding speed ranges.

Figure 9 shows frequency spectra of user T2 on the 6 surfaces. Frequency peaks can be distinguished for the textured surfaces. This result is also valid for the other users, although the amplitudes do vary. An additional observation is that, for T1 and T2, glass generates higher vibration amplitudes.

For a grooved pattern, the frequency peaks depend on the pattern wavelength (LSp) and on the sliding speed [34, 35]. For a given surface, it is thus possible to obtain an expected frequency range by dividing the sliding speed range by the LSp. The expected frequency ranges of each user on each surface are displayed in Fig. 10 with green arrow heads. The sampling frequency of Touchy Finger ® is 3200 Hz. When an expected frequency range goes beyond 1600 Hz, the alias frequency range is added to Fig. 10 with blue bars. The peak frequencies from the 24 experiments are represented in Fig. 10 with black dots. Table 1 highlights the important differences of sliding times among the four users. For the sake of clarity, all frequency values in Fig. 10 have been multiplied by their corresponding mean sliding times, i.e. normalized to their corresponding "sliding frequency".

Figure 10 reveals that most of the reported frequency peaks are within – or close to – the expected frequency ranges or their aliased version. Multiple peak values of the same order of magnitude can be due to the fact that each studied spectrum is a mean of 3 spectra. The variations in sliding speed from one measurement to another can affect

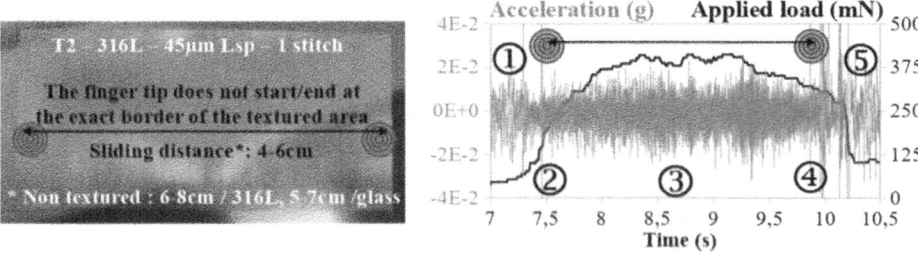

Steps: ①: loading, ②: begining of the sliding, ③: sliding, ④: end of the sliding, ⑤: unloading.

Fig. 8. Different sliding ranges and curves from test s 45 μm 1st. Case (user T2)

Table 1. Mean of the measured sliding times on the 6 surfaces (t_S), for each user

Surfaces (316L = s. glass = g)	Mean t_S - T1	Mean t_S - T2	Mean t_S - T3	Mean t_S - T4
s NT	1.53 ± 0.15	3.18 ± 0.39	0.80 ± 0.03	0.67 ± 0.07
s 30 μm 1 st	1.33 ± 0.06	2.56 ± 0.26	0.74 ± 0.03	0.58 ± 0.04
s 45 μm 1 st	1.23 ± 0.06	2.39 ± 0.04	0.78 ± 0.07	0.71 ± 0.03
s 45 μm 7 st	1.10 ± 0.10	2.00 ± 0.50	0.87 ± 0.04	0.78 ± 0.04
g NT	1.10 ± 0.00	2.18 ± 0.50	0.84 ± 0.05	0.63 ± 0.11
g 45 μm 1 st	1.17 ± 0.12	1.80 ± 0.13	0.84 ± 0.09	0.69 ± 0.07

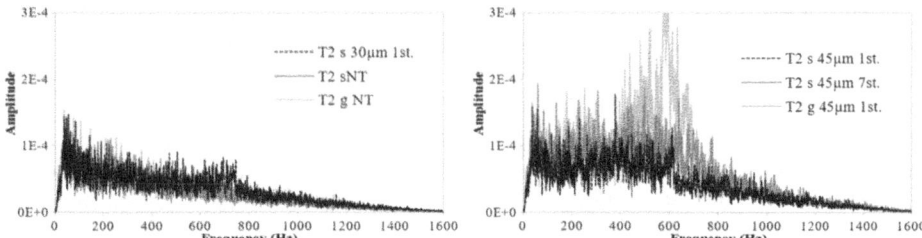

Fig. 9. Frequency spectra for T2 – Each spectrum is the mean of 3 denoised spectra, smoothed with a Savitsky-Golay algorithm (5 points)

the peak frequency. In further studies, it will be necessary to use more samples than 3 per user.

Low frequency peaks could be explained by vibrations due to the finger ridges. Prevost et al. showed that the frequency peaks for an artificial finger sliding on a white noise generating surface was equal to the sliding speed divided by the spacing between finger ridges [36]. With a mean spacing of 350 μm [37], and the different sliding times and distance ranges, these values would be comprised between 45 Hz and 256 Hz for the 24 tests. Once normalized to the "sliding frequencies", this interval becomes 113–229. Apart from these low peaks for T4, no peak is observed for the non-textured surfaces.

The vibration frequency peaks are lower for the three surfaces with 45 μm LSp than for the one with 30 μm LSp. Besides, the frequency peaks of the 3 LSp 45 μm surfaces are close to one another. This consistency confirms the suitability of LST to generate fine textures on 316L and glass. Moreover, the s LSp 45 7st. surface displays the least dispersion around the expected ranges. This could be explained by the smallest disturbance caused by the better stitches on this surface. This difference only occurs for the vibration interaction. Indeed, among the four users, none could feel any stitch, even when knowing what they were and where they were located on the surfaces.

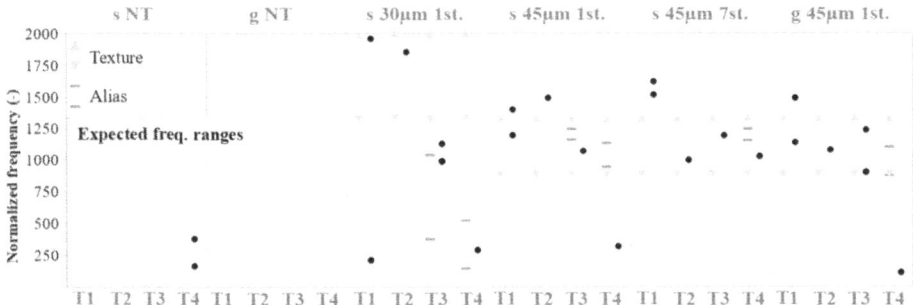

Fig. 10. Expected frequency and alias ranges (green and blue) calculated from Table 1, frequency peaks for each user and surface – frequencies were multiplied by the mean sliding time (normalized to the "sliding frequency")

Finally, Table 1 shows that for every surface, T2 slid more slowly than the other users. This resulted in frequency peaks located in the detection range of Pacinian corpuscles [38]. Abdouni et al. showed that the density of mechanoreceptors is higher for female groups than for male groups. This result is significant for groups below 40y old [4]. T2 is thus expected to have a better tactile perception than the other users. This better tactile perception could explain an involuntary choice to slide slower than the other users, in order to optimize perception. Deeper grooves could help the users to get an improved sense of the patterned surfaces and to adapt their sliding speed.

4   Conclusions

Laser Surface Texturing of fine textures on 316L and glass for tactile perception tests was studied through a systematic process parameter variation on more than 200 surfaces. The most relevant parameters to reduce unwanted surface features in an industry acceptable process duration were selected. The manufacturing of sufficiently large surface areas was then studied through early tactile tests. The 2 methods of stitching led to consistent frequency peaks, and the surface with multiple stitches got the most predictable results. Moreover, the stitches were not perceived by the users.

In subsequent studies, more surfaces will be tested on broader and more diverse panels of test subjects, such as different ages and genders. Their finger physico-chemical, mechanical, thermal and topological properties will be examined. The effect of sliding direction on tactile perception will be addressed as well [4]. The most suitable geometry parameters such as LSp, or wave shapes (Fig. 3c) will be studied to define standardized tests for clinical assessment of fine texture perception.

Such a standardized test for fine textures could ultimately be used for other purposes than the initial clinical one. It could be used as a reference to assess the performance of artificial touch sensor in robotics [39], or the user-dependent finger transfer function for vibrotactile surface rendering [40–42].

Acknowledgments. This work was performed within the framework of the LABEX MANUTECH-SISE (ANR-10-LABX-0075) of Université de Lyon, within the Plan France 2030 operated by the French National Research Agency (ANR). We thank Pauline Paiva of Manutech-USD for her supervision of the laser surface texturing operations. We thank Dr Saman Hosseinpour of Outokumpu for the design and provision of steel samples suitable for study.

Disclosure of Interests. The authors have no competing interests.

Ethical Statement. All the procedures performed in this study were in accordance with the ethical standards of the 1964 Declaration of Helsinki. All the participants were adequately informed of the aims, methods used and results presented in this paper. They were informed they could stop the tests at any time. They gave their written informed consent to the protocol.

References

1. Wickremaratchi, M.M., Llewelyn, J.G.: Effects of ageing on touch. Postgrad. Med. J. **82**, 301–304 (2006)
2. Skedung, L., et al.: Mechanisms of tactile sensory deterioration amongst the elderly. Sci. Rep. **8**, 5303 (2018)
3. Abdouni, A., Djaghloul, M., Thieulin, C., Vargiolu, R., Pailler-Mattei, C., Zahouani, H.: Biophysical properties of the human finger for touch comprehension: influences of ageing and gender. Roy. Soc. Open Sci. **4**, 170321 (2017)
4. Abdouni, A., Moreau, G., Vargiolu, R., Zahouani, H.: Static and active tactile perception and touch anisotropy: aging and gender effect. Sci. Rep. **8**, 14240 (2018)
5. Skedung, L., Buraczewska-Norin, I., Dawood, N., Rutland, M.W., Ringstad, L.: Tactile friction of topical formulations. Skin Res. Technol. **22**, 46–54 (2016)
6. Aimonetti, J.-M., Deshayes, C., Crest, M., Cornuault, P.-H., Weiland, B., Ribot-Ciscar, E.: Long term cosmetic application improves tactile discrimination in the elderly; a new psychophysical approach. Front. Aging Neurosci. **11** (2019)
7. Samain-Aupic, L., Gilbert, L., André, N., Ackerley, R., Ribot-Ciscar, E., Aimonetti, J.-M.: Applying cosmetic oil with added aromatic compounds improves tactile sensitivity and skin properties. Sci. Rep. **13**, 10550 (2023)
8. Skedung, L., Arvidsson, M., Chung, J.Y., Stafford, C.M., Berglund, B., Rutland, M.W.: Feeling small: exploring the tactile perception limits. Sci. Rep. **3**, 2617 (2013)
9. Cesini, I., Ndengue, J.D., Chatelet, E., Faucheu, J., Massi, F.: Correlation between friction-induced vibrations and tactile perception during exploration tasks of isotropic and periodic textures. Tribol. Int. **120**, 330–339 (2018)
10. Hollins, M., Risner, S.R.: Evidence for the duplex theory of tactile texture perception. Percept. Psychophys. **62**, 695–705 (2000)

11. Arvidsson, M., Ringstad, L., Skedung, L., Duvefelt, K., Rutland, M.W.: Feeling fine - the effect of topography and friction on perceived roughness and slipperiness. Biotribology **11**, 92–101 (2017)
12. Faucheu, J., Weiland, B., Juganaru-Mathieu, M., Witt, A., Cornuault, P.-H.: Tactile aesthetics: textures that we like or hate to touch. Acta Physiol (Oxf.) **201**, 102950 (2019)
13. van Kuilenburg, J., Masen, M.A., Groenendijk, M.N.W., Bana, V., van der Heide, E.: An experimental study on the relation between surface texture and tactile friction. Tribol. Int. **48**, 15–21 (2012)
14. Kawasegi, N., Fujii, M., Shimizu, T., Sekiguchi, N., Sumioka, J., Doi, Y.: Evaluation of the human tactile sense to microtexturing on plastic molding surfaces. Precis. Eng. **37**, 433–442 (2013)
15. Massimiani, V., Weiland, B., Chatelet, E., Cornuault, P.-H., Faucheu, J., Massi, F.: The role of mechanical stimuli on hedonistic and topographical discrimination of textures. Tribol. Int. **143**, 106082 (2020)
16. Tang, W., Liu, R., Shi, Y., Hu, C., Bai, S., Zhu, H.: From finger friction to brain activation: tactile perception of the roughness of gratings. J. Adv. Res. **21**, 129–139 (2020)
17. Sahli, R., et al.: Tactile perception of randomly rough surfaces. Sci. Rep. **10**, 15800 (2020)
18. Harris, K.L., Collier, E.S., Skedung, L., Rutland, M.W.: A sticky situation or rough going? Influencing haptic perception of wood coatings through frictional and topographical design. Tribol. Lett. **69**, 113 (2021)
19. Zhang, S., et al.: Selection of micro-fabrication techniques on stainless steel sheet for skin friction. Friction **4**, 89–104 (2016)
20. Zhang, S., Rodriguez Urribarri, A., Morales Hurtado, M., Zeng, X., van der Heide, E.: The role of the sliding direction against a grooved channel texture on tool steel: an experimental study on tactile friction. Int. J. Solids Struct. **56–57**, 53–61 (2015)
21. Zhang, S., et al.: Texture design for light touch perception. Biosurf. Biotribol. **3**, 25–34 (2017)
22. Zhang, S., et al.: Finger pad friction and tactile perception of laser treated, stamped and cold rolled micro-structured stainless steel sheet surfaces. Friction **5**, 207–218 (2017)
23. Zhang, S., Zeng, X., Igartua, A., Rodriguez-Vidal, E., van der Heide, E.: Texture design for reducing tactile friction independent of sliding orientation on stainless steel sheet. Tribol. Lett. **65**, 89 (2017)
24. Birnbaum, M.: Semiconductor surface damage produced by ruby lasers. J. Appl. Phys. **36**, 3688–3689 (1965)
25. Oron, M., Sorensen, G.: New experimental evidence of the periodic surface structure in laser annealing. Appl. Phys. Lett. **35**, 782–784 (1979)
26. Bulgakova, N.M., et al.: Impacts of ambient and ablation plasmas on short- and ultrashort-pulse laser processing of surfaces. Micromachines. **5**, 1344–1372 (2014)
27. Sun, H., Li, J., Liu, M., Yang, D., Li, F.: A review of effects of femtosecond laser parameters on metal surface properties. Coatings **12**, 1596 (2022)
28. Ruf, A., Dausinger, F.: Interaction with metals. In: Dausinger, F., Lubatschowski, H., Lichtner, F. (eds.) Femtosecond Technology for Technical and Medical Applications. Topics in Applied Physics, vol. 96, pp. 105–114. Springer, Heidelberg (2004). https://doi.org/10.1007/978-3-540-39848-6_8
29. Förster, D.J., Jäggi, B., Michalowski, A., Neuenschwander, B.: Review on experimental and theoretical investigations of ultra-short pulsed laser ablation of metals with burst pulses. Materials **14**, 3331 (2021)
30. Di Maio, Y.: Etude de l'interaction laser-matière en régime d'impulsions ultra-courtes : application au micro-usinage de matériaux à destination de senseurs, Ph.D. thesis, Université Jean Monnet - Saint-Etienne (2013)
31. Zahouani, H., Vargiolu, R.: Appareil de mesure d'une force exercée par la face palmaire de la phalange d'un doigt. Patent FR3099359 (2021)

32. Zahouani, H., Mezghani, S., Vargiolu, R., Hoc, T., El Mansori, M.: Effect of roughness on vibration of human finger during a friction test. Wear **301**, 343–352 (2013)
33. Blackman, R.B., Tukey, J.W.: The measurement of power spectra from the point of view of communications engineering—part II. Bell Syst. Tech. J. **37**, 485–569 (1958)
34. Bensmaïa, S.J., Hollins, M.: The vibrations of texture. Somatosens. Mot. Res. **20**, 33–43 (2003)
35. Fagiani, R., Massi, F., Chatelet, E., Berthier, Y., Akay, A.: Tactile perception by friction induced vibrations. Tribol. Int. **44**, 1100–1110 (2011)
36. Prevost, A., Scheibert, J., Debrégeas, G.: Effect of fingerprints orientation on skin vibrations during tactile exploration of textured surfaces. Commun. Integr. Biol. **2**, 422–424 (2009)
37. Maeno, T., Kobayashi, K., Yamazaki, N.: Relationship between the structure of human finger tissue and the location of tactile receptors. JSME Int J. Ser. C **41**, 94–100 (1998)
38. Bolanowski, S.J., Gescheider, G.A., Verrillo, R.T., Checkosky, C.M.: Four channels mediate the mechanical aspects of touch. J. Acoust. Soc. Am. **84**, 1680–1694 (1988)
39. Bounakoff, C., Hayward, V., Genest, J., Michaud, F., Beauvais, J.: Artificial fast-adapting mechanoreceptor based on carbon nanotube percolating network. Sci. Rep. **12**, 2818 (2022)
40. Felicetti, L., Chatelet, E., Latour, A., Cornuault, P.-H., Massi, F.: Tactile rendering of textures by an electro-active polymer piezoelectric device: mimicking friction-induced vibrations. Biotribology. **31**, 100211 (2022)
41. Felicetti, L., Chatelet, E., Bou-Saïd, B., Latour, A., Massi, F.: Investigation on the role of the finger transfer function in tactile rendering by friction-induced-vibrations. Tribol. Int. **190**, 109018 (2023)
42. Weiland, B., Leclinche, F., Kaci, A., Camillieri, B., Lemaire-Semail, B., Bueno, M.-A.: Tactile simulation of textile fabrics: design of simulation signals with regard to fingerprint. Tribol. Int. **191**, 109113 (2024)

Tactile Clip: A Wearable Device for Inducing Softness Illusion Through Skin Deformation

Hikari Yukawa[1]([⊠])[ID], Natsuno Asano[1], Arata Horie[2][ID], Kiryu Tsujita[2][ID], Takatoshi Yoshida[2][ID], Kouta Minamizawa[2][ID], and Yoshihiro Tanaka[1,3][ID]

[1] Nagoya Institute of Technology, Nagoya, Japan
{yukawa.hikari,asano.natsuno,tanaka.yoshihiro}@nitech.ac.jp
[2] Keio University Graduate School of Media Design, Tokyo, Japan
{a.horie,dxkiryu,yoshida,kouta}@kmd.keio.ac.jp
[3] Inamori Research Institute for Science, Kyoto, Japan

Abstract. We proposed the "Tactile Clip," a device to induce an illusion of softness, and verified its effectiveness in augmenting softness perception. A Tactile Clip is a wearable device that provides circumferential-force stimulation to the feet. This force results in the deformation of the foot skin, consequently influencing the perception of softness. We conducted a psychophysical experiment using the interleaved staircase method. Twelve participants assessed foot softness by comparing the stepping sensation of the reference sample with the Tactile Clip to that of the test samples without the Tactile Clip. From the response rate, we derived a psychometric curve and bias value. The results showed a significantly positive bias in the perception, suggesting that the Tactile Clip made the flooring material feel softer than the actual softness. We consider the factors contributing to this phenomenon as a slight increase in the foot's thickness by deformation of sole skin and/or cognitive effects due to changes in force stimulation on the side of the foot in response to stepping.

Keywords: Illusion of softness · Skin deformation · Foot · Wearable device

1 Introduction

Humans interact with their external environment through sensory modalities, in which tactile sensations play a pivotal role in understanding the physical attributes of objects. Many researchers have developed numerous haptic interfaces, which emulate tactile sensations through diverse stimuli, including force, vibrotactile feedback, and electrical stimulation [1,2]. Studies on these haptic technologies have primarily focused on the hands and fingers, whereas tactile receptors are dispersed across the human body, enabling a comprehensive perception of environmental states. In particular, humans engaging in bipedal locomotion spend considerable time perceiving tactile information from the ground

H. Kajimoto et al. (Eds.): EuroHaptics 2024, LNCS 14768, pp. 252–261, 2025.
https://doi.org/10.1007/978-3-031-70058-3_20

surfaces. For instance, the tactile sensation of flooring materials such as carpets is often experienced more as stepping on the soles of the feet than as a feeling of the hands or fingers.

Research on reproducing the feeling of a stepping device can be divided into platform and wearable. Platforms are known to generate geometric terrain using shape-changing displays based on terrain expressions [3] and special mechanical platforms [4,5]. In addition, haptic feedback devices with vibrotactile stimulation [6,7], motor torque [8], and large airbags [9], depending on the user's movement, have been developed to express the form of the ground and the feeling of collision. Recent research has developed a system in which users can perceive terrain materials with large compliance deflections without real-time sensing by the user by manipulating the compression ratio of a spring due to the pre-load force [10]. The wearable type, typically shoe-shaped, expresses the sensation of stepping on a material. These studies attempted to express a broader range of materials using various stimulations, such as vibrotactile stimulation [11], air bladders [12], magnetic fluids [13], and mechanical structures [14]. These wearable devices usually desire to be small and lightweight.

Another method to produce tactile sensation is to change the state of the skin since our tactile sensation is derived by mechanical interaction between our skin and object. This method allows to touch the object directory. Previous research shows that changing the properties of the skin induces a change in our tactile perception [15]. We focused on softness as a crucial factor as well as texture in tactile perception and attempted to enhance the softness by skin deformation. Tao et al. [16] proposed a device with a circular ring to inflate a finger pad by applying pressure. This ring suppresses deformation when it comes in contact with an object. It was demonstrated that the perceived softness of an object can be improved by increasing its contact area. However, this approach is limited to objects smaller than the finger pad.

We proposed a wearable device, "Tactile Clip," that modulates the perceived softness of flooring materials by foot. The skin shape and physical properties are mechanically changed by applying circumferential pressure to the foot. A Tactile Clip does not require a complex structural framework because it does not replicate object softness but generates a softness illusion through skin deformation. Furthermore, it facilitates free walking and compatibility with real-world materials and other rendering systems because its lightweight and wearable design attaches to the top and around the foot without covering the sole surface. In this study, we developed a prototype device that presents force stimuli and assessed the induced softness perceptual bias.

2 Proposed Tactile Clip

The Tactile Clip consists of two parts, as shown in Fig. 1(a). Part A is worn to cover the top of the foot and compresses both sides of the ball and small ball of the foot. Part B is attached to the back of the heel, and applies force around the heel. Both parts are fabricated from an elastic material (Onyx, Markforged Inc.)

made using a 3D printer, and apply elastic force around the foot. When the user steps on the floor, it results in the deformation of the device that applies the force to the foot. Anti-slip materials are incorporated into both parts to enhance stability during use.

Figure 1(b), (c), and (d) show the foot deformation with/without wearing the Tactile Clip. In Fig. 1(b), the Part A clip shows that attaching a Tactile Clip reduces the width of the foot by applying a lateral force. Consequently, the ball of the foot and the small ball become inflated, and the foot becomes thicker. Figure 1(c), which depicts the sole, shows increased wrinkles, indicating increased foot thickness owing to skin gathering and raised sections. There are no visible changes in the rearview in Part B (Fig. 1(d)). However, a slight increase in wrinkles on the soles suggests that lateral forces by Part B also alter skin properties.

3   Psychophysical Experiment

3.1   Setup

We investigated the bias in softness perception induced by a Tactile Clip. We conducted a psychophysical experiment using the interleaved staircase method. The experimental setup is shown in Fig. 2(a). The participants assessed softness by comparing the sensation with one foot attached to the Tactile Clip against the other without it. The experiments were conducted using the prototypes of the Tactile Clips described in the second section, Parts A and B. We prepared three sizes (S, M, and L) of Part A, considering individual differences in foot width around the ball of the foot, and participants chose the optimal size by themselves. Part B was one size because the individual differences in the heel size were small relative to those around the ball of the foot. The size of each Tactile Clip is shown in Fig. 2(b).

3.2   Participants

The experiment was conducted with twelve healthy university students (ten males and two females, age: 22.83 (sd = 2.12)). The experimental protocol followed the Declaration of Helsinki and was approved by the Nagoya Institute of Technology ethics committee. Before the experiment, the content of the experiment was explained, and informed consent was obtained. All participants were naïve to the purpose of the experiment and recruited as paid volunteers.

3.3   Stimuli and Conditions

Polyethylene foam flooring samples (PE Light, INOAC CORPORATION) with six different softness levels were used, each measuring 50 cm square and 3 cm thick. A hardness tester (GS-621R-G, TECLOCK) was used to measure the hardness of the samples. Table 1 shows the measured durometer and compliance values, which are the reciprocals of the durometer value. The stimulus to be

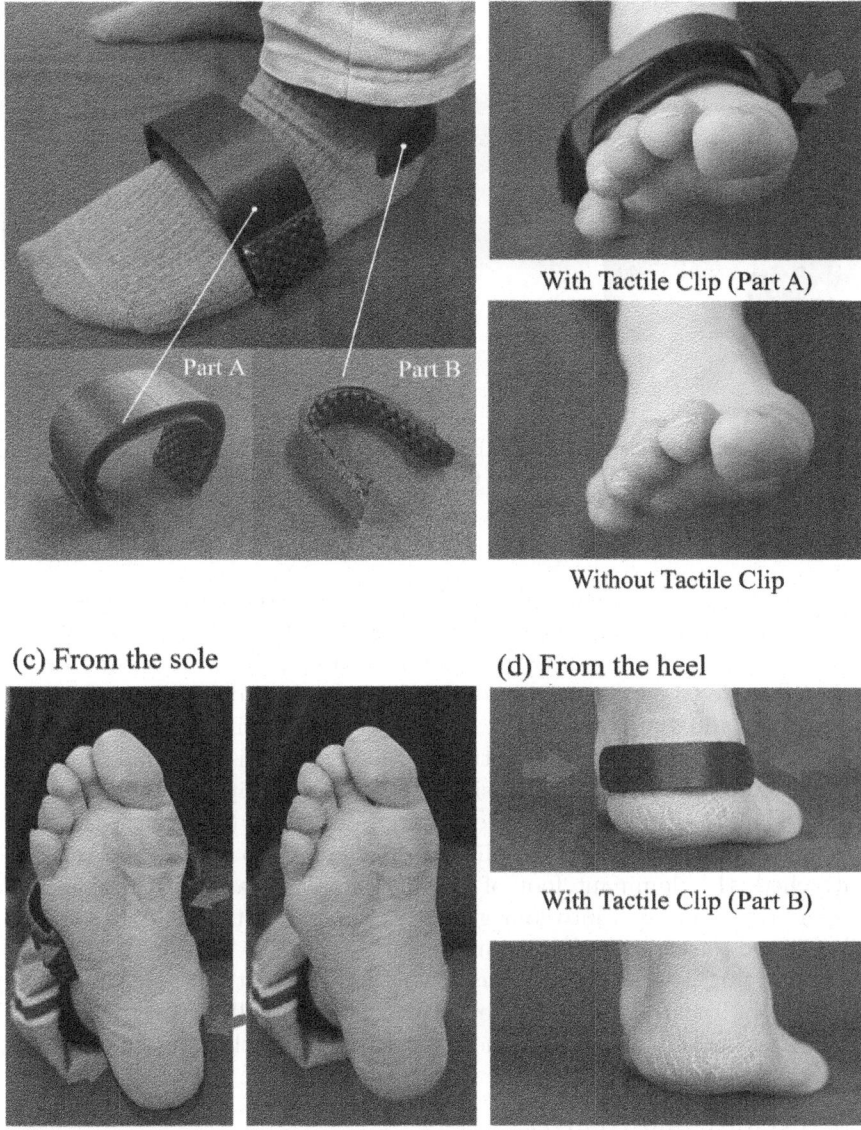

(a) Tactile Clip usage

(b) From the toe

Part A   Part B

With Tactile Clip (Part A)

Without Tactile Clip

(c) From the sole

(d) From the heel

With Tactile Clip (Part B)

Without Tactile Clip

With Tactile Clip   Without Tactile Clip   Without Tactile Clip
(Part A and B)

Fig. 1. The prototype of Tactile Clip. (a) shows the usage of Tactile Clip. Part A is attached around the ball of the foot, and Part B is attached to the back of the heel. (b) shows the deformation of the front of the foot by Part A, (c) shows the deformation of the sole by Part A and B, and (d) shows the deformation of the heel by Part B.

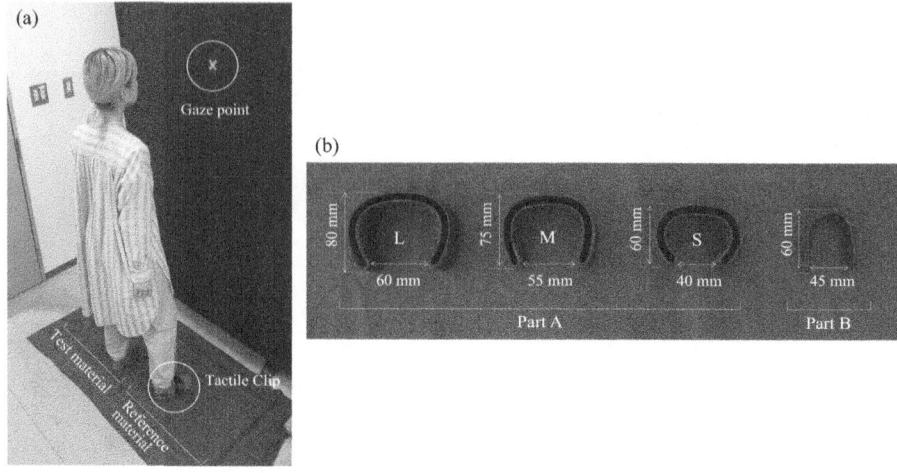

Fig. 2. (a) Experimental setup. The participant compares the softness of the reference sample with the Tactile Clip with the test samples. (b) The size of the Tactile Clip. Part A has three sizes (S, M, and L).

stepped on with the attached Tactile Clip was always the same hardness sample (reference sample (R): the second-hardest sample). Test samples (samples A to F. Sample B were the same as the reference sample.) were used for the foot without a Tactile Clip. The reference and test samples were positioned adjacent to each other, and the participants stepped on them with one foot each. The two samples were covered with a cotton cloth to avoid visual differences.

The assignment of the attached side of the Tactile Clip and the order of test sample presentation in the interleaved staircase method (starting with samples A or F) were counterbalanced. We did not fix the foot to which the Tactile Clip was attached, the dominant foot of the participants was not considered. The participants were divided into four groups of three: (i) starting with sample A, Tactile Clip on the right foot; (ii) starting with sample A, Tactile Clip on the left foot; (iii) starting with sample F, Tactile Clip on the right foot; and (iv) starting with sample F, Tactile Clip on the left foot. Therefore, all the participants wore a Tactile Clip on only one side of their feet throughout the experiment.

3.4   Procedure and Analysis

The experimental procedure was detailed to the participants, and consent was obtained. All participants were asked to wear the same type of socks throughout the experiment. Before the psychophysical experiment, the participants experienced six types of stimuli (samples A to F) with both feet to become accustomed to the softness of the samples used in the experiment. They compared samples without the Tactile Clip and ranked their softness within 5 min. Participants were presented with the correct answer after the evaluation and experienced the correct

Table 1. Compliance value

Sample	Durometer	Compliance
A	60	0.017
B, R	57	0.018
C	47	0.021
D	41	0.024
E	39	0.026
F	24	0.042

order if there were discrepancies. The instructions and practice for attaching and using the Tactile Clip were followed. The participants were instructed to avoid applying pressure to only one part of the foot and to use the entire sole during stepping. The participants were instructed to maintain a fixed gaze point and consistent posture. Although no specific stepping speed was set to preserve natural motion, a minimum of 50 bpm was suggested to prevent excessively slow stepping.

In a psychophysical experiment using the interleaved staircase method, the participants stepped on reference and test samples for up to 10 s. They identified the softer sample (reference sample with Tactile Clips or test sample). When starting with hard test sample (sample A) and the reference sample being judged to be softer, a one-level softer sample (sample B) was used in the subsequent trial. When starting with a soft test sample (sample F) and the test sample was judged softer, a one-level harder sample (sample E) was presented in the subsequent trial. For each participant, 20 trials were performed, alternating between 10 trials starting with sample A and 10 trials starting with sample F. For the analysis, the proportion of participants who perceived the test sample as softer than the reference with the Tactile Clip was calculated for each sample. These values were fitted to a cumulative Gaussian distribution as follows:

$$f(x) = \frac{1}{2}(1 + erf(\frac{x - \mu}{\sigma\sqrt{2}})), \tag{1}$$

where σ is the measure of the discrimination at 84 % of each participant, and μ is the Point of Subjective Equality (PSE). For this fitting, weighting was applied according to the number of responses. The bias value was derived by subtracting the compliance value of the reference sample from the PSE. We examined the normality and homoscedasticity of the bias-value data. A one-sample t-test was conducted if the data followed a parametric distribution, and a one-sample Wilcoxon signed-rank test was used if the data followed a nonparametric distribution.

4   Result

Figure 3 shows an example of the staircase method and psychometric curve results. The bias values for each participant are shown in Fig. 4. For participants other than participant g, the bias was positive, suggesting that wearing the Tactile Clip enhanced softness. There were differences in the magnitude of bias between the participants, and participants b, f, and k had small bias values. Average bias is 0.0024 (sd = 0.0021). Subsequently, we performed a one-sample t-test because normality and homoscedasticity of the data were not rejected. The results confirm a significant positive bias ($t(11) = 3.86$, p = 0.0026). The bias value differed significantly from zero, indicating that the reference sample was perceived as softer than the actual material by attaching the Tactile Clip. And the average measurement of the discrimination at 75 % is 0.0014 (sd = 0.0011).

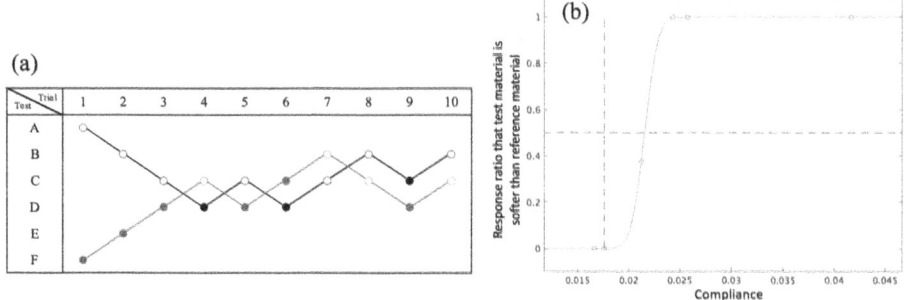

Fig. 3. Examples of (a) staircase method and (b) psychometric curve of one participant. (a) shows the example of the staircase method. The colored circle denotes the response that the reference sample is softer than the test sample, and the white circle denotes the response that the test sample is softer than the reference sample. (b) shows the psychometric curve calculated by the response ratio to response test sample is softer than the reference sample.

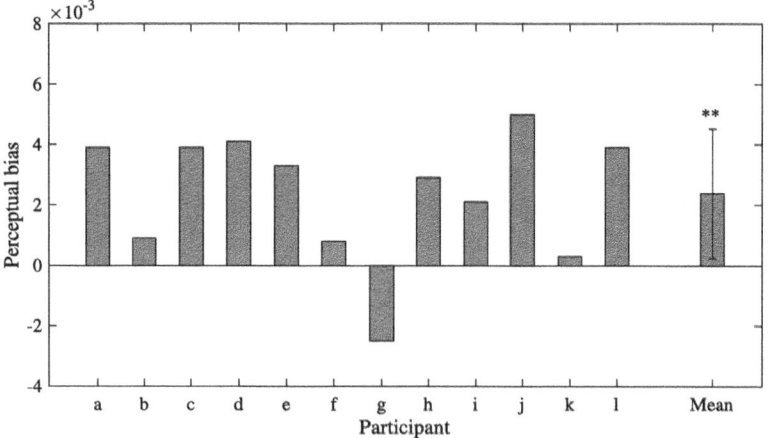

Fig. 4. Results of the bias value of 12 participants. ** denotes p ¡ .01.

Without Tactile Clip

With Tactile Clip

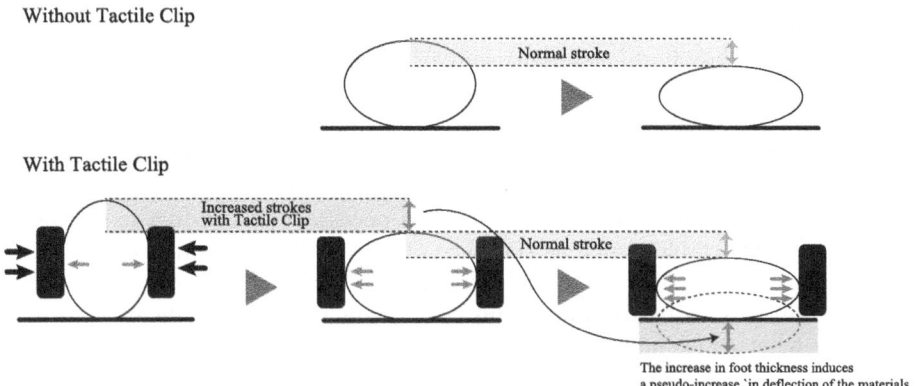

The increase in foot thickness induces
a pseudo-increase 'in deflection of the materials.

Fig. 5. Hypothesis of the principle of perceptual bias. The skin deformation by force increases the stroke of stepping, and the lateral stimulation emphasizes the foot's spreading

5   Discussion

5.1   Principle of perceptual bias

The experiment confirmed the effect of the Tactile Clip, which made the flooring material feel softer than the actual material. We discuss two possible factors that attributed to this effect as illustrated in Fig. 5.

The first is the mechanical deformation of the skin. The relationship between deflection and counterforce is vital for perceiving the softness of an object deformed under applied pressure, such as a flooring material [17]. As shown in Fig. 1(b), the Tactile Clip can increase the foot thickness, allowing for a longer stepping stroke than usual. Although the applied force was constant (equivalent to the body weight), the increased stroke might caused a pseudo-increase in the deflection of the material. This phenomenon may lead to an illusion that the flooring material is softer than the actual material.

The second factor is the cognitive effect. Users wearing the Tactile Clip reported a feeling in their feet gradually spreading. It is possible that the user felt resistance to the expansion of their foot by force, and this resistance might emphasize the sensation of foot expansion. Figure 1(d) shows that the effect of Part B on the shape of the foot appeared minimal, while many users reported feeling the illusion of softness contribution of Part B as well as Part A. This suggests that the cognitive effect of lateral stimulation emphasizing foot spreading contributed to the illusion of softness.

5.2   Limitation and Future Work

Most participants experienced a softness illusion when wearing the Tactile Clip. However, the participant g did not perceive this effect. This may be attributed to the individual physical characteristics. The participant g was a relatively large man with flat feet. Flat feet, which are thinner, cannot stabilize the device attachment, and less fleshiness can result in minimal mechanical changes.

While we have proposed hypotheses regarding the mechanism behind the Tactile Clip illusion, the current lack of supporting evidence necessitates further exploration. In this experiment, participants were instructed to walk naturally to assess the efficacy of the Tactile Clip. However, to achieve a more comprehensive understanding of its mechanism, it is necessary to conduct experiments with controlled variables, such as the pace of stepping, foot dominance, and the specific points at which force is applied. Furthermore, it is essential to investigate the limitations of this simple and convenient device. Specifically, we need to explore whether changing the grip strength can manipulate the intensity of the softness illusion, whether the user's movements affect the effect of Tactile Clip, and whether this system could be applied to other body parts besides the feet, such as the fingers.

In the future, we envision the Tactile Clip illusion being applied in various fields. Its mechanism and expandability could be harnessed to reproduce textures in virtual spaces and enhance online shopping experiences. For instance, it could offer a shopping experience where, within a limited space, consumers can feel various materials of different softness. The Tactile Clip's simple configuration makes it easily implementable, providing a high impact when used in conjunction with actual materials and other systems.

6   Conclusion

This study proposed the "Tactile Clip," a device to induce an illusion of softness in flooring materials by deforming the skin. The Tactile Clip is a wearable device attached to the foot and applies force stimulation from around the foot, thereby deforming the sole and controlling perception.

We conducted a psychophysical experiment using an interleaved staircase method. Participants compared the softness by stepping with one foot on the reference sample with the Tactile Clip and the other on the test samples without the Tactile Clip. The results showed that the perceptual bias values were significantly positive. This suggests that the Tactile Clip has an effect in enhancing the perception of softness. We discussed the factors contributing to this effect, which are mechanical skin deformation and/or a cognitive effect.

Acknowledgments. This work was supported by JST Moonshot R&D Program "Cybernetic being" Project (Grant number JPMJMS2013) and Inamori Research Institute for Science. We gratefully acknowledge HOTTA CARPET CO.,LTD, JAPAN CRAFT&LOCALITY ASSOCIATION, and Mitsubishi Research Institute, Inc.

References

1. Giri, G.S., Maddahi, Y., Zareinia, K.: An application-based review of haptics technology. Robotics **10**(1), 29 (2021)
2. Adilkhanov, A., Rubagotti, M., Kappassov, Z.: Haptic devices: wearability-based taxonomy and literature review. IEEE Access **10**, 91923–91947 (2022)

3. Je, S., et al.: Elevate: a walkable pin-array for large shape-changing terrains. In: Proceedings of the 2021 CHI Conference on Human Factors in Computing Systems, pp. 1–11. Online Virtual Conference (originally Yokohama, Japan) (2021). https://doi.org/10.1145/3411764.3445454

4. Iwata, H., Yano, H., Fukushima, H., Noma, H.: Circulafloor [locomotion interface]. IEEE Comput. Graph. Appl. **25**(1), 64–67 (2005)

5. Hollerbach, J. M., Xu, Y., Christensen, R. R., Jacobsen, S. C.: Designspecifications for the second generation sarcos treadport locomotion interface. In: ASME 2000 International Mechanical Engineering Congress and Exposition, vol. 69, no. 2, pp. 1293–1298. Orlando, Florida, USA (2000). https://doi.org/10.1115/IMECE2000-2446

6. Visell, Y., Giordano, B.L., Millet, G., Cooperstock, J.R.: Vibration influences haptic perception of surface compliance during walking. PLoS ONE **6**(3), e17697 (2011)

7. Blom, K.J., Haringer, M., Beckhaus, S.: Floor-based audio-haptic virtual collision responses. In: Joint Virtual Reality Conference ICAT - EGVE - EuroVR, pp. 57–64. Madrid, Spain (2012). https://doi.org/10.2312/EGVE/JVRC12/057-064

8. Otaran, A., Farkhatdinov, I.: Haptic ankle platform for interactive walking in virtual reality. IEEE Trans. Vis. Comput. Graph. **28**(12), 3974–3985 (2021)

9. Teng, S., Lin, C., Chiang, C., Kuo, T., Chan, L., Huang, D.: TilePop: tile-type pop-up prop for virtual reality. In: The 32nd Annual ACM Symposium on User Interface Software and Technology, pp. 639–649. New Orleans, LA, USA (2019). https://doi.org/10.1145/3332165.3347958

10. Chang, W., Je, S., Pahud, M., Sinclair, M.J., Bianchi, A.: Rendering perceived terrain stiffness in VR via preload variation against body-weight. IEEE Trans. Haptics **16**(4), 616–621 (2023)

11. Strohmeier, P.R., Güngör, S., Herres, L., Gudea, D., Fruchard, B., Steimle,J.: bARefoot: Generating virtual materials using motion coupled vibration in shoes. In: The 33rd Annual ACM Symposium on User Interface Software and Technology, pp. 579–593. Virtual (previously Minneapolis, Minnesota, USA) (2020).https://doi.org/10.1145/3379337.3415828

12. Wang, Y., Truong, T.E., Chesebrough, S.W., Willemsen, P., Foreman, K.B., Merryweather, A.S., Hollerbach, J.M., Minor, M.: Augmenting virtual reality terrain display with smart shoe physical rendering: a pilot study. IEEE Trans. Haptics **14**(1), 174–187 (2020)

13. Yang, T., Son, H., Byeon, S., Gil, H., Hwang, I., Jo, G., Choi, S., Kim, S., Kim, R.: Magnetorheological fluid haptic shoes for walking in VR. IEEE Trans. Haptics **14**(1), 83–94 (2020)

14. Yokota, T., Ohtake, M., Nishimura, Y., Yui, T., Uchikura, R., Hashida, T.: Snow walking: motion-limiting device that reproduces the experience of walking in deep snow. In: 6th Augmented Human International Conference, pp. 45–48. Singapore, Singapore (2015).https://doi.org/10.1145/2735711.2735829

15. Tanaka, Y., Sano, A., Ito, M., Fujimoto, H.: A novel tactile device considering nail function for changing capability of tactile perception. In: 6th International Conference, EuroHaptics 2008. Lecture Notes in Computer Science, vol. 5024, pp. 543-548, Madrid, Spain (2008)

16. Tao, Y., Teng, S., Lopes, P.: Altering perceived softness of real rigid objects by restricting fingerpad deformation. In: The 34th Annual ACM Symposium on User Interface Software and Technology, pp. 985–996. Virtual (2021).https://doi.org/10.1145/3472749.3474800

17. Friedman, R.M., Hester, K.D., Green, B.G., Lamotte, R.H.: Magnitude estimation of softness. Exp. Brain Res. **191**(2), 133–142 (2008)

SENS3: Multisensory Database of Finger-Surface Interactions and Corresponding Sensations

Jagan K. Balasubramanian⬮, Bence L. Kodak⬮, and Yasemin Vardar(✉)⬮

Delft University of Technology, Mekelweg 2, 2628 CD Delft, The Netherlands
y.vardar@tudelft.nl

Abstract. The growing demand for natural interactions with technology underscores the importance of achieving realistic touch sensations in digital environments. Realizing this goal highly depends on comprehensive databases of finger-surface interactions, which need further development. Here, we present SENS3—*www.sens3.net*—an extensive open-access repository of multisensory data acquired from fifty surfaces when two participants explored them with their fingertips through static contact, pressing, tapping, and sliding. SENS3 encompasses high-fidelity visual, audio, and haptic information recorded during these interactions, including videos, sounds, contact forces, torques, positions, accelerations, skin temperature, heat flux, and surface photographs. Additionally, it incorporates thirteen participants' psychophysical sensation ratings (rough–smooth, flat–bumpy, sticky–slippery, hot–cold, regular–irregular, fine–coarse, hard–soft, and wet–dry) while exploring these surfaces freely. Designed with an open-ended framework, SENS3 has the potential to be expanded with additional textures and participants. We anticipate that SENS3 will be valuable for advancing multisensory texture rendering, user experience development, and touch sensing in robotics.

Keywords: Surface · Dataset · Haptic · Multisensory · Sensation

1 Introduction

Recent trends in interactive system development emphasize a naturalistic approach to replicating human engagement with their physical surroundings in digital environments, promising enhanced user experience, accessibility, and improved digital communication [11]. While everyday tasks in the physical world involve engaging with objects through multiple senses [16], transferring this rich sensory information to the digital domain remains a challenge despite advancements in hardware and algorithms for capturing and recreating real-world sensory experiences [28] and understanding multisensory integration [5].

Besides device and algorithm design, achieving naturalistic human-technology interactions relies on one more essential element: data. Thanks to

H. Kajimoto et al. (Eds.): EuroHaptics 2024, LNCS 14768, pp. 262–277, 2025.
https://doi.org/10.1007/978-3-031-70058-3_21

the advancements in camera and microphone technologies, one can effortlessly capture and share their surrounding's audio and visual information. Unfortunately, the same practicality does not apply to tactile data. When humans touch an object, they feel a rich array of tactile cues revealing its distinct surface properties, such as friction, roughness, thermal, and compliance [20], depending on the applied exploratory procedures, e.g., sliding, static contact, and pressing [15]. High-fidelity data collection for each property and exploratory procedure requires specialized expertise and recording technology. This situation makes it challenging for regular users to record the tactile feel of every encountered surface instantly. A fundamental issue is that tactile cues, unlike light and sound waves, require physical contact. The resulting skin deformations depend on finger and surface properties and applied normal force and speed [21]; these can vary substantially even for the same user and surface, hugely influencing the recorded tactile data [26]. Due to these reasons, only a few available databases [6, 21, 24, 25] include tactile recordings from interactions with surfaces. Moreover, existing databases often concentrate on single or dual sensory cues, e.g., vision and tactile, audition and tactile, or tactile only, insufficient for lifelike digitization of surfaces or neglect bare-finger interactions, the most natural way of interacting with our surroundings [17]. Finally, these databases mainly do not include psychophysical sensations humans perceive upon interaction such as rough–smooth, flat–bumpy, sticky–slippery, hot–cold, regular–irregular, fine–coarse, hard–soft, and wet–dry; providing such information can tremendously help understand human texture perception or generate algorithms for machine perception.

To address the above issues with the existing datasets, we propose a novel database, SENS3[1], encompassing all necessary multisensory cues for naturalistic texture digitization. Our open-access database includes visual, auditory, and tactile data recorded while two participants explored 50 surfaces with their fingertips. Data collection measurements were conducted with a custom-designed apparatus, and they consisted of four exploratory procedures: static contact for thermal aspects, pressing and tapping to record compliance and hardness properties, and sliding to capture roughness and friction. Additionally, SENS3 includes psychophysical sensations rated by thirteen participants while freely interacting with the selected surfaces. We also introduce a user-friendly website that describes data collection procedures and collected data for a broad audience. We envision that our database will significantly impact various fields, such as human-machine interaction and robotics, providing a comprehensive and rich resource for researchers and developers to enhance the realism and authenticity of texture rendering and perception and, ultimately, to improve the user experience.

---

[1] https://www.sens3.net.

Table 1. Comparison of attributes of available haptic texture databases with SENS3. Star, *, indicates a not freely accessible database

Dataset	Audio/ Visual	Tactile	Exploration	Interaction	Sensations	Texture count	Participant count
HaTT [6]	Image	3D contact forces & accelerations, finger speed	Sliding	Tool-tip	No	100	1
CBSMR [24]	Audio, image & video	3D contact forces & accelerations, reflectance, conductivity	Tapping & sliding	Tool-tip	Pairwise similarities	108	1
LMT [25]	Audio & image	3D accelerations	Sliding	Bare-finger & tool-tip	No	184	1
		Reflectance	Contour following				
		Thermal conductivity	Static touch	Standalone sensors			
		2D contact forces	Lateral motion	Sensors mounted under finger pad			
		Pressure	Pressing				
		Mass & volume	Holding	Standalone sensors			
Haptex* [14]	No	3D contact forces & torques	Sliding	Bare-finger	No	120	1
Learn2feel [21]	Image	3D contact forces & accelerations, finger speed	Tapping & sliding	Bare-finger	Pairwise similarities	10	10
Concurrent* [9]	Image & video	3D contact forces torques, & accelerations, finger speed	Sliding	Bare-finger	No	10	1
Haptic library [12]	Image	No	Free exploration	Bare-finger	Cluster sorting	84	10
SENS3	Audio, image & video	Skin temperature, heat flux, 3D contact forces & torques	Static contact	Bare-finger	Adjective ratings	50	2/ 13
		3D contact forces & torques, indentation depth	Pressing				
		3D contact forces & accelerations	Tapping				
		3D contact forces, torques & accelerations finger position & speed	Sliding				

2   Existing Haptic Texture Databases

Several databases have been made to date for digitizing tactile textures; see Table 1 for a list of these prior work. The first comprehensive one, the Penn Haptic Texture Toolkit (HaTT), featured unconstrained tool-surface interactions with 100 surfaces, their models for haptic rendering, and high-resolution images of each surface [6]. The tactile data included contact forces, accelerations, and scan speed and was recorded with a custom-designed tool. The authors later demonstrated that the recorded contact accelerations could be used to re-create the vibratory feels of surfaces via a voice-coil actuator [8].

HaTT mainly focused on roughness and friction modalities, overlooking capturing multisensory aspects of texture perception [16]. To address this gap, Strese et al. developed two texture databases [24,25]. They recorded data using multiple custom recording devices while exploring more than 100 surfaces via sliding, tapping, contour following, static contact, pressing, and holding. Their recording devices allowed mainly tool- or sensor-surface interactions. The databases contained contact accelerations, forces, thermal conductivity, reflectance, mass, volume, and interaction sounds and videos; check Table 1 for detailed distribution of collected data in each database.

Although tool/sensor-surface interactions were shown to effectively recreate the vibratory feel of surfaces and machine perception applications, the recorded data still does not represent the information obtained during bare-finger explorations [25]. To combat this, recent studies [9,14,21] introduced databases for finger-surface interactions. These databases include contact forces, accelerations, videos, and sounds recorded during sliding or tapping interactions; see Table 1 for more details of each dataset. The studies also showcased different uses of their data, such as bare-finger texture rendering [14], texture classification [9], or understanding human perception of surfaces [21]. Although these databases provide the much-needed information on finger-surface interactions, they only consider a subset of exploratory procedures and omit other important tactile properties, such as compliance and thermal.

Finally, only some of the databases [12,21,27] include perception data along with the physical one, often focusing on pairwise similarities. While these similarities provide insights into how interaction data shapes human perception, articulating precise statements on how changes in contact data correlate with sensations remains challenging and necessitates adjective ratings for each sensation [21].

3   SENS3 Database

We aimed to overcome the limitations of previous databases summarized in Sect. 2 with SENS3. We recorded the physical interaction data while participants explored the surfaces with their *bare* fingers. As human fingertips show variety in geometry and mechanical properties [18,22], we collected this data from *two* people. Moreover, as previous studies [16] showed evidence that all tactile, visual, and auditory cues contribute to surface perception, we recorded *multisensory* information that spans these three senses. Additionally, we collected sensation ratings from *13* people to complement the physical recordings to understand human texture and machine perception and develop algorithms for effective texture rendering.

For **tactile data**, we aimed to capture perceptually relevant multi-modal material properties: *warmth, compliance, friction*, and *roughness* [20]. Previous works indicated that optimal exploratory procedures differ for the information to be gathered [15]. These procedures include lateral motion for roughness, pressure for compliance/hardness, and static contact for thermal properties. Furthermore, later studies showed that tapping is also suitable for hardness discrimination [7,21]. Therefore, we collected information with four distinct exploratory procedures: static contact, pressing, tapping, and sliding (lateral motion).

Because human fingertips are soft and show nonlinear behavior with applied pressure and moving speed, the interaction data also vary as a function of these parameters [26]. Therefore, we collected contact forces and torques for all interactions. Due to the same reason, the data for the sliding experiments covered a spectrum of speeds (ranging from 0 to 200 mm/s) and applied forces (ranging from 0 to 1 N). Nonetheless, the rest of the interactions covered only one or two selected pressure ranges. In addition, we collected finger vibrations for active dynamic interactions (tapping and sliding), indentation depth for pressing, heat flux, and skin/surface temperature for static contact. See Table 1 for a detailed list of collected data in prior databases and SENS3.

We took top-view high-resolution surface images for **visual data**. We also recorded dynamic finger-surface interactions with two cameras, one from the top and one from the side. Utilizing images and videos is essential, offering valuable perspectives from static and dynamic viewpoints. For **auditory data**, we recorded sounds of active dynamic finger-surface interactions (tapping and sliding).

For the **perceptual data**, we recorded human participants' haptic psychophysical sensations, i.e., adjective ratings, resulting from their free multisensory (visual, audio, and haptic) exploration of the surfaces.

3.1 Selected Surfaces & Data Collection Apparatus

We selected 50 distinct homogeneous surfaces across *ten* material categories: wood, metal, fabric, paper, rubber, plastic, sandpaper, leather, foam, and vinyl (see Fig. 1). These material categories encompass most of the encountered surfaces in daily life, and at least one texture is present per category. Each surface was cut to dimensions of 100×100 mm and stuck on acrylic plates (3 mm thickness for each) using double-sided tape (Tesa PRO double-sided tape, 50 mm $\times 25$ m).

3.2 Data Collection Apparatus

We built a custom data collection apparatus (see Fig. 2 and Supplementary Video), which was placed on an aluminum optical breadboard table (MB60120/M, Thorlabs) and mounted onto a robust supporting frame (PFM52502, Thorlabs). During the experiments, a monitor (UltraSharp 24, Dell) displayed graphical user interfaces (GUI).

The finger-surface contact forces and torques were measured via a 6D force sensor (Nano17 Titanium, ATI) placed on an acrylic plate fixed on the breadboard. We utilized the same sensor for all force measurements to ensure consis-

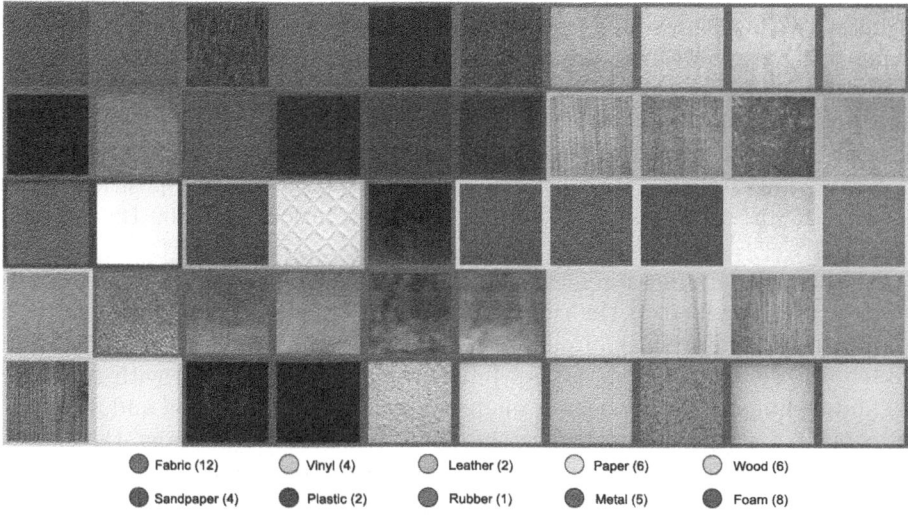

Fabric (12)   Vinyl (4)   Leather (2)   Paper (6)   Wood (6)

Sandpaper (4)   Plastic (2)   Rubber (1)   Metal (5)   Foam (8)

Fig. 1. Recorded surfaces in SENS3 database. The material categories are color-coded; each category's surface count is indicated in the brackets.

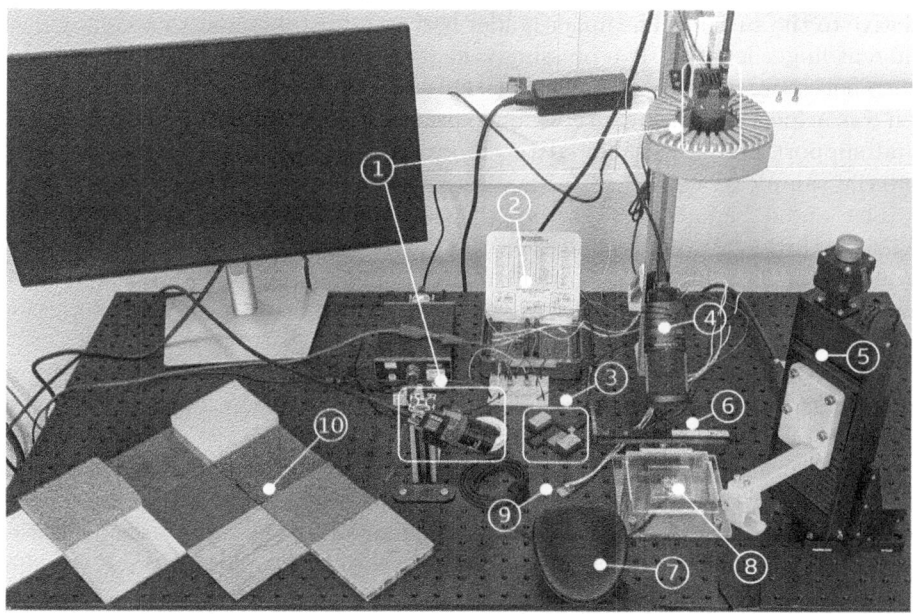

Fig. 2. Data recording apparatus: 1. Cameras, 2. data acquisition board, 3. heat flux sensor, 4. microphone, 5. linear stage, 6. position sensor, 7. armrest, 8. force sensor, 9. accelerometer, 10. selection of surfaces.

tency in the gathered data. Above the sensor, we put another acrylic plate (3 mm thickness) with a 3D-printed side to place the surfaces. The finger accelerations during active explorations were captured via a high-resolution, high-bandwidth, three-axis analog accelerometer (ADXL356, Analog Devices) attached to the index fingernail via double-sided tape. A data acquisition board (PCIe-6323, NI) collected the data from the force and acceleration sensors with a 10 kHz sampling rate. The finger position was measured with a 2D infrared position sensor (NNAMC1580PCEV, Neonode) with 0.1 mm resolution and a 60 Hz sampling rate.

For thermal data measurements, a miniature thermistor (223Fu3122-07U015, Semitec) was attached to the center of the participants' index finger pad with a 1 mm thin strip of insulation tape, and a heat flux sensor (FHF05-15X30, Hukseflux) was mounted on a 3D-printed frame. When touched with the index and middle fingers, the thermistor made contact with the material and measured the contact temperature, whereas the heat flux sensor lay under the middle finger. To ensure full coverage of the finger pad, we chose the FHF-05 15×30 mm heat flux sensor. The initial surface temperatures were measured with an infrared thermometer. The data from the heat flux sensor and thermistor were sampled at 100 Hz using the data acquisition board (PCIe-6323, NI).

A motorized linear stage (NRT100/M, Thorlabs) was equipped for the pressing measurements. A 3D-printed hand support was mounted on the linear stage to keep the participant's index finger in place and provide a 20° contact angle relative to the surface. The finger holder had an adjustable bolt to accommodate different finger lengths. The pressing velocity was controlled by a stepper motor controller (BSC201, Thorlabs) along the vertical axis. The vertical configuration was achieved by using a right-angle bracket (NRT150P1/M, Thorlabs). The hand support and stage allowed the index finger to land on the center of the material sample.

The audio signals were recorded via a cardioid condenser microphone (AT2020 USB+, Audio Technica) in the 20 Hz–20 kHz range with a sampling frequency of 44.1 kHz. As cardioid microphones are unidirectional and most sensitive at the front, we placed the microphone front close to the interaction area.

The visual data was captured by two USB machine vision cameras (Alvium 1800 U-508c). These cameras recorded the interactions from both top and side views at 44 FPS with a resolution of 1200×1200 pixels. The top-view camera was also used to take high-quality images of each material sample. To have enough material exploration space for participants, we placed the top-view and side-view cameras at 450 mm and 170 mm working distances. We chose the 16 mm C series fixed focal length lens (Techspec, Edmund Optics) to address these working distances. Additionally, we used an LED ring light (EFFI-RING, Effilux) to provide illumination in combination with a polarizer (EFFI-RING-POL, Effilux) to eliminate glare and suppress the reflections from the illumination. The two cameras and the LED ring light were mounted to custom-made, partly 3D-printed stands.

We implemented appropriate solutions for synchronization among the recorded multisensory data. A sound cue signaled the start of each experiment for participants. Subsequently, software triggers activated the data acquisition card and microphone, while hardware triggers initiated the cameras. Timestamps were assigned to samples from these devices, ensuring synchronous data recording.

3.3   Data Collection Procedure

Finger-Surface Interaction Data: Two males, the first two authors of this paper, with an age of 26 years, participated in the finger-surface interaction data collection. Both participants gave informed consent and agreed to share their finger-interaction data publicly. Before each measurement, they washed their hands with soap and dried them using a towel. Then, each participant underwent an instruction and training process. They sat on a chair and put their arm on the armrest, facing the monitor. Afterward, their finger-surface interactions were collected while they explored the surfaces by applying static contact, pressing, tapping, and sliding. The detailed procedures for each measurement are described in the following, and Supplementary Video 1 visualizes each interaction. Recording all types of physical interaction data from one participant for fifty surfaces took approximately (not consecutively) 24 h.

a) Static Contact: In these measurements, our goal was quantifying thermal interactions occurring during finger-surface contact. Previous works on thermal perception and rendering determined two major thermal parameters for discriminating materials by touch as the heat flux conducted out of the skin and the corresponding skin temperature [13].

Following a similar approach to [4], we first measured the initial temperature of a surface sample via an infrared thermometer. Afterward, the thermistor was attached to the participant's index finger pad, whereas the 3D-printed frame with the heat flux sensor was positioned on the sample surface and then stabilized with weights. The measurement started with a sound cue while the participants' fingers lay just over the samples, enabling the thermistor to measure the initial finger temperature. After hearing a sound cue, the participant was instructed to place their index and middle finger on the frame. They maintained a constant contact force of 3 N for 60 s for each surface with both fingers to keep the contact area constant.

b) Pressing: With these measurements, we aimed to determine how different materials respond to pressure from the index finger by measuring the force-indentation depth relationship when pressed with the index finger. We chose to record normal force and indentation depth, as recent studies have shown that finger contact area alone is not a distinguishable metric when determining the compliance of objects, and kinesthetic cues likely augment our judgments of compliance [23].

First, the participant mounted their right hand onto the hand support. Then, the linear stage moved in the vertical direction with the participant's finger until reaching 3 N normal force. After staying there for two seconds, the finger was moved to the start position.

c) Tapping: By tapping measurements, we sought to capture the surface hardness. Previous works showed that we can discriminate hardness by tapping on a surface [7], and tap spectral centroid of the contact vibrations is a large contributor in material discrimination [21, 27].

Before tapping measurements, the accelerometer was placed on the participant's fingernail. After the calibration process of the accelerometer, the participant tapped six times on the center of the surface sample using their dominant hand's index finger. The start of each experiment was indicated with a sound cue. They were instructed to apply a maximum force of 1 N in the first three taps and then use a force level that surpasses 2 N for the rest. The desired force level and real-time measured contact force were shown to participants via a Matlab GUI and the force sensor.

d) Sliding: In these measurements, we aimed to capture the friction and roughness properties of the surfaces. Since measured surface friction and roughness features change with applied force and exploration speed [7, 17], our experiments comprehensively explored surfaces across a spectrum of speeds and applied forces.

Similar to tapping, the accelerometer was placed on the fingernail of the participant before these experiments. After the calibration process, the participant was instructed to follow a custom GUI shown on the monitor; refer to SENS3 website for visualization of GUI. The GUI showed a 5×6 matrix of UI elements, each corresponding to a specific finger force and speed range, shown in red color initially. The force ranges of elements corresponded 0–0.2, 0.2–0.4, 0.4–0.6, 0.6–0.8, 0.8–1 N, whereas the speed ranges represented 0–33, 33–66, 66–99, 99–132, 132–165, and 165–200 mm/s. The maximum force and speed were limited to 1 N and 200 mm/s, respectively, as beyond values were challenging to maintain for long. After hearing a sound cue, the participant explored the surface through unconstrained sliding, as this method enables capturing the interaction data more efficiently than constrained exploration [8]. When the participant achieved a specific force-speed combination, the color of the corresponding UI element shifted to green. Each force-speed range required an unconstrained finger-surface exploration with a duration of five seconds. The participants were free to achieve any force-speed combination randomly. The recording from a surface sample was completed when all the UI elements of the matrix turned green. The participant's applied force and speed were shown at the top of the GUI.

Perceptual Data: Through the perceptual measurements, we aimed to record haptic psychophysical sensations felt by participants during the free multisensory exploration of surfaces using their dominant index finger. For this, we used the semantic differential method [20] with 15 points, where participants rated the surfaces based on eight opposing adjectives by moving sliders (implemented in MATLAB's GUI) for each adjective. We selected a subset of surface adjectives used by [1]: rough–smooth, flat–bumpy, sticky–slippery, hot–cold, regular–

irregular, fine–coarse, hard–soft, and wet–dry. Before the experiment, we asked the participants to wash and dry their fingers first and then explore all the surfaces without any sensory restrictions—they were free to interact with surfaces by seeing, hearing, and touching—to adjust their adjective limits mentally. Following this, we conducted mock trials with five random surfaces to acclimate the participants to the study. During the experiments, we randomly presented the surfaces to the participants. We asked participants to explore the surface freely using all their three senses (visual, audio, and haptic) for fifteen seconds after they heard the sound cue. Although they could interact with the surfaces as they wished, we encouraged them to use all haptic exploratory procedures explained during mock trials. After the exploration, the participants rated the surface feels by adjusting the scales in the GUI. The experimental procedure followed the Declaration of Helsinki and was approved by TU Delft's ethics committee with application number 3469. Three women and ten men with an average age of 26.84 years (standard deviation, SD: 2.034) participated in the experiment. All participants gave informed consent. The perceptual data collection procedure took 2 h per participant.

3.4   Collected data

SENS3.net hosts our database. The website was designed to provide a comprehensive resource for the recorded data and the recording procedures. It consists of three pages:

1. 'Home' page: provides a general overview of the database, featuring informative figures and videos that enable users to grasp the essence of our database and the recording techniques.
2. 'Recording Setup' page: includes an overview of the recording setup, data recording procedures, GUI, hardware list, and a figure depicting the camera and microphone arrangement.
3. 'Surfaces' page: hosts the collection of finger-surface interaction data organized into ten distinct categories. Users can click on a category to view the photos of each surface within it and download the associated data. The surface data are further aggregated based on different participants. The 'Surfaces' page also includes a 'Metadata' section, offering details about available files along with their descriptions and further information about surface physical properties; see Table 2).
4. 'About' page: contains citation detail for the paper and additional information about the authors.

Figure 3 showcases the physical data collected from two participants while interacting with two distinct surfaces (metal and foam) with static contact, pressing, tapping, and sliding (only a snippet of five seconds), as well as their adjective ratings collected from psychophysical experiments. The variations in data are visible for different surfaces and participants and demonstrate the significance of variety in surface and participants for the texture databases. For example,

Table 2. Metadata for the recorded data files. (num) represents the allocated number tag of the surface

File	Description
Tapping	
forces.csv	contact forces and torques in 6 columns - Fx, Fy, Fz, Tx, Ty, Tz [N and Nmm]
accelerations.csv	accelerations in 3 columns -Ax, Ay and Az [g]
audio.wav	audio data
video(num)_1.mp4	side view video
video(num)_2.mp4	top view video
Pressing	
forces.csv	contact forces and torques in 6 columns - Fx, Fy, Fz, Tx, Ty, Tz [N and Nmm]
stageposition.csv	position of the linear stage [mm]
video(num)_1_2mms.mp4	side view video
video(num)_2_2mms.mp4	top view video
Static Contact	
forces.csv	contact forces and torques in 6 columns - Fx, Fy, Fz, Tx, Ty, Tz [N and Nmm]
temperature.csv	index finger temperature [°C]
material_temperature.csv	surface temperature of the materials [°C]
heatflux.csv	heat flux between middle finger and material [W/m^2]
Sliding	
Material_sensor_(num).csv	contact forces and torques in 6 columns - Fx, Fy, Fz, Tx, Ty, Tz [N and Nmm], accelerations in 3 columns, Ax, Ay and Az [g]
Material_IR_pos_(num).csv	4 columns of data - elapsed time [s], speed [mm/s], position X [mm], position Y [mm]
Material_(num).mp4	top view video
Material_(num).wav	audio data
Sensation rating	
Order_par(num).csv	order of material presented to participant
Ratings_par(num).csv	adjective ratings
Video(num).mp4	isometric view video
material(num).tif	top view image
thickness.csv	thickness of the materials

there are visible differences between changes in skin temperature, heat flux, compliance hysteresis, and contact accelerations between metal and foam. Similarly, the interaction data from both participants discriminate from each other; these differences could be caused by variations in their exploratory behavior (compare finger position graphs at the last row) and their skin properties [2].

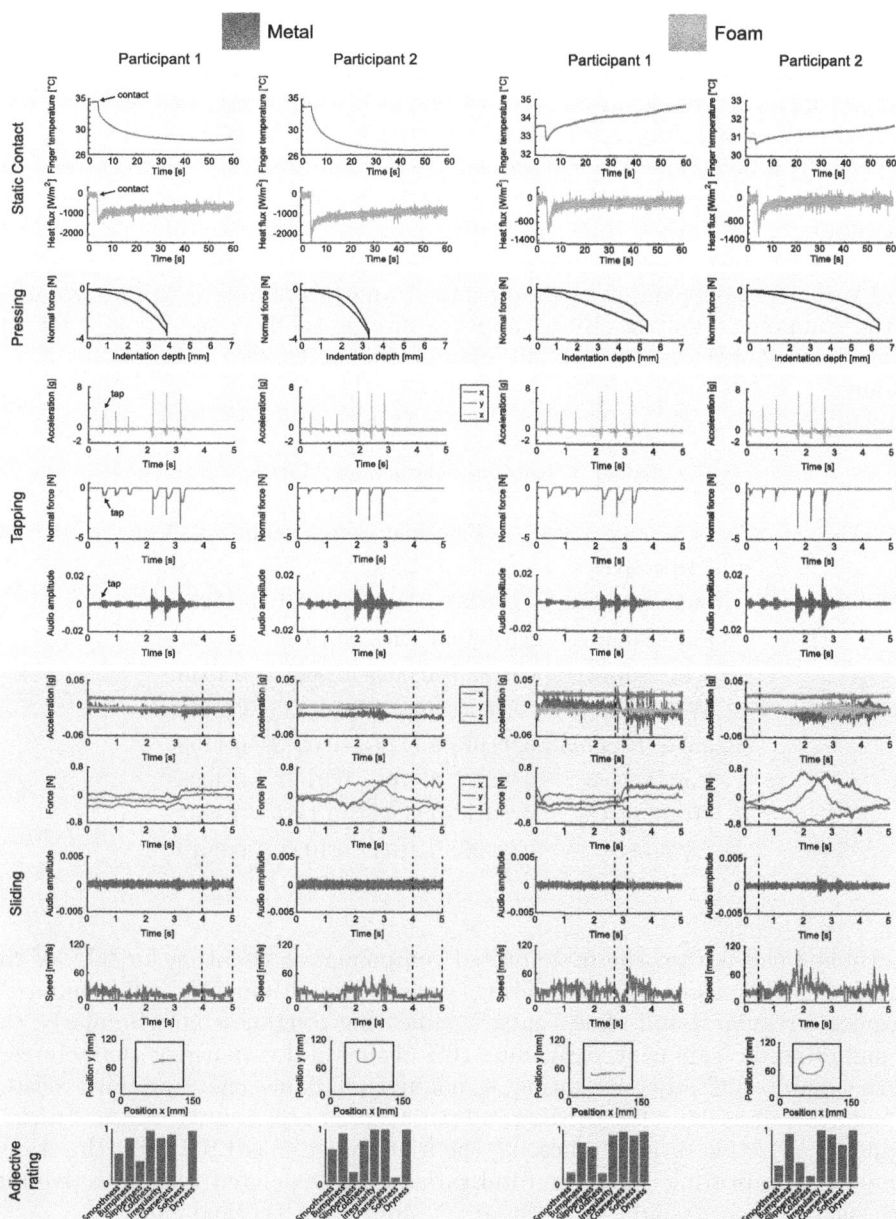

Fig. 3. Example recordings from two participants while interacting with two different surfaces: metal and foam. The sliding data comprises only a 5-s segment from the complete recording for a clear visualization. The dashed lines on the sliding plots highlight one [0.2–0.4 N, 0–33 mm/s] force-speed pair region for each of the measurements. The adjective ratings are normalized based on the maximum and minimum value reported by the participant across all surfaces.

4   Results of the Perceptual Experiments

We used principal component analysis (PCA) to analyze the adjective ratings, first normalizing them using z-scores within each participant. The resulting normalized ratings were averaged across all participants, yielding a 50 × 8 matrix (surfaces by adjective pairs). Subsequently, we computed the eigenvectors of the covariance of the normalized adjective matrix, sorting the diagonal elements in descending order to determine the optimal number of dimensions that capture the data's variance. Four dimensions were chosen as they represent 95% of the total variance. Following this, we did factor analysis through varimax rotation of the component matrix [10] to improve interpretability; see Table 3 for the rotated component matrix with four components and their corresponding factor loadings.

Table 3. Rotated Component Matrix

Adjectives	Principal Components			
	1	2	3	4
Rough–Smooth	−0.7878	−0.0689	0.1380	0.5593
Flat–Bumpy	0.9263	0.2261	0.1576	−0.1397
Sticky–Slippery	0.0183	−0.0629	0.9605	−0.1643
Warm–Cold	−0.2606	−0.4775	−0.1735	0.7923
Regular–Irregular	0.9246	0.1174	−0.0213	−0.1153
Fine–Coarse	0.9237	0.1741	0.0140	−0.3093
Hard–Soft	0.2604	0.8683	−0.1229	−0.3898
Wet–Dry	0.2543	0.2105	0.1923	−0.8912

Table 3 shows that the first rotated component, accounting for 66% of the variance, encompasses adjective ratings such as "rough–smooth", "flat–bumpy", "regular–irregular", and "fine–coarse", indicating roughness cue. Similarly, the second rotated component, explaining 16% of the total variance, is characterized by the "hard–soft" adjective rating, signaling compliance cue. The third rotated component, describing 9% of the total variance, is represented by the "sticky–slippery" adjective rating, indicating the friction cue. Lastly, the fourth rotated component, capturing 4% of the total variance, is associated with adjective ratings such as "warm–cold" and "wet–dry", shinting at thermal cues.

5   Discussion

We presented SENS3, an open-source multisensory database of finger-surface interactions and corresponding psychophysical sensations. This database contains recordings of when two participants interacted with 50 surfaces by static contact, pressing, tapping, and sliding. Our database captures a wide range of sensory information, such as surface photographs, video and sounds of the interactions, contact forces and torques, finger accelerations and positions, skin temperature, and heat flux transfer. In addition to the physical recordings, we captured the psychophysical sensations generated during free exploration through adjective ratings given by thirteen participants. We hope that *SENS3* will provide the necessary data for content design for multisensory user interfaces, understanding human touch and multisensory integration, and providing robots with humanlike touch sensations.

Despite our diligent efforts, the current dataset version exhibits some limitations. It comprises finger-surface interaction recordings from two trained male participants and perceptual ratings from thirteen participants, all collected by exploring fifty surfaces. This limited participant pool hampers our ability to capture the full spectrum of finger biomechanics and human perception. Furthermore, the surface diversity within the dataset represents merely a subset of surfaces encountered in daily life. Additionally, separating physical interaction recordings and adjective ratings poses a challenge in establishing direct correlations between sensations and physical data. Given the interconnected nature of adjective ratings and the inability of a single exploratory method to encapsulate all these sensations, we opted for free exploration, integrating all four exploratory procedures, to gather comprehensive perceptual data [3,19] and high-quality finger-surface interaction data separately with distinct exploratory procedure.

We designed SENS3 as open-ended so that we can address the shortcomings mentioned above. We envision augmenting the dataset with data collected from a more diverse range of surfaces and participants, encompassing variations in age, gender, and finger biomechanics. Our expansion plans also incorporate surface characteristics, finger biomechanical measurements, and simultaneous perceptual data collection separately during each exploratory procedure.

We aim to make *SENS3.net* an open-source platform intended to be a collaborative hub where laboratories, companies, or designers from diverse backgrounds add their new collected data or texture rendering models designed for various interfaces, such as touch-based devices and wearables. This collective effort will accelerate progress in the field of multisensory user interfaces but also empower a wide range of applications, from enhancing accessibility for individuals with sensory impairments to revolutionizing entertainment experiences.

Acknowledgements. This publication is part of the project "From signal-based modeling to sensation-based modeling" with project number 19153 of the research programme Veni partly financed by the Dutch Research Council (NWO) and Huawei Technologies. JK. Balasubramanian and B. Kodak contributed equally to this work.

References

1. Baumgartner, E., Wiebel, C.B., Gegenfurtner, K.R.: Visual and haptic representations of material properties. Multisens. Res. **26**(5), 429–455 (2013)
2. Callier, T., Saal, H.P., Davis-Berg, E.C., Bensmaia, S.J.: Kinematics of unconstrained tactile texture exploration. J. Neurophysiol. **113**(7), 3013–3020 (2015)
3. Cavdan, M., Doerschner, K., Drewing, K.: Task and material properties interactively affect softness explorations along different dimensions. IEEE Trans. Haptics **14**(3), 603–614 (2021)
4. Choi, H., Cho, S., Shin, S., Lee, H., Choi, S.: Data-driven thermal rendering: an initial study. In: 2018 IEEE Haptics Symposium (HAPTICS), pp. 344–350. IEEE (2018)
5. Cornelio, P., Velasco, C., Obrist, M.: Multisensory integration as per technological advances: a review. Front. Neurosci. **15**, 652611 (2021)
6. Culbertson, H., Delgado, J.J.L., Kuchenbecker, K.J.: One hundred data-driven haptic texture models and open-source methods for rendering on 3D objects. In: 2014 IEEE Haptics Symposium (HAPTICS), pp. 319–325. IEEE (2014)
7. Culbertson, H., Kuchenbecker, K.J.: Importance of matching physical friction, hardness, and texture in creating realistic haptic virtual surfaces. IEEE Trans. Haptics **10**(1), 63–74 (2016)
8. Culbertson, H., Unwin, J., Goodman, B.E., Kuchenbecker, K.J.: Generating haptic texture models from unconstrained tool-surface interactions. In: 2013 World Haptics Conference (WHC), pp. 295–300. IEEE (2013)
9. Devillard, A., Ramasamy, A., Faux, D., Hayward, V., Burdet, E.: Concurrent haptic, audio, and visual data set during bare finger interaction with textured surfaces. In: 2023 IEEE World Haptics Conference (WHC), pp. 101–106 (2023)
10. Drewing, K., Weyel, C., Celebi, H., Kaya, D.: Systematic relations between affective and sensory material dimensions in touch. IEEE Trans. Haptics **11**(4), 611–622 (2018)
11. Giri, G.S., Maddahi, Y., Zareinia, K.: An application-based review of haptics technology. Robotics **10**(1), 29 (2021)
12. Hassan, W., Abdulali, A., Abdullah, M., Ahn, S.C., Jeon, S.: Towards universal haptic library: library-based haptic texture assignment using image texture and perceptual space. IEEE Trans. Haptics **11**(2), 291–303 (2017)
13. Ho, H.N., Jones, L.A.: Contribution of thermal cues to material discrimination and localization. Percept. Psychophys. **68**(1), 118–128 (2006)
14. Jiao, J., Zhang, Y., Wang, D., Guo, X., Sun, X.: Haptex: A database of fabric textures for surface tactile display. In: 2019 IEEE World Haptics Conference (WHC), pp. 331–336. IEEE (2019)
15. Lederman, S.J., Klatzky, R.L.: Hand movements: a window into haptic object recognition. Cogn. Psychol. **19**(3), 342–368 (1987)
16. Lederman, S.J., Klatzky, R.L.: Multisensory Texture Perception. MIT Press (2004)
17. Lederman, S.J., Klatzky, R.L., Hamilton, C.L., Ramsay, G.I.: Perceiving surface roughness via a rigid probe: effects of exploration speed and mode of touch. Electron. J. Haptics Res. **1**(1), 1 (1999)
18. Manfredi, L.R., Saal, H.P., Brown, K.J., Zielinski, M.C., Dammann, J.F., III., Polashock, V.S., Bensmaia, S.J.: Natural scenes in tactile texture. J. Neurophysiol. **111**(9), 1792–1802 (2014)
19. Okamoto, S., Nagano, H., Yamada, Y.: Psychophysical dimensions of tactile perception of textures. IEEE Trans. Haptics **6**(1), 81–93 (2012)

20. Okamoto, S., Nagano, H., Yamada, Y.: Psychophysical dimensions of tactile perception of textures. IEEE Trans. Haptics **6**, 81–93 (2013)
21. Richardson, B.A., Vardar, Y., Wallraven, C., Kuchenbecker, K.J.: Learning to feel textures: predicting perceptual similarities from unconstrained finger-surface interactions. IEEE Trans. Haptics **15**(4), 705–717 (2022)
22. Serhat, G., Vardar, Y., Kuchenbecker, K.: Contact evolution of dry and hydrated fingertips at initial touch. PLoS One **17**(7), e0269722 (2022)
23. Srinivasan, M.A., LaMotte, R.H.: Tactual discrimination of softness. J. Neurophysiol. **73**(1), 88–101 (1995)
24. Strese, M., Boeck, Y., Steinbach, E.: Content-based surface material retrieval. In: 2017 IEEE World Haptics Conference (WHC), pp. 352–357. IEEE (2017)
25. Strese, M., Brudermueller, L., Kirsch, J., Steinbach, E.: Haptic material analysis and classification inspired by human exploratory procedures. IEEE Trans. Haptics **13**(2), 404–424 (2019)
26. Tanaka, Y., Bergmann Tiest, W.M., Kappers, A.M., Sano, A.: Contact force and scanning velocity during active roughness perception. PLoS One **9**(3), e93363 (2014)
27. Vardar, Y., Wallraven, C., Kuchenbecker, K.J.: Fingertip interaction metrics correlate with visual and haptic perception of real surfaces. In: 2019 IEEE World Haptics Conference (WHC), pp. 395–400. IEEE (2019)
28. Xia, P.: New advances for haptic rendering: state of the art. Vis. Comput. **34**, 271–287 (2018)

Variable Curvature and Spherically Arranged Ultrasound Transducers for Depth-Adjustable Focused Ultrasound

Shoha Kon[✉], Eifu Narita, Izumi Mizoguchi, and Hiroyuki Kajimoto

The University of Electro-Communications, 1-5-1 Chofugaoka, Chofu, Tokyo, Japan
{shoha.kon,narita,mizoguchi,kajimoto}@kaji-lab.jp

Abstract. In haptic feedback using focused ultrasound, a technique has been proposed that involves creating an ultrasound focal point using a spherical surface without the need for phase control. This method is efficient as each transducer faces directly toward the focal point, but it has a limitation – the focal point remains fixed due to the ultrasonic transducers being attached to a static sphere. This study introduces a method to move the focus in the depth direction by dynamically changing the curvature of the spherical surface to which the transducers are attached. We built a device that can adjust curvature using a servo motor, confirming its capability to relocate the focus position.

Keywords: Curvature Control · Focal Point Control · Mid-Air Haptics

1 Introduction

Airborne ultrasound tactile display can present tactile sensations without directly touching human skin. A typical setup includes ultrasound transducers arranged in a regular pattern on a flat surface. The phase control of each transducer focuses sound waves at any position within the presentation area.

Numerous studies have explored ultrasound focus using phase control [1–7]. Various control methods have been proposed, including the simultaneous formation of multiple focal points [4] and the generation of a strong sensation through fine reciprocating vibration or rotational movement of the focus along the skin while maintaining a constant ultrasound output [5]. In research on user perception, a study [6] investigated the relationship between the number of focal points and their speed. Furthermore, a graphical design tool for the quick presentation of ultrasound tactile sensation [7] has been proposed.

However, phase control represents a relatively costly hardware method. The typical frequency of ultrasound used is 40 kHz, necessitating a control period on the order of MHz for accurate phase control. High-speed control hardware, such as a field programmable gate array (FPGA), is required to control several hundred transducers simultaneously.

Physical methods to form an ultrasound focus without phase control have also been explored [8–10]. For example, methods involving the formation of a focus at the center

H. Kajimoto et al. (Eds.): EuroHaptics 2024, LNCS 14768, pp. 278–284, 2025.
https://doi.org/10.1007/978-3-031-70058-3_22

point of a sphere by arranging ultrasound transducers on the sphere [8], the formation of a focus using the phase difference of sound waves output from a meandering path [9], and the use of a circular deflection diaphragm [10] have been proposed. These methods can create a focal point using ultrasound waves without circuit-based phase control.

However, these methods have limitations in moving the ultrasound focus. For instance, an airborne ultrasound tactile display with ultrasound transducers arranged on a spherical surface has a single fixed focus. While introducing a pan-tilt mechanism to this display facilitates the horizontal or vertical movement of the focus position, moving the focus in the depth direction with this configuration is challenging.

In this study, we propose a method to move the focal point in the depth direction by arranging transducers on a spherical surface and dynamically changing the curvature of the sphere. The contribution of this method is the following two points:

- It simplifies control compared to conventional phase control methods.
- It offers high energy efficiency as the center of the ultrasound transducers consistently aligns with the focal direction.

2 Proposed Methods

2.1 Hardware

Figure 1 illustrates the proposed method. The system utilizes a parametric speaker experiment kit (K-02617, Akizuki Denshi Tsusho) to establish an ultrasound focus. The resonant frequency of the ultrasound transducers is 40 kHz, and FM modulated signals are simultaneously output from all transducers in response to audio input. The 50 transducers attached to the device, spaced approximately 1 cm apart, are arranged radially on a deformable plastic sheet (K-A-PET-BR, AMHA), which undergoes deformation through a servo motor (MG996R, Towerpro) driven by wires.

This results in a group of transducers positioned on a spherical surface with variable curvature. The ultrasound waves output from each transducer converges at the center of the sphere, i.e., vertically above the device. When the curvature of the sphere changes by wire traction, the ultrasound focus moves in the depth direction accordingly.

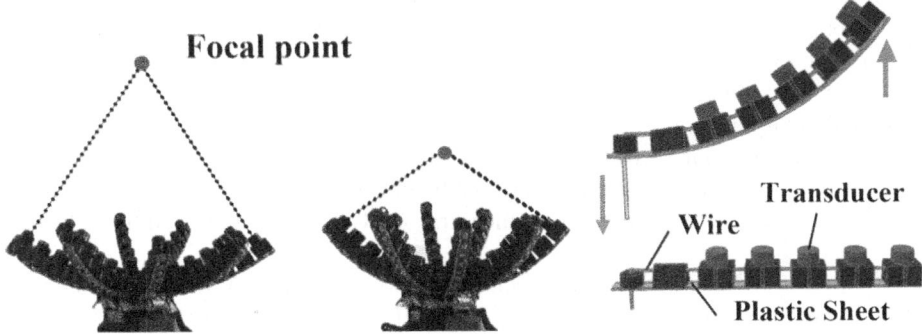

Fig. 1. Focal point movement by changing curvature

2.2 Device Response

To assess the device's responsiveness, we conducted measurements on the time it takes for the servo motor to transition from 0° (when the plastic sheet is parallel to the ground) to a specific angle for each servo motor setting. We conducted five measurements for each angle and present the data in Fig. 2. The offset delay time is approximately 400 ms, with the maximum time required to reach the intended angle being about 1.2 s. Despite this, the system exhibits smooth operation, allowing, for example, the focal point to be adjusted seamlessly in response to the hand's up-and-down motion.

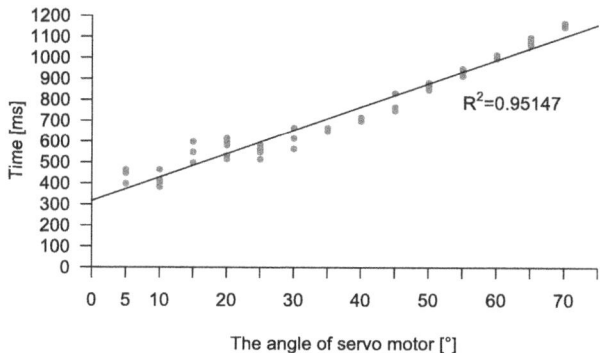

Fig. 2. Curvature formation time for each servo motor angle

3 Experiment: Focal Point Movement with Curvature Change

3.1 Experimental Procedure

This experiment aimed to investigate how the acoustic radiation pressure at the ultrasound focus changes with the traction angle of the servo motor in the depth direction. A mesh screen was put above the transducers, and the change of temperature of the screen that represents acoustic radiation pressure was measured [11]. The experimental setup is illustrated in Fig. 3.

The servo motor varied the traction angles from 5 to 70° in 5-degree increments. For each traction angle, the position of the mesh screen in the depth direction was adjusted by 1 cm, ranging from 9 cm to 42 cm. Subsequently, the surface temperature of the center of the mesh screen was recorded. The height of the mesh screen, corresponding to the maximum surface temperature at the ultrasound focus for each traction angle, was then measured.

The mesh (N No. 420S, Japan Special Fabrics) was arranged in two layers, and a thermal imaging camera (FLIR-E6390) measured the surface temperature. The initial mesh screen position was set at 0 cm when the plastic sheet was parallel to the ground. Temperature and humidity in the room were recorded before the measurement.

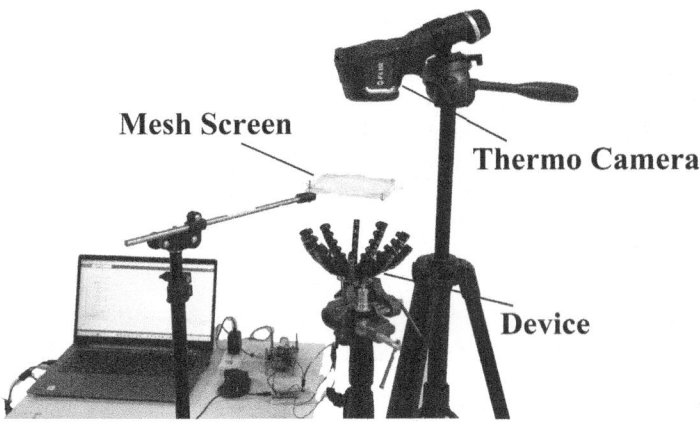

Fig. 3. Experimental setup

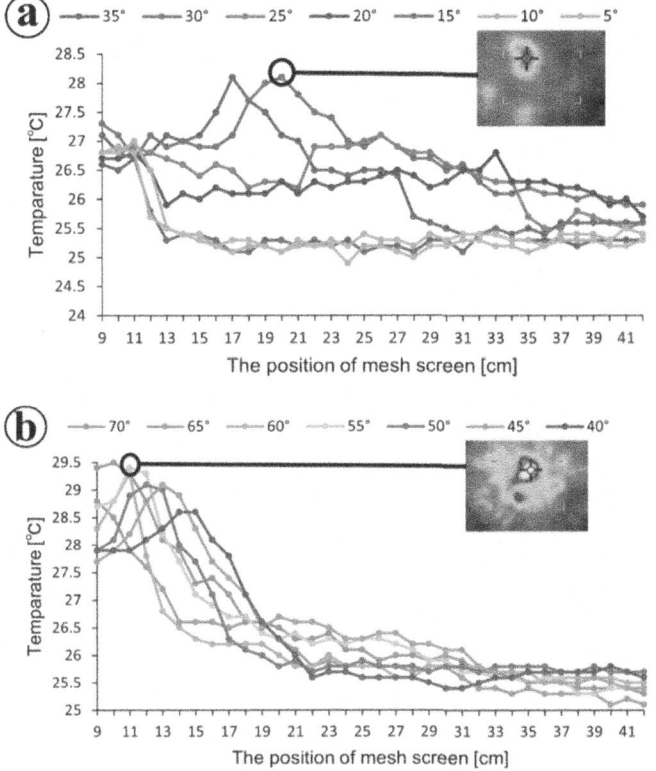

Fig. 4. Temperature changes in distance for each servo motor angle. The figure includes thermal images of the mesh screen surface captured at 30 and 55° as examples.

3.2 Results

During the measurement, the room temperature was 24.0 °C with 25% humidity. Figure 4 illustrates the temperature variations in the depth direction for each traction angle of the servo motor (a: servo motor angles 5° to 35°, b: servo motor angles 40° to 70°). Furthermore, Fig. 5 depicts the changes at the highest temperature point.

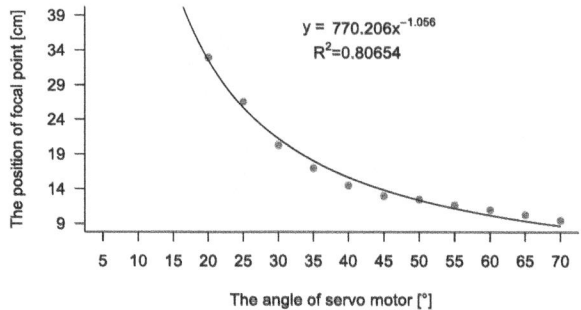

Fig. 5. Dependence of focus position on servo motor angles

4 Discussion

In Fig. 4, local temperature rise is evident at various servo motor traction angles. At 65°, for instance, the temperature peaks at 29.5 °C at 10 cm, and at 30°, it reaches 28 °C at 20 cm. These temperature changes are ascribed to the ultrasound focus. As the servo motor traction angle decreases, temperatures at these focal points decline. This decline may be due to attenuation resulting from reduced curvature as the traction angle decreases and the ultrasonic waves focus over a longer distance, or because the focal point becomes more spread out in the depth direction. Notably, no local temperature increase was observed at traction angles of 15, 10, and 5°. Additionally, the accuracy of the device might contribute to these effects. The device's design introduces minute differences in the curvature of the 10 frames on which the transducers are mounted. The longer the focal distance, the more pronounced the difference in curvature of each frame, potentially causing a shift in the focusing of the sound waves.

A power approximation for nonlinear regression was applied to analyze Fig. 5. It is evident that as the traction angle of the servo motor decreases, the position of the focal point extends in the depth direction. The relationship between the traction angle and the radius of curvature (focal point position) is nonlinear. The focal point shifts approximately 24 cm, moving from 34 cm to 10 cm. As illustrated in Fig. 2, the duration of this transition is less than 1 s. In essence, the system demonstrates sufficient responsiveness to respond to subtle up-and-down hand movements.

When this system is installed on an inclined surface, the plastic sheet to which the ultrasound transducers are attached may bend due to gravity. Therefore, there is a possibility that the acoustic radiation pressure of the focal point formed will weaken or that the focal point itself will not form.

While a detailed subjective evaluation was not conducted, we verified that the tactile sensation is distinctly perceptible when the hand is placed over the system. For instance, at an angle of 70°, tactile sensations were experienced in the range of 9 cm to 10 cm, and at 55°, in the range of 11 cm to 12 cm. These observations are consistent with the physical measurements.

5 Conclusion

This research investigates a device created to deliver ultrasound tactile sensations through a simple control method. The device comprises ten spherical frames with attached ultrasound transducers. The curvature of the spherical surface adjusts according to the servo motor's traction angle, enabling the formation of ultrasound focal points at different positions in the depth direction. Acoustic radiation pressure at the focus was observed for each traction angle using temperature measurements on a mesh screen. The results confirmed that the focus position in the depth direction varies with the servomotor's traction angle in the range of 70 to 20°, indicating that the focus movement in the depth direction can be easily controlled.

Based on the results of this experiment, we will conduct psychological evaluation experiments on human subjects to examine the effectiveness of the proposed method. For example, in the experiment concerning the perception of softness using this device, a hand tracking device (LeapMotion, UltraLeap) is placed nearby to determine the position of the user's hand. By adjusting the angle of the servo motor and the intensity of the ultrasound focus according to the user's finger movement, we can evaluate the sensation of softness when the user presses in with their fingers.

Acknowledgments. This research was supported by JSPS KAKENHI Grant Number JP20H05957.

References

1. Hoshi, T., Takahashi, M., Iwamoto, T., Shinoda, H.: Noncontact tactile display based on radiation pressure of airborne ultrasound. IEE Trans. Haptics 3(3), 155–165 (2010)
2. Suzuki, S., Inoue, S., Fujiwara, M., Makino, Y., Shinoda, H.: AUTD3: scalable airborne ultrasound tactile display. ITEE Trans. Haptics **14**(4), 740–749 (2021)
3. Howard, T., Marchal, M., Lécuyer, A., Pacchierotti, C.: PUMAH: pan-tilt ultrasound mid-air haptics for larger interaction workspace in virtual reality. IEEE Trans. Haptics 13(1), 38–44 (2020)
4. Carter, T., Seah, S.A., Long, B., Drinkwater, B.W., Subramanian, S.: UltraHaptics: multi-point mid-air haptic feedback for touch surfaces. In: UIST 2013, pp. 505–514 (2013)
5. Takahashi, R., Hasegawa, K., Shinoda, H.: Tactile stimulation by repetitive lateral movement of midair ultrasound focus. IEE Trans. Haptics **13**(2), 334–342 (2019)
6. Shen, Z., Vasudevan, M.K., Kučera, J., Obrist, M., Plasencia, D.M.: Multi-point STM: effects of drawing speed and number of focal points on users' responses using ultrasonic mid-air haptics. In: CHI 2023, no. 83, pp. 1–11 (2023)
7. Seifi, H., Chew, S., Nascè, A.J., Lowther, W.E., Frier, W., Hornbæk, K.: Feellustrator: a design tool for ultrasound mid-air haptics. In: CHI 2023, no. 266, pp. 1–16 (2023)

8. Hoshi, T.: Could not help making do-it-yourself acoustic levitation device (2nd report). In: Entertainment Computing Symposium, pp. 100–106 (2015). (in Japanese)

9. Memoli, G., Caleap, M., Asakawa, M., Sahoo, D.R., Drinkwater, B.W., Subramanian, S.: Metamaterial bricks and quantization of meta-surfaces. Nat. Commun. **8**, 14608 (2017)

10. Ito, Y.: High-intensity aerial ultrasonic source with a stripe-mode vibrating plate for improving convergence capability. Acoust. Sci. Technol. **36**(3), 216–224 (2015)

11. Onishi, R., et al.: Two-dimensional measurement of airborne ultrasound field using thermal images. Phys. Rev. Appl. **18**(4), 44–47 (2022)

The HapticSpider: A 7-DoF Wearable Device for Cutaneous Interaction with the Palm

Lisheng Kuang[1]([⊠])[iD], Monica Malvezzi[2][iD], Domenico Prattichizzo[2,3][iD],
Paolo Robuffo Giordano[1][iD], Francesco Chinello[4][iD], and Claudio Pacchierotti[1][iD]

[1] CNRS, University of Rennes, Inria, IRISA, Rennes, France
kuanglisheng@gmail.com
[2] University of Siena, Siena, Italy
[3] Italian Institute of Technology, Genova, Italy
[4] Aarhus University, Herning, Denmark

Abstract. This paper introduces a 7-degrees-of-freedom (7-DoF) hand-mounted haptic device, the "HapticSpider". It is composed of a parallel mechanism characterised by eight legs with an articulated diamond-shaped structure, in turn connected to an origami-like shape-changing end-effector. The device can render surface and edge touch simulations as well as apply normal, shear, and twist forces to the palm. This paper presents the device's mechanical structure, a summary of its kinematic model, actuation control, and preliminary device evaluation, characterizing its workspace and force output.

Keywords: Wearable haptics · Cutaneous feedback · Kinematics

1 Introduction

Haptic technology facilitates remote touch in many scenarios, including robotic teleoperation and virtual interaction, but its mainstream use is limited due to non-portable and non-wearable systems, hindering the field's growth [12,13]. In this respect, recent advancements aim to create more wearable and portable haptic solutions, designed for comfort and ease of use [14]. However, while an increasing number of applications leverage haptic capabilities, the stimuli provided by current wearable interfaces remain limited. Balancing realistic touch with cost, wearability, and portability is crucial for advancing haptic technology toward more immersive remote interactions [10,13].

Wearable cutaneous interfaces have been historically designed for the hand, as it is the most sensitive part of our body and the one that is most often used for grasping, manipulation, and probing the environment, especially targeting the fingertip [3,5,8,9,15] or the palm [2,7,16]. The primary benefit of such wearable haptic technology lies in its compact size and lighter weight compared to grounded solutions. These features enable the potential for engaging in various

H. Kajimoto et al. (Eds.): EuroHaptics 2024, LNCS 14768, pp. 285–292, 2025.
https://doi.org/10.1007/978-3-031-70058-3_23

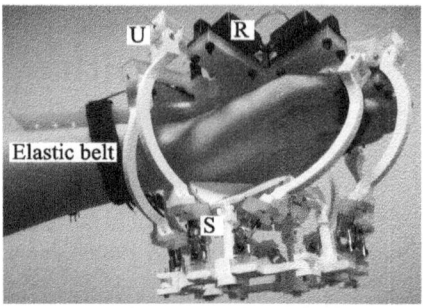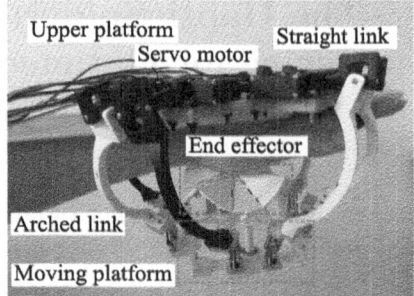

Fig. 1. The proposed 7-DoF haptic device for the palm. Eight actuated legs move the end-effector in all directions/rotations with respect to the palm as well as fold/unfold the origami structure under the user's palm. The fingers are supposed to be straight to bond the strip and to avoid contact with the device legs.

types of haptic interactions in a portable and lightweight manner. However, it is rare to find wearable devices capable of providing multiple haptic sensations, mainly due to limitations in size and weight.

This paper presents a wearable haptic device for the palm, shown in Fig. 1. It features a 7-degrees-of-freedom (7-DoF) end-effector that can move towards/away from the palm, move and rotate on the plane parallel to the user's palm, and fold along a pre-defined origami-like structure, as shown in Fig. 3. The device is therefore able to elicit the sensation of interacting with a wide range of slanted surfaces and curvatures through combined localized pressure (normal force from the generated flat surfaces and sharp edges), and skin stretch sensations (torsional and shear force).

With respect to other wearable haptic solutions for the palm, such as [1,2,7, 17], the proposed device is able to provide a (much) broader range of stimuli, at the cost—however—of a bulkier design.

For example, Malvezzi et al. [3,8] presented a delta-like device for the fingertip that moves a rigid mobile platform thanks to three articulated legs actuated by three servomotors. Williams et al. [18] devised a finger-mounted haptic device with 4-DoF using origami fabrication. This innovative mechanism delivers normal, shear, and torsional haptic feedback to the fingertip. Unlike traditional methods, their approach simplifies joint manufacturing for small devices, reducing complexity and size through origami principles. Addressing the palm, Trinitatova and Tsetserukou [16,17] proposed a 3-DoF inverted delta mechanism consisting of three identical kinematic limbs to move a pin-like end-effector across the palm. Similarly, Dragusanu et al. [4] used a 3-DoF parallel tendon-based mechanical structure to actuate an interchangeable palmar end-effector. Finally, Minamizawa et al. [11] used two motors to actuate two belts moving a flat end-effector on the palm, providing normal and shear forces.

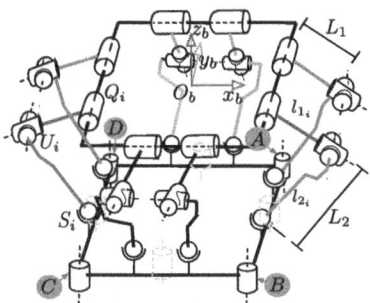

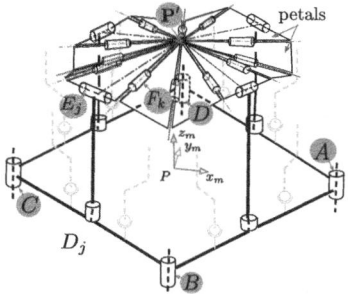

(a) Kinematic structure overview (end-effector not shown).

(b) Detail of the origami end-effector (up).

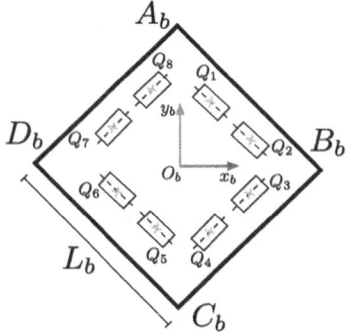

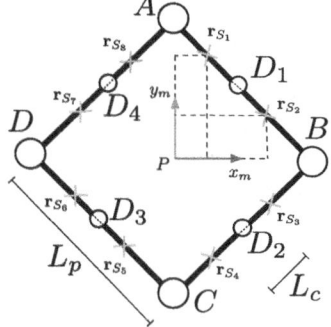

(c) Planar representation of the static base (upper platform).

(d) Planar representation of the moving platform.

Fig. 2. Device structure. (a, c) The static upper base is square, with edge length L_b and vertices A_b, B_b, C_b, D_b. It houses the revolute joints Q_i connected to links $l_{1,i}$, whose length is L_1. The links are in turn connected via universal joints U_i to links $l_{2,i}$ (length L_2). The spherical joints in S_i connect each leg (composed of $l_{1,i}$ and $l_{2,i}$ links) to the moving platform, which has a diamond shape with vertices A, B, C, D. Such points correspond to revolute joints with parallel axes that make the moving platform articulated. A base reference frame with origin in O_b and axes x_b, y_b, z_b is defined. (b, d) The moving platform houses an origami-like end-effector with vertex P' composed of twelve petals held together with twelve hinge joints F_k ($k = 1, \ldots, 12$). The origami is connected to four additional links with the joints E_j ($j = 1, 2, 3, 4$), which are furthermore connected to the moving platform with joints D_j. An auxiliary reference frame with origin in the center P on the moving platform plane and axes x, y, z is defined to identify the moving platform configuration.

2 Device Design and Analysis

2.1 The Device Structure

The device is composed of a static base (upper platform, on the back of the hand) and a moving platform, housing the end-effector in contact with the palm, as shown in Fig. 2a. The upper and moving platforms are connected by eight legs

Table 1. Technical specifications of the HapticSpider

Weight	450 g	L × W × H	12 × 12 × 15 cm
Operating voltage	6.0 V	Control system	ESP32
Max normal force	6.0 N	Joint speed	0.14 s/60°
Max $x \times y$ displacement	10 × 10 cm	Max z displacement	8 cm

(hence, the "spider" reference). These eight identical legs actuate the position, orientation, and configuration of the end-effector mounted on the moving platform.

Each leg is linked to the static upper platform with a revolute joint (Q_i, where $i = \{1, 2, 3, ..., 8\}$) actuated by a servomotor (Hitec HS-85MG). Two rigid links, one straight and shorter ($l_{1,i}$) and one arched and longer ($l_{2,i}$), are serially connected from Q_i to the moving platform through a 2-DoFs universal joint (U_i, between $l_{1,i}$ and $l_{2,i}$) and a final 3-DoF spherical joint S_i, as shown in Fig. 2a. Both the spherical and the universal joints are passive.

The main device features are indicated in Table 1. A video is available as supplemental material, CAD design of the device will be made available on request.

2.2   The Origami End-Effector

The moving platform is composed of four links with length L_p (see Fig. 2d), interconnected between each other through four passive revolute joints, whose axes pass through A, B, C, D. This structure resembles a diamond-shaped frame whose vertex angles can be varied, i.e. a planar four-bar mechanism.

At the mid-point of each link, a passive revolute joint (D_j with $j = \{1, 2, 3, 4\}$, see Fig. 2b), coaxial with the joints in the vertices, connects the moving platform to an additional link, in turn connected to the origami end-effector through the revolute joint in E_j. The rotation axes of the joints in D_j and E_j are orthogonal. The origami end-effector is the part of the device that contacts the palm, inspired from [6,7]. It is composed of 12 elements, that we call "petals", connected by twelve revolute hinge joints F_k, $k = \{1, 2, 3, ..., 12\}$. Joints F_k are passive, and their rotation angle depends on the distance $\overline{AC}$, and thus $\overline{BD}$, as controlled by the underlying spider structure. In other words, the configuration of the origami structure depends on the deformation of the diamond-shaped frame defined by A, B, C, D joints.

2.3   Inverse Kinematics

We indicate with $\mathbf{q} = [q_1, ..., q_8]^{\mathrm{T}}$ the actuators' rotation angles connected to joints Q_i. From Fig. 2d, we indicate with $S_b < O_b, x_b, y_b, z_b >$ the base reference frame and with $S_m < P, x_m, y_m, z_m >$ the local reference frame of the moving platform. The moving platform configuration is defined by the position of point

P, $\mathbf{r}_P = [x_P, y_P, z_P]^T$, and the orientation angles, i.e., roll, pitch, yaw angles, collected in the vector $\boldsymbol{\phi} = [\alpha, \beta, \gamma]^T$, both expressed in the base reference frame S_b. Since the moving platform is not a rigid body, we need a further parameter to represent its opening/closure, due to its diamond-shaped articulated structure. We introduce a further parameter ρ to represent the moving platform shape, defined as half of the diamond diagonal aligned with y_m-axis. The moving platform configuration can be therefore described by the 7-dimensional vector $\mathbf{u} = [x_P, y_P, z_P, \alpha, \beta, \gamma, \rho]^T$. The inverse kinematics defines the relationship mapping $\mathbf{u}$ into $\mathbf{q}$, i.e., $\mathbf{q} = f_d(\mathbf{u})$. According to the above definitions, the coordinates of the diamond vertices A, B, C, D with respect to S_m are

$$\mathbf{r}_A^m = [0, \rho, 0]^T, \ \mathbf{r}_B^m = [\sqrt{L_p^2 - \rho^2}, \ 0, \ 0]^T, \ \mathbf{r}_C^m = -\mathbf{r}_A^m, \ \mathbf{r}_D^m = -\mathbf{r}_B^m. \quad (1)$$

For example, when the angles $\angle ABC = \angle BCD = \angle CDA = \angle DAB = \pi/2$ rad (the moving platform is a square), then $\rho = L_p\sqrt{2}/2$. To calculate the coordinates of the spherical joints S_i connecting the moving platform to the legs (see Fig. 2d), we introduce two bi-dimensional auxiliary vectors $\mathbf{a}$ and $\mathbf{b}$:

$$\mathbf{a} = [a_1, a_2] = 1/L_p \left[L_c \sqrt{L_p^2 - \rho^2} + L_0\rho, \quad (L_p - L_c)\sqrt{Lp^2 - \rho^2} + L_0\rho \right]$$

$$\mathbf{b} = [b_1, b_2] = 1/L_p \left[L_c\rho L_0\sqrt{L_p^2 - \rho^2}/L_p \quad (L_p - L_c)\rho L_0\sqrt{L_p^2 - \rho^2} \right].$$

Thanks to $\mathbf{a}$ and $\mathbf{b}$, with simple geometrical considerations, it is possible to express the coordinates of S_i points with respect to S_m, i.e. $\mathbf{r}_{S_i}^m$. Given $\mathbf{R}_{S_m}^{S_b}$ the rotation matrix from the moving platform to the static base, expressed as a function of orientation angles $\boldsymbol{\phi}$ and the position of P, defined by vector $\mathbf{r}_P$, we can easily evaluate the coordinates of the spherical joints with respect to the S_m frame: $\mathbf{r}_{S_i} = \mathbf{r}_P + \mathbf{R}_{S_m}^{S_b} \mathbf{r}_{S_i}^m$.

On the static upper base, the coordinates of the Q_i points, $\mathbf{r}_{Qi}$ can be evaluated in the S_b frame with simple geometrical considerations, as a function of base parameters L_b and L_p indicated in Fig. 2c and 2d. We introduce eight auxiliary reference frames S_{qi}, with origins in points Q_i, z_i axes parallel to revolute joint axes, and x_i axis on the $x_b y_b$ plane. The rotational matrix $\mathbf{R}_{q_i}$ represents the orientation of S_{qi} with respect to S_b. We can then express the coordinates of S_i points for such auxiliary reference frames

$$\mathbf{r}_{S_i} = [s_{p_{x_i}}, s_{p_{y_i}}, s_{p_{z_i}}]^T = \mathbf{R}_{q_i}^T [\mathbf{r}_{S_i} - \mathbf{r}_{Qi}]^T.$$

This enables us to evaluate the components q_i of the vector joint variables $\mathbf{q}$ as:

$$q_i = 2\arctan\left(\frac{s_{p_{y_i}} + \sqrt{s_{p_{x_i}}^2 + s_{p_{y_i}}^2 - w_i^2}}{w_i + s_{p_{x_i}}}\right)$$

with w_i being an auxiliary variable defined as $w_i = \left(|\mathbf{r}_{S_i}| + L_1^2 - L_2^2\right)/(2L_1)$, where L_{1_i} and L_{2_i} are the same as in Fig. 2a.

| (a) | (b) | (c) |

Fig. 3. By changing the configuration of the moving platform, we change the structure of the origami end-effector. For example, in (a), as the distance between the A and C vertices is equal, the origami bends exposing both G_1 and G_2 edges; in (b), as the distance between A and C is maximised while that between B and D is minimised, edge G_2 is more exposed; conversely, in (c), G_1 is more exposed.

As described before, the moving platform is connected to the origami end-effector, which is capable of folding/unfolding, rendering different local shapes on the palm. The overall structure is visible in Figs. 2b and 3, and it can be changed by controlling the shape of the underlying moving platform. By controlling position vector $\mathbf{r}_P$ and ϕ orientation angles, the position $\mathbf{r}_{P'}$ and orientation of the origami structure can be controlled. By reducing or increasing the distance of the moving platform vertices, through the control of ρ parameter in configuration vector $\mathbf{u}$, it is possible to achieve the two configuration limits shown in Fig. 3b and 3c. In Fig. 3b, the distance between B and D is maximum, so the edge G_2 is sharp and contacts the palm. On the other hand, in Fig. 3c the distance between A and C is maximized, so this time edge G_1 is sharp and contacts the palm. Figure 3a shows an intermediate configuration. Specifically, the position of the origami vertex, indicated with P' in Fig. 3b is directly related to P position and moving platform configuration, represented by ρ parameter: $\mathbf{r}_{P'}^m = [0, 0, z_{P'}^m(\rho)]^{\mathrm{T}}$.

2.4 Pilot Evaluation

We measured the device performance in providing forces. The device was rigidly fixed to a table and a 6-DoF force-torque sensor (ATI-Nano43) was placed below the static upper base, where the palm should be. The origami end-effector was configured such that $||AC|| = ||BD||$, as in Fig. 3a. Using the inverse kinematics described above, we moved the platform ten times in and out of contact with the ATI sensor, applying a maximum vertical force along the z axis. Results show that the device can apply up to 6 N along this direction.

3   Conclusions

This paper presents a novel wearable haptic device for the palm. It features a 7-DoF end-effector capable of diverse movements. Illustrated in Fig. 1, the device can move towards/away from the palm, traverse and rotate on the plane parallel to the user's palm, and fold along a pre-defined origami-like structure. These capabilities enable the device to simulate interactions with various slanted surfaces and curvatures by inducing localized pressure and skin stretch sensations. Moreover, as it is very easy to replace end-effector (e.g., using a different foldable origami structure, see also [7]), the device can be quickly adapted to render custom shape sensations.

Compared to existing wearable haptic solutions for the palm, this device offers a significantly wider range of stimuli, rarely seen in a palmar haptic device. However, it does so at the expense of its wearability, suffering from a rather bulky design that severely limits the portability and comfort of the device. We believe that the device could be anyway a useful reference and benchmark system for developing simpler and lighter wearable devices for the hand palm.

Future work will focus on optimising the design by improving the origami structure's mobility, especially for G1 and G2 edges. Mechanical improvements are crucial to optimise the size and shape of the device to reduce the overall encumbrance, reducing the mass, improve the portability and the mechanical stability while providing the desired haptic feedback. We will also try to organise a proper experimental scenario to validate the work, so as to better evaluate the use of the device in relevant cases.

Acknowledgement. We gratefully acknowledge the funding provided by the project "HARIA - Human-Robot Sensorimotor Augmentation - Wearable Sensorimotor Interfaces and Supernumerary Robotic Limbs for Humans with Upper-limb Disabilities" (EU Horizon Europe, GA No. 101070292).

References

1. Altamirano Cabrera, M., Tsetserukou, D.: Linkglide: a wearable haptic display with inverted five-bar linkages for delivering multi-contact and multi-modal tactile stimuli. In: Proceedings of International AsiaHaptics Conference, pp. 149–154 (2018)
2. Cabrera, M.A., Tirado, J., Heredia, J., Tsetserukou, D.: Linkglide-s: a wearable multi-contact tactile display aimed at rendering object softness at the palm with impedance control in VR and telemanipulation (2022). arXiv preprint arXiv:2208.14149
3. Chinello, F., Pacchierotti, C., Malvezzi, M., Prattichizzo, D.: A three revolute-revolute-spherical wearable fingertip cutaneous device for stiffness rendering. IEEE Trans. Haptics **11**(1), 39–50 (2017)
4. Dragusanu, M., Villani, A., Prattichizzo, D., Malvezzi, M.: Design of a wearable haptic device for hand palm cutaneous feedback. Front. Robot. AI **8**, 706627 (2021)
5. Gabardi, M., Solazzi, M., Leonardis, D., Frisoli, A.: A new wearable fingertip haptic interface for the rendering of virtual shapes and surface features. In: Proceedings of IEEE Haptics Symposium, pp. 140–146 (2016)
6. Kuang, L., Chinello, F., Giordano, P.R., Marchal, M., Pacchierotti, C.: Haptic mushroom: a 3-dof shape-changing encounter-type haptic device with interchangeable end-effectors. In: Proceedings of IEEE World Haptics Conference (WHC) (2023)

7. Kuang, L., Ferro, M., Malvezzi, M., Prattichizzo, D., Giordano, P.R., Chinello, F., Pacchierotti, C.: A wearable haptic device for the hand with interchangeable end-effectors. IEEE Trans. Haptics **17**, 129 (2023)
8. Malvezzi, M., Chinello, F., Prattichizzo, D., Pacchierotti, C.: Design of personalized wearable haptic interfaces to account for fingertip size and shape. IEEE Trans. Haptics **14**(2), 266–272 (2021)
9. Meli, L., Pacchierotti, C., Salvietti, G., Chinello, F., Maisto, M., De Luca, A., Prattichizzo, D.: Combining wearable finger haptics and augmented reality: user evaluation using an external camera and the microsoft hololens. IEEE Robot. Autom. Lett. **3**(4), 4297–4304 (2018)
10. Meli, L., Scheggi, S., Pacchierotti, C., Prattichizzo, D.: Wearable haptics and hand tracking via an RGB-D camera for immersive tactile experiences. In: ACM SIGGRAPH Posters (2014)
11. Minamizawa, K., Kamuro, S., Kawakami, N., Tachi, S.: A palm-worn haptic display for bimanual operations in virtual environments. In: Haptics: Perception, Devices and Scenarios: 6th International Conference, EuroHaptics, pp. 458–463. Springer (2008)
12. Pacchierotti, C., Prattichizzo, D.: Cutaneous/tactile haptic feedback in robotic teleoperation: motivation, survey, and perspectives. IEEE Trans. Robot. **40**, 978 (2024)
13. Pacchierotti, C., Sinclair, S., Solazzi, M., Frisoli, A., Hayward, V., Prattichizzo, D.: Wearable haptic systems for the fingertip and the hand: taxonomy, review, and perspectives. IEEE Trans. Haptics **10**(4), 580–600 (2017)
14. Prattichizzo, D., Otaduy, M., Kajimoto, H., Pacchierotti, C.: Wearable and hand-held haptics. IEEE Trans. Haptics **12**(3), 227–231 (2019)
15. Schorr, S.B., Okamura, A.M.: Fingertip tactile devices for virtual object manipulation and exploration. In: Proceedings of CHI Conference on Human Factors in Computing Systems, pp. 3115–3119 (2017)
16. Trinitatova, D., Tsetserukou, D.: Deltatouch: a 3D haptic display for delivering multimodal tactile stimuli at the palm. In: Proceedings of IEEE World Haptics Conference (WHC), pp. 73–78 (2019)
17. Trinitatova, D., Tsetserukou, D.: Touchvr: a wearable haptic interface for VR aimed at delivering multi-modal stimuli at the user's palm. In: SIGGRAPH Asia 2019 XR, pp. 42–43 (2019)
18. Williams, S.R., Suchoski, J.M., Chua, Z., Okamura, A.M.: A 4-DoF Parallel Origami Haptic Device for Normal, Shear, and Torsion Feedback (2021). arXiv preprint arXiv:2109.12134

Vibrotactile Cues with Net Lateral Forces Resulting from a Travelling Wave

Mondher Ouari$^{(\boxtimes)}$ (ID), Anis Kaci(ID), Christophe Giraud-Audine(ID), Frédéric Giraud(ID), and Betty Lemaire-Semail(ID)

Univ. Lille, Arts et Métiers Institute of Technology, Centrale Lille, Junia, ULR 2697 - L2EP, F-59000 Lille, France
mondher.ouari@univ-lille.fr

Abstract. Ultrasonic travelling waves possess the capability to induce net shear forces on the finger pulp, which finds utility in various applications. This study specifically investigates their potential in generating vibrotactile cues through force modulation. Two experiments are outlined herein. In the first, we assess the detection threshold, estimated at 1.5 µm peak-peak for a modulation frequency of 75 Hz. Remarkably, this threshold closely aligns with that observed in conventional vibrotactile indentation and remains consistent regardless of the force modulation method employed. In the second experiment, we correlate the amplitude of a travelling wave tactile display with the vibration amplitude of a shaker to achieve equivalent vibration intensity. Our findings reveal a substantial amplitude ratio range experienced on the travelling wave device, spanning from 24 to 155, which is pivotal in optimizing vibrotactile stimulator design. Additionally, we demonstrate that vibrations below 1 µm yield negligible net lateral forces on the finger pulp. These results underscore the potential of ultrasonic travelling waves in enhancing tactile feedback systems.

Keywords: Vibrotactile · travelling wave · haptic display · ultrasonic vibration · Net lateral force

1 Introduction

Tactile cues can be conveyed to users through vibrations generated by actuators in contact with the skin, leading to the development of advanced human-computer interaction systems. For example, button clicks can be simulated using electromagnetic [20] or piezoelectric [13] actuators, inducing vibrations on touchscreen displays. Additionally, modulation of vibration patterns enriches user interaction [6,17]. Notably, vibrations can be delivered either in the normal [7] or tangential [19] directions relative to the contact area. Perceptually, sensitivity is higher and more stable for tangential vibrations compared to normal ones [2,15].

A new generation of tactile actuators capable of producing net shear forces on the skin has recently emerged. These actuators employ techniques such as asymmetric friction combined with high-frequency ultrasound [4], elliptical motion

H. Kajimoto et al. (Eds.): EuroHaptics 2024, LNCS 14768, pp. 293–304, 2025.
https://doi.org/10.1007/978-3-031-70058-3_24

of particles induced by travelling waves [3,10], and modal superimposition [8]. Travelling waves and modal superimposition generate net shear forces by inducing elliptical motion of particles on a plate or a ring at ultrasonic frequencies [11]. In both cases, two vibration modes are excited with a ±90° phase shift.

Previous studies on the elliptical motion of particles predominantly focused on the fingertip as the location for force generation. For instance, authors in [10] observed a decrease in force with increasing finger velocity, while [4] noted force occurrence even with a fixed finger.

Travelling wave stimulators offer the advantage of producing low-frequency vibrations (below 1 kHz) with minimal displacement (around 1 μm), ideal for creating vibrotactile patterns on the skin. Modulation of net shear force involves mechanisms such as reducing particle rotational speed by decreasing vibration amplitude [10], or adjusting the standing wave ratio (SWR) through phase shift modification [3,5]. Moreover, reverse motion is achievable by reversing the phase shift from +90° to −90°.

Several applications have been proposed for this type of actuator. For example, [3] aims to create a virtual environment, while [11] achieves a button click sensation by synchronizing the change of direction of the travelling wave with the moment the pressing force exceeds a threshold. This principle was later applied by [8] using modal superimposition on a finite glass plate.

In this paper, we explore the generation of vibrotactile cues using the principle of elliptical motion generated by a travelling wave to stimulate the finger pulp. Devices based on ultra net shear forces have the potential to become more versatile and can offer better haptic feedback by incorporating vibrotactile cues, making this uncharted territory the primary focus of our investigation. Specifically, we investigate how low-frequency vibrations produced by modulated net shear forces can be perceived with the same intensity as normal vibration. To achieve this, we conduct a psychophysical study to determine the minimal detection threshold of vibrotactile cues and compare two methods for modulating net forces to assess sensitivity differences. In the second experiment, participants are tasked with adjusting the level of a normal vibrotactile stimulator to match the intensity of the travelling wave based device.

2   Experimental Setup

The travelling wave is generated by the stator of a Shinsei USR60 ultrasonic motor (Shinsei Corporation, Japan), shown in Fig. 1a, which is composed of a ring to which two sets of piezoelectric actuators are glued. The first set is used to excite a ring bending mode denoted "1" (Fig. 1a), whose resonance frequency is $f = 40$ kHz with a wavelength $\lambda = 21$ mm. The second set is used to excite a twin mode denoted by "2" with the same resonance frequency and the same wavelength, but spatially offset by a quarter of a wavelength $\lambda/4$. The two voltages applied to the sets of piezoelectric actuators are generated by a DSP (Digital Signal Processor, nucleo-STM32F446 for STM) which are amplified via

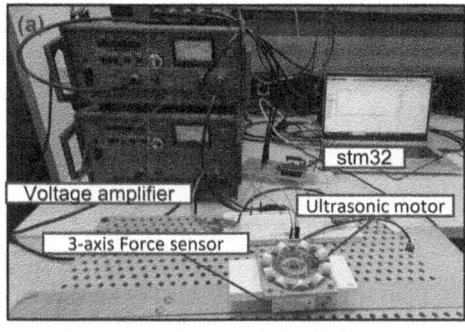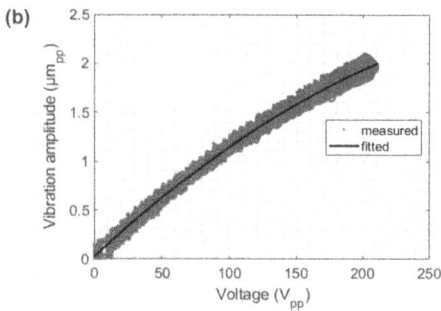

Fig. 1. (a) The experimental setup with the mode shape represented on the USR, (b) Relation between the excitation voltage and the vibration amplitude.

two voltage amplifiers (HSA4051 and HSA4052 from NF JAPAN) to reach up to 300 Vpp. The resulting vibration is written as:

$$w_1(t) = \frac{A}{2} \times cos(2\pi f t) \quad w_2(t) = \frac{A}{2} \times cos(2\pi f t + \phi) \tag{1}$$

with w_1 and w_2 are the deflection of the modes 1 and 2 respectively, A is the vibration amplitude peak to peak and ϕ is the phase shift.

The relationship between the control voltages and the vibration amplitudes of the unloaded device was identified using a laser vibrometer (Polytec OFV 505) and is presented in Fig. 1b for $\phi = 90°$. The curve shows nonlinear behavior, which is taken into account in our study. It is worth noting that the pressing of the finger within the force limits stipulated in this study has a negligible effect on the amplitude.

Finally, a 3-axis force sensor (K3D40 from PM instrument) is mounted under the USR; it monitors the normal force F_N and measures the net lateral force F_L produced by the system.

3 Modulation of the Net Shear Force for Vibrotactile Cue

With a travelling wave, the elliptical motion of the particles of the stator drives the finger pulp due to frictional mechanisms [16]. The model of this interaction is outside the scope of this paper, but it is generally admitted that the force F_L depends on the vibration amplitude A, the phase shift ϕ, and the pressing force F_N. The two methods are implemented in order to modulate the net shear force, and are illustrated in Fig. 2.

The first method modulates the vibration amplitude A (AM) and keeps a phase shift $\phi = \pm\frac{\pi}{2}$. In this way, the travelling wave is maintained, but its amplitude is modulated. The second method keeps a constant vibration amplitude A but modulates the phase shift $-\pi \leq \phi \leq \pi$ (PM). This will modify the standing wave to travelling wave ratio. In the paper, sinusoidal modulations at frequency f_{LF} with vibration amplitude A_0 are used, and we have:

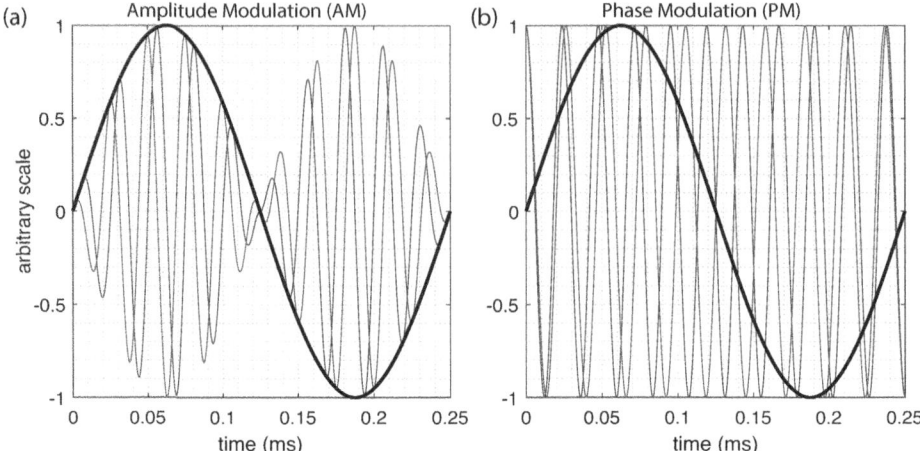

Fig. 2. Proposed methods of modulation, the frequencies are not in scale, they were chosen for illustration. w_1 in blue, w_2 in red and the reference signal in black : (a) Amplitude modulation and (b) Phase modulation. (Color figure online)

- Amplitude Modulation (AM): $A = A_0 \times \sin(2\pi f_{LF} t)$, $\phi = 90°$
- Phase Modulation (PM): $A = A_0$, $\phi = \frac{\pi}{2} \times \sin(2\pi f_{LF} t)$

Since the force generation relies on friction mechanisms, nonlinear behavior may occur, and lead to unwanted additional frequencies. For that purpose, we modulated the net lateral force at different frequencies ($f_{LF} = 25$, 50 and 75 Hz) and same vibration amplitude ($A_0 = 2.6\,\mu m$) according to both methods. A finger is pressing at $F_N = 0.3\,N$. The resulted force in the lateral direction using the 3-axis force sensor is given in Fig. 3. Additional harmonics appear on the spectrum of the measurements, but at very low relative amplitude (20% of the fundamental frequency). Therefore, both methods are suitable to create vibrotactile cues at single frequency.

4   Experiment1: Detection Threshold

In this experiment, we want to measure the detection threshold of vibrotactile stimulation based on net lateral force. To estimate the threshold and slope of the psychometric function, we are using a weighted and transformed 1-up/2-down staircase.

4.1   Participants

A total of twelve participants were initially recruited for the study. However, one participant was excluded due to experiencing loss of touch on the tip of his fingers. Consequently, the final cohort comprised eleven healthy volunteers, consisting of 8 males and 3 females, with an average age of $M = 30$ and a standard

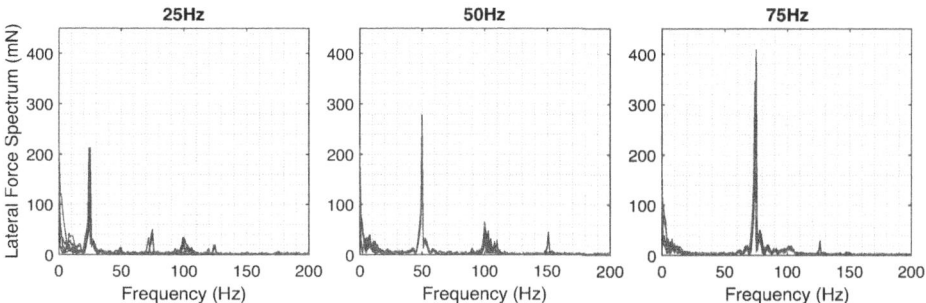

Fig. 3. Spectrum of measured lateral force F_L at $f_{BF} = 25\,\text{Hz}$, $75\,\text{Hz}$ and $150\,\text{Hz}$; AM modulation in blue, PM modulation in red. (Color figure online)

deviation of $SD = 9.16$. Among the participants, 10 were right-handed, while one was left-handed. None of the participants exhibited any motor difficulties or sensory impairments in their dominant hand index finger.

Participation in the experiments was voluntary, and informed consent was obtained from each participant prior to the commencement of the study. The experimental procedures were approved by the ethical committee of Lille University, ensuring adherence to ethical standards.

4.2 Stimuli

Our experiment aims to measure the threshold of vibrotactile feedback based on net lateral force. The two aforementioned methods for generating vibrotactile stimulation (AM and PM), are studied at a frequency $f_{LF} = 75\,\text{Hz}$. Each stimulus is presented for a duration of $2\,\text{s}$.

4.3 Method

Two experiments were conducted to measure the threshold of vibrotactile stimulation and to estimate the psychometric function of the AM stimuli and PM stimuli, respectively. The primary aim was to assess thresholds across the different experimental conditions using a two-alternative forced-choice paradigm.

Each participant underwent both experiments, and we alternated the order to avoid bias due to the presentation order.

To accomplish this task, we employed a weighted and transformed 1-up/2-down staircase method to present the amplitude at each trial. Employing this algorithm, where the upward step size is three times larger than the downward step size, yields a target probability of $p = 0.866$ [9,12].

To evaluate the performance of participants' responses, a subset consisting of a quarter of the trials was randomly chosen as "phantom" trials. In these phantom trials, the haptic device was not actuated, yet participants were instructed to discern and report whether they perceived a vibration.

4.4 Procedure

In both experimental sessions, participants were comfortably seated in a chair facing the device in a quiet room. To ensure that participants become familiar with the experimental setup, a training session consisting of presenting several stimuli helps to ensure a clear understanding of the task ahead.

Following the training session, participants undertook a forced-choice task. In each trial, they were instructed to apply a specific pressing force to the device, ranging between 0.3 N and 0.4 N. To assist them in maintaining the force throughout the trials, a visual cue based on the force sensor measurement was provided, as depicted in Fig. 4. Subsequently, a two-second-long vibration was activated. Upon the vibration's conclusion, participants were tasked with indicating whether they perceived the vibration before advancing to the next trial.

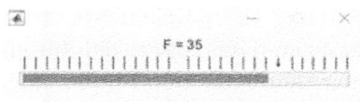

Fig. 4. Visual cue in the form of loading bar that increases with the pressing force (in gramme "gf"). The arrow represent the limit that should not be surpassed.

All the participants performed a total of 100 trials for each experiment. The minimum number of reversal points we achieved with this number of trials during the experiments were 18 (M = 23, SD = 2.32) and 18 (M = 23, SD = 2.69) for AM and PM stimuli, respectively. This number ensures a good estimation of the parameters as we run a pretrial test in order to choose the initial value of the staircase to be close to the expected estimated threshold. We chose a random values between 1.2 μm and 1.5 μm for both conditions.

4.5 Data Analysis

The psychophysical curve was estimated with the weighted and transformed 1-up/2-down staircase. We estimate the threshold and slope of each participant in both experiments separately.

A standard function to predict the psychometric function for a forced-choice experiment is the so-called Weibull cumulative distribution function, defined as follows:

$$W(x) = 1 - (1 - g)e^{-\frac{k \cdot x}{t}^s}, with \ \ k = -\log\left(\frac{1 - a}{1 - g}\right)^{\frac{1}{s}} \tag{2}$$

where g is the chance performance, i.e., the expected performance to be achieved by chance, this will be estimated based on the participant's performance on the phantom trials. t represents the threshold, s the slope of the psychometric function, and a is the performance level or targeted threshold performance (86.6%). Using the log-likelihood algorithm, we estimated the best fitting parameters of the psychometric function [9]. All the data analysis was performed in MATLAB 2022b with the appropriate toolboxes [1]

4.6   Results

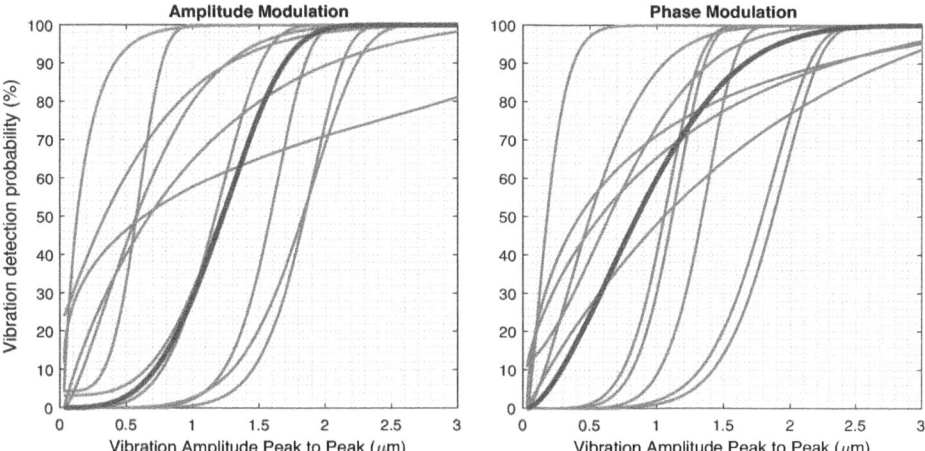

Fig. 5. Vibration discrimination performance using Amplitude Modulation and Phase Modulation. Individual estimated curves are plotted in light gray, and the red curve is based on the median values of threshold and slope (peak to peak vibration amplitude).On the left, the psychometric curve for Amplitude Modulation. On the right, the psychometric curves for phase modulation. (Color figure online)

For each participant, we computed the probability of perceiving the vibration. Subsequently, we estimated the best fitting psychometric function in both experiments by maximizing the log-likelihood function. The psychometric functions are represented in Fig. 5.

For the AM stimuli, the median value of peak to peak threshold and slope estimated were respectively 1.62 μm ($IQR = 1.31-2.02$) and 3.11 ($IQR = 0.95-4.52$) to construct the best-fitting psychometric curve depicted in Fig. 5.a. For the PM stimuli, the median value of peak to peak threshold and slope calculated were respectively 1.54 μm ($IQR = 1.28 - 2.08$) and 1.5 ($IQR = 1.07 - 6.09$. We illustrate the best-fitting using these parameters in Fig. 5b.

Upon visual inspection of best estimated thresholds of both stimuli, as presented in Fig. 6, no distinction between the threshold of both two stimuli is apparent. Additionally, we couldn't find any relation between the same subject threshold in both experiments. To further confirm our observation, we performed an ANOVA to compare between the thresholds of both stimuli (Fig. 7).

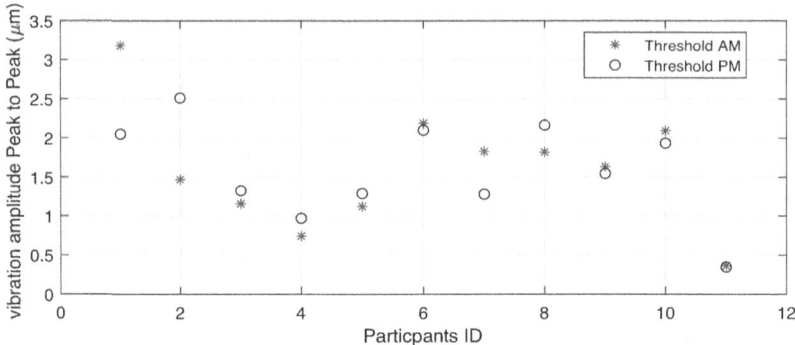

Fig. 6. Best estimated threshold parameters for each participant and both AM stimuli (red stars) and PM stimuli (black circles). (Color figure online)

The analysis shows no statistically significance difference between the thresholds associated with each stimulus. This suggests a comparable impact on perceptual thresholds.

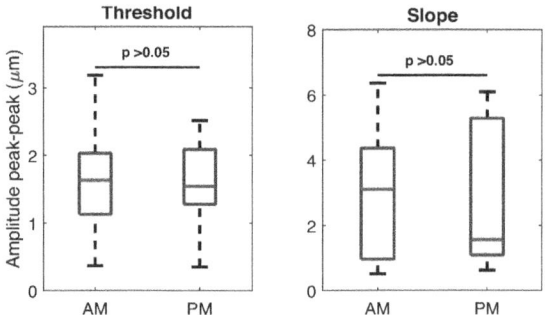

Fig. 7. ANOVA box-plot for threshold and slope for both stimuli.

5 Experiment 2: Stimulus Intensity Comparison with Normal Vibration

The purpose of this experiment is to map the intensity between the normal and the net shear forces vibrotactile stimuli.

5.1 Stimuli and method

The experimental test bench is completed by a shaker, to which an idle stator of USR60 is attached, in order to preserve the same surface properties, as shown in Fig. 8a.

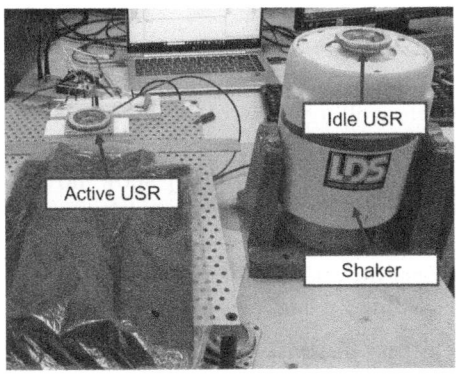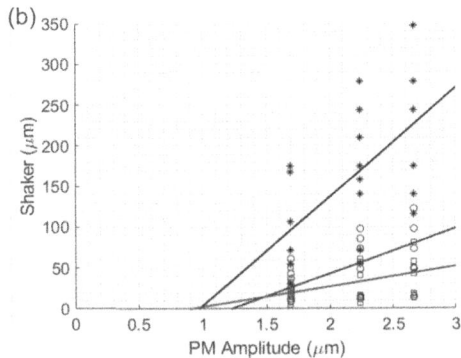

Fig. 8. (a) The experimental setup with an active ultrasonic motor and idle USR placed on the top of the shaker, (b) mapping between normal vibrotactile amplitude and modulated ultrasonic vibration amplitude A_0, for $f_{LF} = 25\,\mathrm{Hz}$ in black, $f_{LF} = 75\,\mathrm{Hz}$ in blue and $f_{LF} = 150\,\mathrm{Hz}$ in red. (Color figure online)

On the travelling wave stimulator, a constant vibrotactile stimulus is presented using the method denoted as PM, with amplitudes A_0 set to $1.69\,\mu\mathrm{m}$, $2.23\,\mu\mathrm{m}$ and $2.66\,\mu\mathrm{m}$, and low-frequency modulation frequencies f_{LF} of $25\,\mathrm{Hz}$, $75\,\mathrm{Hz}$ and $150\,\mathrm{Hz}$. Concurrently, the shaker displays a sinusoidal waveform stimulus of the same low frequency as that of the travelling wave modulation frequency. Participants adjust the stimulation amplitude, denoted as A_S, on the shaker to match the perceived stimulus intensity of the travelling wave stimulator. The amplitude of the shaker stimulus is measured using the same laser micrometer employed previously.

At the end of the experimentation, the participants are asked to report about how difficult it was adjusting the level of vibration on the shaker, and which actuator was more comfortable.

Eight participants (5 male, 3 female, with an average age of 29.75 and standard deviation 8.9) participated in the experiment.

5.2 Results

The results are shown in Fig. 8b. Each symbol (circle, square, star) corresponds to an individual answer by a participant. Each particpants data appeared to follow first order polynome. The lines represent the fitted first-order polynomial across all participants and each frequency, and the linear fit is given by:

- $f_{LF} = 25\,\mathrm{Hz}$: $A_S = 133,64(A_0 - 0.99\,\mu\mathrm{m})$
- $f_{LF} = 75\,\mathrm{Hz}$: $A_S = 55.31(A_0 - 1.21\,\mu\mathrm{m})$
- $f_{LF} = 150\,\mathrm{Hz}$: $A_S = 24.82(A_0 - 0.89\,\mu\mathrm{m})$

Moreover, most participants reported that they preferred the modulated net shear forces, because it is more comfortable, in particular at low frequency.

6   Discussion

In this study, we evaluated the sensitivity of vibrotactile stimulation when produced by net lateral force generated by a travelling wave. We estimated the detection threshold for amplitude modulation (AM) or phase modulation (PM) at 75 Hz; the most likely parameter value (threshold and slope) show no statistical difference between the two experiments. This result leads us to assume that both methods are similar in terms of vibrotactile sensation. Moreover, Fig. 3 indicates no difference in force generation, leading us to infer that vibrotactile sensation may depend more on the force amplitude and frequency rather than the modulation method employed. However, additional work is needed to confirm this trend over a larger span of frequencies.

The value found for the detection threshold is 1.5 µm, and is comparable with the detection threshold obtained with a normal stimulation [14].

However, the results of the second experiment indicate that a much lower vibration amplitude is required for perceiving stimuli with modulated net shear forces generated by the travelling wave compared to normal vibration (shaker). Furthermore, the relationship between A_S (stimulation amplitude) and A_0 (vibrating pot amplitude) seems to be linear. It is noteworthy that one might expect all lines to converge at the origin (0,0); however, interestingly, these lines converge to a point with approximate coordinates of (1 µm, 0). This observation leads to the assumption that the friction mechanisms responsible for creating the net shear forces vanish below a certain vibration amplitude threshold. Indeed, according to [18], a travelling wave can drive objects, including skin, only if there is at least half a wavelength in contact. Therefore, establishing a model of the skin's response to a travelling wave is necessary to further investigate this phenomenon. Additionally, this model would shed light on why the slope of the linear fit depends on frequency.

Finally, the participants reported more comfortable experience with the travelling wave stimulator. Indeed, the device emits no audible noise, even at low frequency, since it is based on ultrasonic vibrations, which represents an advantage over the shaker.

7   Conclusion

Ultrasonic travelling waves have diverse applications in tactile displays, and our paper showcases their effectiveness in delivering vibrotactile cues to a participant's finger. One significant advantage of these devices is their silent operation. Additionally, they require a significantly lower vibration amplitude compared to traditional vibrations-up to 24 to 155 times lower-to achieve the same perceived stimulus intensity. This reduction in amplitude is pivotal and has the potential to drive technological advancements in vibrotactile stimulators, rendering them more compact, versatile and energy-efficient.

However, the mechanisms responsible for producing the net shear forces are still not fully understood. We hypothesize that there exists a minimal vibration amplitude, below which the travelling wave produces no force on the finger pulp. Further investigations are necessary to confirm or refute this assumption and to elucidate the key factors influencing this minimal vibration amplitude. Establishing a model of force generation would be instrumental in designing future prototypes that exploit ultrasonic travelling waves.

Acknowledgments. This project has received support from the French National Agency for Research (ANR) as part of the project HASAMe under agreement ANR-21-CE33-0020. This work is supported by IRCICA (Research Institute on software and hardware devices for Information and Advanced Communication, USR CNRS 3380). The authors acknowledge the use of artificial intelligence for enhancing grammatical accuracy and readability in this work. However, it is important to clarify that the content and ideas presented are solely the creation of the authors. The utilization of AI does not diminish the originality of our work.

Disclosure of Interests. The authors have no competing interests to declare that are relevant to the content of this article

References

1. Alcalá-Quintana, R., García-Pérez, M.A.: Fitting model-based psychometric functions to simultaneity and temporal-order judgment data: Matlab and r routines. Behav. Res. Methods **45**, 972–998 (2013)
2. Biggs, J., Srinivasan, M.A.: Tangential versus normal displacements of skin: relative effectiveness for producing tactile sensations. In: Proceedings 10th Symposium on Haptic Interfaces for Virtual Environment and Teleoperator Systems. HAPTICS 2002, pp. 121–128 (2002). https://api.semanticscholar.org/CorpusID:15300809
3. Cai, Z., Wiertlewski, M.: Ultraloop: active lateral force feedback using resonant traveling waves. IEEE Trans. Haptics **16**, 652 (2023)
4. Chubb, E.C., Colgate, J.E., Peshkin, M.A.: Shiverpad: a glass haptic surface that produces shear force on a bare finger. IEEE Trans. Haptics **3**(3), 189–198 (2010)
5. Dai, X., Colgate, J.E., Peshkin, M.A.: Lateralpad: A surface-haptic device that produces lateral forces on a bare finger. In: 2012 IEEE Haptics Symposium (HAPTICS), pp. 7–14. IEEE (2012)
6. Dharma, A.A.G., Oami, T., Obata, Y., Yan, L., Tomimatsu, K.: Design of a wearable haptic vest as a supportive tool for navigation. In: Human-Computer Interaction. Interaction Modalities and Techniques: 15th International Conference, HCI International 2013, Las Vegas, NV, USA, July 21-26, 2013, Proceedings, Part IV 15, pp. 568–577. Springer (2013)
7. Dhiab, A.B., Hudin, C.: Confinement of vibrotactile stimuli in narrow plates: principle and effect of finger loading. IEEE Trans. Haptics **13**(3), 471–482 (2020)
8. Garcia, P., Giraud, F., Lemaire-Semail, B., Rupin, M., Kaci, A.: Control of an ultrasonic haptic interface for button simulation. Sens. Actuators A **342**, 113624 (2022)
9. García-Pérez, M.A.: Forced-choice staircases with fixed step sizes: asymptotic and small-sample properties. Vis. Res. **38**(12), 1861–1881 (1998)
10. Ghenna, S., Vezzoli, E., Giraud-Audine, C., Giraud, F., Amberg, M., Lemaire-Semail, B.: Enhancing variable friction tactile display using an ultrasonic travelling wave. IEEE Trans. Haptics **10**(2), 296–301 (2016)

11. Gueorguiev, D., Kaci, A., Amberg, M., Giraud, F., Lemaire-Semail, B.: Travelling ultrasonic wave enhances keyclick sensation. In: Haptics: Science, Technology, and Applications: 11th International Conference, EuroHaptics 2018, Pisa, Italy, June 13-16, 2018, Proceedings, Part II 11, pp. 302–312. Springer (2018)

12. Karmali, F., Chaudhuri, S.E., Yi, Y., Merfeld, D.M.: Determining thresholds using adaptive procedures and psychometric fits: evaluating efficiency using theory, simulations, and human experiments. Exp. Brain Res. **234**, 773–789 (2016)

13. Lylykangas, J., Surakka, V., Salminen, K., Raisamo, J., Laitinen, P., Rönning, K., Raisamo, R.: Designing tactile feedback for piezo buttons. In: Proceedings of the SIGCHI Conference on Human Factors in Computing Systems, pp. 3281–3284 (2011)

14. Morioka, M., Whitehouse, D.J., Griffin, M.J.: Vibrotactile thresholds at the fingertip, volar forearm, large toe, and heel. Somatosens. Motor Res. **25**(2), 101–112 (2008)

15. Pra, Y.D., Papetti, S., Järveläinen, H., Bianchi, M., Fontana, F.: Effects of vibration direction and pressing force on finger vibrotactile perception and force control. IEEE Trans. Haptics **16**(1), 23–32 (2023). https://doi.org/10.1109/TOH.2022.3225714

16. Storck, H., Wallaschek, J.: The effect of tangential elasticity of the contact layer between stator and rotor in travelling wave ultrasonic motors. Int. J. Non-Linear Mech. **38**(2), 143–159 (2003). https://doi.org/10.1016/S0020-7462(01)00048-8

17. Terenti, M., Vatavu, R.D.: Measuring the user experience of vibrotactile feedback on the finger, wrist, and forearm for touch input on large displays. In: CHI Conference on Human Factors in Computing Systems Extended Abstracts, pp. 1–7 (2022)

18. Wallaschek, J.: Contact mechanics of piezoelectric ultrasonic motors. Smart Mater. Struct. **7**(3), 369 (1998). https://doi.org/10.1088/0964-1726/7/3/011

19. Wiertlewski, M., Lozada, J., Hayward, V.: The spatial spectrum of tangential skin displacement can encode tactual texture. IEEE Trans. Robot. **27**(3), 461–472 (2011)

20. Zárate, J.J., Shea, H.: Using pot-magnets to enable stable and scalable electromagnetic tactile displays. IEEE Trans. Haptics **10**(1), 106–112 (2016)

Enhancing the Perceived Pseudo-Torque Sensation based on the Distance between Actuators Elicited by Asymmetric Vibrations

Tomosuke Maeda[1]([✉])[iD], Takayoshi Yoshimura[1][iD], Hiroyuki Sakai[1][iD], and Kouta Minamizawa[2][iD]

[1] Toyota Central R&D Labs., Inc., 41-1, Yokomichi, Nagakute, Aichi, Japan
{tmaeda,yoshimura,sakai}@mosk.tytlabs.co.jp
[2] Keio University Graduate School of Media Design, 4-1-1, Hiyoshi Kohoku-ku, Yokohama, Kanagawa, Japan
kouta@kmd.keio.ac.jp

Abstract. The phenomenon of illusory pulling force, induced by asymmetric vibrations, represents a distinct perceptual experience. This pseudo-force sensation has been explored through various techniques involving the deployment of multiple vibrotactile actuators. Although these actuators can elicit multi-dimensional pseudo-forces, the impact of distance between actuators on the rotational pseudo-force (or torque sensation) remains unexplored. This study examines the influence of vibrotactile actuator distance on the pseudo-torque sensation elicited by asymmetric vibrations. We hypothesized that the distance between actuators impacts the perceived pseudo-torque in line with the principle of leverage. Our findings indicate an increase in perceived pseudo-torque up to a 50 mm separation between actuators, with an optimal distance correlating to 0.43 times the hand size for effective pseudo-torque generation. These insights are crucial for designing haptic devices capable of imparting pseudo-torque sensations.

Keywords: Pseudo-Force · Asymmetric Vibration · Perception

1 Introduction

The sensation of a pseudo-force, elicited by asymmetric vibrations and known as the illusory pulling force, represents a distinct perceptual phenomenon. These pseudo-forces create an impression of being pulled, attributable to the disparity between stronger positive and weaker negative accelerations. In contrast to typical vibrotactile stimuli, pseudo-force modalities provide force and directional cues. Asymmetric vibrations that induce pseudo-forces are promising for diverse applications, including navigation (e.g., [2,8]), white cane training (e.g., [10]), and simulating object weight in virtual reality (e.g., [4]).

Supported by Toyota Central R&D Labs., Inc.

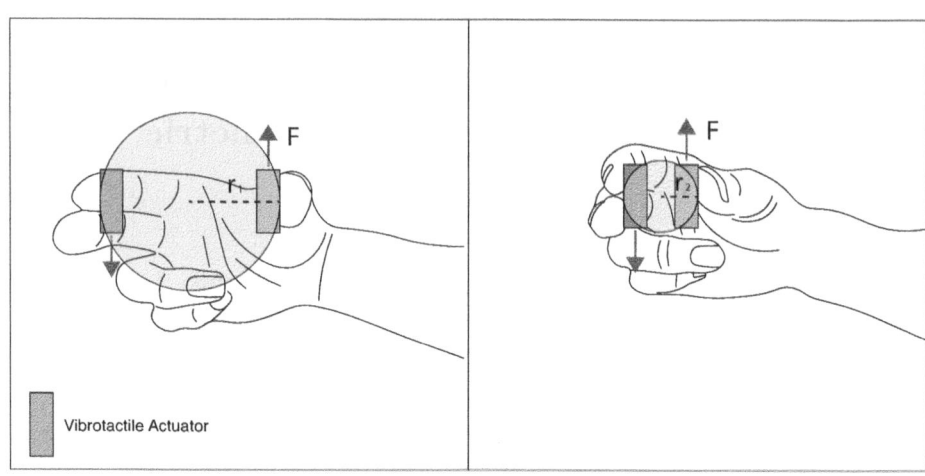

Fig. 1. Conceptual representation of the hypothesis regarding the perceived pseudo-torque sensation induced by asymmetric vibration. This illustration contrasts two scenarios: one with vibrotactile actuators spaced further apart, associated with a stronger perceived pseudo-torque sensation, and the other with actuators spaced closely together, showing a weaker sensation ($F \times r_1 > F \times r_2$).

The sensation of a pseudo-force, elicited by asymmetric vibrations and known as the illusory pulling force, represents a distinct perceptual phenomenon. These pseudo-forces create an impression of being pulled, attributable to the disparity between stronger positive and weaker negative accelerations. In contrast to typical vibrotactile stimuli, pseudo-force modalities provide force and directional cues. Asymmetric vibrations that induce pseudo-forces are promising for diverse applications, including navigation (e.g., [2,8]), white cane training (e.g., [10]), and simulating object weight in virtual reality (e.g., [4]).

Amemiya et al.[1] found that asymmetric vibrations, characterized by rapid positive and slow negative accelerations, stimulate human mechanoreceptors to produce a pulling illusion in the positive direction. Culberson et al. [5] and Imaizumi et al. [7] attributed pseudo-forces to skin displacement resulting from asymmetric vibration. The origins of this sensation via asymmetric vibration have been explored through various methods (e.g., slider-crank [1], LRA [9], voice coil actuator [2], speaker [12]).

Multi-dimensional pseudo-forces can be elicited by deploying multiple vibrotactile actuators with asymmetric vibration. Tanabe et al. [11] and Maeda et al. [8] designed handheld haptic devices that simulate translational and rotational pseudo-forces using two horizontally aligned actuators. Culbertson et al. [6] introduced three-dimensional pseudo-forces, incorporating multiple actuators on the finger to produce both translational and rotational pseudo-force sensations.

Pseudo-torque sensation is critical in navigation systems, haptic rendering, and virtual reality, and has the advantage that stimuli can be presented with small, portable actuators compared to real torque sensation. Investigating rotational pseudo-force (torque sensation) induced by asymmetric vibration is thus

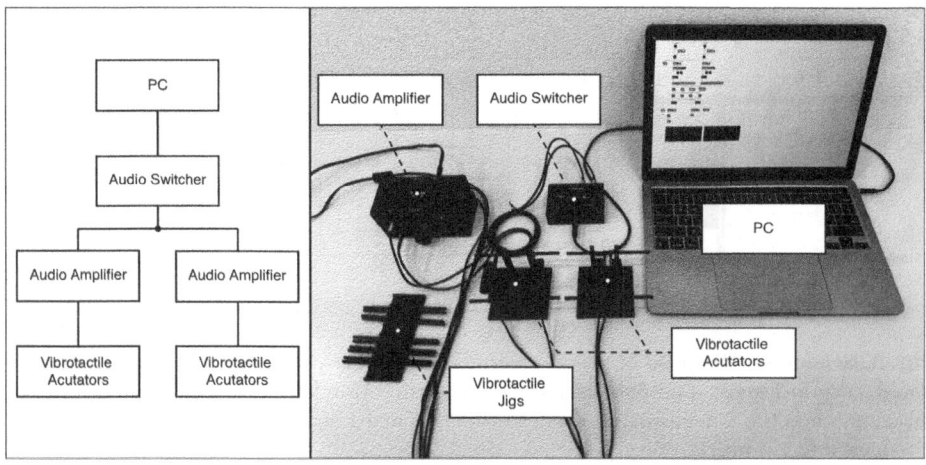

Fig. 2. Experimental setup for pseudo-torque evaluation. (Left) The block diagram shows the signal flow of the PC. The signal is distributed at the switcher among several audio amplifiers. (Right) The figure shows the experimental setup, with the vibrotactile actuators mounted on a fixed jig. This setup enables control of vibration patterns using MAX8 software.

essential. Culbertson et al. [6] observed that accuracy in rotational experiments varied with finger size, suggesting the actuator distance as a key factor, as the intensity of pseudo-torque sensation depends on the lever arm length between the actuator and the finger's rotation center.

This study concentrates on the possible effect of distance between vibrotactile actuators on the pseudo-torque sensation in asymmetric vibrations. We hypothesize that this distance significantly influences the perceived torque (Fig. 1), based on the leverage principle, which posits that increased distance between actuators enhances pseudo-torque sensation. Notably, this sensation is an illusion from asymmetric vibration, not an actual physical torque. This study's findings aim to provide design guidelines for improving pseudo-torque sensation in haptic devices.

2   Materials and Methods

We assessed the impact of distance between actuators on pseudo-force sensation using an experimental protocol adapted from Amemiya et al. [3]. Participants sequentially grasped devices with two distinct actuator distances and subsequently provided a comparative score between the first and second stimuli.

2.1   Participants

The study involved twenty participants (10 females, mean age = 26.6 years, SD = 5.1 years, mean hand size = 182.5 mm, SD = 13.7 mm). The participants' hand

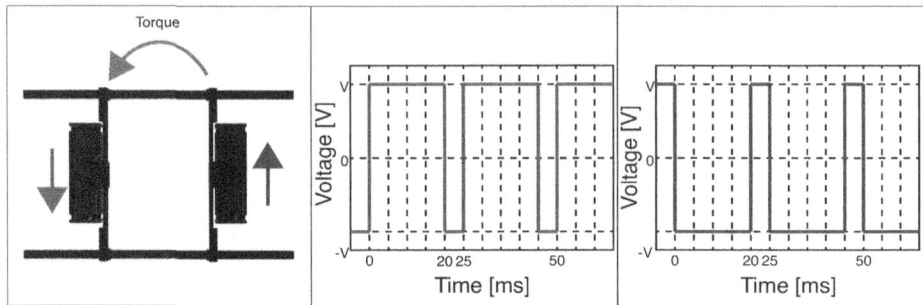

Fig. 3. Schematic Diagram of pseudo-torque feedback method. (Left) Principle of rotational pseudo-forces. The arrows indicate the direction of the pseudo-force, and the color matches the waveform's color. (Center) Upward asymmetric vibrations. (Right) Downward asymmetric vibrations.

size was measured from the tip of the middle finger to the crease under the palm of the hand. All participants received monetary compensation, were unaware of the experiment's purpose, and provided written informed consent. Based on self-reports, one participant was left-handed, while the others were right-handed. The experimental protocol received approval from the Institutional Review Board of Toyota Central R&D Labs., Inc.

2.2 Apparatus and Stimuli

We engineered haptic devices by mounting two vibration actuators (HapCoil One, Tactile Labs.) onto jigs (Fig. 2). Seven jigs, ranging from 30 mm to 90 mm in 10 mm increments, were fabricated using a 3D printer (Mark Two, Markforged) with Onyx filament, which contains microcarbon fiber, to ensure minimal weight. This material is approximately 10× stronger than acrylonitrile butadiene styrene (ABS) plastic. The weight of these haptic devices was approximately 22 g, not including the cable. Each actuator was connected to an acoustic amplifier (AP15, Foster) and could display signal output from a PC (MacBook Pro 13 inch, Apple). The vibration stimulation signals were generated by MAX8 (Cycling '74).

The pseudo-torque sensation was produced by directing the vibrotactile actuators to generate opposing pseudo-forces. By controlling the force direction of each actuator, participants perceived a pseudo-torque sensation. Consistent with prior studies [3,9,11], asymmetric vibrations (square wave, duty ratio = 1:4) at 40 Hz and a maximum acceleration amplitude of approximately $120\,\mathrm{m/s^2}$ were used to induce a pseudo-torque sensation (Fig. 3). This maximum amplitude was set under conditions where an accelerometer (DRV-ACC16-EVM, Texas Instruments Inc.) was mounted on a vibrotactile actuator and the fingertips were held in a pinched position.

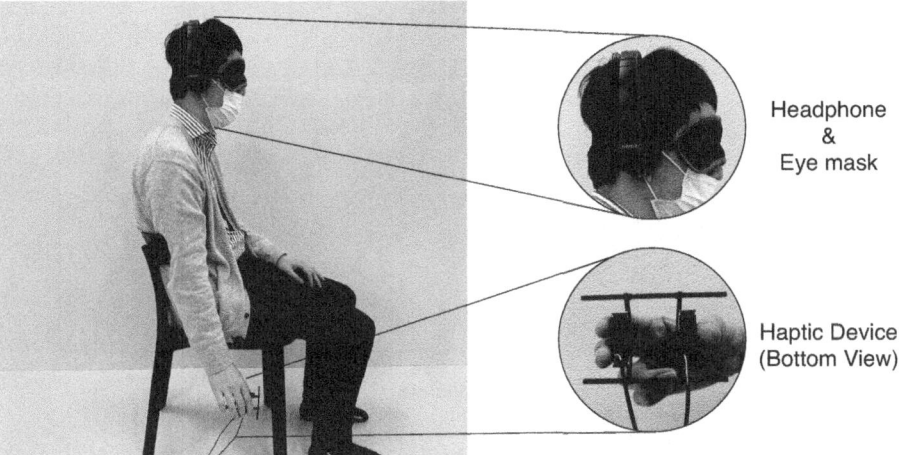

Headphone
&
Eye mask

Haptic Device
(Bottom View)

Fig. 4. Experiment environment. The participant wore headphones and an eye mask and grasped a haptic device.

2.3 Procedure

The experiment employed Scheffé's method [14] of paired comparisons to evaluate the intensity of pseudo-torque sensation as a function of the distance between actuators. Participants seated and equipped with an eye mask and noise-canceling headphones (BOSE, QuietComfort 35) playing white noise were isolated from visual and auditory cues pertaining to the distance between actuators. They were instructed to lightly pinch the haptic device using three fingers (index, middle finger, and thumb) and maintain consistent strength throughout the trials (Fig. 4). Although starting each trial with their hands in a set position, participants could move them during stimulus presentation.

Prior to the main experiment, participants familiarized themselves with the pseudo-torque sensation through a left–right rotational stimulus, with 60 mm between the vibrotactile actuators. Following this learning session, the jigs of two haptic devices were changed by the experimenter to vary the actuator distance. Participants were instructed to hold the haptic device (previous stimulus), via which they were presented with the first pseudo-force stimuli. Each pseudo-torque sensation, alternating between leftward and rightward directions every second, lasted 4 s. They subsequently grasped another haptic device (latter stimulus) with a different actuator distance and experienced a second pseudo-force stimulus similar to the first. Participants verbally assessed the strength of the pseudo-torque sensation using a 5-point Likert scale (-2 = "the previous stimulus felt much stronger", -1 = "the previous stimulus felt stronger", 0 = "both felt the same", 1 = "the latter felt stronger", 2 = "the latter felt much stronger"). The experiment consisted of 42 trials (= $_7P_2$ (combination of condition)), which were presented in random order. The experiment was completed by all participants within approximately 1 h.

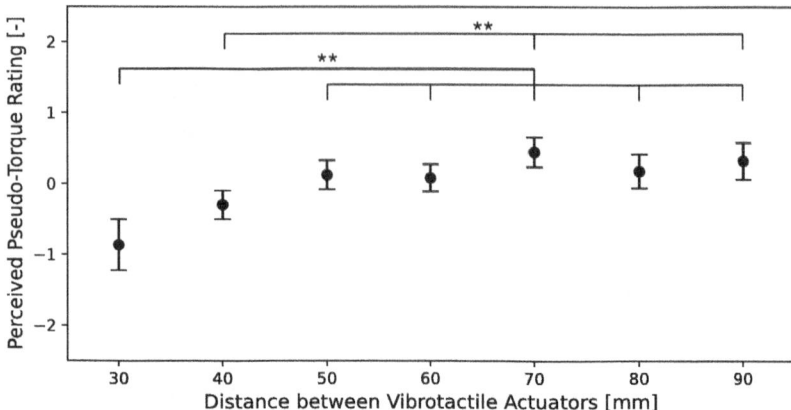

Fig. 5. Mean rating of perceived pseudo-torque sensation versus distance between vibrating actuators for all participants. Error bars indicate 95% confidence interval. (**: p < 0.01)

3   Results and Discussion

Figure 5 shows the relationship between perceived pseudo-torque sensation and distance between actuators. A one-way repeated measures analysis of variance (ANOVA) assessed the impact of distance between actuators on this sensation. The statistical significance threshold was set at $\alpha < 0.05$. Results showed a significant main effect of actuator distance (F(6, 133) = 14.0, p < 0.001, $\eta_p^2 = 0.39$). Post hoc comparisons using Tukey's honestly significant difference (HSD) test revealed significant differences between the 30 mm distance and other distances, and between the 40 mm distance and the 70 mm and 90 mm distances. These findings suggest that pseudo-torque sensation significantly increases with actuator distance and plateaus around 50 mm. This is the first evidence demonstrating the effect of the distance between vibrotactile actuators on the pseudo-torque sensation elicited by asymmetric vibration.

Figure 6 presents the pseudo-torque sensation in relation to the ratio of the distance between actuators to hand size. Considering the physical limitations of hand size, we opted for quadratic polynomial fitting using the least squares method. A polynomial fit yielded a peak value of 0.43, indicating that an actuator distance proportional to 0.43 times the hand size is optimal for generating torque sensation. This insight could inform the design of devices utilizing asymmetric vibration to generate pseudo-torque sensations.

Our study aimed to investigate the relationship between pseudo-torque sensation strength and the distance between vibrotactile actuators. The results confirmed that the perceived sensation increases with distance between actuators, supporting our hypothesis up to a point. The sensation increases with distance, suggesting a virtual center of rotation, aligning with the principle of leverage. Contrary to expectations, the sensation plateaued at distances beyond 50 mm.

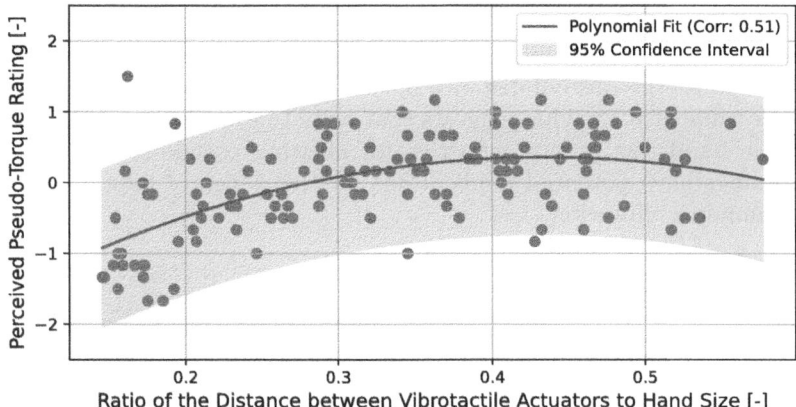

Fig. 6. Mean rating of perceived pseudo-torque sensation versus the ratio of the distance between the vibrotactile actuators to the hand size for all participants. The red line shows polynomial fitting.

A possible explanation is that larger distances hinder appropriate grasping of the actuators, weakening the sensation, as discussed by Teshima et al. [13]. They noted that increased grasping force can augment skin stiffness between the device and finger, potentially influencing pseudo-force sensation. Although participants were instructed to maintain constant grasping force, achieving and maintaining the correct force may have been challenging.

Notably, reduced actuator distances resulted in significant error bars, possibly due to actuator vibrations merging, creating confusion. Our methodology involved activating two actuators simultaneously, complicating the isolation of vibrations.

Future directions include precise measurement of fingertip skin vibrations and deformations, as pseudo-force sensation is influenced by multiple factors, including perceptual characteristics (e.g., [1]) and skin deformation (e.g., [5]). Our experimental framework lacks the capacity for precise assessment of pseudo-force sensation attributes; accurately detecting skin vibration and deformation on the fingertip, particularly when obscured by a device, presents a significant obstacle in haptic research. Furthermore, we did not quantify the physical parameters (e.g., force) linked to the perceived torque sensation. Adapting the methodologies employed in studies such as those by Rekimoto [9] and Tanabe [12] could facilitate the measurement of physical factors associated with torque sensation.

4   Conclusion

In this study, we investigated the perceived pseudo-torque sensations induced by asymmetric vibration as influenced by the distance between vibrotactile actuators. Our findings revealed an enhancement in the perceived pseudo-torque sensation up to a 50 mm actuator distance. This outcome highlights the significance of actuator distance in influencing the perceived pseudo-torque sensation. Furthermore, we applied polynomial fitting to analyze the perceived pseudo-torque sensation against the ratio of distance between actuators to hand size, discovering a peak value at 0.43. This suggests that an actuator distance approximately 43% of the hand size is most effective for eliciting pseudo-torque sensations. These findings have potential implications for the design of haptic devices utilizing asymmetric vibration.

References

1. Amemiya, T., Ando, H., Maeda, T.: Lead-me interface for a pulling sensation from hand-held devices. ACM Trans. Appl. Percept. 5(3), 1 (2008). https://doi.org/10.1145/1402236.1402239

2. Amemiya, T., Gomi, H.: Buru-navi3: Behavioral navigations using illusory pulled sensation created by thumb-sized vibrator. In: ACM SIGGRAPH 2014 Emerging Technologies. SIGGRAPH '14, Association for Computing Machinery, New York, NY, USA (2014). https://doi.org/10.1145/2614066.2614087

3. Amemiya, T., Gomi, H.: Distinct pseudo-attraction force sensation by a thumb-sized vibrator that oscillates asymmetrically. In: Auvray, M., Duriez, C. (eds.) Haptics: Neuroscience, Devices, Modeling, and Applications, pp. 88–95. Springer Berlin Heidelberg, Berlin, Heidelberg (2014). https://doi.org/10.1007/978-3-662-44196-1_12

4. Choi, I., Culbertson, H., Miller, M.R., Olwal, A., Follmer, S.: Grabity: A wearable haptic interface for simulating weight and grasping in virtual reality. In: Proceedings of the 30th Annual ACM Symposium on User Interface Software and Technology, pp. 119–130. UIST '17, Association for Computing Machinery, New York, NY, USA (2017). https://doi.org/10.1145/3126594.3126599

5. Culbertson, H., Walker, J.M., Okamura, A.M.: Modeling and design of asymmetric vibrations to induce ungrounded pulling sensation through asymmetric skin displacement. In: 2016 IEEE Haptics Symposium (HAPTICS), pp. 27–33 (2016). https://doi.org/10.1109/HAPTICS.2016.7463151

6. Culbertson, H., Walker, J.M., Raitor, M., Okamura, A.M.: Waves: A wearable asymmetric vibration excitation system for presenting three-dimensional translation and rotation cues. In: Proceedings of the 2017 CHI Conference on Human Factors in Computing Systems, pp. 4972–4982. CHI '17, Association for Computing Machinery, New York, NY, USA (2017). https://doi.org/10.1145/3025453.3025741

7. Imaizumi, A., Okamoto, S., Yamada, Y.: Friction sensation produced by laterally asymmetric vibrotactile stimulus. In: Auvray, M., Duriez, C. (eds.) Haptics: Neuroscience, Devices, Modeling, and Applications, pp. 11–18. Springer, Berlin Heidelberg, Berlin, Heidelberg (2014). https://doi.org/10.1007/978-3-662-44196-1_2

8. Maeda, T., Yamamoto, J., Yoshimura, T., Sakai, H., Minamizawa, K.: Waylet: Self-contained haptic device for park-scale interactions. In: SIGGRAPH Asia 2023 Emerging Technologies. SA '23, Association for Computing Machinery, New York, NY, USA (2023). https://doi.org/10.1145/3610541.3614567

9. Rekimoto, J.: Traxion: A tactile interaction device with virtual force sensation. In: Proceedings of the 26th Annual ACM Symposium on User Interface Software and Technology, pp. 427–432. UIST '13, Association for Computing Machinery, New York, NY, USA (2013). https://doi.org/10.1145/2501988.2502044

10. Tanabe, T., Fujimoto, Y., Nunokawa, K., Doi, K., Ino, S.: White cane-type holdable device using illusory pulling cues for orientation & mobility training. IEEE Access **11**, 28706–28714 (2023). https://doi.org/10.1109/ACCESS.2023.3259965

11. Tanabe, T., Yano, H., Endo, H., Ino, S., Iwata, H.: Motion guidance using translational force and torque feedback by induced pulling illusion. In: Nisky, I., Hartcher-O'Brien, J., Wiertlewski, M., Smeets, J. (eds.) Haptics: Science, Technology, Applications, pp. 471–479. Springer International Publishing, Cham (2020). https://doi.org/10.1007/978-3-030-58147-3_52

12. Tanabe, T., Yano, H., Iwata, H.: Evaluation of the perceptual characteristics of a force induced by asymmetric vibrations. IEEE Trans. Haptics **11**(2), 220–231 (2018). https://doi.org/10.1109/TOH.2017.2743717

13. Teshima, T., Takamuku, S., Amemiya, T., Gomi, H.: Light touch on pillar array surface greatly improves direction perception induced by asymmetric vibration. In: SIGGRAPH Asia 2015 Haptic Media And Contents Design. SA '15, Association for Computing Machinery, New York, NY, USA (2015). https://doi.org/10.1145/2818384.2818404

14. Ura, S.: An analysis of experiments of paired comparisons. Quality Control **16**, 78–80 (1959)

Presentation of Slip Sensation Using Suction Pressure and Electrotactile Stimulation

Yan Xue Teo[1]([⊠]), Taiga Saito[1], Takayuki Kameoka[2][ID], Izumi Mizoguchi[1][ID], and Kajimoto Hiroyuki[1][ID]

[1] The University of Electro-Communications, 1-5-1 Chofugaoka, Chofu, Tokyo, Japan
teo@kaji-lab.jp
[2] University of Tsukuba, 1-1-1 Tennodai, Tsukuba, Ibaraki, Japan

Abstract. The accurate detection of incipient object slippage is essential for grip control. This study aimed to replicate the sensation of losing grip by simulating a 'partial slip' phenomenon, in which the outer area of a contact point fluctuates slightly while the center of contact stays still. Two types of stimulation methods were used: electrotactile stimulation, chosen for its superior spatial and temporal resolution, which facilitates the precise replication of the partial slip area; and air-suction stimulation selected for its stable pressure sensation. We conducted an experiment to evaluate the effectiveness of each stimulation method, both independently and in combination, in terms of their ability to realistically convey the sensation of slippage, including the perceptual clarity of the slip occurrence and its direction. Results showed that the electrotactile stimulation was proficient in presenting distinct sensations of slippage and its direction while also providing some realism. Moreover, it was observed that the incorporation of suction notably enhanced the realism of the tactile sensation, particularly when used in conjunction with electrotactile stimulation.

Keywords: Electrotactile · Haptics · Partial Slip · Suction Pressure

1 Introduction

The presence of slippage sensations plays a key role in our ability to reliably manipulate objects through grip control with our hands. Such detection enables precise control of grasping force, minimizing it to the requisite level needed to maintain a stable hold on an object. This incipient slip is predominantly perceived through cutaneous sensory feedback. Notably, this type of slippage typically initiates at the periphery of the skin-object contact area while the central region remains stationary, a condition often termed as 'partial slip' [5,7]. Research has posited that the detection and management of partial slip are crucial in facilitating dexterous object manipulation and in the regulation of grip force [7].

H. Kajimoto et al. (Eds.): EuroHaptics 2024, LNCS 14768, pp. 314–320, 2025.
https://doi.org/10.1007/978-3-031-70058-3_26

The objective of this study is to emulate the partial slip phenomenon through the strategic integration of two distinct tactile presentation modalities. The first modality, electrotactile stimulation, entails the application of an electrical current to the skin, thereby activating sensory nerve endings. Its most notable attributes include high spatial and temporal resolution, affording precise control over the localization and timing of stimuli [1,3,8]. This makes the electrotactile stimulation an excellent candidate for presenting slip sensations and could effectively signal the onset of an object's slip and convey directional information about the slip event [6], thereby contributing to the clarity of a slip sensation.

The second modality, suction pressure stimulation, generates a pressure sensation through the application of air suction that deforms the skin. This technique is especially adept at producing sustained pressure sensations [2,4]. It may be occasionally perceived as a sensation of pressure rather than suction [4]. A limitation of air suction stimulation lies in its lower temporal resolution, making it less effective for conveying sensations of rapid impact. Consequently, it is deemed suitable for simulating stable contact in the central region of the grip, thereby augmenting the realism of the slip sensation.

2   Implementation

2.1   Hardware

Figure 1 illustrates the hardware configuration of the apparatus for suction pressure and electrotactile stimulation. For each part, an ESP32 microcontroller was used to communicate with a PC via USB. The apparatus comprises a dual-stimuli head, which accommodates both suction pressure and electrotactile modalities and has a mass of 5 g. The overall weight of the device for a single finger, exclusive of the cables required for connection, is 160 g. For this study, two heads were employed to administer stimuli to the index finger and thumb.

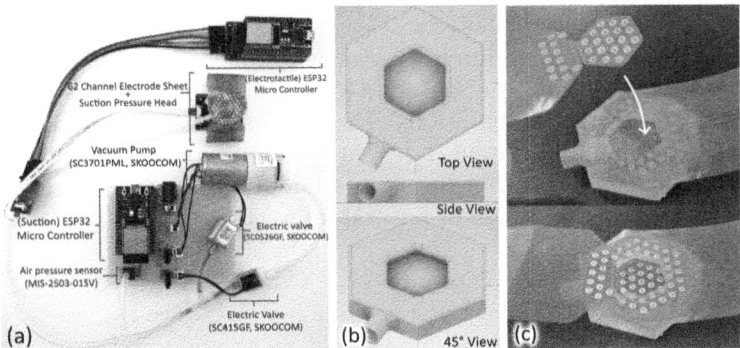

Fig. 1. (a) Suction pressure and electrotactile stimulation hardware setup for a single finger (b) Suction Pressure head (c) Method of combining electrotactile stimulation and suction pressure by pasting the electrodes onto the suction pressure head

The suction pressure module was based on previous work [2], with a newly designed suction pressure head using an optical 3D printer (Form3, FormLabs) and using Elastic Resin as the material with a singular suction hole with a hexagonal shape with equal sides of 6mm. The pressure inside the suction pressure head was controlled by 2 electric valves (SC0526GF, SKOOCOM and SC415GF, SKOOCOM), vacuum pump (SC3701PML, SKOOCOM), and air pressure sensor (MIS-2503-015V).

Two flexible electrode sheets, featuring electrodes arranged in a hexagonal configuration, were affixed to the suction pressure head using double-sided adhesive tape. Each electrode possessed a diameter of 1.5 mm and was positioned at a center-to-center distance of 2.4 mm from adjacent electrodes. The electrical current polarity was configured as anodic, with the pulse width established at 50 μs, and the repetition frequency at 90 Hz, as determined by prior research [1].

2.2 Haptic Rendering Method

The Unity game engine was used to host the simulation of pinching and elevating a virtual object using two fingers. The software operated at a refresh rate of 60 Hz. In this environment, each finger was represented by a set of small rigid bodies that acted as contact sensors in a hexagonal arrangement that curved outward. A dynamic attraction mechanism, employing a 'god-object' (an abstract point in space signifying the sensor's ideal position), was implemented. Here, a spring-like force governed the movement of sensors, intensifying as the distance from the god-object increased.

Each sensor corresponded to its respective electrode, positioned accurately as per the real-world configuration. Sensors representing the electrodes within the suction region contributed collectively to the overall suction pressure. Negative air pressure was modulated based on the spatial deviation between the god-object and the sensor's displaced position, with larger displacements generating greater suction forces. This principle extended to lateral movements, thereby replicating the skin's traction effects.

Electrotactile stimulation intensity was determined by the differential velocity between the object and the sensors. Greater discrepancies in speed resulted in heightened electrotactile sensations. The sensors' spherically convex arrangement, akin to a human fingerpad, inherently allowed for partial slippage at the periphery during contact within the simulation. This organization of sensors to be concave allows for emulation of natural interactions with objects.

3 User Study

The objective of this experiment was to systematically assess the attributes of two stimulation modalities, both in isolation and in combination, focusing on the authenticity of the slip sensation, its perceptual clarity, and the discernibility of the slip's direction.

It was hypothesized that electrotactile stimulation, as a singular modality, could effectively signal the onset of an object's slip and convey directional information about the slip event, which are based on existing examples such as the research of Okabe et al. [6]. Concurrently, judging from the effective pressure sensation from the effective skin deformation produced by suction pressure [2], it was posited that the integration of suction pressure would augment the realism of the slip sensation, primarily by emulating the pressure experienced during object grasping and simulating cutaneous deformation concurrent with the slip.

This experiment encompassed the comparison of three stimulation conditions: suction pressure only, electrotactile only, and a combined application of both stimuli. Additionally, a control scenario, characterized by the absence of any stimulation during contact, was also incorporated. The experiment was approved by the ethics committee of the authors' institution.

3.1 Experiment Setup

The experimental setup employed the Meta Quest 2 Virtual Reality Headset with its right controller to which the suction pressure and electrotactile devices were attached. Participants were able to freely move the sensors using the positional tracking of the controllers and perform a pinching motion by pressing the controller's trigger button.

Prior to the experiment per participant, the intensity of the electrotactile stimulation was calibrated for each finger to the maximum stimulation threshold before pain or discomfort could be felt according to the participant. Steps were also taken to ensure the perceived stimulation strength was uniform, and if an imbalance were felt, the intensity would scale down to match the finger with the lower stimulation strength according to the participant.

The virtual environment was comprised of three distinct cubes, differentiated by their mass. These cubes were assigned mass parameters of 10, 20, and 30 within the Unity's Rigidbody component. However, for enhanced relatability, these were ostensibly labeled as "1 kg", "2 kg", and "3 kg".

Participants were instructed to engage with the three cubes by freely touching and lifting them for a duration of one minute for each of the specified stimulation modes. To obscure any auditory cues emanating from the actuation of suction pressure, white noise was played throughout the interaction. After the one-minute interaction, participants were requested to evaluate their tactile experience on a 7-point Likert scale, with 7 representing a strong agree to the item, while 1 representing a strong disagree to the item, on three aspects: the realism of the tactile sensation, the clarity of the slip sensation, and the distinctness of the slip direction. Each stimulation mode was subjected to four trials, with the sequence being pseudo-randomized to avoid excessive repetition of any one condition. We recruited twelve participants (nine males, three females, average age = 24.75, SD = 2.8) the University of Electro-Communications. The study took about 30 minutes to complete per participant (Fig. 2).

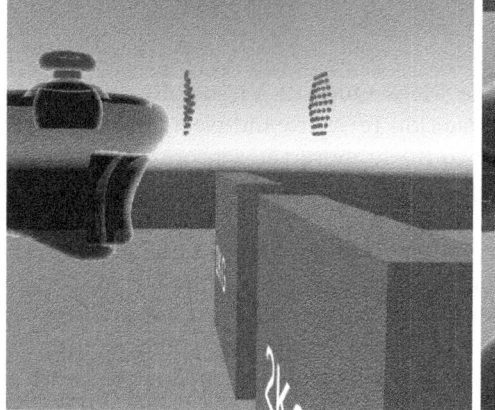

Fig. 2. (Top) Experimental setup using the Meta Quest 2 (Bottom left) Virtual environment with sensor before pinching (Bottom right) Virtual environment with sensor after pinching

3.2   Results and Discussion

The aggregated experiences resulting from four repetitions of each stimulation mode per participant were averaged to generate a composite score for each evaluative criterion, as illustrated in Fig. 3.

Given the ordinal scale format of the questionnaire data, it necessitates treatment as nonparametric. The Friedman test revealed a significant effect of electrotactile stimulation alone compared to no stimulation concerning slip clarity and directionality ($p < 0.001$). Post-hoc analysis, employing Bonferroni correction, further demonstrated significant differences between electrotactile stimulation alone and no stimulation ($p < 0.01$). These findings underscore the efficacy of electrotactile stimulation in enhancing both slip perception clarity and directional accuracy, aligning with our initial hypothesis regarding the conveyance of slip information through electrotactile stimulation. This also corroborates findings from a prior study indicating that electrotactile stimulation induces slip illusion [6].

Additionally, the Friedman test unveiled a significant disparity in the impact of suction pressure on realism compared to the no-stimulation condition ($p < 0.01$), with subsequent Bonferroni correction confirming a significant difference ($p < 0.05$). This suggests that the presence of suction pressure significantly enhances the perception of realism, in line with our hypothesis. However, it's noteworthy that there was no significant difference between sole electrotactile stimulation and the combined stimulation condition. Yet, the considerably stronger significant differences observed between combined stimulations and no-stimulation conditions (Bonferroni corrected values at $p < 0.001$) suggest that electrotactile stimulation also contributes to a sense of realism.

Furthermore, despite the clear enhancement of realism provided by suction pressure stimulation, it did not contribute to the clarity of slip direction. This implies that while directionality may be a crucial cue for object manipulation and preventing objects from slipping, it may not necessarily be pivotal for enhancing realism. The aggregated experiences from the four repetitions of each stimulation mode per participant were averaged, culminating in a composite score for each evaluative criterion, as depicted in Fig. 3.

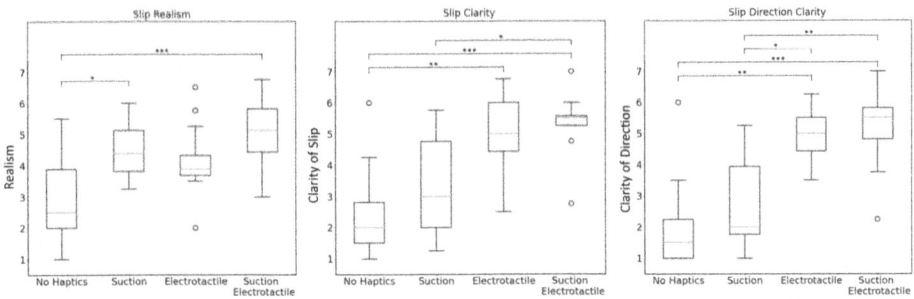

Fig. 3. Averaged Likert scale scores for Slip Sensation with Bonferroni corrected p values. (Left) Realism of each stimulation (Middle) Clarity of each stimulation (Right) Clarity of slip direction of each stimulation. $*$: $p < 0.05$, $**$: $p < 0.01$, $***$: $p < 0.001$

Currently, our work has several limitations. Although the configuration of the virtual sensors was deliberately designed to simulate partial slip, the precise parameters and the extent of their curvature that affect the stimulation strength of the electrodes were not exhaustively explored. The values for these attributes were selected arbitrarily rather than through systematic investigation.

4   Conclusion and Future Works

In summary, this research endeavored to evaluate the efficacy of suction pressure and electrotactile stimulations in replicating a slip sensation. Our findings indicate that suction pressure significantly contributed to the realism of the slip sensation. Conversely, electrotactile stimulation excelled in providing acute awareness of the imminent slip of an object, along with discerning the direction of said slip. While electrotactile stimulation alone imparted a moderate sense of realism, its effectiveness appears to be substantially augmented when combined with suction pressure.

As a progression in our research, we aim to conduct an in-depth investigation into the optimal variation in stimulation intensity between the central and peripheral areas of the contact zone. This study will focus on more accurately simulating the nuances of partial slip sensations.

Acknowledgments. This work was supported by JST A-STEP Grant Number JPMJTR23RC, Japan.

References

1. Kajimoto, H.: Electrotactile display with real-time impedance feedback using pulse width modulation. IEEE Trans. Haptics **5**(2), 184–188 (2011)
2. Kameoka, T., Kajimoto, H.: Design of suction-type tactile presentation mechanism to be embedded in HMD. Front. Virtual Reality **3**, 894873 (2022). https://doi.org/10.3389/frvir.2022.894873
3. Lin, W., et al.: Super-resolution wearable electrotactile rendering system. Sci. Adv. **8**(36), eabp87 (2022). https://doi.org/10.1126/sciadv.abp8738
4. Makino, Y., Asamura, N., Shinoda, H.: Multi primitive tactile display based on suction pressure control. In: 12th International Symposium on Haptic Interfaces for Virtual Environment and Teleoperator Systems, 2004. HAPTICS'04. Proceedings, pp. 90–96. IEEE (2004)
5. Masataka, N., Konyo, M., Maeno, T., Tadokoro, S.: Reflective grasp force control of humans induced by distributed vibration stimuli on finger skin with icpf actuators. In: Proceedings 2006 IEEE International Conference on Robotics and Automation, 2006. ICRA 2006, pp. 3899–3904 (2006). https://doi.org/10.1109/ROBOT.2006.1642299
6. Okabe, H., Fukushima, S., Sato, M., Kajimoto, H.: Fingertip slip illusion with an electrocutaneous display. In: International Conference on Artificial Reality and Telexistence, pp. 10–14 (2011)
7. Schiltz, F., Delhaye, B.P., Thonnard, J.L., Lefèvre, P.: Grip force is adjusted at a level that maintains an upper bound on partial slip across friction conditions during object manipulation. IEEE Trans. Haptics **15**(1), 2–7 (2021)
8. Vizcay, S., Kourtesis, P., Argelaguet, F., Pacchierotti, C., Marchal, M.: Design and evaluation of electrotactile rendering effects for finger-based interactions in virtual reality. In: Proceedings of the 28th ACM Symposium on Virtual Reality Software and Technology. VRST '22, Association for Computing Machinery, New York, NY, USA (2022). https://doi.org/10.1145/3562939.3565634

Applications and Interaction

Blindfolded Operation as a Method of Haptic Feedback Design for Mobile Machinery

Victor Zhidchenko$^{(\boxtimes)}$ (ID), Egor Startcev (ID), and Heikki Handroos (ID)

Department of Mechanical Engineering, LUT University, Lappeenranta, Finland
`victor.zhidchenko@lut.fi`

Abstract. Mobile machinery includes a wide range of machines performing specific operations in off-road environments for agriculture, mining, and construction. Haptic feedback is not widely used in mobile machinery due to several obstacles. One of them is the lack of methods for choosing efficient haptic cues that improve the operation of each type of machine. This paper considers the blindfolded operation of a machine as an approach that facilitates the development of efficient haptic feedback. The method includes the development of a haptic-only human-machine interface that makes possible blindfolded operation of the machine and subsequent interface simplification by excluding the signals, which become unnecessary in the presence of visual information.

Two examples of developing haptic feedback for mobile cranes are presented. Commercially available haptic joysticks were used as the main controls. The haptic cues developed in the study allowed the transmission of 17 signals to an operator through the haptic joysticks, making the blindfolded virtual log crane operation possible. For a limited remote operation of a real crane, haptic cues for transmitting 6 signals were developed with the same approach. The efficiency of the developed haptic feedback was tested with a small group of participants in a real-time simulator of a log crane and in the remote operation of a real crane in a laboratory environment. The participants reported improved depth perception and lower stress levels in remote operation with haptic feedback. The described approach is not limited to mobile cranes and can be used in other types of mobile machinery.

Keywords: Haptic feedback · Mobile machinery · Teleoperation

1 Introduction

Mobile machinery, also known as off-road vehicles (ORV), non-road mobile machinery (NRMM), heavy equipment, and mobile working machines, represent the machines that are widely used in construction, mining, forestry, logistics, and agriculture. They can be defined as machines designed to perform various operations in off-road environments besides transportation. Examples of such machines include excavators, wheel loaders, dozers, mobile cranes, and tractors.

H. Kajimoto et al. (Eds.): EuroHaptics 2024, LNCS 14768, pp. 323–337, 2025.
https://doi.org/10.1007/978-3-031-70058-3_27

Haptic feedback was inherently present in hydraulically actuated machines through the manual control of high-pressure hydraulic valves. Moving the levers directly connected to the valve spools, an operator could naturally feel the forces and vibration existing in the machine. With the introduction of pilot valves integrated into joysticks, the operators did not need to apply large force to operate the machine , but lost information about the pressure in the main valves. Then, electrically controlled valves became widespread in mobile machinery. These devices use solenoids to move the valve spool. An operator controls the voltage signal with electric joysticks, which are totally decoupled from the hydraulic system of the machine. This type of control became predominant in mobile machinery [4]. As a result, haptic feedback has disappeared from most modern machines. Nowadays, new conditions are being observed that are conducive to its return.

Automation is the current trend in mobile machinery design. Manufacturers aim to reduce operator requirements by incorporating automation due to technological advances and the lack of skilled machine operators in many industries. As a result, the operators become even more decoupled from the machines. This situation can cause errors and raise security issues in applications where the level of automation does not allow fully autonomous operation. New information channels are needed to enhance operator situational awareness, providing new opportunities for haptic feedback. Another area that can benefit from the haptic feedback is remote machine operation. This approach improves safety and reduces the requirements for machine operators. When removed from the cabin, an operator loses spatial vision and sound, vibration, smell, and ability to move relative to the machine. Currently, video feed is the main information channel for remote operators.

Haptic feedback creates an additional information channel in teleoperation. This feature is the focus of the current paper. We consider information transmission by haptics more important for remote operation of specific machines (e.g., in load handling) than an ability to "feel" the weight or resistance force.

One of the major problems in employing haptics for mobile machinery is uncertainty in what information to transmit and by which haptic signals. This paper suggests using the blindfolded operation of a machine or its real-time model to answer the aforementioned questions. Blindfolding is a widely used technique in testing haptic devices and evaluating perception in haptics research. Here, we use blindfolding to discover the information required to improve machine operation performance and develop haptic patterns. The essence of the method is simple - a machine operator tries to perform a typical task using the machine while being blindfolded. The operator reports which information is required to operate successfully. Taking into account the movements that the operator must make during the task implementation, haptic patterns that do not hinder those movements are designed. The final step of the method is the elimination of the haptic signals that become unnecessary if the operator can see the working surroundings.

Using this method, the paper presents an example of developing haptic feedback for a mobile crane. We test the approach with two similar machines: a crane mounted on a log truck, which is operated in a real-time simulator, and a real mobile crane operated remotely. To estimate the efficiency of the developed haptic feedback, we performed experiments measuring the performance of operation with the haptic feedback and without it.

2   Studies Related to Haptic Feedback in Mobile Machinery

To our knowledge, the initial research on force feedback related to heavy machinery was conducted by Ostoja-Starzewski in 1989 [15] and Parker in 1992 [16]. Ostoja-Starzewski proposed a prototype of a master-slave force-feedback hydraulic manipulator. Meanwhile, Parker used a force-reflecting master joystick with a CAT 215 log loader in his dissertation. Both of these concepts were later utilized in scientific publications and commercial solutions and paved the road to an ongoing process of developing force-feedback joysticks and manipulators for specific tasks. Researchers such as Kim et al. [9], Heikkinen [7], Chu et al. [3], and Kuang et al. [11] published their studies on this topic. By the beginning of 2000, there were several factory-made manipulators available from companies such as 3D Systems (Phantom), FCS Control Systems (HapticMaster), and Force Dimension (Omega). Various areas of experimental research, such as remote driving, teleoperation for robots, and controlling the hydraulic manipulator of an excavator [6], have made progress with the help of these devices. Hayn et al. found that operators rated haptic assistance as helpful.

In 2012, Villaverde et al. presented a new electromechanical system that allowed for the remote operation of a gantry crane through the Internet using a haptic device [17]. The system addressed stability issues caused by time delays and introduced a virtual environment to improve the interface. However, the authors noted that further tests on a larger scale were necessary to determine the performance improvements of the proposed interface.

In a study by Heikkinen, a virtual environment combined with a gantry crane model and haptic interface was used [7]. The study employed a real-time simulator with a hardware-in-the-loop setup and provided some important findings in the context of the current study. They include the consideration that users should be able to modify force feedback parameters themselves and that the force variables and their directions should be carefully chosen for each application of the haptic controller. These findings raise a question on how to design the haptic feedback for efficient remote operation of mobile machinery.

In 2013, Nitsch et al. conducted a meta-analysis on the impact of haptic interfaces in teleoperation systems [14]. They concluded that haptic interfaces could help tackle challenges such as long task completion times, handling errors, and excessive force application in remote environments, "provided, of course, that they are applied appropriately". However, it is worth mentioning that designing haptic feedback cues for each application case requires a specific approach, and it is impossible to determine whether they were applied appropriately without analyzing performance statistics.

In recent studies, Abdullah discovered that incorporating haptic feedback for simple operations on remote-operated cranes using joysticks does not significantly increase user effectiveness or efficiency [2]. In 2019, Morosi et al. [13] suggested replacing traditional joysticks with a coordinated control paradigm for hydraulic excavators ([10]) using a 3-DoF haptic device developed by Moog FCS Robotics. They used a virtual environment to demonstrate that their system resulted in a 50% reduction in execution time and errors compared to the joystick operation. In 2022, Morosi et al. tested two haptic responses using the same control approach, simulation, and VR environment [12]. After conducting experiments, it was concluded that the operators could not distinguish between the suggested haptic effects. The authors noted that implementing haptic feedback that accurately represents a physical event is challenging.

3   Haptic Joysticks

Some features make mobile machinery distinctive from other types of equipment controlled by haptic devices. First, the controls should follow industry tradition since existing control devices have been refined over decades. Well-established working methods, training schools, and standards are dedicated to existing controls. Second, the devices must allow mass production as several million mobile working machines are produced globally every year [5]. Although some novel concepts have been introduced for haptic control devices for mobile machinery, joysticks remain prevailing in these machines.

Usability tests of several haptic control devices with a virtual real-time model of a rubber tyred gantry crane were performed in [7]. The study concluded that "Using one joint per control direction, as with a joystick, was found to be more natural and less confusing to use".

In the presented study, commercially available haptic joysticks were used, which were provided by Haptronics Oy (Finland) (Fig. 1(a)). The joysticks enable force and vibration feedback on the grip in two axes. The amplitude of the force as well as frequency, amplitude, and duration of vibration can be set in real time. The joysticks are based on the hardware platform manufactured by Brunner Elektronik AG (Switzerland) and offer a customizable communication interface for integration into mobile machinery. This interface is implemented by a controller that serves as an intermediate link between the joysticks and external systems. To use the joysticks for research purposes, a custom communication protocol was developed by Haptronics for LUT University, based on User Datagram Protocol (UDP) running over an Ethernet connection.

In this study, the joysticks controller was connected to the nearby computer by an Ethernet UTP cable. The computer was running real-time simulation software and a program to communicate with a remote computer connected to the real crane. The controller was sending a UDP datagram each 3 ms to the computer. It contained the datagram sequence number and the data from each

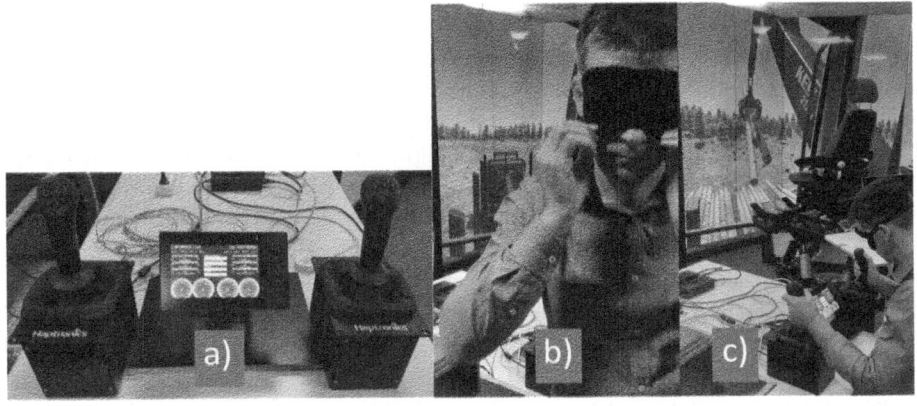

Fig. 1. Haptic joysticks used in the study (a), blindfolding the operator (b), blindfolded operation of a virtual log crane.

joystick, which consisted of the values of joystick output in X and Y directions proportional to the grip angles relative to the corresponding axis, button states, rocker output, and the joystick state. The state contained warnings about dangerous conditions, such as short circuits, extremely high or low voltage, high temperature, and sensor errors. To control haptic feedback, the controller was listening on a separate UDP socket for the datagrams from the computer. These datagrams were generated by Python programs communicating with the simulator or remote crane. They included a datagram sequence number and 16 commands for controlling both joysticks. For each joystick, two commands contained normalized force level (from minimum to maximum) along the X and Y axes, and 6 commands controlled normalized amplitude, normalized frequency, and absolute duration (from 0 to 10 s) of vibration along two axes.

4   Kinesthetic and Tactile Cues for Blindfolded Operation of Log Crane

Mobile cranes are mounted on trucks and tractors. In Finland, a typical example is a log crane mounted on a log truck. This truck moves one or two trailers and delivers the logs from the forest to the wood processing facility. The log crane is used to load the logs into the trailer and to unload them on delivery. Log cranes usually have a cabin from which an operator controls the crane. Removing the crane cabin would increase the payload capacity of the truck and improve safety by eliminating the need for the operator to climb in and out, which can be risky in poor weather conditions. Remote operation of the crane can be performed either from the truck cabin or from a remote operation center (ROC). In these scenarios, operators rely on visual information from cameras, which affects depth perception and hinders the operation.

This study tries to compensate for weaker depth perception using haptic feedback. In log handling, depth information is needed to define the optimal crane position for grabbing and releasing the logs and to avoid collisions. Grabbing the logs at their center of mass leads to the horizontal position of the logs during their movement, minimizes the lifting height and, thus, the time and energy required for operation. If the logs have been grabbed in the center of mass, which usually approximately corresponds to their geometric center, an optimal release position also aligns with the center of unloading space. We completely excluded visual signals to define appropriate ways of delivering information about the position of the crane relative to its surroundings without depth vision. If the goal is to operate the crane with limited visual information (without depth perception), can it be operated without any vision? We asked a person with approximately 20 hours of experience operating the crane in a real-time simulator to consider the blindfolded operation and define the required information. Then, the operator was blindfolded and tried to operate the crane while receiving that information through haptic joysticks (Fig. 1(b),(c)). Several iterations were made to determine the needed information and appropriate ways of transmitting it through haptic joysticks since the initially assumed set of signals was incomplete. The overall process of creating and adjusting the haptic signals and testing them in the real-time simulator took 12 hours. As a result, the operator could perform a log-loading task in the simulator while being blindfolded. The video recording of the operation can be accessed on YouTube [1].

The following features of crane operation with the described haptic joysticks, discovered during their testing in a real-time simulator, were taken into account for the rendering of kinesthetic and tactile cues:

- Force feedback indicating the weight of the load is not needed since the weight is always present in load handling operations. Such a signal would introduce additional fatigue for the operator without providing any help.
- To prevent operator fatigue, force feedback for long periods of time should be avoided.
- It is difficult for the operator to estimate the value of resistance force on the joystick because joystick operation involves dynamic movements and holding the handle in a static position at different angles and for arbitrary periods of time. Considering several hours duration of the work shift and operator fatigue, a feasible approach is to provide a "binary" signal (the joysticks resist or not) or change the resisting force dynamically while the operator holds the joystick in a static position.
- Vibration signals should be used in exceptional cases, for example, to provide warnings, since they can be considered annoying and disturb the operation
- Vibration signals should be repetitive if the haptic joysticks are installed in the crane cabin to prevent hindering the signals by the vibration and shocks produced by the crane.

In total, 17 signals were delivered to the operator through the haptic feedback. The information required for successful log crane operation without visual data and the corresponding tactile cues are presented in Table 1.

Table 1. Information delivered through haptics and corresponding feedback type.

No.	Signal description	Feedback
1	Maximum sufficient height of the grapple above the ground	force
2	Minimum height of the grapple above the ground, sufficient for grabbing the logs	force
3	Maximum angle of rotation of the crane pillar	force
4	Distance to the middle of the logs in the log pile	force
5	Notification when the grapple passes over the right edge of the trailer	force
6	Notification when the grapple passes over the left edge of the trailer	force
7	Minimum allowed angle of the lift boom to avoid collision with the trailer	force
8	Notification when the grapple is close to the vertical passing through the center of the trailer	vibration
9	Grapple state (completely open)	vibration
10	Grapple state (completely closed)	vibration
11	Grapple state (partially closed)	vibration
12	Grapple inclination (less than 30° or more than 30°)	vibration
13	Grapple rotation (less than 3° or more than 3°)	vibration
14	Angles of the crane booms	vibration
15	X coordinate of the crane tip relative to the pillar (in m)	vibration
16	Y coordinate of the crane tip relative to the pillar (in m)	vibration
17	Z coordinate of the crane tip relative to the pillar (in m)	vibration

Force feedback was used to indicate the approaching of the crane tip or boom to a predefined boundary, either a line or a point (signals 1–7 in Table 1). The closer the boundary, the higher was the resistance force of the joystick, with the maximum value at the boundary. The force was applied to the joystick and axis, which produced the command for approaching the boundary. The amplitude of the force was calculated by the following equation:

$$F = \left(1 - \frac{|A - B|}{D}\right) K \qquad (1)$$

where F is the force expressed in relative units ($F \in [0, 1]$, where 0 indicates minimum force, 1 indicates maximum force); A is a value that approaches a boundary expressed in meters or degrees; B is the boundary value expressed in the same units as A; D is the maximum distance from the boundary, at which the force feedback starts, expressed in the same units as A; $K \in (0, 1]$ is a coefficient that limits the maximum force.

Small values of K allowed a small resistance force perceived as an invisible obstacle on the way. Such feedback was used to indicate the fact that the grapple was passing over the edge of the trailer (signals 5 and 6 in Table 1). The values of

K close to 1 were used to limit potentially dangerous movements, for example, when the grapple approached the minimum allowed height above the ground (signal 2 in Table 1).

The opening of the grapple was controlled by the rocker on the right joystick. When the operator was pushing the rocker, the grapple state (signals 9–11 in Table 1) was rendered by the vibration of the right joystick. To obtain this state, the volume flow in the piston chamber of the cylinder moving the grapple and the extension of this cylinder were analyzed. If the grapple was opening or closing, no vibration was generated. Completely open grapple corresponded to the maximum value of the cylinder extension and the volume flow value below a predefined margin. This state was rendered by the vibration signal of 200 ms when the operator was pushing the rocker, trying to open the grapple. The completely closed grapple corresponded to the minimum value of the cylinder extension and the volume flow below the predefined margin. This state was rendered by the vibration signal of 100 ms when the operator was pushing the rocker to close the grapple. The partially closed grapple was detected if the cylinder extension was above the minimum value and the volume flow was below the predefined margin. This state indicated that there were logs (or other objects) in the grapple. It was rendered by the vibration for 1000 ms. These signals helped the operator to detect the successful grabbing and releasing of the logs.

Signals 12–17 (see Table 1) were generated at the operator's request by pressing the joystick buttons. Three buttons located at the left joystick generated the signals 12–14, and three buttons on the right joystick generated the signals indicating the X, Y, and Z position of the joint between the boom and the grapple relative to the pillar (signals 15–17). Signals 12 and 13 used data from the virtual Inertial measurement units (IMUs) located at the grapple and pillar. An inclination of more than 30° was indicated by a short vibration (300 ms) of the left joystick. The smaller inclination angle of the grapple was not indicated, which meant that the grapple was hanging properly (probably swaying). Tracking the rotation of the grapple is important to ensure that the grapple claws are perpendicular to the plane of the booms. It helps with grabbing the logs from the pile and unloading them to the trailer, as in this case, the logs are in the same vertical plane as the booms. The angle of grapple rotation more than 26° relative to the vertical plane of the booms was rendered by a short vibration (100 ms) of the right joystick. Smaller angles were rendered by a long vibration (1000 ms).

These patterns followed the idea that short vibration signals indicated improper states and long vibration or absence of it indicated proper states. If the operator held a button pressed, the vibrations repeated with a 10 ms interval between them. It allowed the detection of improper grapple orientation (rotation relative to the booms or excessive inclination due to collision) and its correction by moving the grapple until the short vibrations disappear. Vibrations were induced at that joystick, which controlled the corresponding correction movement (e.g., right joystick vibration indicated incorrect rotation because the grapple was rotated by the right joystick).

The angle of the lift boom and outer boom (signal 14 in Table 1) was rendered by two consecutive vibrations of the right joystick. The duration of each vibration was proportional to the angle of the corresponding boom (the first signal corresponded to the lift boom, and the second signal - to the outer boom). The similar duration of both signals indicated a regular configuration of the booms. A noticeable difference between the signals indicated that one of the cylinders controlling the booms was abnormally extended or contracted. The signal allowed the operator to quickly check for unusual crane configurations.

The position of the crane tip along a specific axis (signals 15–17 in Table 1) was rounded to meters and rendered by a corresponding number of short vibrations (200 ms) of the left joystick. Since the signals were generated by pressing the right joystick buttons, rendering them at the left joystick allowed the operator to relax the left hand and concentrate attention on the haptic feedback.

As per [8], "if one plans to build a teleoperation, comanipulation or assistive system that adds haptic feedback to interactions that do not already have such, it is probably worthwhile to discuss if all haptic signals have to be measured, transmitted and displayed". In the presence of visual information, some of the signals become unnecessary. Specifically, signals 9–17 were excluded from the final set of haptic cues for [tele]operation of the crane. The effect of the developed haptic feedback on the crane operation in the presence of visual information was tested in a pilot study using the virtual environment and a real teleoperated crane. The experiments are described in the following section.

5   Test of Developed Haptic Cues in Regular Operation

The developed haptic cues were tested using the same operation station in two experiments. The joysticks were installed on a mobile working machine seat mounted on a six-degree-of-freedom motion platform (Fig. 2(a)). The joysticks controller was connected to a computer running the real-time simulator of a log truck and the software communicating with the remote crane. A visualization system comprised 12 55-in. LED panels that formed three flat screens (4 panels each), providing a video wall with a 180-° field of view. The simulator presented a view from the crane cabin and ensured the motion of the real seat corresponding to the movements of the operator seat in the simulation. The motion platform was not actuated in the second experiment during the remote operation of the real crane.

In the first experiment, participants operated a virtual crane. The task was to load the logs into the truck by grabbing them from a log pile located on the right side of the truck and moving them into the trailer (Fig. 2(b)). This is a typical task for a log truck driver. The experiment included four phases: rest for 2 min, crane operation with haptic feedback (5 min), rest for 2 min, and crane operation without haptic feedback (5 min). Experimental data were accumulated in iMotions software (www.imotions.com), which allowed combining the data from the simulator, joysticks, and external video camera on a single time axis. In each operation phase, the following performance parameters were measured:

the number of logs delivered to the trailer in each movement, the total number of logs loaded into the trailer during the phase, the number of collisions with the truck, and the duration of the time periods required to grab the logs and to move them into the trailer.

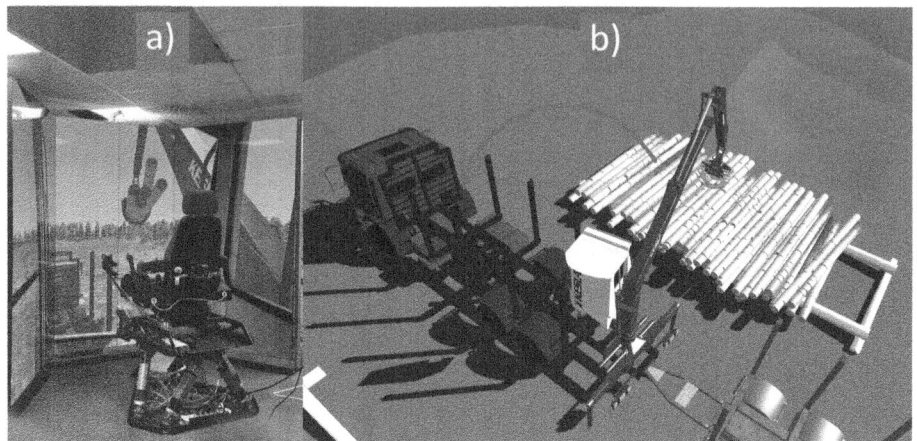

Fig. 2. Operation station (a) and the task performed by the participants in the simulator (b).

Four voluntary male participants (two of them aged 20–30 and two aged 40–50) took part in the pilot study. They were first instructed about the purpose of the experiment, the experimental protocol, the meaning of the joystick commands, and the meaning of the kinesthetic and tactile cues. Then they tried crane operation with haptic feedback and without it. The instruction and practice phase lasted approximately 30 min until the subject reported familiarity with the task and controls. After the experiment, each subject was asked to complete a questionnaire. There were no statistically significant differences in the performance measures between the results of the log crane operation with and without haptic feedback in the real-time simulator. In total, 97 logs were moved to the trailer by all participants with haptic feedback and 99 logs without it. The total number of collisions made by all participants was 1 for the operation with haptic feedback (in total, 18 load/unload cycles) and 3 without haptic feedback (17 load/unload cycles).

In the second experiment, participants remotely controlled a real mobile crane from the same operation station (Fig. 3(a)). The crane was located in another building of the LUT University campus and connected to the operation station by the local area network (Gigabit Ethernet). Visual information was delivered by a camera located behind the crane, which provided a view similar to local operation. The camera view was presented to the operator in a 1.1 × 0.7 m window with a resolution of 1280x720 pixels.

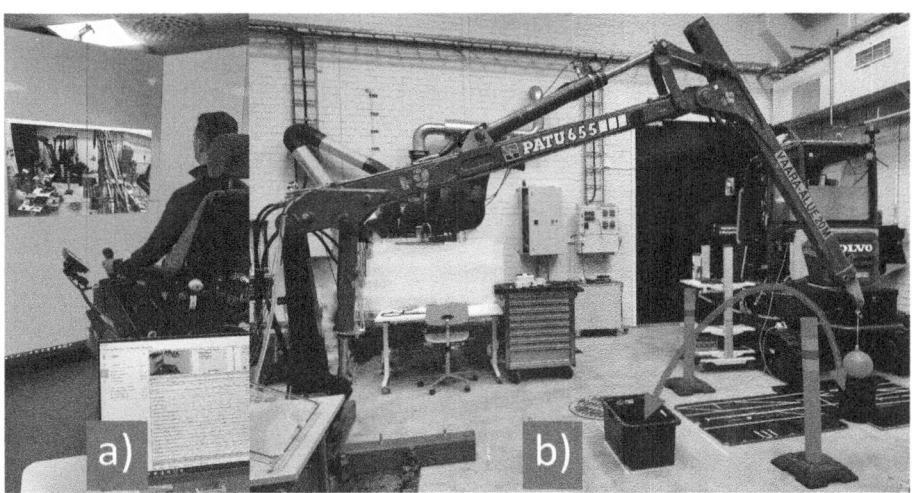

Fig. 3. Operator view in remote operation (a) and the task performed by the participants in remote operation (b).

For safety reasons, we simplified the task implemented with the real crane, trying to retain the features important for real operation. The task was to move a rubber ball hanging on a rope under the crane tip between two points: a basket located at a distance of 4.5 m from the point of the camera installation and a box located at 2.5 m from the same point (Fig. 3(b)). A chain stretched between two poles at the height of 1 m represented an obstacle. The chain was located in the middle of the way between the final points. The 2D video stream from the camera reduced depth perception, making it difficult to hit the box and basket.

Haptic feedback in remote operation experiment was represented by the varying resisting force of the joysticks according to equation (1). As the predefined boundaries, five lines were used, namely, a vertical line crossing the center of the box, a vertical line crossing the center of the basket, a vertical line crossing the chain, and two horizontal lines indicating minimum and maximum allowed height of the crane tip. For crossing the chain, haptic feedback was only enabled if the ball was below the chain. Approaching the chain from both sides was considered separately so that the haptic feedback was enabled only when the ball was moving towards the chain. This approach reduced physical load on the operator by eliminating unnecessary resisting force when the crane was moving away from the chain.

Participants first performed four test movements between the box and basket without haptic feedback to get familiar with the system. Then, they were instructed about the meaning of the haptic cues and performed four movements with the feedback. The protocol was the same as in the first experiment. The number of collisions with the chain, the number of completed movements between the end points, and the time needed for each movement were used as performance measures. The results are presented in Fig. 4.

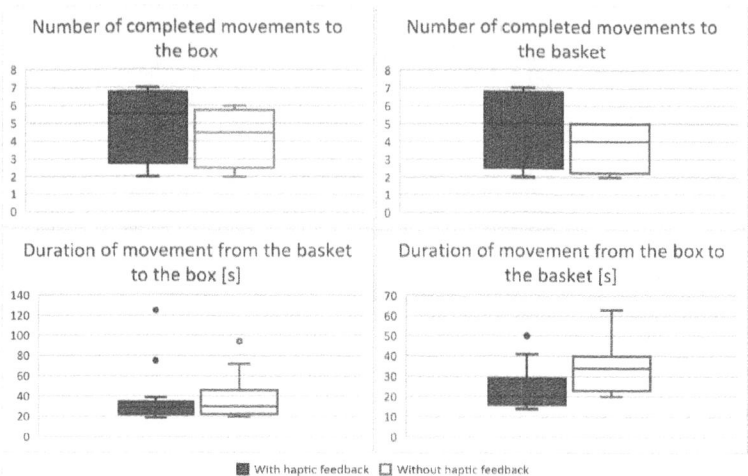

Fig. 4. Measurement results for remote crane operation.

In total, 20 hits to the box and 19 hits to the basket were made with the haptic feedback (accordingly, 17 and 15 hits without haptics). The total number of collisions was equal in both scenarios (5 collisions). The most significant difference was observed for the movement duration from the box to the basket (mean value 24.3 s with haptic feedback and 34.0 s without it). For the opposite movement, mean values were close (45.9 s with haptics and 45.5 s without it).

The results of the questionnaire are presented in Table 2. Qualitative questions were open-ended. To save space in the table, we categorized similar answers and presented them as a single line.

Table 2. Answers provided to the questionnaire after the experiments.

No.	Question						
					Answers from the participants		
	Participant	I	II	III	IV	mean	median
1	Was your position comfortable during the operation?	Yes	Yes	Yes	Yes		
2	How did you recover during the rest periods (1 = very bad; 10 = very good)?	10	10	10	10	10	10
3	How do you estimate your stress level in the simulator environment without any feedback? (1 = not stressful at all; 10 = very stressful)	7	7	6	5	6.25	6.50
4	How do you estimate your stress level in the simulator environment with **haptic feedback**? (1 = not stressful at all; 10 = very stressful)	5	4	4	3	4.00	4.00
5	What was the reason for the higher stress level in the corresponding task?						
	- It was hard to find the center of the logs when grabbing without feedback		X	X	X		
	- It was more difficult to predict the result of an action in the simulator	X					
6	Which phase of the task was the most difficult?						
	- Grabbing the logs from the pile			X	X		
	- Releasing the logs to the truck	X	X				
7	Do you find haptic feedback annoying?	No	No	No	No		
8	How do you estimate your stress level in remote operation without any feedback? (1 = not stressful at all; 10 = very stressful)?	5	5	4	7	5.25	5.00
9	How do you estimate your stress level in remote operation with **haptic feedback**? (1 = not stressful at all; 10 = very stressful)?	3	2	3	5	3.25	3.00
10	What was the reason for higher stress level in the corresponding task in remote operation?						
	- Poor depth perception	X	X		X		
	- It was more difficult to predict the result of an action in remote operation				X		
11	Which phase of the task was the most difficult?						
	- Passing the chain	X			X		
	- Hitting the ball into the box			X			
	- Hitting the ball into the basket				X		

6   Conclusion and Future Work

The presented study demonstrated an approach for building haptic cues for mobile machinery operation by using blindfolded operation as a goal. The cues developed with this approach for a mobile log crane controlled with commercially available joysticks allowed the blindfolded operation of the machine in a real-time simulator. This result exhibits the capability of haptic feedback to serve as a convenient and versatile information channel for the operation of mobile machinery. In the presented setup, 17 signals were transmitted to the operator through haptics.

An advantage of haptic feedback in relation to other information channels, such as visual and auditory, is the ability of a machine operator to receive information without being distracted from the task. Developing haptic cues for blindfolded operation is beneficial for another reason. In the experiments, we noticed that with haptic feedback, some operators start to rely on it so that they pay less attention to visual information. This effect shifts the operator's attention from the visual to the tactile channel so that the operator stops looking attentively and waits for a haptic signal. As a result, haptic cues designed to provide information about the machine and its surroundings for blindfolded operation, with the visual channel totally excluded, become more relevant. They can better compensate for the loss of visual information compared to the haptic cues developed with other methods. Further studies with various types of machines and work tasks are required to determine which tactile signals can be safely excluded without losing this capability.

Although no significant effect of haptic cues on work performance in the simulator was observed in the pilot study, such an effect was present in the remote operation of the real crane. The experimental results highlight the teleoperation of mobile machinery as a promising application area for haptics. Remote operation station isolates the operator from the shocks and vibrations inherent in in-cabin work, which improves sensitivity to haptic signals and makes them more feasible compared to traditional operation.

In the future, we plan to conduct experiments involving more participants and compare several types of feedback in the remote operation of mobile machinery (haptic, auditory, and visual feedback). In addition to the crane, we will test haptic feedback using joysticks in the remote operation of a mini-excavator.

Acknowledgments. Research reported in this publication was financially supported by Business Finland, project name "SANTTU".

Disclosure of Interests. Author 3 has shares in the company Haptronics Oy, whose equipment was used in the study. Haptronics Oy is a spin-off company of LUT University.

References

1. Blindfolded operation of a log crane in a real-time simulator using haptic joysticks - YouTube. https://www.youtube.com/watch?si=92TkAs7SF1itI-uE&v=5ecD7s9CZGM&feature=youtu.be. Accessed 02 Feb 2024
2. Abdullah, U.N.N.: Novel Methods for Assessing and Improving Usability of a Remote-operated Off-Road Vehicle Interface. Ph.D. thesis, Lappeenranta University of Technology (2019)
3. Chu, Y., Zhang, H., Wang, W.: Enhancement of virtual simulator for marine crane operations via haptic device with force feedback. In: Haptics: Perception, Devices, Control, and Applications, pp. 327–337. Springer International Publishing (2016)
4. Duffy, O.C., Wright, G., Heard, S.A.: Fundamentals of Mobile Heavy Equipment. Jones & Bartlett Learning, CDX Learning Systems (2017)

5. Geimer, M.: Mobile Working Machines. SAE International (2020)
6. Hayn, H., Schwarzm, D.: A haptically enhanced operational concept for a hydraulic excavator. In: Advances in Haptics. InTech (2010)
7. Heikkinen, J.: Virtual Technology and Haptic Interface Solutions for Design and Control of Mobile Working Machines. Ph.D. Thesis, Lappeenranta University of Technology, Lappeenranta (2013)
8. Kern, T.A., Hatzfeld, C., Abbasimoshaei, A.: Engineering Haptic Devices. Springer International Publishing, Cham (2023). https://doi.org/10.1007/978-3-031-04536-3
9. Kim, D., Oh, K.W., Hong, D., Park, J.H., Hong, S.H.: Remote control of excavator with designed haptic device. In: 2008 International Conference on Control, Automation and Systems, pp. 1830–1834. IEEE (2008)
10. Kontz, M.E.: Haptic Control of Hydraulic Machinery Using Proportional Valves. Ph.D. Thesis, Georgia Institute of Technology (2007)
11. Kuang, L., Marchal, M., Giordano, P.R., Pacchierotti, C.: Rolling Handle for Hand Motion Guidance and Teleoperation. In: EuroHaptics 2022 - International Conference on Haptics: Science, Technology, Applications, pp. 1–3 (2022)
12. Morosi, F., Caruso, G.: Configuring a VR simulator for the evaluation of advanced human-machine interfaces for hydraulic excavators. Virtual Reality **26**(3), 801–816 (2022)
13. Morosi, F., Rossoni, M., Caruso, G.: Coordinated control paradigm for hydraulic excavator with haptic device. Autom. Constr. **105**, 102848 (2019)
14. Nitsch, V., Färber, B.: A meta-analysis of the effects of haptic interfaces on task performance with teleoperation systems. IEEE Trans. Haptics **6**(4), 387–398 (2013)
15. Ostoja-Starzewski, M., Skibniewski, M.: A master-slave manipulator for excavation and construction tasks. Robot. Auton. Syst. **4**(4), 333–337 (1989)
16. Parker, N.R.: Application of Force Feedback to Heavy Duty Hydraulic Machine. Ph.D. Thesis, University of British Columbia (1992)
17. Villaverde, A.F., Raimúndez, C., Barreiro, A.: Passive internet-based crane teleoperation with haptic aids. Int. J. Control Autom. Syst. **10**(1), 78–87 (2012)

Effects of Rendering Discrete Force Feedback on the Wrist During Virtual Exploration

Samet Mert Ercan$^{(\boxtimes)}$ [ID], Ayoade Adeyemi[ID], and Mine Sarac[ID]

Kadir Has University, Istanbul, Turkey
{sametmert,ayoade.adeyemi}@stu.khas.edu.tr, mine.sarac@khas.edu.tr

Abstract. Relocating the haptic feedback from the fingertip to the wrist is a trendy topic in haptic-assisted virtual interactions, and finding its best practices still requires a lot of research. In this paper, we investigate the perceptual and performance differences while rendering haptic feedback on the wrist in single-bump, discrete force feedback (through custom voice coil actuation of CoWrHap) or continuous force feedback (through linear DC actuation of LAWrHap). We conducted a user study experiment where participants interacted with identical-looking virtual objects with different stiffness properties and identified the ones with a higher stiffness level based on the haptic feedback they received. Our results indicate that participants performed the tasks *(i)* with higher sensitivity (higher JND), with more confidence (Number of Taps), and with better user experience using LAWrHap compared to using CoWrHap, and *(ii)* with no difference in terms of task accuracy (PSE), exploration and interaction time between using LAWrHap and CoWrHap.

Keywords: Haptic Interfaces · Virtual Reality Interactions

1 Introduction

Haptics is crucial to improving the quality of user performance and experience during interactions in Virtual and Augmented Reality (VR/AR) environments [24]. Rendering haptic information based on physical properties is useful for various applications, e.g., medical simulators [18] or exploration tasks [6] – significantly increasing the task performance [22] or immersion [16].

Most tactile haptic devices are designed for the fingertips [3,11] due to the highest mechanoreceptor intensity and increased realism [9]. However, rendering haptic feedback on the fingertips comes with drawbacks too. To increase comfort and wearability, they must be designed in small sizes; thus must be equipped with powerful but minimized actuators – increasing the overall cost. Even with the smallest design possible, they might limit the transparency of natural hand movements. Finally, they might prevent the interaction capabilities with virtual objects – especially during AR interactions.

H. Kajimoto et al. (Eds.): EuroHaptics 2024, LNCS 14768, pp. 338–351, 2025.
https://doi.org/10.1007/978-3-031-70058-3_28

In summary, haptic feedback on the skin around the fingers and hands through wearable haptic devices might impose different challenges during VR and AR interactions, which can be addressed by leaving the hands free of mechanisms. Relocating the haptic feedback from the fingertips to an alternative body location (like the wrist) can address these issues and challenges by freeing the hands and still rendering useful (and believable) information regarding virtual interactions. In such relocated scenarios, users perform exploration or manipulation tasks actively using their fingers in virtual or real environments while perceiving related haptic sensations in a remote body location other than the fingers or hands [2,5,12,13,15,19,25].

Despite the potential of wrist-worn devices, the most effective practices of such relocation are yet to be explored – especially in connection with VR/AR interactions. Understanding the human perception to discriminate objects is crucial to get their thresholds in exploratory tasks by utilizing different haptic devices. Several experimental studies have developed techniques to investigate the stiffness discrimination abilities using vibrotactile actuators [10], rubber and spring cells [21], and linear actuator [17].

Conventionally, mechanical properties of virtual objects are rendered through continuous force feedback (e.g., servo actuators [14,15] or linear actuators [17, 19]. Such continuous force feedback might help participants exhibit better user performance and experience since the useful information is rendered following conventional exploration mechanisms in the real world. For example, as the user squeezes an object, interaction forces occur based on the stiffness property of the object – thus, the stiffer the object, the faster and higher the interaction forces are perceived. Because all the haptic information is perceived with a similar logic to the real-world exploration tasks, participants might find this form of haptic rendering easier to interpret and more *believable*.

Relocating the haptic feedback to the wrist simplifies the rendering quality from *highly realistic* to *believable*. We previously hypothesized that haptic rendering could be simplified even further with a single-bump (discrete) force feedback instead of a continuous-level force feedback while still helping participants discriminate different mechanical properties of the virtual objects. We previously designed a novel custom-made voice coil device (CoWrHap) Fig. 1 (b) that can provide single-bump force feedback on the wrist [1], inspired by fingertip devices based on custom voice-coil actuation that would render force feedback [4,23]. CoWrHap might provide better user performance or experience since the discrete force feedback is rendered to the user through event-based triggers, i.e., whenever the user interacts with the surface of the object. Even if not realistic, such an event-based trigger might simplify the process of exploration in a more comparable way. Because all the haptic information is perceived in the same amount of time for all comparison pairs, how much time spent to complete the exploration might be shorter with discrete force feedback than continuous.

In summary, CoWrHap has great potential due to its low cost, compact size, and adjustability for force and frequency ranges as needed. It features a custom-made actuator that allows precise modifications of wire diameter and magnet selection. These design enhancements enable us to precisely control the size, skin deformation, and interaction forces, tailoring the device to specific needs [1,20]. This adaptability and customization highlight CoWrHap's advantages, and its true performance over conventional force feedback devices that can render continuous force feedback is still unknown.

In this study, we empirically compare the effect of rendering discrete and continuous force feedback through wrist-worn haptic devices while exploring the mechanical properties of virtual objects, an approach not previously explored in haptic research for VR/AR applications. We are committed to answering the research question (*RQ*): *are there perceptual and performance differences between discrete and continuous force feedback rendered at the wrist while participants discriminate stiffness properties of virtual objects?* With this motivation, we performed a user study experiment to validate that participants can use such controversial but observable and believable force feedback to discriminate different levels of mechanical properties of objects in a VR environment. The paper is structured as follows: Sect. 2 will detail the experiment setup including the haptic devices to render discrete and continuous force feedback separately, the VR environment, and the experiment protocol, Sect. 3 will present the experiment results, Sect. 4 will detail our discussions on the results, and Sect. 5 will offer the conclusion.

2   Experiment Setup

Figure 1 (a) shows the experiment setup: participants sit on a chair with arm supports and wear Oculus Quest 2 headset for (i) real-time hand tracking through the Hand Physics Toolkit (HPTK) [8] and (ii) visual representation of the VR environment. Once an interaction occurs between the real-time tracked hand avatar and the virtual objects, a haptic bracelet renders discrete or continuous information on the wrist based on the mechanical property of the object. Based on previous studies, we have observed that participants perform well in discriminating between different mechanical properties of virtual objects while *(i)* interacting with virtual tools using their dominant hand and *(ii)* receiving stationary feedback on their wrists [1]. The literature also includes successful examples of delivering haptic feedback to the non-moving wrist while interacting with dominant hand, included in psychophysical studies [7]. Consequently, we will have participants wear the haptic devices on their non-dominant wrists and interact with the virtual environment using their dominant hand. Noise-cancellation headphones with white noise minimize the environment and actuator noise.

2.1   Haptic Rendering Scenarios Through Wrist-Worn Devices

We are motivated to investigate the effect of rendering haptic feedback discretely or continuously on user performance, behavior, and experience during virtual exploration tasks. We used two different haptic devices in this study: a voice-coil based wrist-worn haptic device (CoWrHap) and a linear actuated wrist-worn haptic device (LAWrHap).

Voice-Coil Based Wrist-Worn Haptic Device (CoWrHap): CoWrHap in Fig. 1 (b) consists of a permanent magnet and a voice coil built by wrapping a copper wire around a cylindrical base [20]. By changing the levels of current passing through the wire, a magnetic field is created inside the cylinder, which ultimately creates a displacement for the overall coil since the magnet is attached to the base. In the meantime, this deformation causes a corresponding interaction force on the user's skin. It is driven by an L293B motor driver and controlled with a Raspberry Pi Pico microcontroller. The current passing through the coil can be changed through levels of duty cycle. Despite the unusual mapping between a stiffness exploration task in a VR setting and the discrete single-bump force feedback rendered by CoWrHap, its potential to help participants discriminate different levels of stiffness on virtual objects has been shown previously [1].

Linear Actuated Wrist-Worn Haptic Device (LAWrHap): LAWrHap in Fig. 1 (c) consists of a commercial linear DC actuator placed at the ventral side of the wrist through custom-designed 3D parts. We chose Actuonix PQ12-P linear actuator due to its low weight (15 g), maximum stroke (20 mm), high output force (18 N), and straightforward control using an integrated position sensor. It is driven by Actuonix Linear Actuator Controller (LAC) unit, which can perform PI controller with on-board controller gains. Users' comfort and ergonomy are improved through a silicone pad between the plastic and the skin and wide Velcro straps to keep the grounding stable. Further details about the design and performance are described in [19].

Calibration: The biggest challenge of conducting a user study with alternative haptic devices is to ensure that the intensity levels of haptic stimuli are the same, such that their impact on the user performance and experience can be

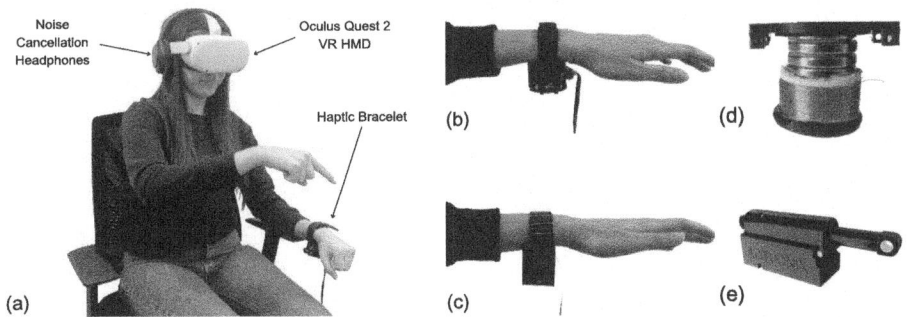

Fig. 1. Experiment setup: (a) The participant wears an Oculus headset to experience VR, haptic feedback, and noise-canceling headphones to minimize the environment and actuation noise. The haptic bracelet renders (b) discrete, single-bump force feedback through (d) voice coil actuation – CoWrHap or (c) continuous, linear force feedback through (e) DC actuation – LAWrHap.

compared. Thus, we measured forces with a force-sensing resistor (FSR 402) on users' skin while wearing CoWrHap and LAWrHap separately in a total of 15 repetitive tests for each device. Figure 2 (a) shows the force measurements taken at the duty cycles of CoWrHap in a previous study of stiffness discrimination [1] (from 55% to 80%) with red line showing the linear behavior connecting the mean values. We then explored different actuator displacements for LAWrHap that would yield similar force measurements with CoWrHap in a linear pattern, as indicated in Fig. 2 (b).

2.2   Virtual Environment

In this study, we utilized a VR environment that was previously developed and described in our research [1]. We implemented this environment using Unity 2021.3.14f1. The environment includes a white desk, a hand avatar, and two identical-looking, red, rigid boxes as shown in Fig. 3, to be displayed on the Oculus Quest 2 headset. HPTK tracks the user's real hand movements and visualizes them in the virtual environment through a black hand avatar. The black hand avatar strictly follows the actual hand movements as long as there is no interaction or collisions with other rigid objects in the environment. In the event of an interaction, the movements of the black hand avatar are limited by the surface of these rigid objects while an invisible hand avatar (i.e., a proxy avatar) continues following the actual hand movements. The difference between the visible and the invisible hand avatars helps us evaluate how much the object has been deformed and calculate the expected interaction forces depending on the mechanical properties of the object. When there is no interaction, these two hands superimpose each other. An interaction with a rigid object causes these hands to be separated and triggers the haptic stimuli to be rendered according to the mechanical properties of the object.

Figure 3 shows the experiment flow. Each trial starts with two identical-looking red boxes with 7 possible comparison pairs: one of the boxes was set with a constant (reference) stiffness value (104 N/m) and the other box with a varying

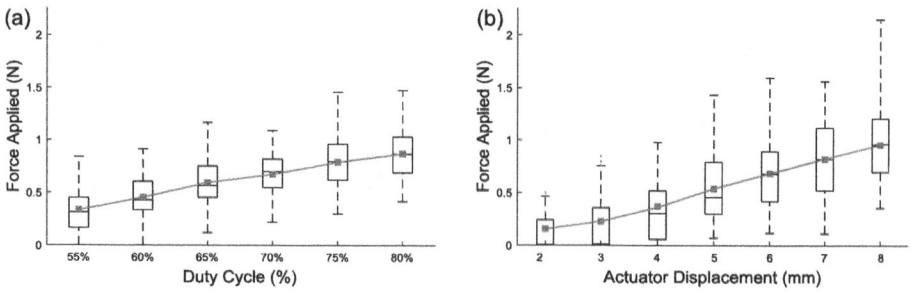

Fig. 2. Box plots of results obtained from fifteen repetitive force sensing measurements (a) CoWrHap data, (b) LAWrHap data.

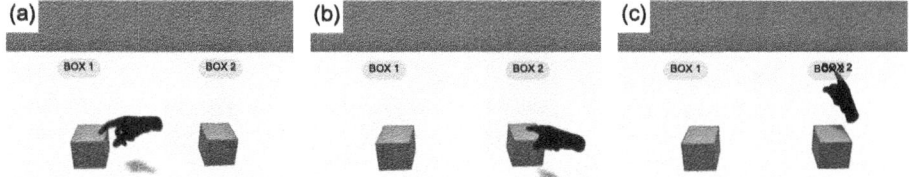

Fig. 3. Virtual task and the environment: the participant (a) explores Box 1 by pushing from the top, (b) explores Box 2 by pushing from the top, and (c) chooses the "stiffer" box which feels stiffer by clicking on the related button.

stiffness value $(50, 68, 86, 104, 122, 140, 158$ N/m$)$. The participant interacts with both objects as needed while receiving haptic feedback about their stiffness:

- **With CoWrHap:** Each stiffness level has a corresponding duty cycle to be sent to CoWrHap for haptic rendering. The event of virtual interaction triggers this communication, which renders the haptic feedback in a single-bump, discrete manner. The reference value was presented with a duty cycle of 65% while varying stiffness values were presented with duty cycles of $50, 55, 60, 65, 70, 75, 80\%$.
- **With LAWrHap:** Each stiffness value is used to compute the real-time force following Hook's Law based on the level of displacement after the first interaction with the surface of objects. These forces are then mapped to the actuator displacements to render haptic feedback in a continuous manner. The continuous integration forces (or actuator displacements) are saturated based on the actuator displacements that were calibrated based on Fig. 2 such that the maximum forces that can be perceived while exploring are the same with both devices. The maximum actuator displacement was set to 5 mm as the reference and $2, 3, 4, 5, 6, 7, 8$ mm as the varying stiffness values.

Based on the haptic feedback they received, participants determined which box was stiffer by clicking on the "Box 1" or "Box 2" button on top of the selected box. Each participant interacted with each comparison pair with 10 repetitions while wearing each bracelet. Ultimately, they performed the experiments with both devices (with 70 repetitions each) – resulting in 140 repetitions. The order of the comparison pairs and the order of the haptic bracelets were randomized.

Three breaks of 2 min each were given while wearing both devices to prevent fatigue. While changing the first haptic device with the second one, they were given a longer break of approximately 5 min. The overall experiment took approximately 30 min.

During the trials, we recorded participants' responses to the "stiffer" box, how many times they tapped on the boxes, how long it took participants to make a decision, their interaction time with the objects, and how deep they moved from the surface of the object to analyze their performance. They also filled out a post-experiment questionnaire about their overall experience regarding the VR environment and their haptic perception.

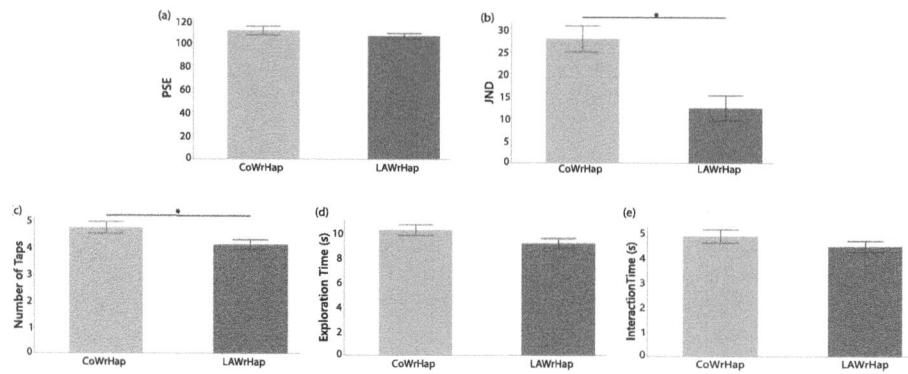

Fig. 4. Bar plots of results obtained from experiments considering device condition by calculating averages for (a) PSEs, (b) JNDs, (c) the number of exploration taps, (d) exploration time, and (e) total interaction time with boxes.

2.3   Participants

We conducted a user study with 16 participants from the local university (7 males and 9 females) with ages ranging between 19 and 29 years (22.0 ± 2.5). The University Review Board approved the experimental protocol, and all participants gave informed consent. In terms of hand dominance, 15 participants reported being right-handed, and 1 reported being left-handed. We asked about their prior experiences with VR, 5 participants stated no prior experience, 4 participants one to three times, and 7 participants more than five times.

3   Results

The data were processed on JMP and analyzed on SPSS 27. We analyzed the data in terms of discrimination performance (through a psychometric curve), discrimination efforts, exploration behavior, and user experience. Figure 4 and Fig. 5 show the mean in the graphs, and the error bars represent the standard error of the mean.

3.1   Discrimination Performance

Participants' discrimination performance was evaluated based on their accuracy in correctly choosing the stiffer object for each stiffness pair. With this accuracy information, we formed a psychometric curve that summarizes their discrimination performance. This curve was formed using the sigmoid function technique presented in a previous study [19]. We then analyzed the results using a t-test with the haptic device type as a within-subject factor, as summarized in Table 1.

Point of Subjective Equality (PSE) values indicate changes in stiffness intensity as they relate to changes in perceived sensation. Closer PSEs to the actual reference indicate better discrimination performance. Figure 4 (a) shows the average PSEs for CoWrHap (66.0 ± 3.11) and LAWrHap (65.4 ± 2.46). The results of the t-test indicated no statistically significant difference between rendering continuous and discrete haptic feedback on the wrist.

Just Noticeable Difference (JND) values indicate the smallest change in stiffness that can be detected by participants. Lower JNDs indicate better discrimination performance. Figure 4 (b) shows the average JNDs for CoWrHap (7.4 ± 2.78) and LAWrHap (3.4 ± 3.10). The results of the t-test indicated that participants performed the discrimination task statistically significantly better with LAWrHap than with CoWrHap ($p < 0.001$).

3.2 Discrimination Efforts and Exploration Behavior

We also focused on the level of effort raised by participants while receiving both haptic rendering strategies (i.e., discrete or continuous). Discrimination efforts are evaluated through *(i)* how many times participants interacted with the boxes and *(ii)* how much time it took them to complete each trial. We are also motivated to investigate whether or not receiving discrete haptic feedback would change the exploration behavior through and *(i)* how much time they spent in each trial in contact with the boxes. We analyzed the results via a two-way repeated-measures analysis of variance (RM-ANOVA) considering two within-subject factors as summarized in Table 1. The columns denote the statistical metrics for each factor: "Device" indicates the type of wrist-worn haptic device used (CoWrHap or LAWrHap), "Stiffness Pairs" refers to the comparison between reference stiffness and varying stiffness values, and "Device X Stiffness Pairs" represents the interaction effect between the device and stiffness pair.

Number of Taps is how many times participants touched virtual objects with hand avatars related to their level of confidence or confusion: a lower number indicates more confident decisions. Figure 4 (c) shows the average number of taps for all participants, all stiffness pairs, and Box 1 and Box 2 – since we observed that they explored both objects in almost equal amounts. We observed

Table 1. Statistical results from user study.

	Device	Stiffness Pairs	Device X Stiffness Pairs
Number of Taps on Boxes	$F(1,6) = 8.063$, $p = 0.012$, $\eta^2 = 0.163$	$F(1,6) = 9.384$, $p<0.001$, $\eta^2 = 0.076$	$F(1,6) = 9.445$, $p<0.001$, $\eta^2 = 0.076$
Exploration Time	$F = 3.640$, $p = .076$, $\eta^2 = 0.123$	$F(1,6) = 13.161$, $p<0.001$, $\eta^2 = 0.079$	$F(1,6) = 6.828$, $p<0.001$, $\eta^2 = 0.050$
Interaction Time	$F(1,6) = 1.308$, $p = .271$, $\eta^2 = 10.028$	$F(1,6) = 8.546$, $p = 0.001$, $\eta^2 = 44.336$	$F(1,6) = 5.628$, $p = 0.002$, $\eta^2 = 16.763$

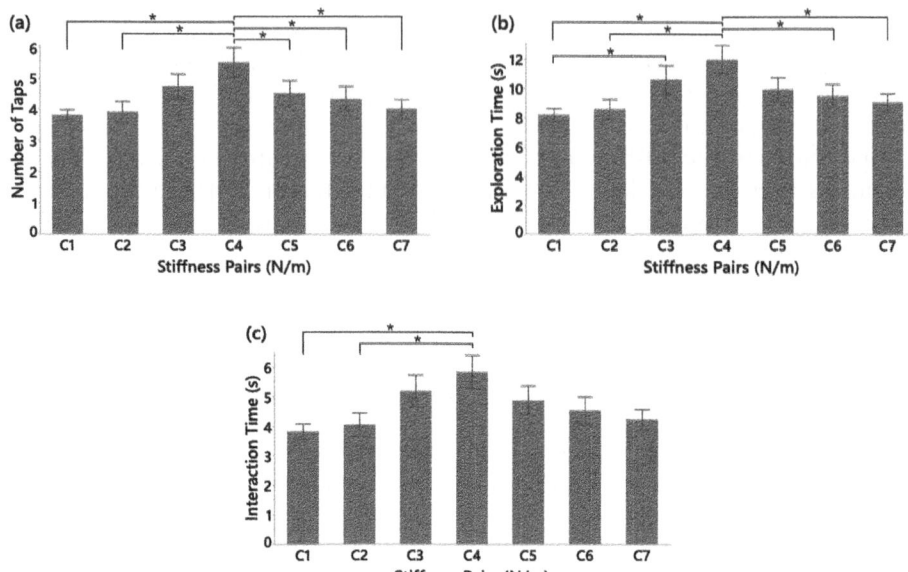

Fig. 5. Bar plots of results obtained from experiments considering stiffness pairs condition by calculating averages (a) number of taps with respect to stiffness pairs per task, (b) exploration time with respect to stiffness pairs per task, (c) interaction time with respect to stiffness pairs per task considering both devices.

that participants interacted with the boxes before making a decision statistically significantly higher while using CoWrHap compared to using LAWrHap (Table 1). In terms of stiffness pairs, the number of taps on each trial was found to be statistically significantly higher when varying stiffness is close to the reference value (Fig. 5(a)), as revealed by the follow-up post-hoc analysis.

Exploration Time refers to the total duration participants require to complete each trial including any periods of inactivity or decision-making, which is influenced by their level of confidence or confusion. A shorter exploration time typically indicates that participants are more confident in their decisions, as they spend less time deliberating over the choices presented. Figure 4 (d) shows the average exploration time for all participants and all stiffness pairs. We observed that participants completed the trials with no significant difference while receiving the haptic feedback in a discrete or continuous way – i.e., while using CoWrHap or LAWrHap – (Table 1). In terms of stiffness pairs, exploration time on each trial was found to be statistically significantly higher when varying stiffness is close to the reference value (Fig. 5(b)), as revealed by the follow-up post-hoc analysis.

Interaction Time refers to the duration of direct engagement between the hand avatar and the virtual boxes within a trial. This metric focuses on the active period of interacting with the boxes, reflecting the participants' exploration behavior. Figure 4 (e) shows the average interaction time for all participants, all stiffness pairs, and Box 1 and Box 2 – since they interacted with both objects in almost equal amounts of time. We observed that participants interacted with objects during the trials with no significant difference while receiving the haptic feedback in a discrete or continuous way – i.e., while using CoWrHap or LAWrHap. In terms of stiffness pairs, interaction time on each trial was found to be statistically significantly higher when varying stiffness is close to the reference value (Fig. 5(c)), as revealed by the follow-up post-hoc analysis.

3.3   Subjective Questionnaire

Participants completed a post-experiment questionnaire on their experience and preferences. 16 participants reported that they preferred receiving continuous force feedback through LAWrHap. We then asked them to explain the reason behind their preferences in their own words. They reported "I had to think more while using (CoWrHap) and try it multiple times to find the difference", "the increase and decrease of the haptic feedback (with LAWrHap) made it easier to make the difference", "(LAWrHap) led to a smoother decision-making process. I did not need to try multiple each time as I was able to feel the difference easier relatively to the other one (CoWrHap).", "Its effect is easier to understand the pressure", "LAWrHap feels more sensitive to me", "The impact is more noticeable on the (LAWrHap) as it exhibits a higher push-like sensation" and so on.

   We then asked participants to rate both haptic rendering modes (i.e., CoWrHap and LAWrHap, respectively) in terms of different user experience concepts on a 7-point Likert scale. Table 2 summarizes the results: participants reported the task to be easier, more pleasant, more believable, and less tiring mentally and physically while using LAWrHap compared to CoWrHap. These ratings are also in line with their explanations reported above.

Table 2. Post experiment questionnaire results.

	CoWrHap	LAWrHap
Task Ease	3.3 ± 0.7	5.9 ± 0.3
Pleasantness	3.9 ± 1.8	5.9 ± 1.0
Believability of Interactions	3.4 ± 1.2	5.6 ± 1.2
Mental Fatigue	3.3 ± 1.8	2.6 ± 1.3
Physical Fatigue	3.4 ± 1.3	3.1 ± 1.5

4   Discussions

In this paper, we follow up on our previous work showed that rendering single-bump force feedback on users' wrists could help participants perform virtual stiffness discrimination tasks successfully by comparing its effect with conventional force feedback strategies where continuous forces are rendered on users' wrists through linear actuators. We used two separate devices to render such different rendering scenarios: discrete force feedback was rendered through a custom voice-coil based wrist-worn haptic device (CoWrHap), and continuous force feedback was rendered through a linear actuated wrist-worn haptic device (LAWrHap).

The first step of such an empirical comparison was to calibrate the interaction forces. We measured the forces applied by CoWrHap during our previous stiffness discrimination study for rendering each stiffness value to the user through FSR sensors integrated between the user's skin and CoWrHap. We specifically chose the duty cycles for CoWrHap to indicate different levels of stiffness rendering by ensuring that the rendered forces result in a linear behavior. Then, the same calibration was needed to map the required interaction forces to the desired actuator displacements that would be saturated after a certain level of object deformation in the virtual environment. We showed that such a calibration was possible by presenting the results as in Fig. 2.

We then conducted a user study where 16 participants were asked to explore two identical-looking objects with different stiffness values in a virtual environment and choose the "stiffer" object based on haptic perception. We first evaluated users' discrimination performance through a psychometric curve. Our results showed that participants performed the experiment with statistically significantly higher sensitivity (lower JNDs) using continuous force feedback (i.e., LAWrHap), indicating that participants could detect relatively closer variations in stiffness when the force feedback was continuous. On the other hand, they exhibited similar accuracy (PSEs) while using discrete or continuous force feedback, meaning participants could correctly identify the stiffer object at similar rates regardless of the feedback mechanism we utilized. Reflecting on RQ; these indications suggest that *continuous force feedback may offer a more varied and clear haptic experience, helping to the perception of subtle differences in virtual object mechanical properties.*

We extended our data analyses to discrimination efforts and exploration behavior (e.g., number of taps, exploration time, interaction time). Participants interacted with the boxes significantly more with discrete force feedback (using CoWrHap) than with continuous force feedback (using LAWrHap), as expected. Interestingly, the interactions between the stiffness pairs and the haptic devices were also found to be statistically significant – thus, it is possible that interaction and exploration times show the benefit of LAWrHap for certain stiffness pairs even if not for all. Additionally, the exploration time and number of taps varied significantly across different stiffness pairs, indicating that the perceived similarity or difference in stiffness influenced the exploration strategy. Reflecting on RQ; these indications suggest that *continuous force feedback may help participants discriminate stiffness properties of virtual objects easier and faster.* On the other hand, we did not observe any statistical significance between the two devices in terms of the interaction and exploration times.

Finally, we conducted a post-experiment questionnaire to investigate the user experience through the post-experiment questionnaire with 7-point Likert scale. Participants' subjective comments showed that (interestingly) all participants preferred continuous force feedback (LAWrHap) over discrete feedback (CoWrHap). Reflecting on **RQ**; these comments suggest that *continuous force feedback may help participants enjoy the haptic sensation received during the virtual interactions better.*

Ultimately, while LAWrHap shows better potential in discrimination tasks in general, there are certain metrics in which both feedback modalities show similar performance (e.g., discrimination sensitivity, exploration time, or interaction time). We believe that CoWrHap can also be utilized to benefit from its other advantages in applications where these factors are more crucial than factors such as accuracy or number of taps, which were found to be statistically better with LAWrHap.

5   Conclusions

In this paper, we investigated the perceptual and performance differences while rendering haptic feedback on the wrist in single-bump force feedback (through voice coil actuation) or in continuous force feedback (through linear DC actuation) and conducted a user study experiment. Overall, the findings of this study revealed that rendering continuous force feedback through LAWrHap offers benefits in terms of discrimination sensitivity, discrimination confidence, and user experience. While we could not find any measure that would highlight the benefit of using discrete force feedback through CoWrHap, these two rendering strategies were found to be not different than each other in terms of discrimination accuracy, interaction time, and exploration time.

In the future, we will explore how the user behavior changes in terms of interacting with the objects during the VR explorations. Furthermore, while these results are specific to the stiffness discrimination task, other exploration scenarios might indicate different results. We will also explore other VR interaction scenarios to explore the differences between discrete and continuous force feedback rendering strategies, such as rendering forces for event-based interactions.

References

1. Adeyemi, A., Sen, U., Ercan, S.M., Sarac, M.: Hand dominance and congruence for wrist-worn haptics using custom voice-coil actuation. IEEE Robot. Autom. Lett. (RA-L) (2024)
2. Aggravi, M., Pausé, F., Giordano, P.R., Pacchierotti, C.: Design and evaluation of a wearable haptic device for skin stretch, pressure, and vibrotactile stimuli. IEEE Robot. Autom. Lett. **3**(3), 2166–2173 (2018)
3. Bortone, I., et al.: Wearable haptics and immersive virtual reality rehabilitation training in children with neuromotor impairments. IEEE Trans. Neural Syst. Rehabil. Eng. **26**(7), 1469–1478 (2018)

4. Camardella, C., Gabardi, M., Frisoli, A., Leonardis, D.: Wearable haptics in a modern VR rehabilitation system: design comparison for usability and engagement. In: Seifi, H., et al. (eds.) EuroHaptics 2022. LNCS, vol. 13235, pp. 274–282. Springer, Cham (2022). https://doi.org/10.1007/978-3-031-06249-0_31
5. Clark, J.P., Lentini, G., Barontini, F., Catalano, M.G., Bianchi, M., O'Malley, M.K.: On the role of wearable haptics for force feedback in teleimpedance control for dual-arm robotic teleoperation. In: IEEE International Conference on Robotics and Automation (ICRA), pp. 5187–5193 (2019)
6. Gaffary, Y., Le Gouis, B., Marchal, M., Argelaguet, F., Arnaldi, B., Lecuyer, A.: AR feels "Softer" than VR: haptic perception of stiffness in augmented versus virtual reality. IEEE Trans. Visual Comput. Graph. 23(11), 2372–2377 (2017)
7. Gaudeni, C., Meli, L., Jones, L.A., Prattichizzo, D.: Presenting surface features using a haptic ring: a psychophysical study on relocating vibrotactile feedback. IEEE Trans. Haptics 12(4), 428–437 (2019)
8. González, J.J.: Hand physics toolkit (2020). https://github.com/jorgejgnz/HPTK
9. Johansson, R., Vallbo, A.: Tactile sensibility in the human hand: relative and absolute density of four types of mechanoreceptive units in glabrous skin. J. Physiol. 286, 283–300 (1979)
10. Maereg, A.T., Nagar, A., Reid, D., Secco, E.L.: Wearable vibrotactile haptic device for stiffness discrimination during virtual interactions. Front. Robot. AI 4, 42 (2017)
11. Maisto, M., Pacchierotti, C., Chinello, F., Salvietti, G., De Luca, A., Prattichizzo, D.: Evaluation of wearable haptic systems for the fingers in augmented reality applications. IEEE Trans. Haptics 10(4), 511–522 (2017)
12. Miyatake, Y., Hiraki, T., Iwai, D., Sato, K.: Haptomapping: visuo-haptic augmented reality by embedding user-imperceptible tactile display control signals in a projected image. IEEE Trans. Visual Comput. Graph. 29(4), 2005–2019 (2023)
13. Moriyama, T., Kajimoto, H.: Wearable haptic device presenting sensations of fingertips to the forearm. IEEE Trans. Haptics 15(1), 91–96 (2022)
14. Palmer, J.E., Sarac, M., Garza, A.A., Okamura, A.M.: Haptic feedback relocation from the fingertips to the wrist for two-finger manipulation in virtual reality. In: IEEE/RSJ International Conference on Intelligent Robots and Systems (IROS), pp. 628–633 (2022)
15. Pezent, E., Agarwal, P., Hartcher-O'Brien, J., Colonnese, N., O'Malley, M.K.: Design, control, and psychophysics of tasbi: a force-controlled multimodal haptic bracelet. IEEE Trans. Rob. 38(5), 2962–2978 (2022)
16. Sallnäs, E.L., Rassmus-Gröhn, K., Sjöström, C.: Supporting presence in collaborative environments by haptic force feedback. ACM Trans. Comput. Hum. Int. 7(4), 461–476 (2000)
17. Sarac, M., Di Luca, M., Okamura, A.M.: Perception of mechanical properties via wrist haptics: effects of feedback congruence. In: IEEE/RSJ International Conference on Intelligent Robots and Systems (IROS), pp. 620–627 (2022)
18. Sarac, M., Hallett, K., Saunders, J., Makled, B., Okamura, A.M.: Augmented needle decompression task with a wrist-worn haptic device. In: IEEE World Haptics Conference (WHC), pp. 873–873 (2021)
19. Sarac, M., Huh, T.M., Choi, H., Cutkosky, M.R., Luca, M.D., Okamura, A.M.: Perceived intensities of normal and shear skin stimuli using a wearable haptic bracelet. IEEE Robot. Autom. Lett. (RA-L) 7(3), 6099–6106 (2022)
20. Sen, U., Sarac, M.: Design for wrist-worn haptic device with custom voice coil actuation. In: IEEE World Haptics Conference (Work-in-Progress) (2023)

21. Srinivasan, M.A., LaMotte, R.H.: Tactual discrimination of softness. J. Neurophysiol. **73**(1), 88–101 (1995)
22. Swapp, D., Pawar, V., Loscos, C.: Interaction with co-located haptic feedback in virtual reality. Virtual Reality **10**(1), 24–30 (2006)
23. Tanacar, N.T., Hudhud, M., Batmaz, A.U., Leonardis, D., Sarac, M.: The impact of haptic feedback during sudden, rapid virtual interactions. In: IEEE World Haptics Conference, pp. 1 – 7 (2023)
24. Wee, C., Yap, K.M., Lim, W.N.: Haptic interfaces for virtual reality: challenges and research directions. IEEE Access **9**, 112145–112162 (2021). https://doi.org/10.1109/ACCESS.2021.3103598
25. Young, E.M., Memar, A.H., Agarwal, P., Colonnese, N.: Bellowband: a pneumatic wristband for delivering local pressure and vibration. In: IEEE World Haptics Conference, pp. 55–60 (2019)

Viscous Damping Displayed by Surface Haptics Improves Touchscreen Interactions

Zhaochong Cai[✉][iD] and Michaël Wiertlewski[iD]

TU Delft, 2628 Delft, CD, The Netherlands
{z.cai-1,m.wiertlewski}@tudelft.nl

Abstract. Virtual targets on touchscreens (e.g., icons, slide bars, etc.) are notoriously challenging to reach without vision. The performance of the interaction can fortunately be improved by surface haptics, using friction modulation. However, most methods use position-dependent rendering, which forces users to be aware of the target choice. Instead, we propose using tactile feedback dependent on users' speed, providing a viscous feeling. In this study, we compared three viscous damping conditions: *positive damping*, *negative damping*, and *variable damping* (viscosity was high during slow movements and low during fast movements), against a baseline condition with no tactile feedback. These viscous fields are created by changing net lateral forces based on velocity. Results indicate that, during the initial phase of movement when the finger approaches the target, various viscous feedback has an insignificant impact on targeting trajectories and movement velocity. However, positive damping and variable damping significantly influence behavior during the selection phase by reducing oscillation around the target and completion time. Questionnaire responses suggest user preference for viscous conditions and disapproval of negative viscous forces. This study provides insights into the role of viscous resistance in touchscreen interactions.

Keywords: surface haptics · viscous forces · pointing tasks

1 Introduction

Because of the lack of tactile cues, users interacting with touchscreens and touchpads have to interact using only visual cues. The need for visual attention can be dangerous in situations such as driving or walking. While some manufacturers are reverting back to physical buttons, they are also losing the flexibility that touchscreens provide. Programmable tactile feedback, where feedback is provided to the user's bare finger, circumvents all these limits and reduces the need for visual attention. Moreover, the feedback enhances the performance of the interaction and improves the overall user experience. The standard approach to implement programmable tactile feedback, commonly found in consumer electronics, uses vibrotactile feedback to inform users with vibrations [13]. While

© The Author(s), under exclusive license to Springer Nature Switzerland AG 2025
H. Kajimoto et al. (Eds.): EuroHaptics 2024, LNCS 14768, pp. 352–364, 2025.
https://doi.org/10.1007/978-3-031-70058-3_29

effective at signaling the user, vibrotactile feedback only provides transient or periodic stimulation. In contrast, friction modulation offers finer and continuous stimuli providing a natural physical rendering of a target. It has been shown that a simple binary friction profile reduces pointing task completion time by providing more intuitive guidance to users [4, 16, 23].

However, all existing methods that employ position-based feedback require knowledge of the target location and, consequently, must predict the user's intention in selecting their target. The position-based approach can be effective when the interface has only a few targets but may be impractical when localized targets do not exist. To facilitate interaction across complex interfaces, we need to implement a target-independent rendering strategy, for example, velocity-dependent forces that feel similar to viscous elements to guide the user on the surface.

Pointing tasks, where the finger reaches a target on the screen, are fundamental in human-computer interaction. Fitts demonstrated that the time taken to reach a target during these pointing tasks depends on its distance and width [8, 9]. The kinematics are governed by the principle of minimum variance control [12, 20, 21], suggesting that users minimize target variance by slowing down when approaching a target. The velocity profile forms a bell shape, dividing into two phases: an approaching phase, resembling a ballistic movement with minimal sensory feedback [5, 17], and a subsequent slower adjustment phase to pinpoint the exact location using continuous sensory feedback.

Therefore, we postulate that task completion time can be reduced by modulating feedback along these phases. This translates to accelerating the ballistic movement for a quicker approach and slowing down the adjustment phase for

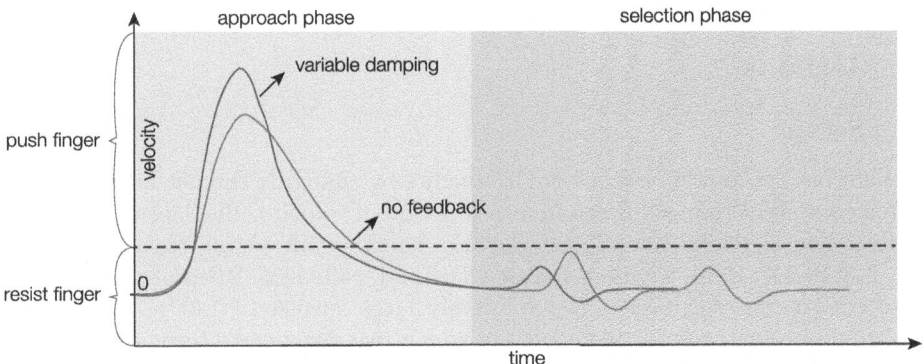

Fig. 1. Typical velocity profiles observed during a pointing task. The cyan curve represents a typical velocity profile without tactile feedback, where the finger moves back and forth when selecting a target. The purple curve illustrates the proposed approach that varies damping. We hypothesize that, during the approach to the target, low viscosity (or even negative damping) can accelerate finger movement, while during the target selection, high viscous viscosity helps locate the target with fewer back-and-forth movements.

finer control, as illustrated in Fig. 1. We implemented this strategy by changing the damping coefficients as a function of the user's velocity. For example, we can display negative damping at high speeds to increase speed and positive damping at low speeds to dampen the approach. We expect that this feedback provides better performance with a shorter time to completion compared to scenarios without tactile feedback.

Viscous-based assistance has been implemented in the past using force feedback. In contrast to surface haptics, users interact with force feedback devices through a handle rather than their bare fingers. Despite these differences, the findings offer insights into expected behavior. Notably, it has been observed that adding constant static friction leads to a decrease in reaching times and improves accuracy when moving low-mass objects [1, 6, 19]. This improvement is attributed to friction forces decelerating and filtering out jittery movements. Keemink et al. demonstrated that both constant damping and position-dependent damping reduce movement time and increase endpoint accuracy [15]. These findings suggest that viscous damping in force feedback operations is beneficial to human operators. The impact of such forces on bare fingertip interactions remains unexplored, in part due to the lack of a surface haptic device capable of providing the desired lateral force.

In this paper, we investigated the effects of variable viscous damping on users' targeting strategies using a novel surface haptic device called Ultraloop [3]. This device generates lateral forces based on the user's velocity. We compared users' performance when reaching targets when presented with a positive damping, a velocity-dependent damping, a negative damping, and a control condition without any damping. We found that viscous conditions do not significantly affect the movement trajectories during the approach phase but notably decrease the back-and-forth movements during the selection phase.

2   Methods

2.1   Setup

In this work, we render viscous environments by changing the net lateral forces, with a consistent reduced friction, as a function of velocity. The lateral forces are produced by active surface haptic devices that use ultrasonic traveling waves, e.g. [2, 3, 10, 11]. Here, we use a haptic touchpad, called the Ultraloop, which can deliver active lateral forces on a relatively large surface of $140 \times 30 \ mm^2$. It has an aluminum ring-shaped cavity in which two degenerate resonant standing wave modes are excited at approximately 40 kHz with a 90° phase shift. These standing waves superimpose into either a counter-clockwise (when the phase is 90°) or a clockwise (when the phase is -90°) traveling wave that propagates around the ring. The traveling wave interacts with the skin and produces a net lateral force that can push or pull fingertips. The direction and magnitude of the force can be modulated by varying the amplitude and phase shift of these standing waves. To create lateral forces as a function of velocity, we used a Teensy 3.6 microcontroller to program the phase of two driving voltages in response to

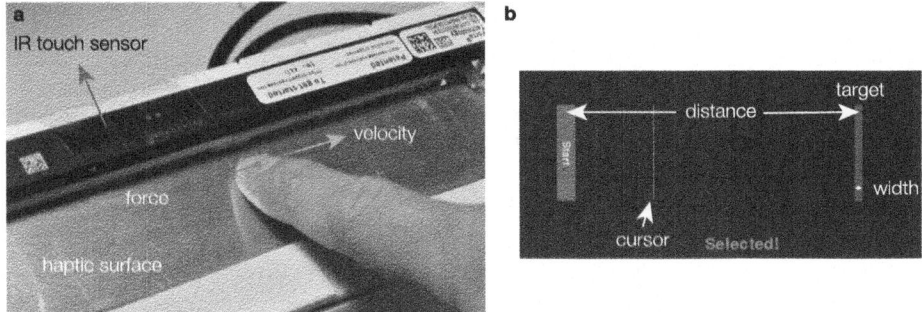

Fig. 2. a, Experimental setup: Participants slide their index finger on the touch surface of the Ultraloop while experiencing lateral forces generated as a function of the measured velocity. **b**, Graphical user interface displaying visuals for a one-dimensional reciprocal targeting task.

finger velocity, derived from the first-order backward difference of the position tracked by an infrared sensor (Neonode, NNAMC1580PCEV) (Fig. 2a).

2.2   Experimental Conditions

In the experiments, participants were asked to reach for a target while being assisted by three different viscous environments or not assisted at all (baseline condition). In the baseline condition, the surface had uniform low friction with no externally applied lateral forces. In the viscous conditions, net lateral forces generated by the Ultraloop were a function of finger movement speed, formulated as $F = -bv$, while the strength of friction reduction remains the same as the baseline condition. We designed three experimental conditions:

1. *Positive damping*: Here, b is a constant positive value, creating a viscous resistance similar to what can be experienced in daily life.
2. *Negative damping*: In this condition, b is a negative value, and the faster the users go the stronger the lateral forces push.
3. *Variable damping*: b varies linearly with velocity, turning negative when the finger moves faster than 0.08 m/s.

Due to the limitations of the Ultraloop, the net forces plateau at approximately 300 mN. Therefore, the damping force cannot increase beyond a certain finger speed. Figure 3 illustrates the proposed damping coefficient and phase shift profiles as a function of finger speed for each condition. It is important to note that the amplitude of ultrasonic vibration remains constant across all feedback conditions to minimize variations in the strength of friction reduction. The phase of the driving signals is the only parameter tuned, based on velocity.

2.3   Protocol and Design

The graphical user interface used for the experiment is shown in Fig. 2b. Participants conducted one-dimensional reciprocal targeting tasks. Twelve successive

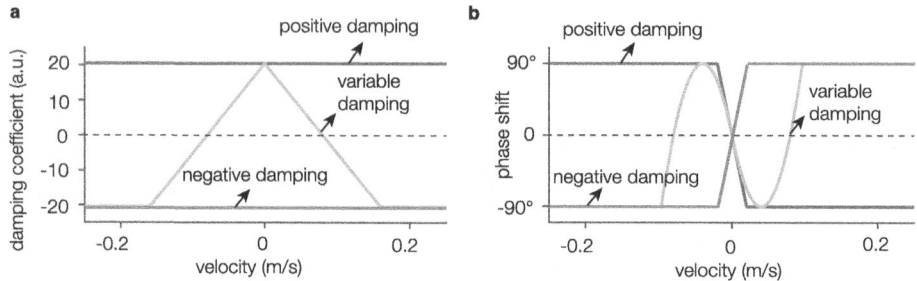

Fig. 3. a, Damping coefficients as a function of finger velocity. **b,** Phase shift between the two channels of driving signals. The maximum lateral forces are generated with a phase shift of ±90°.

tasks with the same target width and viscous condition were grouped together as a block. At the start of each block, they placed their index finger on the active surface of the Ultraloop and slid the cursor to the start area. After holding the cursor in the start area for 0.2 s, the first trial of this block started, and the participant slid the cursor to the target and pressed the "ctrl" key with their non-dominant hand to confirm the acquisition. Next, a new target appeared on the other side of the user interface. Participants were instructed to complete the tasks both quickly and accurately, aiming for a success rate of approximately 96 %. If a participant missed more than one target in a block, a message on the user interface would prompt them to slow down for greater accuracy. Conversely, if they completed one block without any misses, they were encouraged to increase their speed.

We used a repeated within-subject design, with independent variables as viscous environments and target widths. These widths were set at 8, 16, 24, and 32 pixels, equivalent to 0.8, 1.6, 2.4, and 3.2 mm on the touch surface. The target distance is fixed at 7.5 mm. Before experiments, participants spent ten minutes familiarizing themselves with the Ultraloop and the interface.

The experiment consisted of 32 blocks, with each block containing 12 trials with the same target and feedback condition. These 32 blocks were divided into four sessions, each dedicated to one of the viscous conditions. We applied a Latin Square design to counterbalance the presentation order of viscous conditions among participants (Fig. 4). Each session had eight successive blocks, with target widths presented in a descending order, and grouped by the same width. Participants were allowed a one-minute break after each block to rest their hands and fingers. In summary, each participant completed 384 trials, calculated as 4 sessions × 4 widths × 2 repeats × 12 targets.

2.4 Participants

Nine individuals from TU Delft participated in the experiments (seven males, and two females; aged 22–32, average age 25.4). All participants were right-handed, had no tactile impairments, their fingers were free of cuts and calluses,

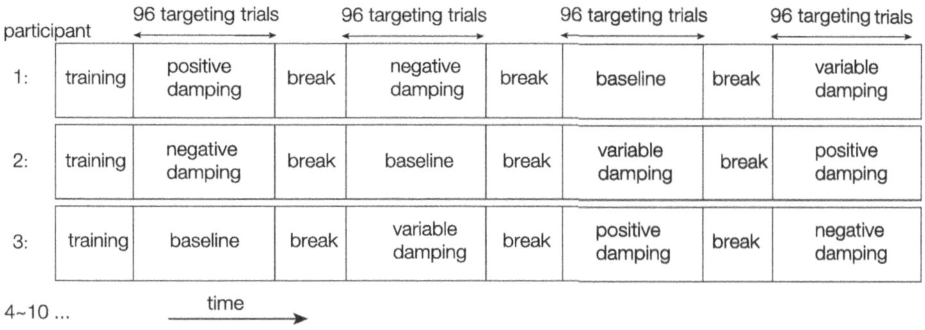

participant	96 targeting trials			96 targeting trials			96 targeting trials			96 targeting trials
1:	training	positive damping	break	negative damping	break	baseline	break	variable damping
2:	training	negative damping	break	baseline	break	variable damping	break	positive damping
3:	training	baseline	break	variable damping	break	positive damping	break	negative damping

4~10 ...   time →

Fig. 4. Experimental procedure overview. Targeting tasks are organized into sessions based on viscous conditions. The sequence in which these conditions are presented to participants is determined by a Latin Square design.

and were unaware of the aim of the study. Every participant provided informed consent before the experiments. The study received ethical approval from the ethics committee of Delft University of Technology, complying with the Declaration of Helsinki.

2.5 Data Processing

The area where participants began the task had a specific width, so we mitigated variations in trajectory timings by aligning the traces to a common reference time point. The velocity profile as a function of position was obtained by interpolating the time-domain position data from each trial. Additionally, we excluded data from the first session of one participant who could not perform movements as fast and accurately as possible, resulting in considerably slower movement compared to the rest of the cohort. In a set of successive trials organized into two blocks of 24 trials, which have the same viscous conditions and target widths, we excluded the first four trials from the first block and the first two trials from the second block to allow for adaptation. The exclusion removed the trial where the learning effect was present, in turn providing focus to the data where the performance was stable.

3 Results

Movement time for each trial is defined as the time between the onset of movement to the selection of a target. A repeated measures ANOVA analysis revealed a significant impact of viscous conditions on the average movement time ($F_{3,21}$ = 13.392, p < 0.001). Notably, both *variable damping* and *viscous damping* conditions exhibited shorter average movement time (mean = 1.41 s and 1.46 s) compared to the baseline condition (mean = 1.54 s). In contrast, *negative damping* increased the completion time of the pointing task (mean = 1.74 s).

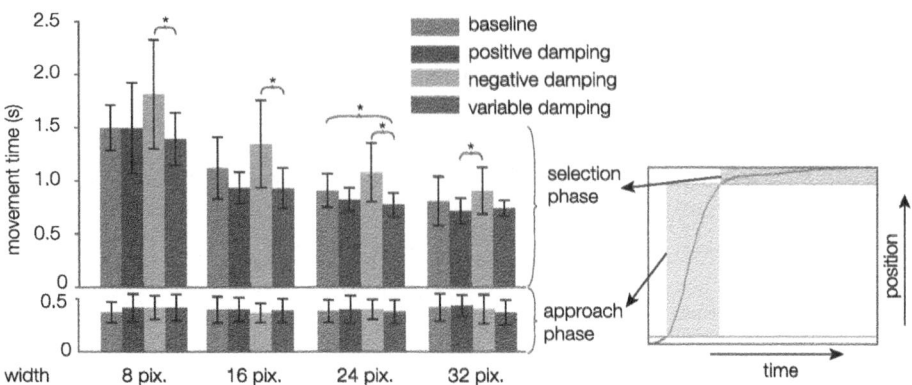

Fig. 5. Mean movement time during the selection phase and approach phase across all target width conditions. Bar charts for these two phases use the same scaling in the y direction. Stars "*" indicate the significance of $p \leq 0.05$. The inset illustrates a typical movement trajectory, with the two phases indicated by the shaded areas.

To explore the potential source for the variance in movement time across viscous conditions, we divided the movement of a trial into two phases: the approach phase and the selection phase. The approach phase spans from the moment the finger exits the start area to when 90% of the target distance is covered. The selection phase comprises the remaining time until task completion. We chose the divide point at 90 % following preliminary observations indicating that the phase before this point exhibits a rapid, monotonous movement towards the target, often described by a bell-shaped velocity profile [17]. Beyond this point, user movements become non-monotonous and involve corrective motions, indicating a shift from rapid approach to precise target alignment.

These phases were represented in the inset of Fig. 5 and statistically analyzed separately to quantify their distinct contributions to task performance. Notably, significant differences in movement times were predominantly observed during the selection phase. Figures. 6 a and b showed that significant differences in movement time were not observed during the approach phase ($F_{3,21} = 1.72$, p $= 0.19$), but during the selection phase ($F_{3,21} = 12.826$, p < 0.001). Specifically, *negative damping* recorded the longest selection time (mean $= 1.29$ s), while the *variable damping* recorded the shortest (mean $= 0.96$ s). Further analysis of movement times in different target widths indicated that the primary difference occurred in the selection phase, with minor variations in the ranking of viscous conditions (Fig. 6c).

We further analyzed the averaged movement profiles under the same viscous condition and target width, as depicted in Fig. 7. Across all width conditions, position and velocity profiles showed small differences between viscous conditions, considering notable standard deviations. Additionally, the averaged peak velocities are similar across different viscous conditions. It further suggests that varying viscous resistance does not effectively speed up or slow down user move-

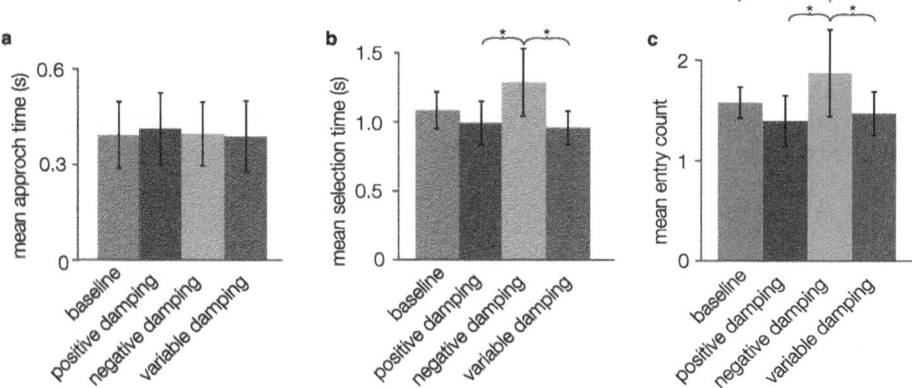

Fig. 6. a and **b**, the mean duration during the approach phase and selection phase. **c**, Mean entry count across different viscous conditions. Standard deviations across participants are indicated by error bars. Stars "*" indicate the significance of $p \leq 0.05$.

ment during the approach. Furthermore, averaged velocity versus position profiles during the approach phase also exhibit small deviations from each other for widths of 8, 16, and 24 pixels, yet a relatively larger deviation was observed for a width of 32 pixels. In contrast, the selection phase was notably affected by the viscous conditions. We observed large variations in the number of oscillations around the target and selection duration. The condition *negative damping* significantly increased the average entry count (mean $= 1.875$), which is the number of times the finger moves into the target area, while both *positive damping* and *variable damping* conditions effectively reduced the number of oscillations around the target (mean $= 1.397$ and 1.47), compared to the baseline condition (mean $= 1.578$). Interestingly, despite the opposite viscosity in the *positive damping* and *variable damping* conditions during the approach phase, participants obtained similar entry counts, as indicated by the pairwise post-hoc analyses. It suggests that the selection behavior is primarily influenced by the viscous conditions in the low-speed regime.

After each viscous condition, participants were asked to respond to three questions, with scores ranging from 1 (strongly disagree) to 10 (strongly agree). The questions were as follows: Q1:"I performed well", Q2: "I enjoyed the tactile feedback when interacting with the touchpad", and Q3: "It is easy to hit the target". Responses were collected from eight participants. One participant did not comply with the requirement to complete tasks both quickly and accurately at the first experimental session and was excluded to ensure the reliability of the data. The average scores for all three questions followed the same order across conditions: *positive damping > variable damping > baseline > negative damping*, as displayed in Fig. 8. After completing the experiment, participants were asked to select their most and least preferred conditions. Four out of eight preferred *positive damping*, three *baseline*, and one *variable damping*. In contrast, seven

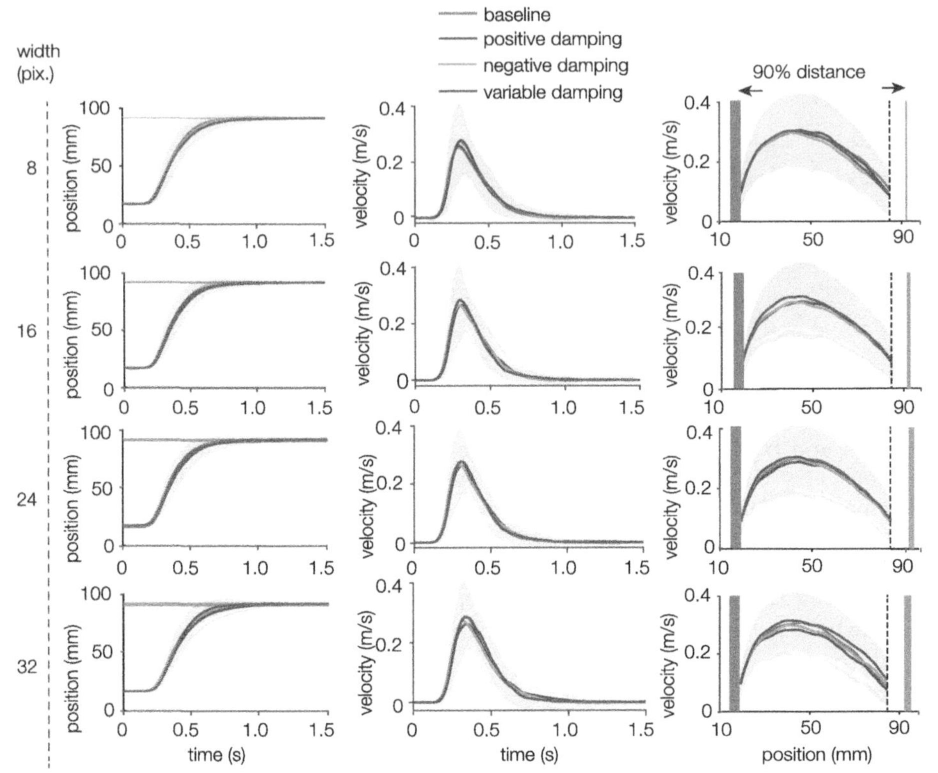

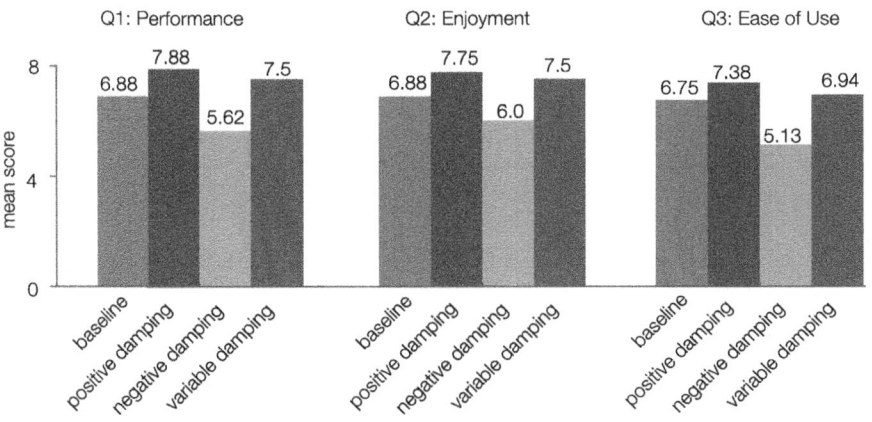

Fig. 7. Averaged movement profiles. Left and middle panels: Averaged movement and velocity profiles of nine participants. Light red bars indicate the targets. Right panels: Averaged velocity as a function of position during the approach phase. Standard deviations are indicated by the shaded areas. (Color figure online)

Fig. 8. Mean questionnaire responses, with 10 = strongly agree and 1 = strongly disagree.

out of eight participants chose *negative damping* as the one to not use, with only one choosing *positive damping*.

4   Discussion and Conclusion

We demonstrated a new method to guide users on haptic touchpads and touchscreens. The guidance is created using velocity-dependent forces on the finger, which produce a low damping effect when the finger is moving fast and a high damping effect when it is moving slowly. The user studies indicate that the viscous forces applied to fingertips affect the performance when reaching for a target. The gains in performance are mostly in the later phase of the movement when the user selects the target instead of in the phase when approaching the target. Notably, even when comparing two opposite viscous conditions, i.e., *negative damping* and *variable damping*, their velocity profiles follow similar bell-shaped trajectories, with comparable peak velocities. This observation seems inconsistent with studies using force feedback devices, which reported significant changes in approach trajectories [15]. Two alternative explanations for this inconsistent behavior during the approach phase can be raised.

First, the inconsistency may be attributed to the small variations in the magnitude of the applied force. Hand-operated force feedback devices typically exert forces in an order of 10 N, which are sufficient to impact the limb dynamics. In comparison, the Ultraloop produces much smaller net lateral forces, approximately 0.2 N. The interaction forces at the fingertip —the combination of net lateral force and sliding friction— may differ by a maximum of 0.4 N across different feedback conditions. This variation is notable between *negative damping* and *positive damping*, which produce net tangential forces of opposite signs, with sliding friction consistently opposing movement. These small variations in interaction forces do not significantly accelerate or decelerate finger movement. For instance, in *negative damping*, the forward forces are neutralized by friction forces, possibly leading to resistive interaction forces [10]. The movement during the approach phase likely follows a feedforward behavior, unimpeded by the level of resistance generated by the device.

Second, the short interaction distance in our study was only 7.5 cm, ensuring that participants consistently had a clear visual target throughout the movement. The distance is notably shorter than similar experiments using force feedback devices, such as the 23 cm mentioned in [15]. The salience of the visual cue likely led to a dominance of visual stimuli over haptic stimuli. It is well accepted under the multisensory integration framework that when visual information has a minimal variance, it becomes the primary component of the perceived stimuli. We hypothesize that in this task, participants primarily relied on visual cues, which provided consistent positional feedback. By contrast, the velocity-dependent haptic feedback, which in principle does not infer the target location, played a lesser role. This visual dominance likely explains why variations in viscous damping had minimal impact on the trajectories during the approach phase. This hypothesis is in line with a study by Levesque et al. [16], where the authors report no significant difference in movement speed when using constant low or constant high friction conditions.

Conversely, in the selection phase, where the finger is approaching the target, positive damping gives an advantage to the user for positioning at the right location. We measure the advantage by the reduced oscillations around the target and shorter selection times. The observations align with findings from force feedback device studies [14], where the authors attributed this benefit to haptic damping forces mitigating motor noise during positioning. The positive damping, which creates an energy-dissipative environment, helps dampen unintentional small movements of the user's finger. Our experiments with negative damping show that this effect reverses when the environment is generating energy, creating more oscillations and longer selection times. In addition, participants also described it as the "most challenging," with "unpredictable" movement.

The results regarding active surface haptics can be compared to previous studies that use passive surface haptics with friction modulation. With friction modulation, the target is represented with a low or high friction part, and everything outside is high or low friction, respectively. In both conditions, the friction pattern provides a distinct sensation upon touching the virtual target, and the additional tactile feedback can effectively reduce the need for visual attention. However, the discontinuity in resistance may not be preferred by users [16], especially if it conflicts with other feedback channels or tasks. In contrast, creating viscous damping environments using ultrasonic traveling waves induced active force does not involve a discontinuity in friction or lateral force, which assists in targeting continuously. We believe it improves the movement by attenuating motor noise during the precision epoch of the movement. This feedback scheme smoothly updates the lateral forces, and as a consequence, feels continuous, free from irregularities, and does not interfere with the visual channel. Therefore, it can be an effective complement to the screen in visual-dominant tasks or shared control tasks.

In conclusion, our investigation focused on the effects of viscous forces using active lateral force feedback in touch interactions. Results reveal that viscous forces do not significantly change targeting strategies during the approach to the target but help in positioning toward the target. However, it should be noted that the insights were conducted with only nine participants, which may potentially affect the generalizability of our findings and an improvement could involve a larger participant pool. Moreover, future work could exploit the potential benefits of viscous damping in tasks where moving targets are tracked. With the right design, viscous damping environments may enhance these dynamic tasks that involve frequent acceleration and deceleration [7,18]. Another avenue is to explore how humans adapt to viscous environments created through pure friction modulation. This setting may yield different observations, as humans can perceive friction change before sliding occurs [22].

Acknowledgments. The authors thank participants for contributing to the user study. The work of Zhaochong Cai was supported by China Scholarship Council under Grant 202006320048.

Disclosure of Interests. The authors have no competing interests to declare that are relevant to the content of this article.

References

1. Berkelman, P., Ma, J.: Effects of friction parameters on completion times for sustained planar positioning tasks with a haptic interface. In: 2006 IEEE/RSJ International Conference on Intelligent Robots and Systems, pp. 1115–1120 (2006)
2. Biet, M., Giraud, F., Martinot, F., Semail, B.: A piezoelectric tactile display using travelling lamb wave. In: Proceedings of Eurohaptics, pp. 567–570 (2006)
3. Cai, Z., Wiertlewski, M.: Ultraloop: active lateral force feedback using resonant traveling waves. IEEE Trans. Haptics **16**(4), 652–657 (2023)
4. Casiez, G., Roussel, N., Vanbelleghem, R., Giraud, F.: Surfpad: riding towards targets on a squeeze film effect. In: Proceedings of the SIGCHI Conference on Human Factors in Computing Systems, pp. 2491–2500. CHI 2011, New York, NY, USA (2011)
5. Chen, Y., Hoffmann, E.R., Goonetilleke, R.S.: Structure of hand/mouse movements. IEEE Trans. Human-Mach. Syst. **45**(6), 790–798 (2015)
6. Crommentuijn, K., Hermes, D.J.: The effect of coulomb friction in a haptic interface on positioning performance. In: Kappers, A.M.L., van Erp, J.B.F., Bergmann Tiest, W.M., van der Helm, F.C.T. (eds.) EuroHaptics 2010. LNCS, vol. 6192, pp. 398–405. Springer, Heidelberg (2010). https://doi.org/10.1007/978-3-642-14075-4_59
7. De Winter, J., Dodou, D., De Groot, S., Abbink, D., Wieringa, P.: Hands-on experience of manual control in a human-machine systems engineering course. In: Proceedings of the 37th Annual Conference of the European Society for Engineering Education SEFI, (MCM) (2009)
8. Fitts, P.M.: The information capacity of the human motor system in controlling the amplitude of movement. J. Exp. Psychol. **47**(6), 381–391 (1954)
9. Fitts, P.M., Peterson, J.R.: Information capacity of discrete motor responses. J. Exp. Psychol. **67**(2), 103–112 (1964)
10. Ghenna, S., Vezzoli, E., Giraud-Audine, C., Giraud, F., Amberg, M., Lemaire-Semail, B.: Enhancing variable friction tactile display using an ultrasonic travelling wave. IEEE Trans. Haptics **10**(2), 296–301 (2017)
11. Gueorguiev, D., Kaci, A., Amberg, M., Giraud, F., Lemaire-Semail, B.: Travelling ultrasonic wave enhances keyclick sensation. In: Haptics: Science, Technology, and Applications, pp. 302–312 (2018)
12. Harris, C.M., Wolpert, D.M.: Signal-dependent noise determines motor planning. Nature **394**(6695), 780–784 (1998)
13. Hoggan, E., Brewster, S.A., Johnston, J.: Investigating the effectiveness of tactile feedback for mobile touchscreens. In: Proceedings of the SIGCHI Conference on Human Factors in Computing Systems, pp. 1573–1582. CHI 2008, New York, NY, USA (2008)
14. , Keemink, A.Q., Beckers, N., van der Kooij, H.: Resistance is not futile: haptic damping forces mitigate effects of motor noise during reaching. In: 2018 7th IEEE International Conference on Biomedical Robotics and Biomechatronics (Biorob), pp. 357–363 (2018)
15. Keemink, A.Q., et al.: Using position dependent damping forces around reaching targets for transporting heavy objects: a fitts' law approach. In: 2016 6th IEEE International Conference on Biomedical Robotics and Biomechatronics (BioRob), pp. 1323–1329 (2016)

16. Levesque, V., et al.: Enhancing physicality in touch interaction with programmable friction. In: Proceedings of the SIGCHI Conference on Human Factors in Computing Systems. CHI 2011, pp. 2481–2490 (2011)
17. Lin, R.F., Tsai, Y.C.: The use of ballistic movement as an additional method to assess performance of computer mice. Int. J. Ind. Ergon. **45**, 71–81 (2015)
18. McRuer, D., Jex, H.: A review of quasi-linear pilot models. IEEE Trans. Hum. Fact. Electr. **3**, 231–249 (1967)
19. Richard, C., Cutkosky, M.: The effects of real and computer generated friction on human performance in a targeting task. In: Proceedings of the ASME Dynamic Systems and Control Division, pp. 1101–1108 (2021)
20. Todorov, E.: Optimality principles in sensorimotor control. Nat. Neurosci. **7**(9), 907–915 (2004)
21. Todorov, E.: Stochastic optimal control and estimation methods adapted to the noise characteristics of the sensorimotor system. Neural Comput. **17**(5), 1084–1108 (2005)
22. Willemet, L., Kanzari, K., Monnoyer, J., Birznieks, I., Wiertlewski, M.: Initial contact shapes the perception of friction. Proc. Natl. Acad. Sci. U.S.A. **118**(49), e2109109118 (2021)
23. Zhang, Y., Harrison, C.: Quantifying the targeting performance benefit of electrostatic haptic feedback on touchscreens. In: Proceedings of the 2015 International Conference on Interactive Tabletops & Surfaces, pp. 43–46. ITS 2015, New York, NY, USA (2015)

Latency Compensation in Ultrasound Tactile Presentation by Linear Prediction of Hand Posture

Atsushi Matsubayashi$^{(\boxtimes)}$ ⓘ, Yasutoshi Makino ⓘ, and Hiroyuki Shinoda ⓘ

The University of Tokyo, 7-3-1 Hongo, Bunkyo, Tokyo 113-8654, Japan
matsubayashi@hapis.k.u-tokyo.ac.jp

Abstract. Interaction systems using ultrasound haptics technology present tactile stimuli in fixed coordinates in space; hence, system delays cause not only temporal differences in the stimuli but also spatial shifts in presentation points when targets are moving. In particular, if the transducer array is placed surrounding the hand workspace, a shift in ultrasound focus could result in providing strong tactile stimuli to unintended parts of the hand, such as the opposite side of the fingers. In this study, we examine the feasibility of mitigating the delay effect by predicting the surface shape of the hand. The verification system fits the hand surface shape acquired by a depth camera with a hand model represented by low-dimensional posture parameters, and then performs Kalman prediction on the parameter transitions. The results of the user study show that for finger contacts under constant velocity motion conditions, the prediction method can mitigate the decrease in perceived intensity due to ultrasonic focus shift and increase in perceived intensity at undesirable areas.

Keywords: Ultrasound haptics · Human–computer interaction

1 Introduction

Ultrasound haptics has recently emerged as a technology capable of high-resolution tactile presentation in mid-air [4,6]. The absence of contact sensation renders this technology suitable for being applied to interactive systems, and various studies have proposed systems that combine it with auto-stereoscopic displays [9,10] or head-mounted displays [7,12,15]. However, this advantage also exacerbates the impact of delay, which is a common problem in many interactive systems. Many systems using ultrasound tactile presentation sense the position of the user's hand, and then converge ultrasound waves emitted from a transducer array onto a target position set on its surface. Therefore, in contrast to wearable devices, the tactile presentation points are displaced owing to the hand movements between measurement and tactile presentation. For instance, if there is a delay of 50 ms (equivalent to the latency in a typical commercial device [2]) between the sensing of the hand and arrival of the ultrasound when

H. Kajimoto et al. (Eds.): EuroHaptics 2024, LNCS 14768, pp. 365–377, 2025.
https://doi.org/10.1007/978-3-031-70058-3_30

the hand movement speed is 100 mm/s, then the ultrasound focus position will
be shifted 5 mm from the target point on the skin. Considering that the diameter of the ultrasound focus is approximately 5 mm to 10 mm, the target point
would be displaced from the high-sound-pressure area, and the resulting tactile sensation would be considerably smaller than expected. Latency causes even
more problems when the transducer array is also positioned in the direction in
which the ultrasound focus is shifted. Several systems have been proposed that
enable tactile interaction from various directions by arranging the array around
the workspace [8,9]. In such cases, shifting the focus can lead to not only a
decrease in pressure at the target location but also an increase in pressure on
the opposite side as shown in Fig. 1.

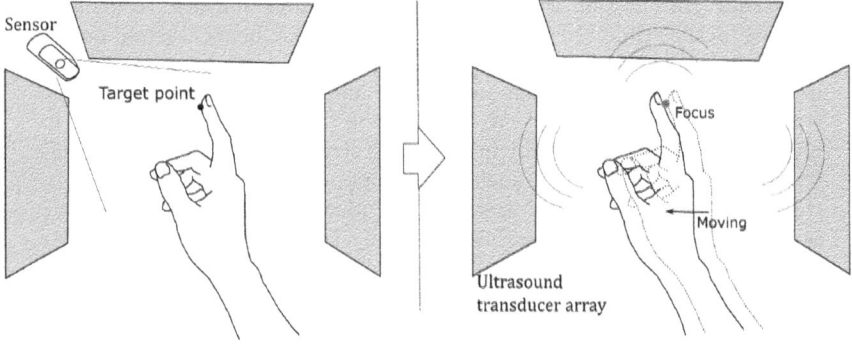

Fig. 1. Latency between hand sensing and ultrasound arrival causes the displacement
of the position at which the tactile sensation is presented from the target.

In this study, we examine the feasibility of mitigating the delay effect by
predicting future hand postures based on hand tracking results. Hand posture
tracking is often used in mid-air haptic interaction systems [7,12,15]. In such systems, a hand mesh model is created based on the estimated hand posture, which
is used to determine contact characteristics with virtual objects and resulting
tactile presentation positions. The main advantage of this method compared to
contact detection using point cloud data acquired with depth cameras is that
it can use the information regarding the temporal transitions, for example, the
velocity of the vertices of the hand model. This is useful for simulating the physical response of the virtual object during hand contact, which can also be used
to predict future hand posture. Focus shifts due to latency can be compensated
by estimating the future hand shape based on the posture parameters of the
hand model and determining the corresponding tactile presentation positions.
This study examines the effect of Kalman prediction, which assumes a simple
linear model for the transition of hand posture, on the intensity of tactile presentation. Numerical simulations were performed to investigate the changes in
the intensity of stimuli perceived on the side where the tactile sensation was
desired as well as on the opposite side, caused by the shift of the ultrasound

focus. The user study also examined the variation of the perceived intensity with the speed of finger movement at a constant velocity and the extent of its improvement by the compensations made based on the predictions.

2 Experiment Setup

To investigate the effects of latency, we constructed the apparatus shown in Fig. 2(a). 20 array units, each consisting of 249 transducers [14], were placed surrounding the workspace, with depth cameras (Intel RealSense D415) placed at the four corners to capture the surface geometry of hands. The depth cameras were operated at 60 Hz with 640 × 360 pixels, An auto-stereoscopic display (SONY ELF-SR2) was placed at the back of the array to display virtual objects to be touched during the user study.

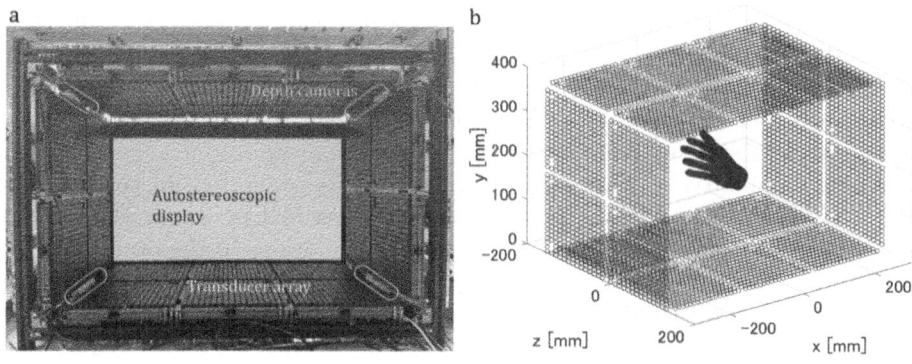

Fig. 2. a) Experiment setup: 4980 transducers surround the workspace. b) Simulation setup: The transducer array is configured in the same way as that implemented in the experiment setup.

2.1 Numerical Simulation of Ultrasound Focus

We numerically simulated the sound pressure distribution over a hand surface generated by the ultrasound focus using the experimental setup. The simulation configuration is shown in Fig. 2(b). A hand mesh model consisting of 170599 faces was placed at the center of the transducer array, and the center of the index finger pad was set as the ultrasound focusing target. We assumed that sound pressure of the incident wave $p_{inc}(\boldsymbol{r})$ from each transducer was a spherical wave with directivity D:

$$p_{inc}(\boldsymbol{r}) = \sum_n D(\boldsymbol{r} - \boldsymbol{x}_n)\frac{e^{-(jk+\alpha)\|\boldsymbol{r}-\boldsymbol{x}_n\|}}{\|\boldsymbol{r} - \boldsymbol{x}_n\|}q_n, \tag{1}$$

where $\alpha \in \mathbb{R}$ is the attenuation coefficient, and $q_n \in \mathbb{C}$ and $\boldsymbol{x}_n \in \mathbb{R}^3$ are the complex amplitudes and positions of the transducer $n \in \{1, \cdots N\}$, respectively. Given the position $\boldsymbol{x}_{focus}$ to generate the focus, q_n was set as follows:

$$q_n = ae^{jk\|\boldsymbol{x}_{focus} - \boldsymbol{x}_n\|}, \tag{2}$$

where $a \in \mathbb{R}$ is the maximum amplitude of transducers. The focal point $\boldsymbol{x}_{focus}$ was assumed to be shifted from the target position owing to latency. The direction of shift was defined as the direction opposite to the normal of the mesh model at the target point. Along this direction, the distance to the opposite surface was 13.8 mm. Simulations were based on the collocation method with the fast multipole boundary element method using Burton and Miller's integral equation [3]. The hand is regarded as completely sound hard. The directivity function D was determined based on the datasheet of the transducer (NIPPON CERAMIC T4010A1) used in the device and $a = 28.5$ kPa based on prior measurements. Figure 3 shows the change in sound pressure at the target position and opposite position with respect to focus shifts. Figure 4 also shows the pressure distribution over the hand surface for focus shifts of 0, 6.9, and 13.8 mm. As the focus shift increases, the sound pressure at the target position decreases while that at the opposite position increases. However, even when the focal point is shifted to the opposite side, a sound pressure of approximately 8 kPa is generated at the target. Similarly, at zero shift some pressure is generated on the opposite side. The perceptibility of this sound pressure is verified by a user study. Note that these results are based on sound pressure calculations in the range of linear acoustics. It has been observed that with the high sound pressure range, the increase in sound pressure is suppressed due to nonlinear effects [14]. Therefore, the actual sound pressure plots exhibit a more gentle curve and lower peaks.

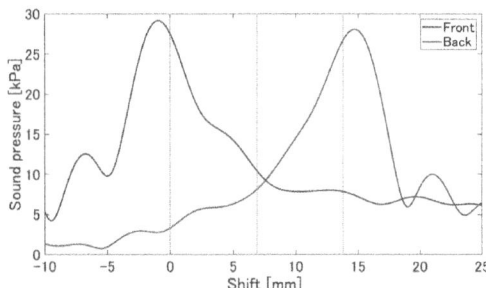

Fig. 3. Change in sound pressure at the target point (front) and the point on the opposite side (back) with respect to the shift in focus.

Notably, the peak of the sound pressure is shifted from the hand surface by approximately 1 mm. This seems to be caused by the diffraction of the sound waves from the arrays placed parallel to the normal direction of the target; however, a detailed verification is beyond the scope of this study.

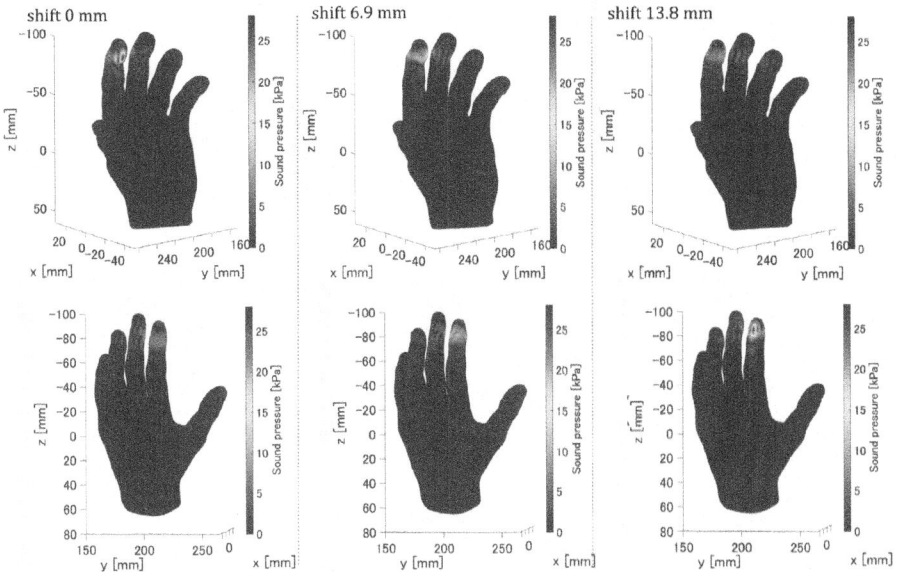

Fig. 4. Sound pressure distribution over the hand surface at focus shifts of 0 mm (on the target point), 6.9 mm, and 13.8 mm (on the opposite point).

3 Implementation

3.1 Hand Tracking

To create a mesh model that follows the surface shape of the hand, posture parameter optimization is performed to fit the point cloud data acquired in four directions using the depth cameras. As a template model for fitting, we used a hand model learned from various hand postures of multiple subjects called MANO [11], which captures the changes in a non-rigid shape due to pose. In MANO, the hand mesh model is determined by 10 parameters representing the individual differences in hand shape and 51 posture parameters (3 DOFs of rotation at 16 joints + 3 DOFs of translation; see Fig. 5(a) for the joint positions). In our system, each parameter is optimized with respect to every frame to fit the point cloud obtained from the depth cameras; hence the update rate for parameter optimization is 60 Hz. For a detailed method, see the appendix. An example of the fitted models is shown in Fig. 5(b).

3.2 Prediction of Hand Pose

Future hand postures are predicted based on the posture parameters obtained by optimization. In this study, we assume that the state transitions are represented by a linear model considering acceleration as noise. Let $\boldsymbol{x}_t \in \mathbb{R}^{51}$ be the parameter vector at a given frame t in which the 1–48th elements of this vector

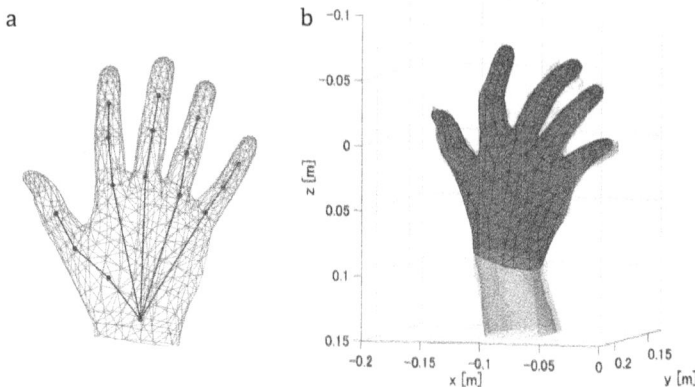

Fig. 5. a) Template model for fitting to a hand surface profile. b) Example of fitting to a point cloud obtained from the depth cameras.

denote the rotational parameters of the joint (the concatenation of 16 rotation vectors), and the 49–51st elements denote the translational parameters. Moreover let $v_t \in \mathbb{R}^{51}$ be rate of change in x_t, then in the next frame $t + 1$, the transition is expressed as follows:

$$\begin{bmatrix} x_{t+1} \\ v_{t+1} \end{bmatrix} = G \begin{bmatrix} x_t \\ v_t \end{bmatrix} + u_t, \tag{3}$$

$$G = \begin{pmatrix} 1 & & & \Delta t & & 0 \\ & \ddots & & & \ddots & \\ & & 1 & & & \Delta t \\ & & & 1 & & \\ & & & & \ddots & \\ 0 & & & & & 1 \end{pmatrix}, \tag{4}$$

where Δt represents the interval between frames; $\Delta t = 16.667$ ms in our system. u_t follows a normal distribution and adds noise to the velocity components. Assuming that the noise for each parameter is independent, let σ_r^2 be the variance of the noise in the velocity component of the joint rotation parameter and σ_r^2 be the variance of the noise in the translation velocity:

$$u_t \sim N(0, U), \tag{5}$$

$$U = \mathrm{diag}([0, \cdots, 0, \sigma_r^2, \cdots, \sigma_r^2, \sigma_t^2, \sigma_t^2, \sigma_t^2]^T), \tag{6}$$

where $\mathrm{diag}(b)$ denotes a matrix having vector b as its diagonal components. The parameters y_t obtained as a result of the optimization at the frame t is assumed to be x_t with noise s_t, which are expressed as follows:

$$\boldsymbol{y}_t = F \begin{bmatrix} \boldsymbol{x}_t \\ \boldsymbol{v}_t \end{bmatrix} + \boldsymbol{s}_t, \quad \boldsymbol{s}_t \sim N(0, S), \tag{7}$$

$$F = \begin{pmatrix} 1 \\ & \ddots & 0 \\ & & 1 \end{pmatrix}, \tag{8}$$

$$S = \mathrm{diag}([\sigma^2_{obs\_r}, \cdots, \sigma^2_{obs\_r}, \sigma^2_{obs\_t}, \sigma^2_{obs\_t}, \sigma^2_{obs\_t}]^T), \tag{9}$$

In our system, we set $\sigma^2_r = 10^{-1}\ \mathrm{rad}^2/\mathrm{s}^2$, $\sigma^2_t = 10^{-4}\ \mathrm{m}^2/\mathrm{s}^2$, $\sigma^2_{obs\_r} = 10^{-2}\ \mathrm{rad}^2$, and $\sigma^2_{obs\_t} = 10^{-7}\ \mathrm{m}^2$.

The above assumptions yield the maximum likelihood estimate $\boldsymbol{m}_t$ of the state $[\boldsymbol{x}_t \boldsymbol{v}_t]^T$ under the observation $\boldsymbol{y}_0 \ldots \boldsymbol{y}_t$. Without getting into details, $\boldsymbol{m}_t$ is updated as follows:

$$\boldsymbol{m}_t = G\boldsymbol{m}_{t-1} + K_t(\boldsymbol{y}_t - FG\boldsymbol{m}_{t-1}), \tag{10}$$

$$K_t = R_t F^T (F R_t F^T + S)^{-1}, \tag{11}$$

$$R_t = G C_{t-1} G^T + U, \tag{12}$$

$$C_t = (I - K_t F) R_t. \tag{13}$$

$\boldsymbol{m}_t$ and C_t represent the mean and variance-covariance matrix of the filter distribution for the state $[\boldsymbol{x}_t \boldsymbol{v}_t]^T$ under $\boldsymbol{y}_0 \ldots \boldsymbol{y}_t$, respectively.

The above expressions detail the procedure implemented to obtain the estimated values of the parameters at the current frame using the Kalman filter. Assuming that the change rate of the delay parameters is constant, the parameters of the future state $\boldsymbol{x}^{pred}_t$ in Δt_{pred} are predicted as follows:

$$\boldsymbol{x}^{pred}_t = \begin{pmatrix} 1 & & \Delta t_{pred} & & 0 \\ & \ddots & & \ddots & \\ & & 1 & & \Delta t_{pred} \end{pmatrix} \boldsymbol{m}_t. \tag{14}$$

Our system reduces the effect of delay by creating a hand mesh model with predictive parameters $\boldsymbol{x}^{pred}$ and setting the tactile presentation position to its surface.

An example of a hand movement prediction made using our system is shown in Fig. 6, in which the motion of grasping and releasing while moving the hand from right to left was predicted. The hand was moved for 3 s (180 frames) and stopped at the end. We assumed that the latency (and thus the prediction time) is equal to 4 frames ($\Delta t_{pred} = 66.67\mathrm{ms}$). Because it is difficult to accurately measure the latency of a system, especially a shutter timing of the camera, we chose the number of frames close to the measured computation time for hand tracking and prediction (approximately 15 ms) plus the time of flight of the sound wave (approximately 1 ms) and the delay time of the depth camera (approximately 50 ms, determined with reference [1]). The hand models at frames 1, 31, 61, 91, 121, and 151 are depicted in the figure. The observed hand model for the current frame is shown by the red line, the predicted hand model by the

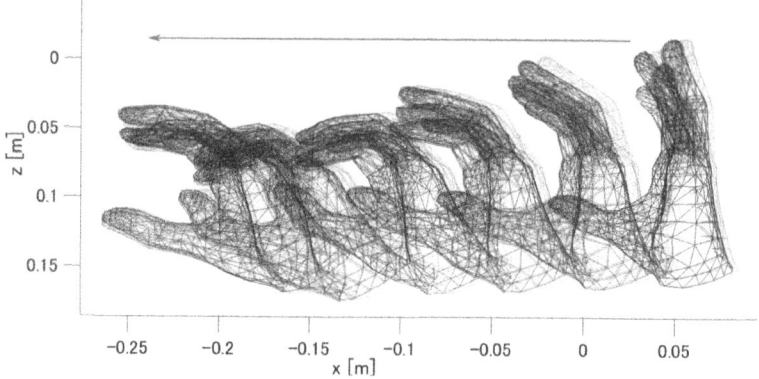

Fig. 6. An example of hand posture prediction. The green and blue lines are the predicted and observed data four frames after the red line, respectively. (Color figure online)

green line, and the observed hand model four frames ahead by the blue line. Although there is some error, the blue and green lines mostly overlap, indicating that the predictive model functions properly. Figure 7(a) presents a plot of the predicted vs. 4-frame-ahead observed rotation parameters at the first joint of the index finger for the same movement.

Because the prediction is based on the assumption that the speed remains constant, it is accurate in areas where the gradient changes moderately, whereas overshoot is observed in areas where the speed changes significantly, such as when transitioning from grasp to release, and in areas where the movement completely stops. Figure 7(b) shows the average error of the hand model vertices for each case with and without prediction, i.e., the difference between the prediction or current frame observation and four-frame-ahead observation. Compared to the no-prediction condition, the error has improved, with the smallest error being approximately 1 mm. However, the error remains larger in areas where there are significant speed changes, particularly when stopping the motion.

4   User Study

We experimentally investigated the extent to which focus shift due to latency affects perception, and whether this can be compensated by prediction. First, we tested the intensity of the unwanted stimulus produced due to latency at the opposite side of the target point. During the experiment, participants put their hands inside the apparatus as shown in Fig. 8(a) and touch the image on the auto-stereoscopic display. The image displayed is shown in Fig. 8(b). The participants touched the image with their index fingers from right to left so that they penetrated the lower rectangle. To dictate the touching speed, a white marker that moved at a constant velocity was displayed. The marker started

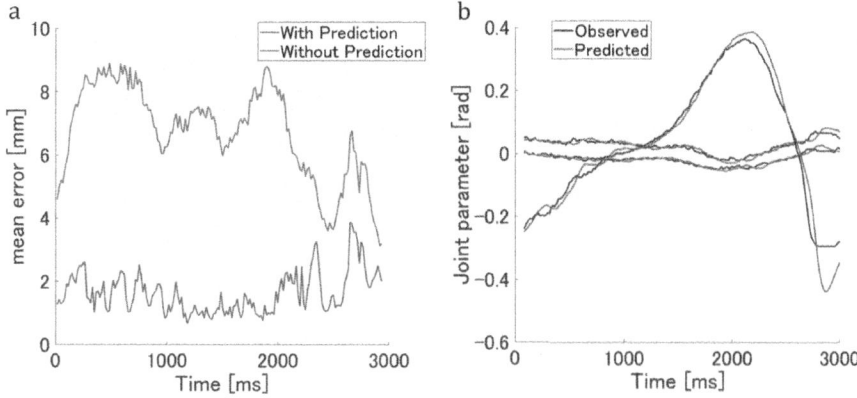

Fig. 7. a) Predicted and observed rotation parameters at the first joint of the index. b) The average of the error from the vertices of the hand model 4 frames ahead, with and without prediction.

5.5 cm to the right of the cuboid, moved 16 cm to the left, and then moved again from the initial position. The position of focus generation while touching the cuboid is determined as shown in Fig. 9(a). First, the centroid of the vertices of the predicted hand model that are inside the cuboid is calculated. Then, the target point was determined to be the location where the centroid was projected onto the surface of the hand model along the direction of touch (x-axis negative direction). Participants were asked to report the tactile intensity they felt on the back of their fingers. As a reference, a thin cuboid was shown at the top. When this was touched, the focus was generated on the back of the finger, the location of which was determined by the projection along the opposite direction (x-axis positive direction) to that traversed when touching the lower cuboid (Fig. 9(b)). The participants were asked to describe the intensity of the stimulus with a number 0–9 after touching the lower cuboid, with the reference stimulus value being set to 8. When touching the reference cuboid, the participants were instructed to keep their hands still to avoid delay effects. To increase the perceived intensity, a circular spatiotemporal modulation [5] was applied to the focus. The focus was moved at 200 Hz on a circle of 2 mm radius along a plane perpendicular to the direction of touch (x-axis). A filter was also applied to the array to gradually change the phase of the transducers to suppress any audible sound [13]. The speed at which the markers moved was either 50 mm/s, 100 mm/s, or 150 mm/s, while the prediction time Δt_{pred} was either 0, 20, 40, $\cdots$, 120, or 140 ms. Participants performed the task three times under each condition, $3 \times 8 \times 3 = 72$ times in total. The order of tasks was randomized.

Subsequently, the participants repeated the same experiment, this time assessing the intensity of the tactile stimulus perceived on the front surface of the finger. In this experiment, the position of focus generation in the reference cuboid was determined in the same manner as that implemented with in

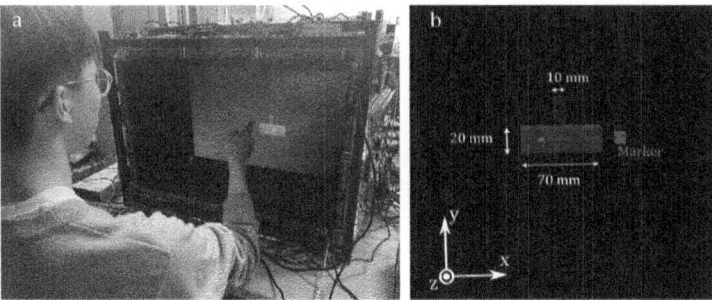

Fig. 8. a) User study. Participants repeat the task of touching the images on the auto-stereoscopic display. b) Image displayed on the auto-stereoscopic display during the user study.

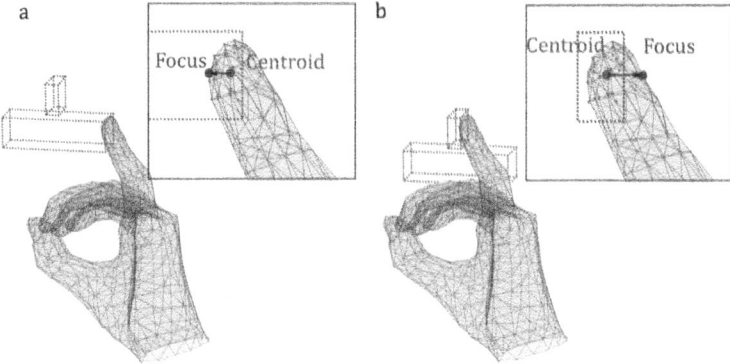

Fig. 9. Algorithm used for tactile presentation in the user study. a) When the lower cuboid is touched. b) When the upper cuboid is touched.

the lower cuboid. Participants similarly performed a randomized task 72 times. The reason for splitting the procedure into two parts, rather than getting two responses on the same task, was to make the participants focus on the stimuli to one side and prevent confusion. The participants were 10 male subjects aged between 23 to 32 years old.

Results. Figures 10(a) and (b) show the averages of the relative intensity felt by the participants at the back and front sides of their fingers, respectively. Although there are large individual differences between participants, the averages exhibit certain trends. The intensity felt on the back decreases as the prediction time increases, while the intensity felt on the front reaches a maximum at prediction times of approximately 60 to 80 ms. The perceived intensity on the back also intersects when comparing the plots of the three speeds in the same area. These results suggest that subjects perceive changes in intensity due to shifts in focus, hand posture prediction increases perceived intensity, and latency in this system

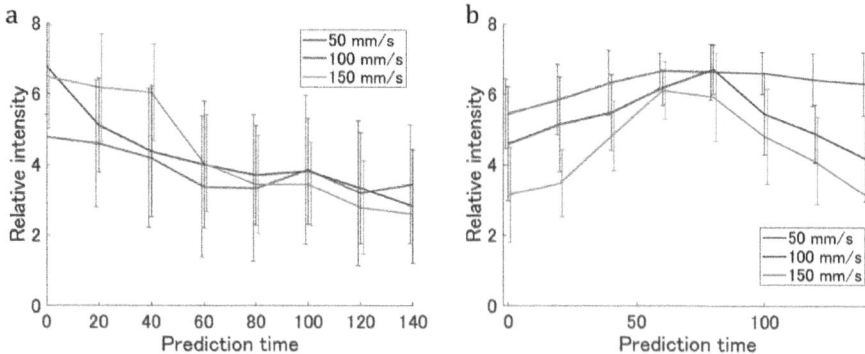

Fig. 10. Plots of perceived intensity relative to the reference stimulus versus prediction time. a) Perceived intensity on the back side of the finger, b) perceived intensity on the front side of the finger.

is approximately 60–80 ms. Assuming a latency of 70 ms, the focal point would be shifted by 10.5 mm when the velocity is 150 mm/s without prediction. This is close to the thickness of a finger; thus, the intensity plot drawn for the back side at a velocity of 150 mm/s is a curve that peaks near 0 ms.

Both front and back results show that the relative intensity average is greater than 3, even when the focal point is completely on the opposite side. As observed from the simulation results (Fig. 4), a small sound pressure is generated on the opposite side, and this could have been perceived. Therefore, to eliminate the unwanted stimulus, it may be necessary to not only compensate for the delay but also selectively drive the transducers on the array according to the direction in which the ultrasound is emitted. It is also possible that although participants were instructed to move their fingers at a constant velocity, some moved and stopped their fingers discretely, which may have led to poor prediction results.

Because the effect of delay increases with an increase in the speed of touch, the 150 mm/s result on the front side depicts a higher effect of delay compensation but with a peak value that is lower compared to those of the other speeds. this could be due to the fact that the prediction error increases with faster motion or the update speed of the focal position is relatively slow. In our system, the projected hand model is updated at the same rate as the hand tracking (60 Hz). Therefore, the change in stimulus point due to finger movement during 16.7 ms could lower the perceived intensity. Performing multiple predictions per tracking and updating the focal point more continuously could solve this problem, although the computational cost would be higher.

5   Conclusion

In this study, we experimentally demonstrate that the delay in focus generation has a detrimental effect on the perceived intensity of stimuli in mid-air tactile presentation systems, which can be improved by the prediction of hand postures. The system used a simple linear transition model to make predictions, which is suitable for slowly changing hand postures, but cannot adequately predict complex motions. The user study verified the effectiveness of the prediction under the simplest scenario of constant velocity finger movement, but deeper verification is needed for its effectiveness in more natural touch. Recently, neural network-based approaches have been widely implemented for predicting human behavior. Applying these to hand models could yield highly accurate predictions. A future research direction for us is to develop a new prediction algorithm based on results reported in this study and provide a more robust tactile presentation.

Appendix

The hand-tracking algorithm implemented in our system is briefly described below. First, downsampling is performed on the point cloud data obtained from the depth cameras. Then, the workspace was divided into voxels, and the point cloud was downsampled to the group of centroids of the points within the voxels. The posture parameters x and shape parameters β are optimized to fit the downsampled point cloud. Accordingly, the following cost function is minimized:

$$E(x, \beta) = \lambda_{point} E_{point}(x, \beta) + \lambda_{rot} E_{rot}(x) + \lambda_{shape} E_{shape}(\beta). \qquad (15)$$

$E_{point}(x, \beta)$ is the sum of the distances between points and their nearest vertices in the hand model. However, if the inner product of the normals of a point and its closest vertex is less than a given threshold, then the pair of distances is not included in the sum.

$E_{rot}(x)$ represent the constraints on the direction and magnitude of the rotation vector. When the rotation parameter at joint i is represented as $\theta_i \in \mathbb{R}^3$,

$$E_{rot}(x) = \sum_{i=1}^{16} \|A_i \theta_i\|_2^2. \qquad (16)$$

$A_i \in \mathbb{R}^{3 \times 3}$ is a matrix that constrains the direction of the rotation vector, which is different at each joint.

$E_{shape}(\beta)$ controls individual differences in hand shape. Because in MANO, shape parameters β are normalized quantities based on principal component analysis,

$$E_{shape}(\beta) = \|\beta\|_2^2. \qquad (17)$$

In our system, the above cost function is minimized using the Levenberg–Marquardt method to estimate the mesh model along the point cloud.

References

1. High-speed capture mode of intel® realsense™ depth camera d435 (2023). https://dev.intelrealsense.com/docs/high-speed-capture-mode-of-intel-realsense-depth-camera-d435

2. Abdlkarim, D., et al.: A methodological framework to assess the accuracy of virtual reality hand-tracking systems: a case study with the meta quest 2. Behav. Res. Meth. **56**(2), 1052–1063 (2024)

3. Burton, A., Miller, G.: The application of integral equation methods to the numerical solution of some exterior boundary-value problems. Proc. Royal Soc. London Math. Phys. Sci. **323**(1553), 201–210 (1971)

4. Carter, T., Seah, S.A., Long, B., Drinkwater, B., Subramanian, S.: UltraHaptics: multi-point mid-air haptic feedback for touch surfaces. In: Proceedings of the 26th Annual ACM Symposium on User Interface Software and Technology, pp. 505–514. ACM (2013)

5. Frier, W., et al.: Using spatiotemporal modulation to draw tactile patterns in mid-air. In: Prattichizzo, D., Shinoda, H., Tan, H.Z., Ruffaldi, E., Frisoli, A. (eds.) EuroHaptics 2018. LNCS, vol. 10893, pp. 270–281. Springer, Cham (2018). https://doi.org/10.1007/978-3-319-93445-7_24

6. Hoshi, T., Takahashi, M., Iwamoto, T., Shinoda, H.: Noncontact tactile display based on radiation pressure of airborne ultrasound. IEEE Trans. Haptics **3**(3), 155–165 (2010)

7. Hwang, I., Son, H., Kim, J.R.: AirPiano: enhancing music playing experience in virtual reality with mid-air haptic feedback. In: 2017 IEEE World Haptics Conference (WHC), pp. 213–218. IEEE (2017)

8. Makino, Y., Furuyama, Y., Inoue, S., Shinoda, H.: HaptoClone (haptic-optical clone) for mutual tele-environment by real-time 3d image transfer with midair force feedback. In: Proceedings of the Conference on Human Factors in Computing Systems, pp. 1980–1990. ACM (2016)

9. Matsubayashi, A., Makino, Y., Shinoda, H.: Direct finger manipulation of 3D object image with ultrasound haptic feedback. In: Proceedings of the Conference on Human Factors in Computing Systems. ACM (2019)

10. Morisaki, T., Fujiwara, M., Makino, Y., Shinoda, H.: Midair haptic-optic display with multi-tactile texture based on presenting vibration and pressure sensation by ultrasound. In: SIGGRAPH Asia 2021 Emerging Technologies, pp. 1–2 (2021)

11. Romero, J., Tzionas, D., Black, M.J.: Embodied hands: modeling and capturing hands and bodies together. ACM Trans. Graph. (Proc. SIGGRAPH Asia) **36**(6) (2017)

12. Sand, A., Rakkolainen, I., Isokoski, P., Kangas, J., Raisamo, R., Palovuori, K.: Head-mounted display with mid-air tactile feedback. In: Proceedings of the 21st ACM Symposium on Virtual Reality Software and Technology, pp. 51–58. ACM (2015)

13. Suzuki, S., Fujiwara, M., Makino, Y., Shinoda, H.: Reducing amplitude fluctuation by gradual phase shift in midair ultrasound haptics. IEEE Trans. Haptics **13**(1), 87–93 (2020)

14. Suzuki, S., Inoue, S., Fujiwara, M., Makino, Y., Shinoda, H.: AUTD3: scalable airborne ultrasound tactile display. IEEE Trans. Haptics **14**(4), 740–749 (2021)

15. Villa, S., Mayer, S., Hartcher-O'Brien, J., Schmidt, A., Machulla, T.K.: Extended mid-air ultrasound haptics for virtual reality. Proc. ACM Hum. Comput. Inter. **6**(ISS), 500–524 (2022)

Pseudo-Frequency Modulation: A New Rendering Technique for Virtual Textures

Paras Kumar[(✉)] and Rebecca F. Friesen

Texas A&M University, College Station, TX 77801, USA
paras.kumar@tamu.edu

Abstract. Creating virtual textures with speed-invariant identities requires spatial constancy, attained by changing temporal frequency with a finger's sliding speed. Implementing sensations of continuous frequency change, however, is non-trivial for low-cost wearable vibrotactile displays, as the amplitude of commonplace voice coil motors and linear resonant actuators varies strongly with driving frequency. In this paper, we present Pseudo-Frequency Modulation (PFM), a technique to continuously change the perceived frequency by modulating only the amplitudes of two single-frequency components in response to changes in sliding speed. Results of a psychophysical study with a wrist-worn vibrotactile display show that people perceive smooth changes in frequency with PFM, and this technique can be reliably used to display distinguishable textures which differ in terms of perceived surface properties such as coarseness.

Keywords: Vibrotactile display · Perception and psychophysics · Virtual reality

1 Introduction

Diverse texture rendering can add a new dimension to virtual experiences by producing a wide range of distinguishable haptic sensations. In this work, we seek to preserve both a diversity of texture sensations and a fundamental perception of "texture" (rather than "buzzing") using as simple as possible vibrotactile rendering. Our technique continuously modulates a perceived temporal frequency in response to finger sliding speed. This is done in order to preserve spatial constancy [8] and realism [3] when actively exploring a virtual texture. While actual preservation of spatial constancy requires continuous changes in temporal frequency, our technique is significantly simpler. It renders continuous shifts in perceived temporal frequency by superimposing two single-frequency components and modulating only their vibration *amplitudes* as a function of user's sliding speed [13].

A vibrotactile texture display provides direct vibration to the skin as a substitute for texture-elicited skin vibrations. A real textured surface has a spatial frequency, defined by the spacing between its individual asperities. A uniform

H. Kajimoto et al. (Eds.): EuroHaptics 2024, LNCS 14768, pp. 378–390, 2025.
https://doi.org/10.1007/978-3-031-70058-3_31

texture, for example, would have a constant spatial frequency in a given direction. When we slide our skin across such a real textured surface, contact with surface asperities evoke vibrations in our skin. These vibrations have a temporal frequency, which depends not only on the texture's spatial frequency, but also on our sliding speed [8]. These temporal changes in skin vibrations form the basis of our texture perception. If a real texture has a constant spatial frequency, our perception of its identity also remains constant regardless of our sliding speed [2]. At the neural level, our sense of movement is fused with a proportional change in the temporal frequency of the texture-elicited vibration, preserving spatial constancy. Extending this understanding of texture perception to virtual environments, vibrotactile feedback provided by texture displays should also respond dynamically to the user's speed by proportionally changing the temporal frequency.

Many previous studies have employed this speed-responsiveness in texture displays. For example, Konyo et al. [12] continuously modulated the vibration frequency in response to the hand velocity. However, they had to continuously apply phase adjustments to produce smooth frequency-modulated outputs. Similarly, rendering of a coarse texture was perceived as more real when frequency was continuously modulated as a function of the sliding speed [6]. Here, the driving voltage to the broadband Voice Coil Motor (VCM) had to be filtered to achieve matched displacement amplitude at all the frequencies. This was necessary to compensate for the non-flat frequency response of a VCM. Such examples are evidence for high real-time processing demands of continuous frequency modulation.

Researchers have also explored data-driven techniques to achieve high realism in texture rendering. They recorded acceleration signals while actively exploring a real texture to generate texture models. These models were used to synthesize vibration signals as a function of user's normal force and sliding speed. These techniques, however, have only been assessed for tool-texture interactions; further, it requires extensive data acquisition and processing for generation of texture models [4,24].

Aiming to simplify the process of texture rendering, while preserving a fundamental perception, we looked at superimposed vibrations. In a past study, it was shown that superposition of two frequency components is perceived as a single pitch equivalent, where the pitch equivalent correlates to the amplitude ratio of individual single-frequency components [5]. Therefore, to circumvent the real-time processing demands of continuous frequency modulation, we asked: can we continuously modulate a perceived frequency by using superposition of just two single-frequency signals?

Indeed, superimposed vibrations have been considered previously in texture rendering [12]. Researchers have also used dual-frequency vibrations in audio-to-haptic conversion with the same principle of reducing the amount of information while still preserving a basic expression [10,15]. Some studies have extensively characterized the perceived properties of dual-frequency vibrations for hand-

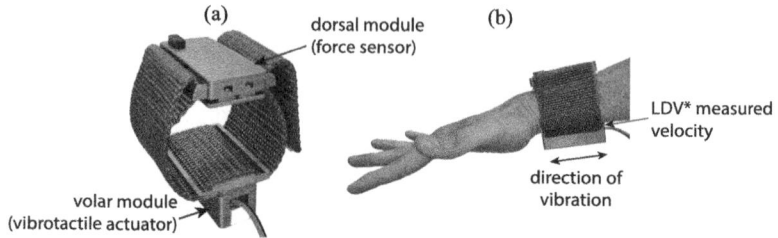

Fig. 1. (a) Wrist strap with two 3D printed modules in gray. The volar and dorsal modules house the vibrotactile actuator and the force sensor, respectively. (b) A user wearing the wrist strap and actively exploring a virtual texture in mid-air using their index fingertip. The directions of actuator vibration and LDV measured velocity are also shown. (*LDV: Laser Doppler Vibrometer)

held vibrotactile devices [11,29]. However, to our knowledge, none have used superimposed vibrations to preserve spatial constancy of virtual textures.

Based on these observations, we propose a technique to continuously modulate the perceived frequency of a vibrotactile display as a function of finger sliding speed. We are initially exploring the simplest possible signals, two superimposed sinusoids, but anticipate studies with more complex vibrations in the future. The key contributions of this work include (1) an experimental setup to evaluate vibrotactile signals in wrist displays, and (2) preliminary results which validate the use of two single-frequency components to continuously modulate the perceived frequency and preserve spatial constancy.

2   Methods

2.1   Wrist Strap Design

The desired form factor of a tactile display is a critical early design decision. Wearable vibrotactile displays can be broadly classified into two types. The first renders sensations directly to the fingertips and takes the form of thimbles [27] and gloves [9,23]. The second relocates feedback to proximal body sites, e.g. the ventral side of the distal [17] and proximal phalanx [6,25] or around the wrist [7,21]. While lower tactile acuity could be a drawback of relocating sensation away from the fingertips [7,14], relocating feedback remains of interest as it frees the fingertips to dexterously interact with the real world objects in augmented or mixed reality applications. A wrist form factor, in particular, offers low occlusion of the hand, which improves seamless hand tracking using camera-based trackers. It also allows the use of larger broadband actuators due to increased real estate on the wrist. Additionally, changes along the roughness-smoothness dimension can still be encoded via changes in frequency at locations away from the fingertips, including the wrist [7]. Based on these considerations, we designed a wrist strap. It accommodates two components: an actuator to facilitate vibration, and a force sensor to monitor the strap tightness. These components are housed in 3D printed modules, as shown in Fig. 1a.

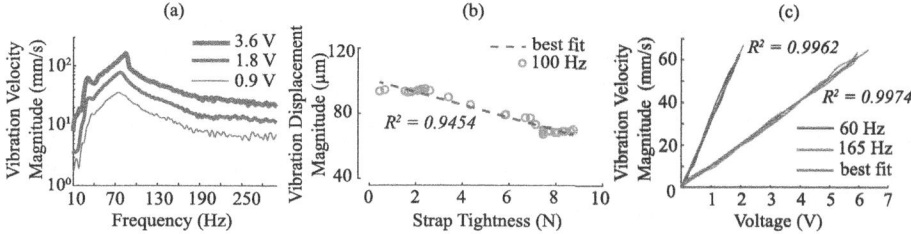

Fig. 2. (a) Frequency response of the wrist strap actuated by a Voice Coil Motor (VCM). The vibration intensity is strong for a broad range of frequencies, but the response does not exhibit a flat characteristic. (b) Effect of strap tightness on displacement magnitude of a 100 Hz vibration - evidence for the need to ensure consistency of strap tightness in psychophysical experiments. (c) The relationship between measured vibration velocity magnitude and actuator input voltage is linear up to 2 V and 6 V for 60 Hz and 165 Hz signals, respectively, as shown by the lines of best fit. The vibration velocity magnitude of both the single-frequency components was kept under 36 mm/s, and hence in the linear range.

Vibrotactile Actuator. The volar module consists of a vibrotactile actuator and a cover, both snap-fitted to a base plate. The module's position can be adjusted along the strap to accommodate for different wrist sizes. The actuator vibrates parallel to the skin, in the distal-proximal direction, as shown in Fig. 1b. These vibrations are transmitted to the volar skin through the base plate and the strap. We chose the volar side for tactile feedback due to its higher tactile acuity than the dorsal side [1,16,19]. Additionally, this keeps feedback on the same side of the forearm as the finger pad, where sensations will appear to occur during virtual interactions. We selected a VCM (Hapcoil-One, Actronika) for vibrotactile actuation owing to its strong vibrations at a broad range of frequencies. Figure 2a shows the non-flat frequency response of the strap, obtained by playing a chirp (10 Hz to 300 Hz) on the actuator at three different voltage levels. All vibration measurements were made in the direction of the actuator vibration using a single point Laser Doppler Vibrometer (LDV Polytec VFX-F-110, VFX-I-160 sensor head). The strap was worn with a tightness of 2.6 N and the laser was focused at the strap edge, which directly contacts the volar side, as depicted in Fig. 1b. Data was acquired using National Instruments USB-6211 at a sampling rate of 10 kHz, and high pass filtered with a 20 Hz cutoff frequency to remove variations caused by low frequency hand tremors.

Force Sensor. The dorsal module monitors the strap tightness, which is known to affect the vibration amplitude of a wearable [22]. Inspired from Tasbi [21], it has a circular capacitive force sensor (SingleTact CS15-4.5N), held in place by a pre-compressed disc spring. The spring is attached to a movable plate, which is in direct contact with the wrist as shown in Fig. 3a. An increase in strap tightness pushes the plate, compresses the spring, and generates a measurable load on the force sensor. To calibrate the force sensor, we removed the

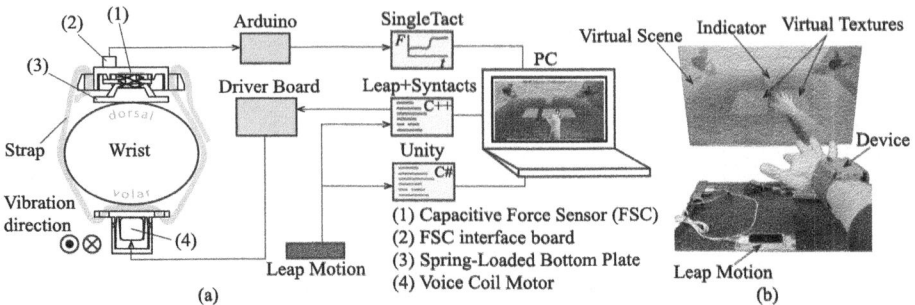

Fig. 3. (a) Schematic of the setup showing the wrist strap's cross-sectional view, sub-components, and peripherals. (b) Picture of a participant wearing the wrist strap and exploring a virtual texture. A virtual scene is also shown on a screen which contains textures and a real-time hand animation, rendered via hand tracking using Leap Motion.

volar module, applied known pull forces to the volar strap worn on the wrist, and measured the force response. Figure 2b shows the effect of strap tightness on the displacement magnitude of a 100 Hz vibration. This provides further evidence for the need to ensure fit consistency during psychophysical experiments.

2.2   Virtual Environment

We acquired index fingertip position data–streamed from a Leap Motion Controller (a stereo camera for hand tracking from Ultraleap) at 120 Hz–to calculate the instantaneous speed at each frame. We then considered six consecutive frames to obtain a weighted moving average and modulated the vibration signals as a function of this average speed (in a C++ application).

We used Leap Motion's Unity plug-in to render a real-time hand animation in virtual environments. These environments included virtual texture surfaces and an indicator light which turns green to indicate user's presence in the texture's interaction zone, as shown in Fig. 3b. The two-dimensional size of each interaction zone differed across the virtual environments; but the height was set to a constant 75 mm, with the texture surface at its center. This gave the users flexibility to deviate slightly from the plane of the texture and still remain in touch with the virtual mid-air textures. We entered the dimensions of each interaction zone manually in both the C++ and Unity applications to toggle the actuator and the virtual indicator, respectively.

2.3   Signal Design

To modulate vibrotactile feedback as a function of a user's finger speed, we need to generate a speed-dependent signal. We achieved this by integrating Leap Motion and Syntacts (an open-source software from Rice University [20]) in a single C++ application. We customized the spatializer feature of Syntacts to modulate the amplitude of sinusoidal audio signals. The processed audio signals

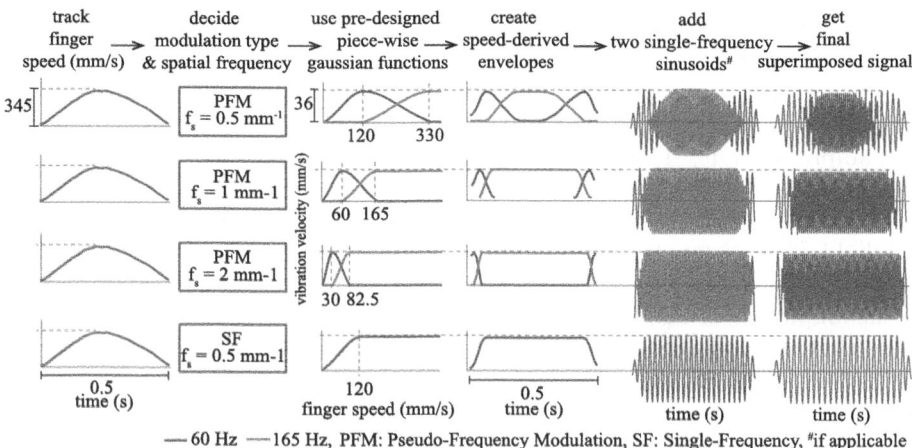

track   decide   use pre-designed   create   add   get
finger   →   modulation type → piece-wise → speed-derived → two single-frequency → final
speed (mm/s) & spatial frequency gaussian functions envelopes   sinusoids# superimposed signal

— 60 Hz —165 Hz, PFM: Pseudo-Frequency Modulation, SF: Single-Frequency, #if applicable

Fig. 4. Signal design procedure for Pseudo-frequency modulated (PFM) and single-frequency (SF) textures used in this study. The PFM signals were designed for three distinct spatial frequencies: 0.5, 1, and 2 per mm. The single frequency signal was designed for only 0.5 per mm.

were converted to Pulse Width Modulated (PWM) signals by a multichannel H-bridge driver board (HSD Mk 1, Actronika) which outputs an amplified signal to power the actuator.

Figure 4 outlines the proposed signal design using a combination of 60 Hz and 165 Hz vibrations. We selected this combination from many possible choices, wherein the amplitudes can be matched without overloading the actuator. During pilot studies, we observed that the proposed technique is more effective with such a fluttery-smooth combination i.e. low frequency in 10–70 Hz range and high frequency above 150 Hz [26]. To predict the physical amplitude of the vibration as a function of the input voltage, we obtained the voltage characteristics by measuring the vibration response to a sinusoidal input voltage at each of the two frequencies. Figure 2c shows that the relationship between vibration velocity magnitude and input voltage is linear up to 2 V and 6 V for 60 Hz and 165 Hz signals, respectively.

Pre-designed piecewise gaussian functions, shown in Fig. 4, are the main component of the proposed signal design for a given design spatial frequency f_s. In the context of our signal design process, spatial frequency is just a design parameter; our intent is to create distinguishable textures, and not to perfectly mimic the textures with such physical spatial frequencies. At sliding speed s under a threshold speed value s_1, we only modulate the 60 Hz vibration as a function of speed, which then saturates to a set level A. As the speed increases further, the amplitude of 60 Hz vibration diminishes, while the 165 Hz vibration ramps up to the set level at the same rate, which is defined by a second threshold speed s_2. Speeds higher than the second threshold will only produce set level 165 Hz vibration. This technique will hereafter be referred to as Pseudo-Frequency Modulation (PFM).

These gaussian functions involve two important design decisions: (1) threshold sliding speeds ($s_i, i = 1, 2$) and (2) the set level or matched maximum vibration amplitude (A). The threshold sliding speed s_i for each single-frequency component is simply the ratio of the temporal frequency f_t and the desired spatial frequency f_s. For example, for a desired spatial frequency of $0.5\,\mathrm{mm}^{-1}$, the threshold sliding speeds are 120 mm/s and 330 mm/s for 60 Hz and 165 Hz components, respectively. The matched maximum vibration amplitude A could be the amplitude of displacement, velocity, or acceleration. Greenspon et al. [8] found that the displacement of texture-elicited vibrations remains constant with the sliding speed. Displacement matching, however, posed a challenge due to the reduced frequency response of the actuator at higher frequencies (Fig. 2a). At actuator-safe input voltage levels, a 60 Hz vibration, displacement matched with a 165 Hz vibration, is perceptually weak. Therefore, we chose to match the velocity amplitudes, with $A = 36$ mm/s, which is in the linear range of the voltage characteristics shown in Fig. 2c. We designed three gaussian functions for design spatial frequencies $f_s = 0.5, 1$, and $2\,\mathrm{mm}^{-1}$. Equations (1–2) show the mathematical expressions for the designed gaussian functions. Here F is a gaussian function that depends on tracked finger speed s, threshold speed s_i, and a variable c. G_1 and G_2 are piecewise gaussian functions (formulated using F) for 60 Hz and 165 Hz components, respectively.

$$F(s, s_i, c) = 0.25A[5e^{\frac{-(s-s_i)^2}{2s_i^2 c^2}} - 1] \tag{1}$$

$$G_i(s) = \begin{cases} G_1 = F(s, s_1, 0.55), \ G_2 = 0 & s \leq s_1, \ c = 0.55 \\ G_1 = F(s_1, 0.97), \ G_2 = F(s, s_2, 0.97) & s_1 < s \leq s_2, \ c = 0.97 \\ G_1 = 0, \ G_2 = A & s > s_2, \end{cases} \tag{2}$$

Using these piecewise gaussian functions G_i in conjuction with tracked finger speed s provides speed-derived envelopes in time domain. We also designed a single frequency signal of 60 Hz vibration, corresponding to a $0.5\,\mathrm{mm}^{-1}$ spatial frequency, for comparison with PFM, as shown in Fig. 4.

3   Psychophysical Experiments

We conducted a psychophysical study (TAMU IRB2022-1042) with 15 participants (one left-handed, seven women, ages 18–34). Participants wore the device on their dominant wrist with a strap tightness of 2.6 N. We instructed them to keep all fingers extended with an open hand posture while exploring the virtual textures, since gestures also affect strap tightness. Pink noise was played through a pair of noise cancelling headphones to mask any audio from the actuator. The experiment spanned two preliminary training tasks followed by four subsequent experiments. Only Experiment 4 had a visual representation of textural features overlaid on the haptic vibrotactile feedback. For all the prior experiments, participants saw only a blank gray surface.

3.1   Training

The objective of these tasks was to introduce participants to the concepts of amplitude and frequency and to assess if they can correctly describe changes in these properties. In the first task, we played a sequence of three distinct single-frequency vibrations on the actuator, each with a different amplitude (30, 50, and 90 μm displacement) but same frequency (100 Hz). We then asked participants to assign labels of low, mid, and high based on the amplitude. The second task involved the same procedure, but the vibrations differed in terms of frequency (60, 120, and 240 Hz) at a constant voltage of 1.8 V (the actuator's frequency response determined the amplitude at each of the three frequencies). Each task consisted of 3 trials per participant (total 45 trials), and the sequence was randomized in each trial. We used simple terms to describe vibration properties to non-STEM participants, e.g. 'loudness' or 'intensity' for amplitude and 'how fast' for frequency.

Results. Figure 5a demonstrates that low, medium, and high amplitudes were correctly identified by almost all participants. When identifying vibrations of different frequency, a subset of participants (3 out of 15) consistently confused high and low frequency values and reported them in the reverse order. Post-training queries indicate that these participants focused only on amplitude.

3.2   Experiment 1: Detection of PFM

Experiment 1 consisted of a 30 cm square texture slab at the center of the virtual environment. We asked participants to practice sliding their finger across the virtual texture in mid-air while being consistently in 'touch', signaled by a green indicator. After necessary practicing, we displayed PFM $f_s = 0.5\,\mathrm{mm}^{-1}$ on the virtual texture slab through the actuator. We then asked participants to slide their fingers in circular motions, spanning the entire square texture slab and at a constant speed, regulated using a metronome. After repeating the procedure for three different speeds (with 40, 69, and 100 beats per minute; 2 beats per circular motion), we asked participants if they sensed a change in the vibration between the three speeds; if yes, then which of two properties changed- frequency or amplitude.

Results. Figure 5b shows that 14/15 participants sensed a change in frequency (Z-test, $p = 0.0004$). When asked to gradually increase their speed in a single circular motion on the virtual texture, 13/15 participants sensed a continuous change in frequency (Z-test, $p = 0.0023$).

3.3   Experiment 2: Comparing PFM with Single Frequencies

Experiment 2 consisted of two 22.5 cm square texture slabs placed side by side. We displayed PFM on one and a single frequency vibration of 60 Hz on the other

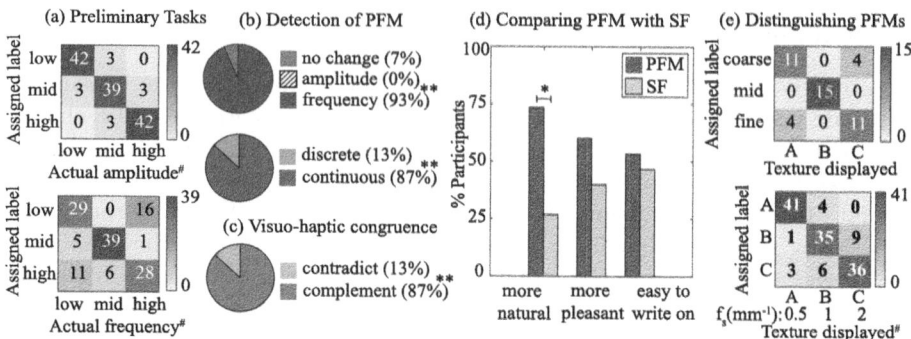

Fig. 5. Results for psychophysical experiments: (a) Preliminary tasks demonstrate ability of participants to discern changes in amplitude and frequency. (b) Experiment 1 shows that participants perceived a continuous change in frequency with PFM. (c) Addition of visuals complemented the haptic texture feedback. (d) PFM is significantly rated as more natural. (e) Participants accurately identified the trend of changing coarseness and distinguished three different PFM textures made with just two single-frequency components. Statistically significant results are marked with asterisks; * and ** mean Z-test $p < 0.05$ and $p < 0.01$, respectively.

slab (both corresponding to $f_s = 0.5\,\mathrm{mm}^{-1}$). The placement order was randomized for each participant. We allowed the participants to actively explore each texture without any speed or time constraints, and asked them which texture feels more natural and pleasant. We also asked them to write invisible alphabets like 'P' or 'M' on each slab using their index finger and choose which texture requires less cognitive effort, i.e. which was easier to write on.

Results. Figure 5d shows that the participants rated PFM texture as more natural (Z-test, $p = 0.0349$). While the other two ratings were also higher for PFM, the difference is insignificant.

3.4   Experiment 3: Distinguishing PFMs

Experiment 3 consisted of three 15 cm square texture slabs placed side by side, as shown in Fig. 3a (see PC). First, we displayed PFM $f_s = 0.5\,\mathrm{mm}^{-1}$, $1\,\mathrm{mm}^{-1}$ and $2\,\mathrm{mm}^{-1}$ on texture slabs left to right, and named them Texture-A, B, and C. We asked participants to get familiar with each of the three textures by active exploration, and arrange them in increasing order of coarseness. Next, we randomized the placement order of these three textures and asked participants to identify and assign labels of A, B, and C, from their memory of each texture. The texture identification task was repeated for 3 trials per participant (total 45 trials).

Results. The top confusion matrix in Fig. 5e shows that participants accurately identified the trend of changing coarseness. However, some confusion occurred between most and least coarse textures due to difference in the notion of coarseness for a few participants. The bottom confusion matrix in Fig. 5e shows that the participants accurately identified each of the three virtual textures, which differed in terms of design spatial frequency.

3.5   Experiment 4: Visuo-Haptic Congruency

Experiment 4 consisted of two 22.5 cm square texture slabs. The left and right slabs had a haptic display of PFM $f_s = 0.5\,\mathrm{mm}^{-1}$ and $1\,\mathrm{mm}^{-1}$, and a visual display of spatial frequency $1\,\mathrm{cm}^{-1}$ and $2\,\mathrm{cm}^{-1}$, respectively. We adopted an order of magnitude higher spatial frequency in visual display to make it easy for the participants to visually distinguish the textures. The participants compared the two slabs by active exploration, and we asked them if the visual representation complements or contradicts the vibrotactile haptic display.

Results. Figure 5c shows that the visual representation had complemented the haptic display for 13/15 participants (Z-test, $p = 0.0023$).

4   Discussion

In this study, we proposed Pseudo-Frequency Modulation (PFM), a technique to continuously modulate the perceived temporal frequency as a function of finger sliding speed by only modulating the *amplitude* of two single-frequency components. We then conducted a psychophysical study to assess the effectiveness of PFM in preserving spatial constancy. To our knowledge, this is the first assessment of such a technique, and hence we implemented a minimal signal design using a single combination of pure sinusoids to prove the concept.

We found that participants overwhelmingly detected a continuous change in frequency with variations in their sliding speed on a PFM texture. When compared to a single frequency (SF) texture, they rated PFM as more natural, associating it with real textured surfaces e.g., tarp, velcro, breadboard, etc. Some participants emphasized that PFM feels the same everywhere i.e., uniform. Participant preference for the SF texture in the pleasantness task, however, may be due to its lower vibration frequency.

During pilot studies, early participants reported greater control over their motion on PFM textures. This allowed them to perform tasks like writing with higher accuracy, which informed our design of Experiment 2. Conversely, the results of the writing task in Experiment 2 did not show a significantly better rating for PFM textures. This could owe to task comprehension issues e.g., some ratings were given based on coarseness under the assumption that larger coarse features are easier to write on.

We also found that participants can reliably distinguish three PFM textures even though they are composed of the same single-frequency components. This has implications for the success of diverse virtual texture designs using only two single-frequency signals, either with broadband VCMs or by using two tiny LRAs with distinct resonant frequencies placed within the two-point discrimination threshold [18]. Finally, visual representation of textural features complemented the PFM based vibrotactile display, suggesting the potential of integrating PFM with visual feedback. Some participants emphasized that while they can describe texture coarseness even without the visuals, the process of distinguishing textures becomes faster with a visual representation.

Nevertheless, the findings of this study are limited to simple sinusoidal single-frequency vibration components. The range of possible diverse textures is also limited with just two single frequency components. More complex signal combinations can potentially create sensation of properties other than coarseness, e.g. stickiness, adding to the diversity. In addition, Leap Motion's low sampling rate and actuator's slow dynamic response contribute to the latency of the system, thereby impacting the realism of the haptic feedback. Faster hand tracking solutions and actuators with improved dynamic response may help alleviate such latency issues in the future. The psychophysical study is also limited in terms of its categorical data, and the use of numeric ratings may help in better quantification of conclusions made in this study. Lastly, the results are limited to wrist displays and the effectiveness of the technique remains unexplored at the fingertips. Considering similar frequency discrimination performance between the two body sites [16] (despite marked differences in detection thresholds), the technique may translate to the fingertips as well.

In the future, we will build on this experimental platform by improving the latency. We will do a comprehensive study of possible frequency combinations to achieve similar effects. We will compare PFM with actual frequency modulation and assess its effectiveness at other body sites, like the fingertips. We will explore the possibility of using beat perception [28] to lower the frequencies below 20 Hz (for low speeds and coarse textures) while keeping the amplitude perceptually strong. We will also evaluate the use of more than two single-frequency components to further broaden the PFM's operating frequency range.

References

1. Bolanowski, S.J., Gescheider, G.A., Verrillo, R.T.: Hairy skin: psychophysical channels and their physiological substrates. Somatosens. Motor Res. 11(3), 279–290 (1994)
2. Boundy-Singer, Z.M., Saal, H.P., Bensmaia, S.J.: Speed invariance of tactile texture perception. J. Neurophysiol. 118(4), 2371–2377 (2017)
3. Culbertson, H., Kuchenbecker, K.J.: Should haptic texture vibrations respond to user force and speed? In: 2015 IEEE World Haptics Conference (WHC), pp. 106–112 (2015). https://doi.org/10.1109/WHC.2015.7177699
4. Culbertson, H., Unwin, J., Kuchenbecker, K.J.: Modeling and rendering realistic textures from unconstrained tool-surface interactions. IEEE Trans. Haptics 7(3), 381–393 (2014)
5. Friesen, R.F., Klatzky, R.L., Peshkin, M.A., Colgate, J.E.: Single pitch perception of multi-frequency textures. In: 2018 IEEE Haptics Symposium (HAPTICS), pp. 290–295. IEEE (2018)

6. Friesen, R.F., Vardar, Y.: Perceived realism of virtual textures rendered by a vibro-tactile wearable ring display. IEEE Trans. Haptics **17**, 1–11 (2023). https://doi.org/10.1109/toh.2023.3304899
7. Gaudeni, C., Meli, L., Jones, L.A., Prattichizzo, D.: Presenting surface features using a haptic ring: a psychophysical study on relocating vibrotactile feedback. IEEE Trans. Haptics **12**(4), 428–437 (2019)
8. Greenspon, C.M., McLellan, K.R., Lieber, J.D., Bensmaia, S.J.: Effect of scanning speed on texture-elicited vibrations. J. R. Soc. Interface **17**(167), 20190892 (2020). https://doi.org/10.1098/rsif.2019.0892
9. HaptX Inc.: Haptx (2023). https://haptX.com. Accessed 29 Sep 2023]
10. Hwang, I., Lee, H., Choi, S.: Real-time dual-band haptic music player for mobile devices. IEEE Trans. Haptics **6**(3), 340–351 (2013). https://doi.org/10.1109/TOH.2013.7
11. Hwang, I., Seo, J., Choi, S.: Perceptual space of superimposed dual-frequency vibrations in the hands. PLoS One **12**(1), e0169570 (2017). https://doi.org/10.1371/journal.pone.0169570
12. Konyo, M., Tadokoro, S., Yoshida, A., Saiwaki, N.: A tactile synthesis method using multiple frequency vibrations for representing virtual touch. IEEE. https://doi.org/10.1109/iros.2005.1545130
13. Kumar, P., Friesen, R.: Wearable multi-frequency vibrotactile display for virtual textures. In: Work-in-Progress Paper (1 page) Presented at the IEEE World Haptics Conference (WHC) (2023)
14. Lederman, S.J., Klatzky, R.L.: Haptic perception: a tutorial. Attent. Percept. Psychophys. **71**(7), 1439–1459 (2009). https://doi.org/10.3758/app.71.7.1439
15. Lee, J., Choi, S.: Real-time perception-level translation from audio signals to vibrotactile effects (2013). https://doi.org/10.1145/2470654.2481354
16. Mahns, D.A., Perkins, N., Sahai, V., Robinson, L., Rowe, M.: Vibrotactile frequency discrimination in human hairy skin. J. Neurophysiol. **95**(3), 1442–1450 (2006)
17. Manus: Manus (2023). https://www.manus-meta.com/. Accessed 24 Sep 2023
18. Martinez, J.S., Tan, H.Z., Cholewiak, R.W.: Psychophysical validation of interleaving narrowband tactile stimuli to achieve broadband effects. In: 2021 IEEE World Haptics Conference (WHC), pp. 709–714. IEEE (2021)
19. Morioka, M., Whitehouse, D.J., Griffin, M.J.: Vibrotactile thresholds at the fingertip, volar forearm, large toe, and heel. Somatosens. Motor Res. **25**(2), 101–112 (2008)
20. Pezent, E., Cambio, B., O'Malley, M.K.: Syntacts: open-source software and hardware for audio-controlled haptics. IEEE Trans. Haptics **14**(1), 225–233 (2020)
21. Pezent, E., Israr, A., Samad, M., Robinson, S., Agarwal, P., Benko, H., Colonnese, N.: Tasbi: Multisensory squeeze and vibrotactile wrist haptics for augmented and virtual reality. In: 2019 IEEE World Haptics Conference (WHC), pp. 1–6 (2019). https://doi.org/10.1109/WHC.2019.8816098
22. Rokhmanova, N., Faulkner, R., Martus, J., Fiene, J., Kuchenbecker, K.J.: Strap tightness and tissue composition both affect the vibration created by a wearable device. In: Work-in-Progress Paper (1 page) Presented at the IEEE World Haptics Conference (WHC) (2023)
23. SenseGlove: Senseglove (2023). https://senseglove.com. Accessed 24 Sep 2023
24. Shin, S., Choi, S.: Hybrid framework for haptic texture modeling and rendering. IEEE Access **8**, 149825–149840 (2020). https://doi.org/10.1109/ACCESS.2020.3015861

25. Talhan, A., Kim, H., Jeon, S.: Tactile ring: multi-mode finger-worn soft actuator for rich haptic feedback. IEEE Access **8**, 957–966 (2019)
26. Tan, H.Z., Durlach, N.I., Reed, C.M., Rabinowitz, W.M.: Information transmission with a multifinger tactual display. Percept. Psychophys. **61**(6), 993–1008 (1999). https://doi.org/10.3758/bf03207608
27. WeArt: Weart (2023). https://weart.it/. Accessed 29 Sep 2023
28. Yang, S., Tippey, K., Ferris, T.K.: Exploring the emergent perception of haptic beats from paired vibrotactile presentation. In: Proceedings of the Human Factors and Ergonomics Society Annual Meeting. vol. 58, pp. 1716–1720. SAGE Publications Sage CA, Los Angeles CA (2014)
29. Yoo, Y., Hwang, I., Choi, S.: Perceived intensity model of dual-frequency superimposed vibration: Pythagorean sum. IEEE Trans. Haptics **15**(2), 405–415 (2022). https://doi.org/10.1109/toh.2022.3144290

Do Vibrotactile Patterns on both Hands Improve Guided Navigation with a Walker?

Inès Lacôte[1]([⊠]), Pierre-Antoine Cabaret[1], Claudio Pacchierotti[2],
Marie Babel[1], David Gueorguiev[3], and Maud Marchal[1,4]

[1] CNRS, Univ Rennes, INSA Rennes, Inria, IRISA, Rennes, France
`lacote@isir.upmc.fr`,
`{pierre-antoine.cabaret,marie.babel,maud.marchal}@irisa.fr`
[2] CNRS, Univ Rennes, Inria, IRISA, Rennes, France
`claudio.pacchierotti@irisa.fr`
[3] CNRS, Sorbonne Université, ISIR, Rennes, France
`david.gueorguiev@isir.upmc.fr`
[4] Institut Universitaire de France (IUF), Rennes, France

Abstract. Current assistive technologies for navigation guidance using non-visual feedback often rely on audiovisual cues only. Haptic cues can also provide rich information while leaving the auditory sensory channel unencumbered. By enhancing the sensory information available for guidance, haptics has the potential to improve the adoption of guiding technologies by people with, e.g., visual impairments. This paper evaluates the capacity of tactile patterns delivered by haptic handles to guide users walking with a walker while they are asked to follow a predefined path. We conducted a user study on 18 participants who used two haptic handles mounted on a walker in actual walking condition. We implemented three types of vibrotactile patterns for guidance: uni-manual (one handle), bi-manual (two handles), and dual (combining one-handle and two-handles patterns) were used depending on the direction. We also compared vibration and tapping stimulation modes to test for their potential influence on the guiding strategy and the user's preference. Results showed no significant effect of the strategy or the stimulation mode on the accuracy of following the target path. However, they showed that bi-manual conditions presented a higher satisfaction rate, gave a sensation of being mentally less demanding to users, improved the confidence rate in succeeding the task, and increased the navigation speed.

Keywords: Vibrotactile feedback · Navigation assistance · Walking · Path following · Haptic handle

1 Introduction

In the realm of navigation assistance, today we encounter a large number of challenges, particularly when considering the diverse needs of the different user

© The Author(s), under exclusive license to Springer Nature Switzerland AG 2025
H. Kajimoto et al. (Eds.): EuroHaptics 2024, LNCS 14768, pp. 391–404, 2025.
https://doi.org/10.1007/978-3-031-70058-3_32

groups. While vehicular navigation systems have advanced notably through turn by turn systems, offering sufficiently efficient guidance for these applications, these tools are not so well suited for pedestrians, who need significantly more detailed information [11]. The limitations of today's tools become even more apparent when we consider the accessibility aspect and the specific needs of, e.g. people with visual impairments. For example, current navigation technologies using non-visual feedback often focus on providing auditory-based navigation, ending up overloading the auditory sense. However, the auditory sense is paramount to acknowledge the environment, move safely, a socially interact with others, especially if other senses are impaired. This limitation is further exacerbated for those users experiencing deafblindness, a condition where both visual and auditory senses are impaired, making traditional navigation systems largely inaccessible. Indeed, nowadays most guidance is provided through visual, auditory, or audiovisual stimuli, which is limiting in many situations [2,3,19].

In this perspective, many works have focused on studying the potential of haptics for navigation assistance, as an alternative to vision and audition. Kappers et al. review of hand-held haptics devices for navigation assistance [11] suggests that vibrotactile cues are an interesting and viable way of conveying directional information. While the use of vibrations is straightforward and widespread, receiving these cues during long periods of time can potentially cause tiredness and create tactile noises [7,24]. In that respect, apart from vibrotactile devices, we can see various types of haptic stimuli emerge, using shape-changing [21,22], force [6], pseudo-force [1], and skin-stretch [4,5,13]. On addressing mobility issues encountered by the elderly, some have focused on the development of on-walker embedded assistance. While a part of those works target rehabilitation of gait mobility [20], using for example rhythmic haptic cues after a stroke [9], others aim at better navigation assistance for people suffering from combined sensory and mobility deficiencies [10,18,26]. In this last case, research that involves haptic feedback mainly focus on active guidance by physically pulling the user towards a direction [18,26].

In this paper, we present an experiment investigating the performance of two haptic handles embedded on a walker in delivering navigation information. The handles use the Apparent Haptic Motion (AHM) illusion to deliver rich directional indications to users asked to follow a predefined path. We create and compare three navigation strategies using one or two handles with different tactile patterns [17] and we compare two different modes of stimulation: vibrations and tapping sensations. The study aimed to identify the most effective and intuitive tactile patterns and thus the preferred navigation strategy, as well as to evaluate potential difference in effectiveness and user experience between the tapping and vibrotactile stimulations during an actual navigation task. As the handles are designed to be adaptable to multiple mobility aids, e.g., walkers, wheelchairs, precanes, white canes, these results might be readily applicable to a much broader range of situations and systems.

2   Experimental Device

We developed a haptic handle equipped with five electromagnetic actuators, as depicted in Fig. 1. Inspired by our earlier work [16], this version of the device incorporates significant improvements, such as an enhanced actuator design and positioning. The device has also undergone miniaturisation for seamless integration onto a walker and, in the future, other mobility aids. The handle, made through 3D printing using TPU soft material (Filaflex 82 A, 0.8 mm thickness), has a cylindrical shape designed for comfortable single-handed grip. Actuators are arranged in a "T"-configuration across the handle to stimulate the metacarpal bones of the palm (three motors), thumb (one motor), and index finger (one motor). This handle is designed to replace the default handles of a standard walker (ErgoClick 4-wheels walker), shown in Fig. 2. The same handles were used in [17] for a preliminary study aiming to evaluate the device performance in a static position before having people actually walking around with the device.

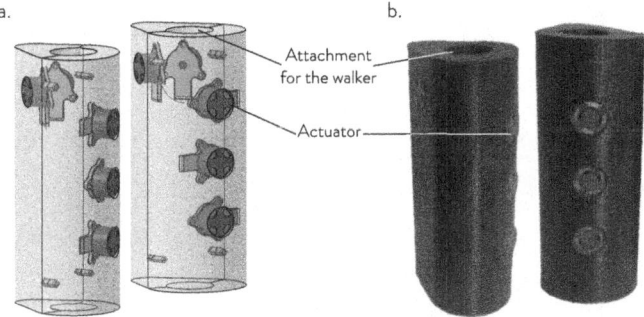

Fig. 1. Right and left haptic handles designed for providing directional tactile sensations. a) Computer-Aided Design (CAD) of the 3D-printed handle, made from flexible TPU material, featuring alcoves to house the actuators in a "T" - shape configuration. This design naturally stimulates the thumb, the second metacarpal bone of the palm, and the index finger. b) Image of the two symmetrically designed handles [17].

The actuators are custom-designed electromagnetic motors, drawing inspiration from the Hapticomm device [8] and previous studies on the Apparent Haptic Motion (AHM) illusion [14–16]. They are capable of conveying both vibratory and tapping sensations, referred to as "taps". The actuators are controlled by M5 Stack controllers and powered with a power bank (capacity: 10 Ah/37 Wh, Output: 6–9 V DC, max 2 A), enabling more than four hours of use. A significant effort has been made for miniaturisation and portability, as it is a crucial aspect for the implementation in a study close to real-life condition and its transition towards an everyday use.

3   User Study

Our objective is to evaluate the effectiveness of providing rich and precise navigation assistance through the Apparent Haptic Motion (AHM) paradigm delivered with our custom hand-held interfaces. More precisely, our focus is to compare the accuracy rates achieved with different tactile patterns using one or both hands/handles to identify the most intuitive ones for the walker users in navigation tasks. To do so, we carried out a human participants study aiming to test our devices and strategies in the context of following unknown trajectories through tactile instructions in actual walking modalities. The study focus on three key factors to be tested: (i) stimulation *Modes*: standard vibrations or discrete "tapping" sensations; (ii) pattern *Strategies*: Bi-manual (using both handles for all directional instructions), Dual (using either one or two handles depending on the instruction), and Uni-manual (only using the handle on the dominant hand side) with patterns distributed around or along the handles; and (iii) complexity of the *Path* to follow. The study received approval from Inria's ethics committee (COERLE, 2021-39).

3.1   Experimental Setup and Methods

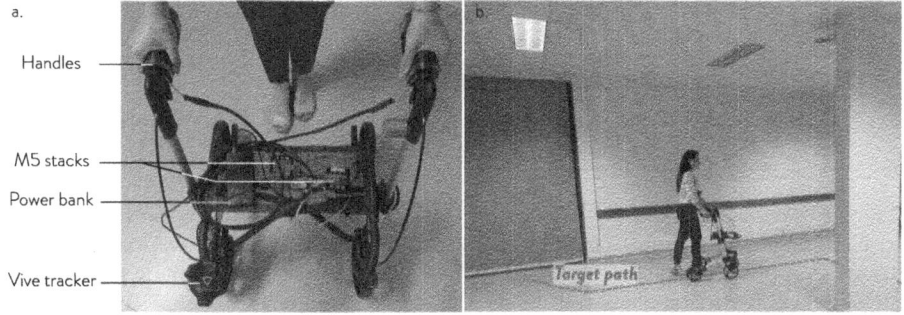

Fig. 2. a) Experimental setup: The handles are mounted on the walker. Signals are amplified and sent to the handles by the M5 Stacks controllers connected via WiFi. The motors are powered through the M5 Stacks cabled to a power bank. b) Navigation: The participant is guided by the tactile patterns along a target virtual path (not visible by the user, dashed blue line in the Figure).

Experimental Setup. The experimental setup featured a standard walker equipped with two haptic handles described in Sect. 2, positioned 48 cm apart with adjustable height for user comfort. Participants, wearing noise-cancelling headphones, stood behind the walker, which housed the handle controlled by the electronic setup. The participants navigated in a 8 × 8 m room in which their position and orientation were tracked using a Vive tracking system, with base stations in the four corners of the room ceiling and a tracker on the walker. The room was empty, with a pillar close to the centre. Participants were asked to follow precisely predefined virtual paths. The position and orientation tracking enabled to evaluate the proper tactile patterns to be given as a directional instruction to the user, according to the experimental condition at hand. No visual information or details about the trajectory to follow were given to the participants, and there were no visual clues except for the starting points marked with tape on the floor. We defined two different starting points diagonally positioned in two of the room's corners. We considered the coordinates of the starting point as the origin of the spatial referential $(x_{start}; y_{start}) = (0; 0)$, and the closest corner $(x_{corn}; y_{corn}) = (-1; -1)$. Figure 3 shows the room dimensions and the *Paths* starting at $(x_{start}; y_{start}) = (0; 0)$. The representation shows the two different starting points in opposite corners. When applying a 180° rotation of the participants trajectories in orange, we obtain consistent representations of both blue and orange *Paths* together with superimposed starting points.

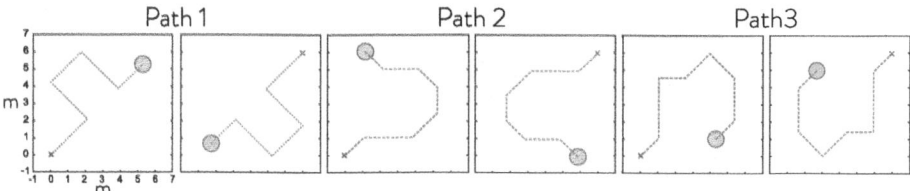

Fig. 3. Paths that participants were tasked to follow. The blue and orange paths are identical but start from two different starting points (× and ×). The coloured discs (● and ◉) represent the arrival area, where the target is considered reached.

Tactile Patterns. Three strategies to provide directional information were considered: Bi-manual, Dual, and Uni-manual. Figure 4 shows the three navigating strategies and the tactile patterns associated. Building on a previous study on directional indication in a static position [17], we selected tactile patterns utilising the Apparent Haptic Motion (AHM) illusion for navigation cues. These patterns, activated through electromagnetic actuators, conveyed directional information either on a single handle (three actuators) or both handles (six actuators). They are distributed around or along the handles. The coloured arrows in Fig. 4 illustrate the activation patterns for each signal. The gradient of colours, ranging from green to red, depicts the temporal activation of the motors.

Type of Haptic Stimuli: Two distinct methods of delivering haptic cues—vibrations and "taps"—were achieved by altering the activation *Mode* of the electromagnetic actuators within the handles. The "tap" stimulations comprised a 220 ms square signal, while the vibration stimulation consisted of a 120 Hz sinusoidal signal within a 220 ms square envelope. In both *Modes*, signals were preceded by a negative impulse to pull the magnets down in the coil before pushing them out. The Stimuli Onset Asynchrony (SOA), the activation delay between two subsequent actuators, was set at 110 ms (see [16]).

Paths to Follow: Participants had to follow virtual *Paths*. We created three of them to vary complexity and prevent participants from learning them. Paths were designed to have multiple turns, between four and six, with at least two right turns and two left turns. We differentiated *Paths* by the angles of their turns. "Path 1" presents only 90° turns, "Path 2" only 45° turns, and "Path 3" both 45° and 90° turns (see Fig. 3).

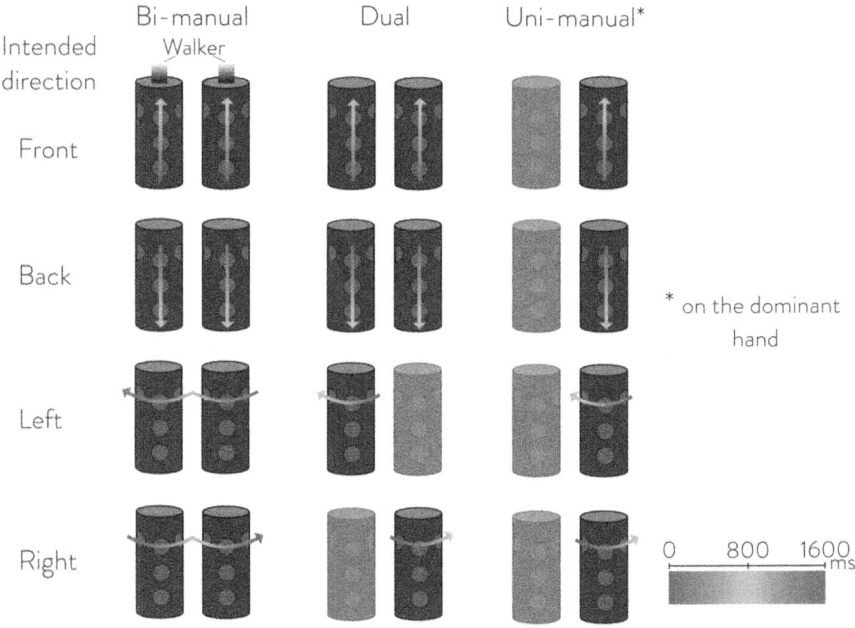

Fig. 4. Directional patterns using the AHM illusion organised in strategies. The coloured arrows depict the pattern of activation, with the colour gradient indicating temporal activation (green to red). Each motor is individually activated, providing tapping or vibratory sensations. For example, when considering the *Front* signal of the *Bi-manual Strategy*, it initiates the progressive activation of the three motors along each handle held by the user's. On the other side, the *Front* signal of the *Uni-manual Strategy* activates the three actuators along the user's dominant side handle. For *Left/Right* indications, the sequential activation stimulates the thumb first, the second metacarpal bone (across the purlicue), and the index finger. (Color figure online)

3.2   Experimental Task and Design

Participants were tasked with manoeuvring the walker based on the directional information provided through the handle(s), with no prior knowledge of the trajectory. Signals were generated at a frequency of 0.5 Hz, offering directional information every two seconds. The tracking system monitored the walker position and rotation, which were used to compute and select the directional pattern to be displayed by the handle(s). Users were guided towards a point one meter ahead on the path, similarly to [12]. The tactile patterns were evaluated depending on the angle computed to reach the target position. If the target locates at $\pm 15°$ compared to the orientation of the user, being the 0° reference, the instruction given to the users is *Front*, creating a cone of 30° of tolerance for *Front* direction, comparable to the parameters used in [25]. The *Back* instruction was sent for a target point located at $\pm 127.5°$ compared to the orientation of the user.

The experiment comprised three blocks (for the three *Strategies*), each featuring two series, one per stimulation *Mode* ("taps" or vibrations). Within each series, participants performed three trials corresponding to the three different target *Paths*. To prevent learning bias, participants alternated between starting points, and the order of presentation of the *Path*, *Mode*, and *Strategy* conditions were counter-balanced. This protocol resulted in 18 navigation trials per participant corresponding to the 18 conditions: $3\,Paths \times 2\,Modes \times 3\,Strategies$, with no repetition.

3.3   Participants and Collected Data

Eighteen participants, aged between 21 and 35 ($mean = 25.2$, $sd = 4.4$), including four left-handed people, six women, eleven men, and a participant identifying as non-binary, took part of the experiment. Participants were naive about the study hypotheses and design. The experiment lasted one hour. They signed an informed consent form and completed demographic questionnaires. Then, they got 2 minutes to experience manoeuvring the walker without receiving tactile patterns on the handles. Before each change of *Strategy* and *Mode*, participants tested the patterns while the experimenter indicated what was the given interpretation of each tactile pattern. Throughout the task, we collected the participant position, the walker orientation, and navigation information, as well as their level of agreement on a 7-point Likert scale to various statements, as detailed in Sect. 4. At the end of the experiment, we collected the user's preferred guiding *Strategy* and *Mode*, as well as open comments and general feedback.

4   Results

Figure 5 shows the data of all eighteen participants' trajectories, categorised according to the *Strategy* and the stimulation *Mode* conditions for the three different target *Paths*. Each graph presents the target *Path* (doted black line) and the eighteen participants courses (coloured lines) for each condition.

4.1 Statistical Analysis on Trajectory Data

For our data analysis we considered the following variables. For the independent variables, we considered the navigation strategies, which lead to different tactile patterns using one or two handles (Uni-manual, Dual, Bi-manual), the stimulation *Mode* ("taps" and vibrations), and target *Paths* (see Fig. 3). For the dependant variables, we considered (i) the Dynamic Time Wrapping (DTW) computed on participants trajectories, (ii) the duration of their courses, and (iii) the total distance covered during each trial.

Dynamic Time Wrapping (DTW). We computed the Dynamic Time Wrapping distance as a measure of the distance between matching pairs of points in the target *Path* and the participant trajectories. More specifically, the DTW identifies a wrapping *Path* between the target signal and the participant trajectory signal so as to find the closest corresponding point on the target, and create the corresponding pairs of points. We used the package "SimilarityMeasures" on RStudio and the function DTW [23]. As the DTW metrics also depends on the number of points over the signal, we normalised the DTW by the number of points in the participant trajectory. We thus obtained a mean value of distance to the target *Path* along the whole trajectory. Data showed overall: a mean value of distance from the target $D_{mean} = 0.58m$; median $D_{med} = 0.49m$ and standard deviation $D_{std} = 0.38m$ over all participants and conditions. As the data distribution does not follow a normal distribution, we ran a Generalised Linear Mixed Model analysis, considering as independent variables the three navigation *Strategies*, the stimulation *Modes*, and the three different *Paths*. From the Wald Chi-square test performed on the GLMM, none of the three factors has an impact on the distance to the target calculated through the DTW ($p > 0.05$).

Trajectories Duration. For the measure of the duration, we looked at the impact of the independent variables *Strategy and Mode*. Because Paths 1, 2 and 3 have different lengths, we separated the data based on the *Path*. Furthermore, as the data did not satisfy the normality requirements to perform an ANOVA, we ran a GLM analysis to investigate the possible effect of the *Strategy* and the *Mode* of stimulation on the time (thus the velocity of participants) and the total distance covered by participants during a trajectory. For the time data, the Wald Chi-square test (*Strategy, Mode*) performed on the GLM showed no impact of the *Mode* ($p > 0.05$). However it showed a significant effect of the *Strategy*. For Path 1: $\chi^2(2) = 13.176$, $p = 0.001$; for Path 2: $\chi^2(2) = 11.523$, $p = 0.003$; for Path 3: $\chi^2(2) = 14.618$, $p < 0.001$. The post-hoc Holm test on the *Strategy* showed that Uni-manual *Strategy* significantly increases the reaching time compared to Bi-manual (p < 0.01) and to Dual (p < 0.01). No significant difference was found between Bi-manual and Dual *Strategy*. Data showed overall: a mean value of time $T_{mean} = \{83; 75; 88\}$ s for *Paths* $\{1; 2; 3\}$ respectively; median $D_{med} = \{74; 63; 74\}$ s and standard deviation $D_{std} = \{34; 36; 41\}$ s over all participants and conditions.

Total Covered Distance for Each Trajectory. We look at the total distance of participants' trajectories, separating once again the data based on the studied *Path* number, also analysing the impact of the *Strategy* and *Mode*. From

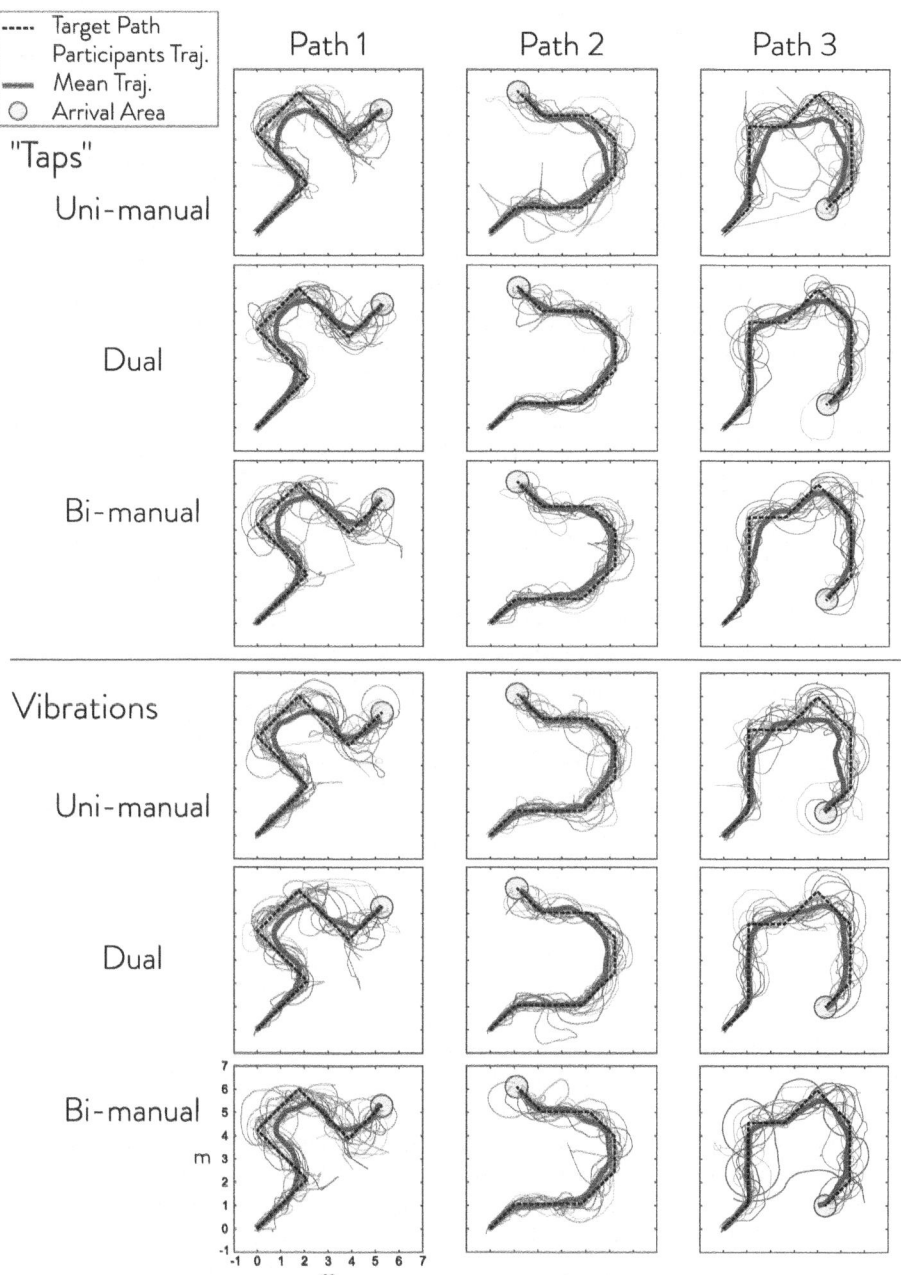

Fig. 5. Collection of all participants trajectories for each condition of guiding: *Strategy*, stimulation *Mode*, and *Path*. The dashed black line represents the target *Path*, the thick green line is the mean trajectory among the 18 participants, and the thin coloured lines are participant trajectories. The light green disc represent a 0.5 m radius area around the arrival point, considered as the arrival area. The axis represent the 2D space of the room, with the origin of the task placed at the same starting location for representation. (Color figure online)

the Wald Chi-square test (Strategy, Mode) performed on the GLM, none of the two factors had an impact on the total distance ($p > 0.05$).

Subjective Evaluation. After each *Path*, *Mode*, and *Strategy*, participants had to evaluate on a 7-point Likert scale how much they agreed with the given statement, 1 being "totally disagree" and 7 "totally agree". The statements are named according to their order of presentation to the participants, from 1 to 14. Statements for which strategies showed no significant effect are reported in Table 1, with the corresponding median value and the standard deviation. Statements on which strategies had an effect are presented in Table 2, showing the p-values of the corresponding strategy effect. The values of median and standard deviations for these statements are then given in Fig. 6.

Table 1. Median and standard deviation of the 7-point Likert scale evaluation where the Post-Hoc Holm tests results showed no significant effect of the *Strategy* ($p > 0.05$). Data for all 18 participants.

Results on a 7-points Likert scale for the three strategies		
	Median	Std
S3: The stimulations are pleasant	6	1.14
S4: The stimulation intensity is appropriate	5	1.64
S5: The use of the device is effortless	6	1.76
S7: The task is physically demanding	1	0.95
S8: I confidently navigated	6	1.26
S9: The haptic assistance is useful to follow the trajectory	6	1.13
S10: The haptic assistance is easy to use	6	1.52
S11: I could use the device without prior instructions	5	1.57
S12: I learnt how to use the device quickly	6	1.43
S13: It is fun to use	6	1.26

For the statements *S1* to *S7*, the same statements are given for vibrations and "taps". Each data is answered according to the three *Paths* covered in the same *Strategy and Mode* conditions. We performed a GLMM with a Poisson model, where *Strategy* and *Mode* are the 2 independent variables. For each answer the participants did 3 trajectories (Path 1-2-3) in each *Mode* and *Strategy*. The Post-Hoc Holm tests results are reported in Table 1 for non significant results and in Table 2 along with the evaluation results in Fig. 6 for results showing an effect of the *Strategy*. The significant effect are shown using "*". No effect of the *Mode* was found on these statements ($p > 0.05$).

For statements *S8* to *S14*, the question/answers were given after performing all trials for a specific *Strategy*, compiling data for "tap" and vibratory *Mode* or the *Path*. Only the *Strategy* is a independent variable. Answers are still given on a 1-7 Likert scale. This type of data required Krustal-Wallis rank sum test as the data did not comply with normal distribution requirements for ANOVA.

Table 2. Post-Hoc Holm tests results for the subjective evaluation of the stimuli showing a significant effect of the navigation strategy. Data for all 18 participants.

	Strategy comparison		
	Dual - Bi	Uni - Dual	Uni - Bi
S1: I think I succeeded in completing the task	$p > 0.05$	$p = 0.032^*$	$p = 0.032^*$
S2: The guiding *Strategy* is easy to use	$p > 0.05$	$p = 0.037^*$	$p = 0.024^*$
S6: The task is mentally demanding	$p > 0.05$	$p = 0.030^*$	$p > 0.05$
S14: I am satisfied with the navigation *Strategy* (tactile patterns)	$p > 0.05$	$p = 0.042^*$	$p = 0.042^*$

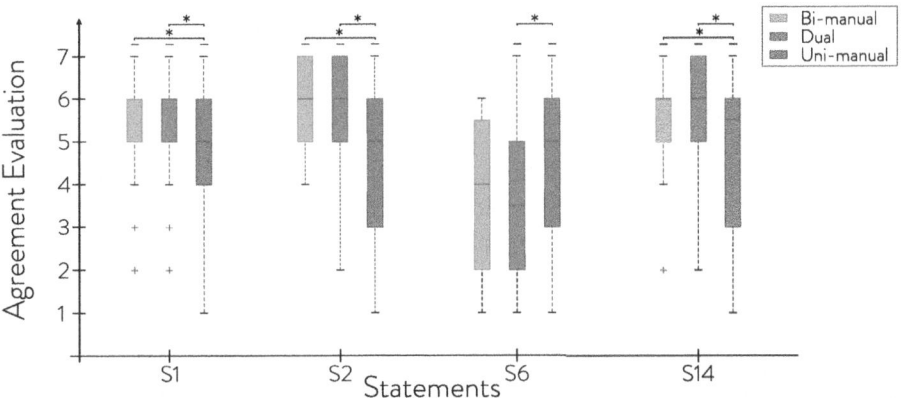

Fig. 6. Participants' evaluation of given statements showing significant effect of the *Strategy*. Answers are given on a 7-point Likert scale. Red markers indicate the median value. The bottom and top edges of the box indicate the 25th and 75th percentiles, respectively. The whiskers extend to the most extreme data points not considering outliers. Outliers are plotted using the $+$.

5   Discussion

We conducted a user study with eighteen participants, who walked on virtual paths with the sole help of tactile indications provided on custom walker haptic handles. We tested three strategies of tactile patterns, using one or two handles. In this study, various factors were examined concerning navigation strategies using different tactile patterns, stimulation modes ("taps" and vibrations), and target paths. Statistical analyses were conducted on dependent variables such as participant trajectories—to evaluate the distance to the target (using the DTW), course duration—to determine impact on intuitiveness and confidence, total distance—to see if participants would over-walk in certain conditions, and Likert scale evaluations—to assess the user experience and cognitive load. The Dynamic Time Wrapping distance was employed for trajectory analysis, revealing no significant impact of *Strategy*, *Mode*, or *Path*. The value of the distance to the target found with the DTW ($mean = 0.58\ m$) is promising. Indeed, it suggests that participants should be guided on a path located about 60 cm away from obstacles, which is compatible with most public and outdoor environments, where the device is meant to be used. The same observation was made on the total distance covered during the trajectories. However, trajectory duration was significantly influenced by strategy, with the Uni-manual approach making the participants' courses longer in time, suggesting the Uni-manual strategy affected the confidence of participants. Likert scale evaluations indeed indicated that the Uni-manual strategy negatively affected participants' confidence, as they perceived their success in the task lower, and found the guiding strategy was less easy to use, mentally more demanding and were less satisfied with this strategy. However it is important to notice that their actual displacements were as good with the Uni-manual strategy as with the other two strategies.

6   Conclusion and Perspectives

We conducted an experiment with eighteen participants to evaluate in close to real-life conditions two haptic handles mounted on a walker for performing navigation tasks. Two stimulations modes ("taps" and vibrations) and three navigation strategies (Bi-manual, Uni-manual, and Dual) were tested. For all strategies and modes, the path was adequately followed by participants with no significant difference. However, Uni-manual was never cited as a preferred navigation strategy: nine participant preferred the Dual strategy and nine preferred the Bi-manual. Similarly, for mode preferences, nine participants preferred "taps" and nine of them preferred vibrations. Overall, these findings highlight the importance of strategy in influencing both objective performance metrics and subjective user experiences during navigation tasks. The design of the handles and the results on participants' capacity to follow a *path* with all three *strategies* makes it easy to adapt to different mobility aids, that will be tested in the future, such as the precanes, the power-wheelchair, the cane, for people with single or multiple impairments.

A limitation of this work is the narrow range of ages that we considered for our participants. Moreover, as mentioned for the Uni-manual strategy, training could have an impact on participants performance, and needs further evaluation. It would be indeed relevant to understand and evaluate how long-term use of the device could impact and improve its effectiveness.

Acknowledgments. This work was supported by Inria Défi Project "DORNELL". This work involved human subjects or animals in its research. Approval of all ethical and experimental procedures and protocols was granted by Inria's ethics committee (COERLE No. 2021-39).

References

1. Amemiya, T., Gomi, H.: Buru-Navi3: behavioral navigations using illusory pulled sensation created by thumb-sized vibrator. In: Proceedings of ACM SIGGRAPH (2014)
2. Azenkot, S., et al.: Smartphone haptic feedback for nonvisual wayfinding. In: Proceedings of ACM SIGACCESS, pp. 281–282 (2011)
3. Bouzbib, E., Kuang, L., Robuffo Giordano, P., Lécuyer, A., Pacchierotti, C.: Survey of Wearable Haptic Technologies for Navigation Guidance (2023). https://inria.hal.science/hal-04356277
4. Chinello, F., Pacchierotti, C., Tsagarakis, N.G., Prattichizzo, D.: Design of a wearable skin stretch cutaneous device for the upper limb. In: Proceedings of IEEE Haptics Symposium (HAPTICS), pp. 14–20 (2016)
5. Chinello, F., et al.: Design and evaluation of a wearable skin stretch device for haptic guidance. IEEE RA-L **3**(1), 524–531 (2018)
6. Choiniere, J.P., Gosselin, C.: Development and experimental validation of a haptic compass based on asymmetric torque stimuli. IEEE Trans. Haptics **10**(1), 29–39 (2017)
7. Devigne, L., Aggravi, M., Bivaud, M., Balix, N., Teodorescu, C.S., Carlson, T., Spreters, T., Pacchierotti, C., Babel, M.: Power wheelchair navigation assistance using wearable vibrotactile haptics. IEEE Trans. Haptics **13**(1), 52–58 (2020)
8. Duvernoy, B., et al.: HaptiComm: a touch-mediated communication device for deafblind individuals. IEEE RA-L **8**(4), 2014–2021 (2023)
9. Georgiou, T., Holland, S., van der Linden, J.: Wearable haptic devices for post-stroke gait rehabilitation. In: Proceedings of the 2016 ACM International Joint Conference on Pervasive and Ubiquitous Computing: Adjunct, pp. 1114–1119 (2016)
10. Jiménez, M., et al.: Assistive locomotion device with haptic feedback for guiding visually impaired people. Med. Eng. Phys. **80**, 18–25 (2020)
11. Kappers, A.M.L., Oen, M.F.S., Junggeburth, T.J.W., Plaisier, M.A.O.: Hand-held haptic navigation devices for actual walking. IEEE Trans. Haptics **15**(4), 655–666 (2022)
12. Kuang, L., Marchal, M., Aggravi, M., Giordano, P.R., Pacchierotti, C.: Design of a 2-DoF haptic device for motion guidance. In: International Conference on Human Haptic Sensing and Touch Enabled Computer Applications, pp. 198–206 (2022)
13. Kuang, L., et al.: Wearable cutaneous device for applying position/location haptic feedback in navigation applications. In: Proceedings of IEEE Haptic Symposium, pp. 1–6 (2022)
14. Lacôte, I., et al.: Speed discrimination in the apparent haptic motion illusion. In: Proceedings of EuroHaptics. Lecture Notes in Computer Science, vol. 13235, pp. 48–56. Springer, Cham (2022)
15. Lacôte, I., et al.: "Tap Stimulation": an alternative to vibrations to convey the apparent haptic motion illusion. In: Proceedings of IEEE Haptics Symposium (2022)
16. Lacôte, I., et al.: Investigating the haptic perception of directional information within a handle. IEEE Trans. Haptics **16**(4), 680–686 (2023)
17. Lacôte, I., et al.: Comparing the haptic perception of directional information using a uni-manual or bi-manual strategy on a walker. In: Proceedings of IEEE Haptics Symposium (2024)

18. Morris, A., et al.: A Robotic Walker that Provides Guidance. In: Proceedings of IEEE ICRA, pp. 25–30 (2003)
19. Nasser, A., et al.: ThermalCane: exploring thermotactile directional cues on cane-grip for non-visual navigation. In: Proceedings of ACM SIGACCESS (2020)
20. Schmidt, H., et al.: HapticWalker –Haptic foot device for gait rehabilitation. In: Human Haptic Perception: Basics and Applications, pp. 501–511. Birkhäuser (2008)
21. Spiers, A., et al.: The S-BAN: insights into the perception of shape-changing haptic interfaces via virtual pedestrian navigation. ACM Trans. CHI **30**(1), 1 (2023)
22. Spiers, A.J., Dollar, A.M.: Outdoor pedestrian navigation assistance with a shape-changing haptic interface and comparison with a vibrotactile device. In: Proceedings of IEEE Haptics Symposium, pp. 34–40 (2016)
23. Toohey, K., Duckham, M.: Trajectory similarity measures. ACM SIGSPATIAL Special **7**(1), 43–50 (2015)
24. Van Erp, J.B.: Guidelines for the use of vibro-tactile displays in human computer interaction. In: Proceedings of Eurohaptics, pp. 18–22 (2002)
25. Wachaja, A., et al.: Navigating blind people with walking impairments using a smart walker. Auton. Robot. **41**(3), 555–573 (2017)
26. Yokota, S., et al.: The assistive walker using hand haptics - the design of the prototype. In: Proceedings of International Conference on Human System Interactions (HSI), pp. 214–218 (2013)

Data-Driven Haptic Modeling of Inhomogeneous Viscoelastic Deformable Objects

Gautam Kumar[1], Shashi Prakash[1], Hojun Cha[2], Amit Bhardwaj[1(✉)], and Seungmoon Choi[2]

[1] Indian Institute of Technology, Jodhpur, India
{kumar.156,prakash.8,amitb}@iitj.ac.in
[2] Pohang University of Science and Technology, Pohang, South Korea
{hersammc,hoism}@postech.ac.kr

Abstract. This work provides a new approach for modeling of inhomogeneous viscoelastic deformable objects. The approach is validated on a dataset collected at multiple locations of three different inhomogeneous deformable objects. The dataset consists of position and force measurements corresponding to single finger normal interaction. The approach, first, employs the principles of feature-based learning and perceptual adaptive sampling mechanisms to reduce the dataset. Then, a single random forest-fractional derivative (RF-FD) based data-driven model is trained on the reduced dataset to estimate a non-parametric relation between position and force samples for each deformable object. Thus, the proposed approach requires just one trained model to predict interactions at unknown locations of the object with good accuracy, unlike the existing clustering based solution in the literature where one model is trained for each cluster. Our results demonstrate that the proposed approach provides a better prediction accuracy in estimating the responses on inhomogeneous objects as compared to the existing solution in the literature in terms of the relative root mean square error (less than 0.15) and maximum error (less than 0.75 N).

Keywords: Data-driven haptic modeling · Deformable object · Clustering · Random forest · Fractional derivatives

1 Introduction

With the availability of large numbers of force feedback haptic devices, we are allowed to interact with an object into a virtual environment and feel the sense of touch of the underlying object. The sense of touch (i.e., the haptic sense) is felt primarily in terms of the interaction force between the device and the object. An algorithm which computes this interaction force is called haptic rendering. Thus, haptic rendering algorithms along with the haptic devices augment the sense of touch into the virtual environment. Bringing the sense of touch into a virtual

© The Author(s), under exclusive license to Springer Nature Switzerland AG 2025
H. Kajimoto et al. (Eds.): EuroHaptics 2024, LNCS 14768, pp. 405–418, 2025.
https://doi.org/10.1007/978-3-031-70058-3_33

environment has many applications in the field of virtual or augmented reality (VR or AR) as it promises to provide more realistic experiences to the users. In particular, haptic rendering has potential in designing medical training setups into virtual environments.

Data-driven haptic rendering has been a focus of the haptic research community over the last one decade as an alternative to conventional physics-based rendering approaches [8]. Unlike the physics-based approach, it is a model free approach and has the capacity to handle complex mechanisms with high perceptual realism. Here the interaction data (position, velocity, acceleration, force, etc.,) are acquired from real objects, and machine learning models are trained on the acquired data to model a non-parametric interpolation function between the input and output. The trained models are then used for haptic rendering of objects in a virtual environment. Data-driven haptic rendering is also named as *Haptic Camera* in the literature in a sense that we capture haptic properties of a real object and play/feel those haptic sensations in the virtual environment with high realism [11].

In the recent past, data-driven approaches have been employed for modeling and rendering various haptic properties (like surface texture, visco-elastic behavior and thermal sensation) in the virtual environment [1,5,8,12–14,17]. Here in this work, we focus on modeling of viscoelastic deformable objects using data-driven approaches. The very first work in this direction employed radial basis functions (RBFs) for estimating the viscoelastic response of a deformable object [7,8]. The goal was modeling of normal interactions. In [14], authors extended the approach to handle slip interactions on the object. In the very recent work [15], authors have proposed a feature-based learning (FB-L) for data-driven haptic rendering to model two-finger normal interactions on a deformable object. The FB-L approach reduces the size of the captured data by retaining only the most informative features. This transforms the discrete-time data into a low-dimensional feature space in the frequency domain. Finally, the RBF-based models are trained on the reduced discrete-time dataset.

In another recent work [1], authors have proposed an alternative data-driven approach for modeling of normal interactions on a viscoelastic deformable object. This approach is based on a random forest (a popular machine learning technique; RF) [2]. A RF is trained on the input feature vectors (consisting of current and past position information) for estimating the response force. Results showed that the RF-based approach required less training data than the RBF-based approach to provide similar modeling accuracy. In [10], authors have proposed a Cat-Boost based haptic modeling method for multi-finger interaction on deformable objects. However, this method has not yet been validated for haptic rendering.

In the literature, modeling of inhomogeneous deformable objects has received little attention so far as compared to homogeneous objects. Modeling inhomogeneity of a deformable object is a complex phenomenon. Unlike the homogeneous modeling case, the measurements need to be collected at multiple locations of the object, and the modeling of this resulting dataset is a challenging task. In [17], Choi et al. incorporated the position of a proxy in the RBF formulation to model frictional responses on an inhomogeneous deformable object. In another study [14], authors employed the RBF-based approach along with a pattern matching algorithm for modeling inhomogeneity. The approach, first, segments an object into homogeneous clusters and then trains a separate RBF model for each cluster. The models trained at different clusters are employed for predicting responses at any unknown location on the object. The resulting algorithm may lead to spatially disconnected homogeneous clusters which makes the interpolation of the forces at unknown locations difficult. In addition, both the works [14,17] have employed the RBF-based data-driven approach which becomes computationally intractable, when learned/trained on large datasets. Hence, we need an alternative computationally efficient mechanism to model and render the inhomogeneity of deformable objects. In this work, we propose an approach for modeling of an inhomogeneous deformable object where a single RF-FD model is trained for the whole object. The approach is inspired from the feature-based learning (FB-L) [15] and the perceptual sampling [4,16]. The results show that the single trained model predicts the interactions at unknown locations with good accuracy.

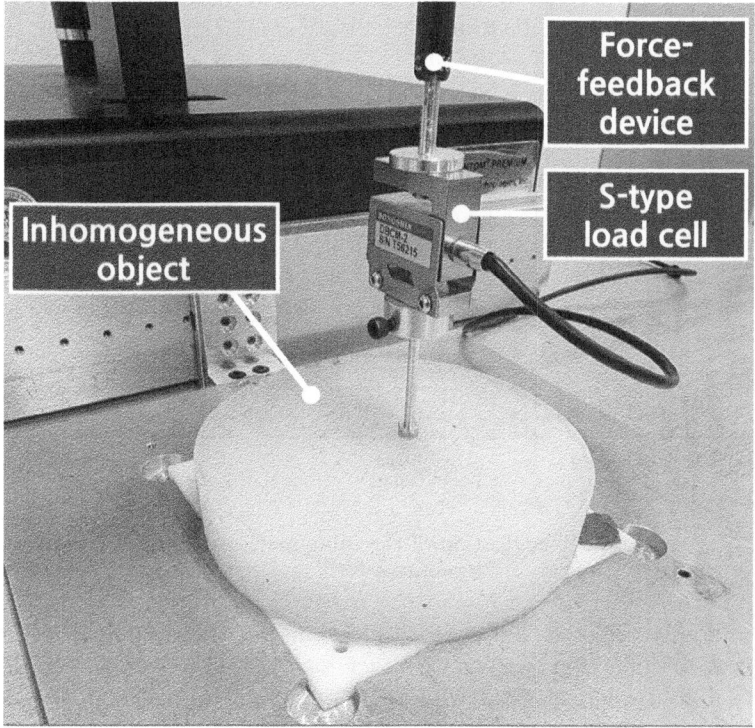

Fig. 1. Data collection configuration for a single inhomogeneous object.

2   Data Collection

We design an experimental setup to record input-output measurements on an inhomogeneous deformable object. The experimental setup consists of a force-feedback device (Phantom Premium 1.5 HF, 3D Systems), the load cell (DBCM-2 kg, Bongshin) and an inhomogeneous deformable object, as shown in Fig. 1. We attached the load cell at the end-effector of the device to measure the response forces. A 6-mm diameter tip is attached at the end of the load cell for interaction with the object.

The data is collected on three physical mockups made of soft silicone as used in [17]. All the mockups were made of Ecoflex 0010 from SmoothOn Inc. The first mockup has a cylindrical shape while the second and third mockup have half-spherical shapes. In the first mockup, we embedded small silicone pieces of different stiffness at its centre to represent inhomogeneity. This mockup will be referred to as Object O_1 in the paper. A void and stiffer spherical rubber ball is embedded at the center of the second and third mockups, respectively (as shown in Fig. 2). The second and third mockups will be referred to as Object O_2 and O_3 in the paper. For data collection at each object, we chose a modeling area (60 mm by 50 mm) which shows a high level of inhomogeneity as depicted in Fig. 2 with a detailed specification.

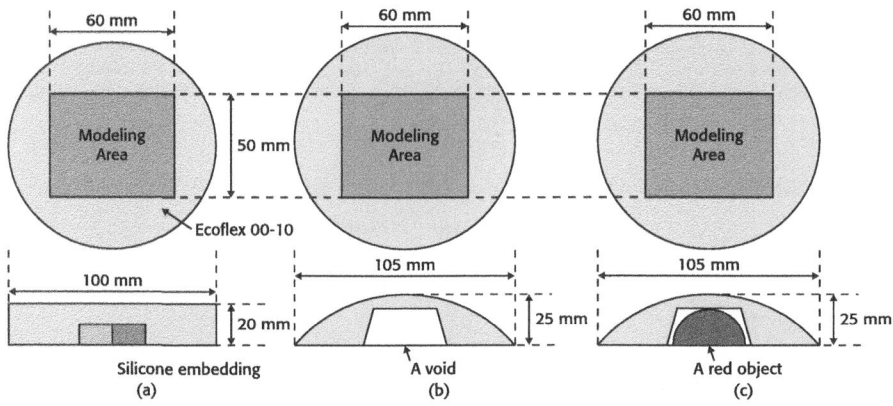

Fig. 2. Specification of the inhomogeneous object.

We randomly selected 200 contact locations on the modeling area of each object for data collection. At each contact location, the data is recorded using the position controlled method where the device is controlled to reach at the target displacement. The position-force measurements are recorded using the force-feedback device by applying an inverted cosine force control signal to the device's tip:

$$d(t) = -\frac{a}{2}(\cos(2\pi\nu t) - 1),   \tag{1}$$

where a is the target (peak-to-peak) amplitude in mm, ν is the indentation frequency in Hz. In order to capture the non-linear and rate-time dependent behavior of visco-elastic deformable objects, the data is recorded for three amplitudes, $a \in \{4, 8, 12\}$ (in mm) and three frequencies $\nu \in \{0.5, 1, 2\}$ (in Hz). For each pair (a, ν), the interaction data (position and force vectors) is collected for a single indentation cycle. We denote the resulting force and position vectors at any typical contact location l as $f_k^l \in \mathbb{R}^{T_k}$ and $d_k^l \in \mathbb{R}^{T_k}$, where $k \in \{1, \cdots, K\}$, $l \in \{1, 2, \cdots, L\}$ and T_k represents the length of the kth interaction. In addition, velocities are also stored for each interaction in v_k^l. Next, we create two global vectors using the obtained vectors: $Z = \{z_k^l | z_k^l \in \mathbb{R}^{2 \times T_k}, \ k = 1, 2, \cdots, K; \ l = 1, 2, \cdots, L\}$, where $z_k^l = [d_k^l, \ v_k^l]$ and $F = \{f_k^l | f_k^l \in \mathbb{R}^{T_k}, \ k = 1, 2, \cdots, K; \ l = 1, 2, \cdots, L\}$. Thus for each object, the interaction data (position, velocity and force measurements) are collected for 1800 ($K = 9$ and $L = 200$) sinusoidal signals. In Fig. 3, we show a typical collected position and force curves for $a = 8$ mm and $\nu = 1$ Hz. Figure 4 shows hysteresis curves (force vs position) at different locations of Object O_1 for a typical fixed indentation. Different shapes of the curves signify towards inhomogeneity of the deformable object.

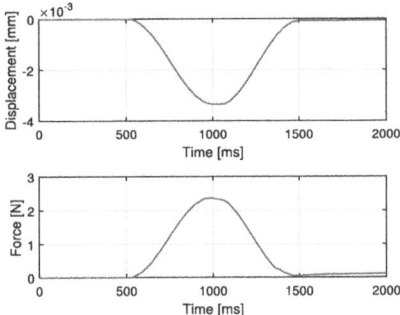

Fig. 3. Measured force and position data at a typical contact location on Object O_1

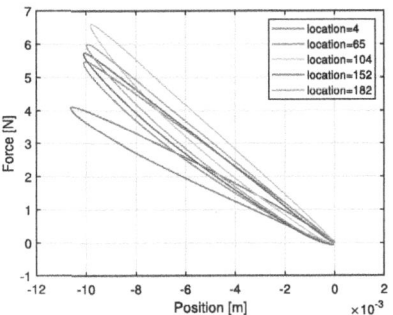

Fig. 4. Illustration of inhomogeneity: hysteresis curves at different locations of Object O_1

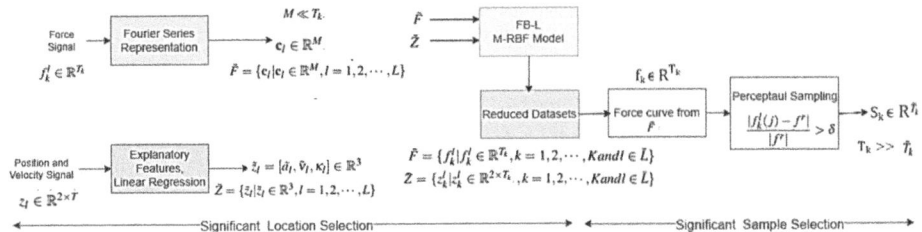

Fig. 5. Overview of data reduction process

3   Methodology

Having collected the interaction data at different locations of each deformable object, the next task is to develop a single machine learning model to predict a mapping function between the recorded position/velocity and force measurements across the whole modeling area of the object. Since the data is collected for $K = 9$ different types of interactions at each $L = 200$ locations, we, first, identify significant locations using the feature-based learning (FB-L) approach from [15]. For further reduction of the data points, a perceptual adaptive sampling mechanism is applied on each interaction of the selected significant locations to identify significant samples. An overview of the data reduction process is shown in Fig. 5. At last, a single RF-FD based data-driven model is trained on the reduced dataset for each object. Below, we discuss each of the above mentioned steps in detail.

3.1   Significant Location Selection

We employ the FB-L approach from [15] to identify significant locations on the object. For the purpose, data points corresponding to a typical k^{th} interaction are considered for each location. That is, the approach is applied on data points of two new global vectors $Z_s = \{z_k^l \mid z_k^l \in \mathbb{R}^{2 \times T_k}, \; k = k_s; \; l = 1, 2, \cdots, 200\} \subset Z$, and $F_s = \{f_k^l \mid f_k^l \in \mathbb{R}^{T_k}, \; k = k_s; \; l = 1, 2, \cdots, L\} \subset F$. For simplicity, we omit the k symbol for the data points of the newly defined global vectors. In the approach, firstly, the recorded position and velocity data points, $z^l = \{d^l, \; v^l\}$, and force values f^l are transformed into lower dimensional explanatory (input) and response features (output), respectively.

To get lower dimensional response features, the force vectors f^l are represented as a Fourier series:

$$f^l \approx c_{0,l} + \sum_{n=1}^{N} \left[c'_{n,l} \cos(n\omega t) + \mathrm{j} \, c''_{n,l} \sin(n\omega t) \right] \tag{2}$$

where $c_{0,l}, c'_{n,l}$ and $c''_{n,l}$ represent the Fourier coefficients of the time domain force curve f^l, and ω corresponds to the angular frequency of the force vector. In order to compute the Fourier coefficients, it has been assumed that the force vectors f^l are periodic and stationary, provided that their energy is localized to the low frequency. The Fourier coefficients are computed using the Discrete Fourier Transform and stacked together into a single response feature vector $\mathbf{c}_l = [c_{0,l}, \; c'_{1,l}, \; c''_{1,l}, \cdots, \; c'_{N,l}, \; c''_{N,l}] \in \mathbb{R}^{M=2N+1}$. For better readability, the elements of the vector $\mathbf{c}_l$ are denoted as $c_{m,l}$, $m = 0, 1, \cdots, 2N$. This vector represents the recorded force vector f^l in the Fourier domain. Thus, this step has transformed a force vector $f_k^l \in \mathbb{R}^{T_k}$ to $\mathbf{c}_l \in \mathbb{R}^M$, where $M \ll T_k$. In this work, we consider $M = 11$ feature components in $\mathbf{c}_l$ for each recorded signal. The obtained vectors $\mathbf{c}_l$ are used to form a new response feature set $\tilde{F} = \{\mathbf{c}_l \mid \mathbf{c}_l \in \mathbb{R}^M, \; l = 1, 2, \cdots, L\}$.

Next, we target to reduce the dimensionality of $z_l \in \mathbb{R}^{2 \times T}$. The force vector f^l depends on the following parameters: amplitude a, angular frequency ν and the object stiffness κ at the corresponding location. We consider these parameters as our explanatory (input features) variables for the recorded displacement and velocity signals and stack them together to generate a explanatory feature vector $\tilde{z}_l = [\tilde{a}_l, \ \tilde{\nu}_l, \ \kappa_l] \in \mathbb{R}^3$. These features are derived from the measured data. To compute object stiffness κ_l at the lth location, we solve a simple linear regression problem between f^l and d^l. All the features in the explanatory feature vector $\tilde{z}_l$ are normalized in the range $[0, 1]$. The resultant explanatory features are used to form a new explanatory feature set $\tilde{Z} = \{\tilde{z}_l| \ \tilde{z}_l \in \mathbb{R}^3, \ l = 1, 2, \cdots, L\}$.

Next, the approach employs a simple greedy search algorithm along with radial basis functions (RBFs) progressively on M sets of explanatory and response features $\{\tilde{Z}, \ c_m\}$ where $c_m = [c_{m,1}, c_{m,2}, \cdots, c_{m,L}] \in \mathbb{R}^L$, $m = 0, 1, \cdots, 2N + 1$. The approach starts with $m = 0$, where we learn the RBF interpolation function on L response feature vectors $\tilde{z}_l$ from the feature set $\tilde{Z}$ to predict response feature vector c_0. The predicted response vector is denoted as $c_0(\tilde{z})$. For the purpose, the RBF interpolation function is, firstly, initialized with a single RBF centre $\tilde{z}_1 = [\tilde{a}_1, \ \tilde{\nu}_1, \ \kappa_1]$ and learned to predict the response feature vector $c_0(\tilde{z})$. The component $l \in L$ for which the absolute relative error between $c_0(\tilde{z})$ and c_0 is maximum is added to the list of RBF centres, and the function is re-trained. This process continues until the absolute relative error between the predicted and actual response feature sample falls below a threshold ϵ. The RBF centres finally obtained for $m = 0$ are retained and used as an initialization for the prediction of $c_1(\tilde{z})$ (i.e., $m = 1$ stage interpolation). This process continues till the RBF models are learned for all m. Finally, the indices of the RBF centres obtained for $m = 2N + 1$ stage RBF interpolation are termed as significant locations for the object, which are stored in a set $\bar{L}$. This way, we obtain the reduced datasets $\bar{Z} = \{z_k^l | z_k^l \in \mathbb{R}^{2 \times T_k}, \ k = 1, 2, \cdots, K; \ l \in \bar{L}\}$ and $\bar{F} = \{f_k^l | f_k^l \in \mathbb{R}^{T_k}, \ k = 1, 2, \cdots, K; \ l \in \bar{L}\}$. Thus, both the reduced datasets consist of the data points for $T_r = (K \times \bar{L})$ number of interactions on each deformable object for further process. Figure 6 shows the illustration of the FB-L approach on Object O_1. Here, the blue points represent the significant locations identified by the FB-L approach. For $M = 11$ and $\epsilon = 0.05$, we get 26, 28, and 27 number of significant locations for Objects O_1, O_2 and O_3, respectively.

3.2 Significant Sample Selection

In the last step, we obtained the reduced datasets $\bar{Z}$ and $\bar{F}$ after identifying the significant locations on the object. But, the length of the respective vectors d_k^l, v_k^l, and f_k^l in the reduced datasets remain unchanged. Thus, we further reduce the data by employing a perceptually adaptive sampling mechanism on force vectors f_k^l. The mechanism has been widely used in haptic data communication [4,16] for data reduction. The mechanism is based on Weber's law in perception which selects only perceptually significant samples using the Weber fraction δ for data transmission. The mechanism is named as perceptual deadband or

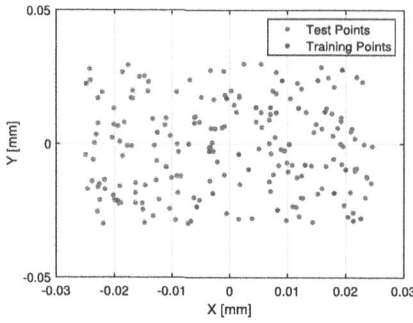

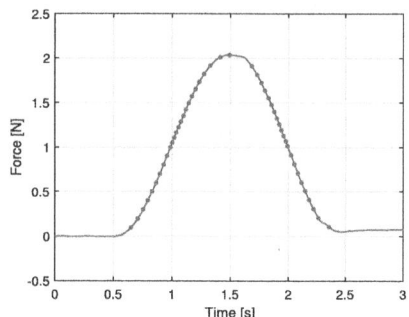

Fig. 6. Illustration of feature-based learning: blue points correspond to the significant locations and red points for test locations

Fig. 7. Illustration of perceptual sampling: red points correspond to the selected sample points

deadzone; see [4] for further details. We employ the same adaptive sampling mechanism for data reduction. The procedure is applied to each sample of the force curve f_k^l. It begins with the first sample $f_k^l(1)$ and considers it as the reference force f^r. We progressively search for the sample index j that satisfies

$$\frac{|f_k^l(j) - f^r|}{|f^r|} > \delta. \tag{3}$$

Since the Weber's law for force defined by Eq. (3) is not effective for very small forces [4], the following condition is searched for the sample index j for forces in $(0, 1)$ N

$$|f_k(j) - f^r| > c \tag{4}$$

where c is a level-crossings constant [4]. This sample $f_k^l(j)$ is considered to have significant information with respect to the reference f^r. The sample index j is stored, and the reference force f^r is replaced with $f_k^l(j)$ for the next search. We repeat this process on each force curve in the dataset separately and store the indices of all selected samples.

This procedure may discard local minima or maxima in the force curves, which are essential for model training. Thus, we also consider the indices of the local minima/maxima. Suppose that the selected sample points from a typical force curve f_k^l are marked in the index sets A_k and the local minima and maxima are in the index set B_k. Then, the reduced data is represented by the union index set $S_k = A_k \cup B_k$ and $\bar{T}_k$ represents the number of samples in the set. Finally, we only consider the sample points belonging to the union index set S in the reduced datasets $\bar{Z}$ and $\bar{F}$ for further process.

Figure 7 shows the results of adaptive sampling on a typical force curve for $\delta = 0.05$ ($\simeq$ force perception JND [9]) and $c = 0.10$ N, where red dots represent the selected samples. Hereafter, the same value of c is used. As illustrated from the plot, the approach can handle the complexities (non-linearity and rate-dependent behavior) of viscoelastic objects. For example, when the signal changes rapidly (slowly), the approach is bound to select more (less) samples. The Weber fraction δ determines the size of the reduced data. The Weber fraction δ for force perception varies up to 15% [16]. In this work, we select the Weber fraction $\delta = 0.05$ considering the threshold of the force perception of the finger. With this threshold, the approach retains only around 4% of the sample points from the force vectors f_k^l.

3.3   Random Forest Modeling

Next, we employ RF-FD [3] based data driven model on the reduced datasets $\bar{Z}$ and $\bar{F}$ to learn a non-parametric mapping function between the input (position) and output (force) samples. The RF-FD model is trained on the input features derived from the fractional derivatives (FDs) of position information. Apart from this, we also include the X-Y position information of the contact location in the input feature vector. The nth input feature vector $I[n]$ for the kth interaction is defined as follows:

$$I[n] = \left(X, Y, D^{r1}d[n], D^{r2}d[n], \cdots, D^{r10}d[n] \right) \tag{5}$$

where X, Y denote the x and y coordinates of the contact location under consideration and $D^r d[n]$ denotes the rth fractional derivative of $d[n]$ ($r_i \in [0, 1]$, $i = 1, 2, \cdots, 10$). Then, the feature vectors for the kth interaction are denoted as

$$\mathbf{I}_k = \{I[i] | I \in \mathbb{R}^{12}, i \in S_k\} \tag{6}$$

$$\mathbf{f}_k = \{f_k[i] | f_k \in \mathbb{R}, i \in S_k\} \tag{7}$$

where S_k represents the union index set for the kth interaction obtained after perceptual sampling. Next, we stack together these feature vectors to form two new input and output vectors:

$$\mathbf{I}_g = \{\mathbf{I}_k | \mathbf{I}_k \in \mathbb{R}^{12 \times \bar{T}_k}, k = 1, 2, \cdots, T_r\} \tag{8}$$

$$\mathbf{f}_g = \{\mathbf{f}_k | \mathbf{f}_k \in \mathbb{R}^{1 \times \bar{T}_k}, k = 1, 2, \cdots, T_r\} \tag{9}$$

where T_r represents the total number of interactions selected for an object across all the locations after applying the FB-L approach. The RF-FD model is trained on the global vector pair $(\mathbf{I}_g, \mathbf{f}_g)$ for each object with the following parameters: ten orders of FDs=$\{0.05, 0.10, \cdots, 0.50\}$, the number of decision trees= 100 and the stopping criteria= minimum five samples at leaf nodes. The parameters may be further optimized for each object.

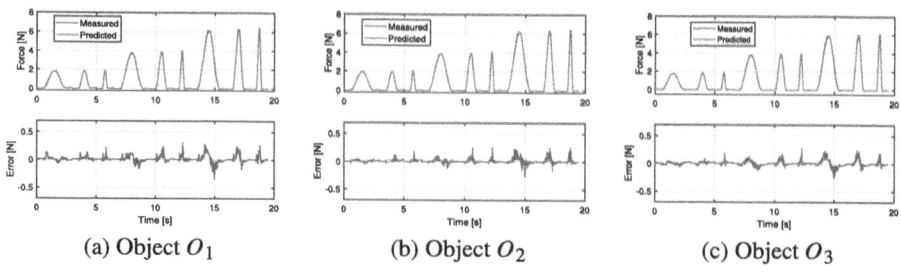

Fig. 8. Validation: measured and predicted force curves for all interactions (combined together) at typical test locations on each object (top panel), and the corresponding errors in the bottom panel

4   Results and Validation

Next, we validate the trained RF-FD model on the remaining locations (not selected by FB-L procedure) of the object. Figure 8 compares the measured and predicted force curves at typical test locations on all three objects. The bottom panel of the figure also shows the error between the measured and predicted curves. The results show that both the curves match quite well for all the objects as the error is limited to 0.5 N.

For quantitative valuation, we compute the relative root mean square error (RMSE) between the measured and predicted forces. Let there are n number of samples in the measured/predicted force curves. Then, the relative RMSE is computed as follows:

$$\text{Relative RMSE} = \frac{\sqrt{\frac{1}{n}\sum_{i=1}^{n}(f_i - \hat{f}_i)^2}}{\sqrt{\frac{1}{n}\sum_{i=1}^{n} f_i^2}} \tag{10}$$

where f_i and $\hat{f}_i$ denote i^{th} measured and predicted force sample. Figure 9 shows the histogram of the relative RMSEs computed at all the test locations of each deformable object. For Objects 1, 2 and 3, the proposed approach predicts more than 90%, 95% and 75%, of the test locations with the relative RMSE less than 0.10, respectively. Further, we also plot the spatial distribution of the relative RMSEs as a function of their test contact locations on each object in Fig. 10. The results show that for all the objects, the relative RMSEs more than 0.10 are observed around the center of the object (i.e., the region of inhomogeneity). If the parameters of the FB-L are optimized for each object, a better performance may also be achieved.

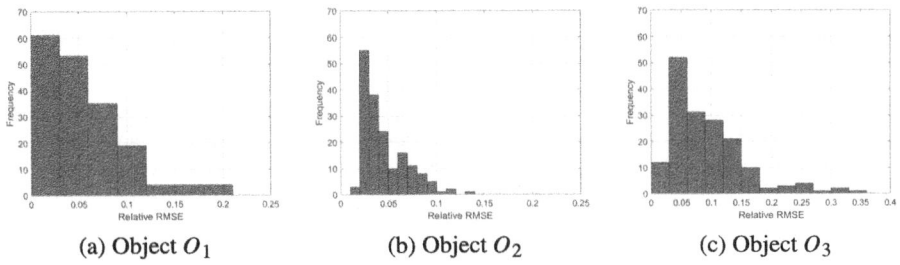

(a) Object O_1 (b) Object O_2 (c) Object O_3

Fig. 9. Quantitative Validation: histograms of the relative RMSEs computed at all test locations of the object

In addition, we also compute the absolute relative error (ARE) between the measured and predicted force curves as follows:

$$\text{ARE} = \frac{1}{n} \sum_{i=1}^{n} \left| \frac{f_i - \hat{f}_i}{f_i + 0.01} \right| \tag{11}$$

Figure 11 shows the ARE as a function of the measured force for Object O_1 corresponding to the test location where the maximum error between the measured and predicted force curves are found. It is generally less than 0.10 for the measured forces more than 2 N and tends to decrease as the the measured force is increased. As the JND for force perception lies between 7–15% [6], the absolute relative error is in the acceptable perceptual limit. The similar behavior is observed for the other test locations of each deformable object. The above results signify that the trained RF-FD based model predicts the forces at unknown locations of an inhomogeneous object with good accuracy.

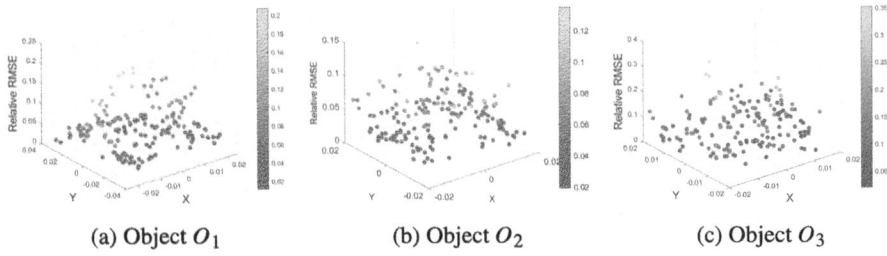

(a) Object O_1 (b) Object O_2 (c) Object O_3

Fig. 10. Quantitative Validation: Spatial distribution of the relative RMSEs computed at all test locations of the object

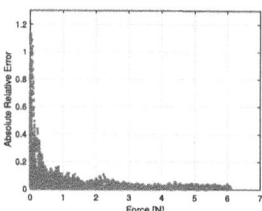
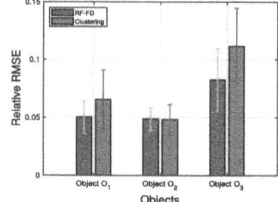
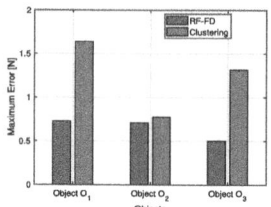

Fig. 11. Absolute relative error vs force for Object O_1

Fig. 12. Relative RMSE across all the objects

Fig. 13. Maximum error across all the objects

4.1 Comparison with the Clustering Approach

We compare the proposed approach with the existing clustering approach in the literature [14] for modeling inhomogeneous deformable objects. As mentioned earlier, the clustering approach segments the object into homogeneous clusters using a widely known K-mean algorithm. Randomly selected force samples are used as features for the clustering algorithm. Having clustered the object into homogeneous clusters, a cluster representative (CR) is identified for each cluster. A CR location represents a location on the cluster that provides the least mean square (measured between the force curves) with respect to all other locations in the corresponding cluster. Finally, a data-driven model is trained at each CR location of the object. For any unknown location on the object, the model trained on corresponding CR location will be employed for predicting the responses. We select five CR locations (identified by elbow method) using the K-means algorithm on each deformable object and train the same RF-FD model on each CR location.

Figure 12 compares the average RMSE (over all test locations) of the proposed approach with respect to the clustering approach for each object. The figure also shows the standard deviation of the RMSEs. If we compare the standard deviation of the RMSEs, the results show that even after training the five RF-FD models, the clustering approach has much higher variation in the RMSEs across all test locations of each object as compared to the proposed approach. The average relative RMSEs provided by our approach lie in the range of [0.05–0.08] across all the objects, whereas it lies in the range of [0.05–0.13] for the clustering approach. For Objects O_1 and O_3, the proposed approach performs much better than the previous approach in terms of the RMSE accuracy. Both the approaches provide the similar RMSEs for Object O_2, even though, the clustering approach requires five trained models for providing that prediction performance. Reason for getting different results for Object O_2 is that it has a void at its centre unlike the other two objects. In Fig. 13, we also compare both the approaches in terms of the maximum error between the measured and predicted forces across all test locations. Across all the objects, the maximum error of the proposed approach is restricted up to 0.75 N and is much lesser than that of the clustering approach, where it is up to 1.5 N. Hence, the above results signify that the proposed single trained RF-FD based data-driven model performs better than the clustering approach in terms of the average RMSEs and maximum errors. In addition, we also compare the performances of both the methods in terms of the training time. The training time for the proposed approach is 44.77 s while the clustering approach requires 238.12 s for training. Thus, the proposed approach is six times faster than the clustering approach in terms of the training time.

4.2 Discussion

The single trained RF-FD model may be finally employed for haptic rendering of the deformable object in a virtual environment. The implementation of the RF-FD model for haptic rendering in CHAI 3D platform is straight forward [3]. In our previous work [3], we showed using human perceptual experiments that the trained RF-FD model provided stable haptic interactions on homogeneous deformable objects in a virtual environment. In future, we intend to validate the proposed approach on actual haptic rendering of inhomogeneous objects using human studies.

The proposed method also has some limitations. First, all data-driven methods have poor extrapolation performance outside the range of training data, and ours is not an exception. Second, the current approach deals with only normal interactions. We plan to extend the approach to perform on interactions in other directions.

5 Conclusions

We have proposed a new computationally efficient approach to model inhomogeneous viscoelastic deformable objects. The approach required to train just one RF-FD based data-driven model on the dataset (reduced by the feature-based learning and perceptual adaptive sampling mechanisms) for modeling an inhomogeneous deformable object. The approach is validated on three different inhomogeneous deformable objects. Experimental results verify that the single trained model predicts interactions at unknown locations of the objects with good accuracy. The proposed approach also requires a much lesser number of trained models as compared to the existing clustering based solution and provides better prediction accuracy than it in terms of the RMSE and maximum error. All of these demonstrate the capability of the proposed approach to model inhomogeneity of viscoelastic deformable objects.

Acknowledgment. This work was supported in part by a grant from i-HUb Drishti, TIH Jodhpur for a project titled as "Haptics based Medical Simulator for Abdomen Palpation and Pulse Behavior" and in part by a grant from IHFC, TIH IIT Delhi for a project titled as "Telepresence and Teleaction System for Robot Assisted Dentistry".

References

1. Bhardwaj, A., Cha, H., Choi, S.: Data-driven haptic modeling of normal interactions on viscoelastic deformable objects using a random forest. IEEE Robot. Autom. Lett. **4**(2), 1379–1386 (2019)
2. Breiman, L.: Machine Learning, chap. 1, pp. 5–32. Springer (2001)
3. Cha, H., Bhardwaj, A., Choi, S.: Data-driven haptic modeling and rendering of viscoelastic behavior using fractional derivatives. IEEE Access **10**, 130894–130907 (2022)
4. Chaudhuri, S., Bhardwaj, A.: Kinesthetic Perception: a Machine Learning Approach, vol. 748. Springer (2017)
5. Choi, H., Cho, S., Shin, S., Lee, H., Choi, S.: Data-driven thermal rendering: an initial study. In: IEEE Haptics Symposium (HAPTICS), pp. 344–350 (2018)
6. Hinterseer, P., Hirche, S., Chaudhuri, S., Steinbach, E., Buss, M.: Perception-based data reduction and transmission of haptic data in telepresence and teleaction systems. IEEE Trans. Signal Process. **56**(2), 588–597 (2008)
7. Höver, R., Harders, M.: Measuring and incorporating slip in data-driven haptic rendering. In: IEEE Haptics Symposium (HAPTICS), pp. 175–182 (2010)
8. Hover, R., Kósa, G., Szekly, G., Harders, M.: Data-driven haptic rendering-from viscous fluids to visco-elastic solids. IEEE Trans. Haptics **2**(1), 15–27 (2009)
9. Jeon, S., Choi, S.: Real stiffness augmentation for haptic augmented reality. Presence Teleoper. Virtual Environ. **20**(4), 337–370 (2011)
10. Kumar, G., Prakash, S., Bhardwaj, A.: Catboost for haptic modeling of homogeneous viscoelastic deformable objects. In: IEEE World Haptics Conference (WHC), pp. 273–278. IEEE (2023)
11. MacLean, K.E.: The 'haptic camera': A technique for characterizing and playing back haptic properties of real environments. In: Proceedings of Haptic Interfaces for Virtual Environments and Teleoperator Systems (HAPTICS), pp. 459–467 (1996)
12. Osgouei, R.H., Shin, S., Kim, J.R., Choi, S.: An inverse neural network model for data-driven texture rendering on electrovibration display. In: IEEE Haptics Symposium (HAPTICS), pp. 270–277 (2018)
13. Shin, S., Choi, S.: Geometry-based haptic texture modeling and rendering using photometric stereo. In: IEEE Haptics Symposium (HAPTICS), pp. 262–269 (2018)
14. Sianov, A., Harders, M.: Data-driven haptics: addressing inhomogeneities and computational formulation. In: IEEE World Haptics Conference (WHC), pp. 301–306 (2013)
15. Sianov, A., Harders, M.: Exploring feature-based learning for data-driven haptic rendering. IEEE Trans. Haptics **11**(3), 388–399 (2018)
16. Steinbach, E., Strese, M., Eid, M., Liu, X., Bhardwaj, A., Liu, Q., Al-Ja'afreh, M., Mahmoodi, T., Hassen, R., El Saddik, A., et al.: Haptic codecs for the tactile internet. Proc. IEEE **107**(2), 447–470 (2018)
17. Yim, S., Jeon, S., Choi, S.: Data-driven haptic modeling and rendering of viscoelastic and frictional responses of deformable objects. IEEE Trans. Haptics **9**(4), 548–559 (2016)

Perception of Paired Vibrotactile Stimulus on the Upper Limb: Implications for the Design of Wearable Technology

Dorine Arcangeli[1,2]($\boxtimes$) ⓘ, Gabriel Arnold[2], Agnès Roby-Brami[3],
Giovanni de Marco[1], Nathanaël Jarrassé[3], and Ross Parry[1] ⓘ

[1] LINP2, UPL, UFR STAPS, Université Paris Nanterre, 200 Avenue de la République, 92001 Nanterre, France
dorine.arcangeli@caylar.net
[2] CAYLAR, 14 Avenue du Québec, 91140 Villebon Sur Yvette, France
[3] ISIR, Sorbonne University, CNRS UMR 7222, ERL INSERM U 1150, 75005 Paris, France

Abstract. The ability to correctly perceive multiple stimuli represents an important barrier to the use of vibrotactile devices in training complex behaviors. The aim of this study was to evaluate how stimulation parameters influence perception of vibrotactile patterns applied to the forearm and upper arm. In this experimental protocol, participants (N = 16) were asked to compare two vibrotactile sequences and indicate whether they were the same or different. We examined the effects of (1) sequential versus simultaneous vibrotactile stimulation; (2) the temporal structure of the vibrotactile stimulus upon subject perception and (3) difference in pattern recognition between the two segments. Our results confirmed that perception was generally superior when the two stimuli were presented sequentially and when there was a marked difference in the temporal structure of the signals. At the same time, participants were highly capable of detecting two identical sequences when presented simultaneously. These findings may have important implications for the design of wearable vibrotactile devices intended for guiding upper limb movement.

Keywords: Haptic Technology · Vibrotactile Stimulation · Tactile Discrimination · Judgement · Tactile Pattern Recognition

1 Introduction

Assistive technology increasingly uses tactile cues to influence or correct motor behavior. Broadly speaking, this involves transmission of a mechanical stimulus (e.g. pressure or vibration) to the user's body. Devices using augmented tactile feedback may take different forms, including instrumented objects or robotic comanipulation devices [1, 2]. Vibrotactile matrices which integrate multiple vibrating actuators into a wearable garment are increasingly popular, with potential applications in navigation, behavioral training, and motor learning [3, 4]. The basic premise of such devices is that the tactile stimulus confers information regarding the user's actions, and how they conform with

H. Kajimoto et al. (Eds.): EuroHaptics 2024, LNCS 14768, pp. 419–427, 2025.
https://doi.org/10.1007/978-3-031-70058-3_34

the desired performance. In motor learning, feedback might indicate whether a particular goal has been achieved (i.e. knowledge of results). Alternatively, feedback may be used to provide knowledge of performance, that is to say, information regarding the coordination of movement and how that might be improved [5].

The appropriation of wearable vibrotactile devices for training motor skills across the health, industrial and sporting sectors, however, remains somewhat limited. In a recent systematic review, van Breda et al. found that learning effects when using these technologies were generally poor [6]. In relatively simple applications, such as using a vibrotactile system at the level of the wrist to guide hand movements in a two-dimensional plane, some benefits may be observed [7, 8]. Similarly, vibrotactile feedback may assist in training the orientation of a given segment (i.e. forearm) in a simulated reaching task [9]. The benefits of vibrotactile feedback for improving knowledge of performance across multiple segments though, are yet to be proven. Certain authors have attempted to train more complex upper limb gestures, including those used in combat sports or use of musical instruments [9, 10]. However, these attempts have often been based on the correction of end-state postures rather than online coordination between the different segments involved. Studies using vibratory stimulus to guide both forearm and upper arm movement simultaneously through the course of a gesture are less present in the scientific literature. Moreover, the results of those studies generally indicate that the positive effects of the vibrotactile system is negligible in gestures involving multiple degrees of freedom [6, 11].

Invariably, one of the key barriers to the use of this technology is the user's ability to interpret multiple tactile cues, particularly when they are presented simultaneously. In effect, subject ability to attend to multiple distinct vibrotactile signals may be associated with both physiological (e.g. masking phenomena associated with mutually inhibitory pathways) [12], as well as higher order cognitive processing capacity [13, 14]. For vibrotactile technology to be effective in more complex motor learning tasks, further work is required to determine specific configurations which favor one's ability to perceive and respond to multiple vibratory stimuli. In the present paper, our aim is to explore how stimulation parameters influence subject perception of vibrotactile patterns applied to the forearm and upper arm. More specifically, we examined the effects of (1) sequential versus simultaneous vibrotactile stimulation; (2) the temporal structure of vibrotactile stimulus upon subject perception; and (3) differences in pattern recognition between the two segments. During the experimental task, subjects were required to compare two vibrotactile sequences with varying levels of synchrony (i.e. temporal patterning between bursts of vibratory stimulus).

2   Materials and Methods

2.1   Participants

Sixteen healthy adult participants (7 males) with an average age of 24 years (SD = 2 years, range = 20–27 years) and no known neurological or orthopedic conditions were recruited for this study. All had either normal or corrected vision.

2.2 Experimental Setup

Vibrotactile stimuli were generated using a customized system comprised of a control unit and two vibratory modules each consisting of twenty-four vibrating actuators arranged in an 8 × 3 matrix (MTX-Lab, Caylar, France). Each vibrating actuator was a rotating eccentric mass encapsulated in a cylindrical tube with a 5 mm diameter and 11 mm length. Vibrators were spaced 10 mm from center to center and were fixed into a silicon support, holding them perpendicular to the surface of the skin. Surface contact between each vibrator and the underlying skin was thus 19.6 mm². Only 6 adjacent vibrators of each module (3 × 2 matrix at the center of the module) were activated (see Fig. 1.1).

The two vibratory modules were placed on the left arm of each participant with Velcro straps (see Fig. 1.2). Each was applied lengthwise along the ventral surface of the arm, with one at the level of the forearm and the other at the level of the upper arm. Participants were seated at a desk. The left arm was placed in a comfortable position before them, the shoulder flexed and slightly abducted, with the hand resting along a raised horizontal surface. A computer was then placed on the desk in front of the participant such that they read instructions on the screen whilst using their keyboard with their right hand to provide responses. This computer piloted the tactile device via ethernet connection and delivered visual instructions to the screen. The participants also wore headphones to render vibrating actuators inaudible during the experiment.

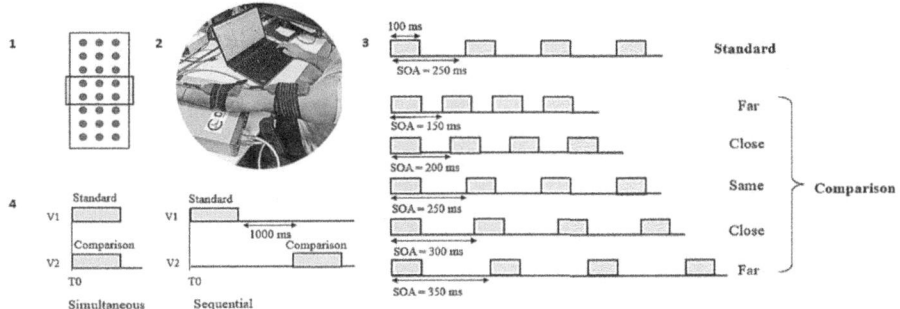

Fig. 1. 1. Vibrators activated during experiment. 2. Experimental Setup. 3. Standard and comparison sequences. 4. Simultaneous and sequential presentation.

2.3 Experimental Procedure

The experimental task required participants to indicate whether the temporal patterning of the stimuli applied to the forearm and upper arm were the same or different. Vibrotactile sequences consisting of 4 or 6 100 ms vibratory bursts (250 Hz) punctuated by specific stimulus onset asynchronies (SOA) were presented. One of the two vibratory sequences, the standard sequence, had a constant SOA of 250 ms. The other vibratory sequence, the comparison sequence, had a varying SOA, ranging between 150 ms and 350 ms (i.e. 150 ms, 200 ms, 250 ms, 300 ms, 350 ms). Therefore, the comparison sequence was composed of either the same temporal patterning as the standard, or a different temporal patterning, that is, with shorter or longer SOA. Also, the difference in temporal patterning between the standard and the comparison could be classified as either far (SOA of 150 ms or 350 ms) or close (SOA of 200 ms or 300 ms).

To prevent response strategies whereby participants might attend primarily to the number of vibratory bursts, or to the total duration of the sequence, the number of bursts (4 or 6) changed between sequences. According to the number of bursts and the SOA, the total duration of the sequences varied between 600 ms and 2100 ms (see Fig. 1.3). The standard and comparison sequences were presented using two different modes: (1) simultaneous presentation, during which both vibrotactile sequences were presented at the same time; and (2) sequential presentation, during which the two vibrotactile sequences were presented one after the other, separated by a 1 s delay (seeFig. 1.4). In this experimental protocol, the standard vibrotactile sequence was always presented first. However, the standard sequence was presented on the forearm or the upper arm, depending upon the experimental block.

During the experimental procedure, visual instructions were provided via the computer screen. Following each trial, participants were prompted to indicate whether the paired vibratory sequences were "the same" or "different" using the associated colored buttons on the keyboard. A time limit of 4 s was imposed for providing responses.

Each participant completed the experimental task for both presentation modes (simultaneous, sequential) and for both positions of the standard vibratory sequence (forearm, upper arm), corresponding to a total of four experimental blocks. The order of the experimental blocks was counterbalanced across participants. Within each experimental block, comparison sequences with SOA different from the standard (i.e. 150 ms, 200 ms, 300 ms, 350 ms) were presented 4 times each. Comparison sequences with the same SOA as the standard (i.e. 250 ms) were presented 8 times. In this way, 8 of the 24 trials in each block comprised two sequences with identical temporal patterning, 8 trials comprised sequences with a shorter temporal patterning than the standard, and 8 trials comprised comparison sequences with a longer temporal patterning than the standard.

2.4 Statistical Analysis

Percentage of correct responses was computed for each participant and each condition. Repeated-measures analysis of variance (ANOVA) was carried out, with Presentation mode (sequential, simultaneous), Comparison sequence SOA (150 ms, 200 ms, 250 ms, 300 ms, 350 ms), and Position of the standard sequence (forearm, upper arm), as within participant factors.

3 Results

Overall, participants correctly perceived whether the two vibrotactile sequences applied to their upper limb were the same or different in 75.5% (SD = 12.9) of the trials. The rate of correct responses was significantly greater for sequential presentation (81.9%, SD = 14.2) than for simultaneous presentation (69.1%, SD = 8.5) (F (1, 15) = 10.90, $p < .01$, $\eta_p^2 = .42$). The position of the standard vibrotactile sequence did not have a significant effect upon perception (F (1, 15) < 1, $p = .68$, NS), with a correct response rate of 74.9% and 76.0% for the forearm and upper arm respectively.

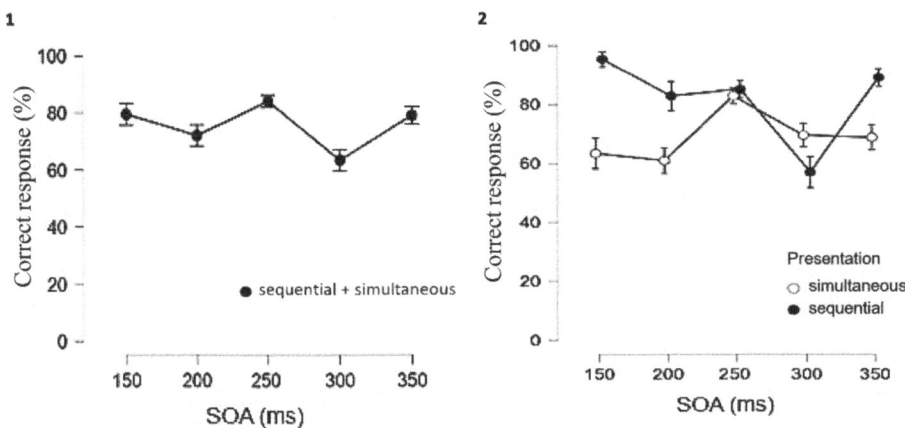

Fig. 2. 1. Percentage of correct answers according to varying stimulus onset asynchrony (SOA). 2. Percentage of correct responses according to SOA for the simultaneous and sequential presentation of vibratory stimulus (error bars represent standard error).

There was a significant effect of SOA (F (4,60) = 12.83, $p < .001$, $\eta_p^2 = .46$) showing an advantage of far over close SOA ($p < .001$) and an advantage of same over close SOA ($p < .01$; see Fig. 2.1). Finally, there was also a significant interaction between Presentation and SOA (F (4, 60) = 7.39, $p < .001$, $\eta_p^2 = .33$) (see Fig. 2.2). This interaction showed two different patterns of the effect of SOA, according to the mode of Presentation. In the sequential presentation, there was an advantage of far over close SOA ($p < .001$) and an advantage of same over close SOA. However, in the simultaneous presentation, there was a global advantage of the same SOA over all different SOA ($p < .001$) but no advantage of far over close SOA.

4 Discussion

The aim of this study was to examine how stimulation parameters influence subject perception of vibrotactile patterns. Using an experimental protocol, participants were required to indicate whether the temporal patterning of the stimuli applied to the forearm and upper arm were the same or different. The results of this experiment indicated that (1) perception was more accurate when the two signals were presented one after the other than simultaneously; (2) temporal patterning strongly influenced the accuracy of responses and (3) placement of the standard sequence on either the forearm or upper arm had no effect on task performance. Here we discuss how these findings may assist in the design of assistive technologies which incorporate multiple vibrotactile signals in shaping movement responses.

Effectively configuring stimulation parameters for wearable vibrotactile devices represents an important challenge when using these types of devices to influence complex behaviors. The results of the present study indicate that user ability to perceive specific patterns of vibrotactile stimulation is generally superior when presented in a sequential, rather than simultaneous manner. This observation appears consistent with recent studies where subjects compared the intensity of a standard vibratory stimulus to comparison stimuli [15] or compared vibratory patterns reproducing Braille alphabet [16]. Superior performance when distinguishing sequentially presented vibratory stimuli is likely attributable to the ability to represent differences between the tactile stimuli in the somatosensory cortex, and to leverage working memory in the prefrontal cortex [15, 17]. Conversely, simultaneously vibratory stimulation implies concurrent transmission and may induce stimulus integration. As such, the use of simultaneous vibratory signals would thus be generally susceptible to interelement masking at lower levels of the perceptual system [18]. In the present study, these effects upon the ability to distinguish the two signals were observed regardless of whether the standard vibratory stimulus was placed on the forearm or upper arm, revealing no asymmetrical masking between the two segments.

Differences in the temporal structure of the vibratory patterns revealed interesting effects upon subject ability to distinguish signals across the sequential and simultaneous conditions. In the sequential presentation of vibratory stimuli, we found a rather classic effect [19], where participant ability to distinguish two signals was greater when faced with markedly different temporal patterns (i.e. far SOA sequences) as opposed to minor temporal differences (i.e. close SOA sequences), and with equally reliable perception when faced with identical temporal patterns. The specific decrease in the accuracy of responses for the 300 ms SOA in the sequential presentation condition is somewhat more difficult to account for. Certain studies have previously suggested the possibility of SOA specific effects in the decay of short-term tactile representations [20]. It may be the case that neural time courses involved in encoding and recall of haptic memory contribute to a specific instance where these two patterns are not easily individuated [21].

On the other hand, when presented simultaneously, participants perceived the condition in which the signals comprised identical temporal patterning significantly better than in situations with mixed temporal patterning. This may be due, in part, to the use of a distinct perceptual strategy. In this situation, the synchronous nature of the vibrations may lead subjects to perceive the stimuli as a unified sensation, rather than as two separate vibrotactile signals. In this manner, participants would attach less importance to the specific temporal structure per se [18]. Taken together, these findings indicate firstly that the level of differentiation between multiple vibrotactile signals must be high for them to be clearly perceived and interpreted when using sequential vibratory feedback; and secondly, that the use of synchronized patterns may be particularly salient for vibrotactile devices which transmit feedback online across multiple segments.

The performance differences observed between the perception of vibrotactile signals in the sequential and simultaneous conditions may provide insight into the use of paired vibratory stimuli in motor learning. For example, it may be pertinent to select the feedback mode according to the movement task. In ballistic movements (e.g. golf swing), the duration of the gesture and the associated motor planning does not lend itself to real-time, simultaneous feedback. Sequential feedback might, however, be harnessed immediately following the gesture to provide information regarding task performance. This might potentially include distinct vibrotactile signals indicating timing or coordination errors [5, 22]. Based upon our results, it would also be advisable to exploit patterns with marked temporal asynchrony to strongly convey the desired message. Conversely, more deliberate actions (e.g. handwriting) which imply greater movement durations could benefit from feedback through the course of the gesture. As indicated above, the key to this type of feedback modality might be exploiting highly synchronous stimulation. In this type of configuration, the resonant property of the paired vibrotactile stimulus might be used to signal compliance with the desired coordination between the two segments (i.e. forearm, upper arm), as opposed to cueing each segment independently.

Following this study, several important questions regarding the perception of multiple vibratory stimuli remain unanswered. As indicated above, it is difficult to determine the exact cognitive and perceptual mechanisms associated with the performance discrepancies observed in the experiments carried out here. The use of specific perceptual strategies when faced with varying patterns of vibrotactile stimuli has been previously suggested in the literature [21]. It might therefore be pertinent to further investigate potential strategies using experimental and/or phenomenological methods. For example, this might involve evaluating discrimination capacities when faced with different degrees of vibrotactile synchrony (i.e. regular vs. chaotic). It should also be recognized that the present study manipulated temporal patterning of feedback signals by varying delays between the bursts of vibrotactile stimulus. Future experiments might also exploit burst time parameters to accentuate potential differences between signals, and thereby render them more perceptible to the user. Finally, perception of the vibrotactile signals here was limited to static postures. It remains to be seen whether response rates would be comparable through the course of movements where there is a continuous flux of tactile and proprioceptive afferents associated with segmental displacement and physical contact with the surrounding environment.

Acknowledgments. Author D.A was supported by an industrial research scholarship supported by the National Defence Innovation Agency (AID-CIFRE N°2022/003).

Disclosure of Interests. The authors have no competing interests to declare that are relevant to the content of this article.

References

1. Basalp, E., Wolf, P., Marchal-Crespo, L.: Haptic training: which types facilitate (re)learning of which motor task and for whom? Answers by a review. IEEE Trans. Haptics **14**(4), 722–739 (2021). https://doi.org/10.1109/TOH.2021.3104518
2. Choi, S., Kuchenbecker, K.J.: Vibrotactile display: perception, technology, and applications. Proc. IEEE **101**(9), 2093–2104 (2013). https://doi.org/10.1109/JPROC.2012.2221071
3. Alahakone, A.U., Senanayake, S. M. N. A.: Vibrotactile feedback systems: current trends in rehabilitation, sports and information display. In: 2009 IEEE/ASME International Conference on Advanced Intelligent Mechatronics, pp. 1148–1153 (2009). https://doi.org/10.1109/AIM.2009.5229741

4. Schönauer, C., Fukushi, K., Olwal, A., Kaufmann, H., Raskar, R.: Multimodal motion guidance: techniques for adaptive and dynamic feedback. In: Proceedings of the 14th ACM International Conference on Multimodal Interaction, pp. 133–140 (2012). https://doi.org/10.1145/2388676.2388706

5. Williams, C.K., Carnahan, H.: Motor learning perspectives on haptic training for the upper extremities. IEEE Trans. Haptics 7(2), 240–250 (2014). https://doi.org/10.1109/TOH.2013.2297102

6. Van Breda, E., Verwulgen, S., Saeys, W., Wuyts, K., Peeters, T., Truijen, S.: Vibrotactile feedback as a tool to improve motor learning and sports performance: A systematic review. BMJ Open Sport Exerc. Med. 3(1), e000216 (2017). https://doi.org/10.1136/bmjsem-2016-000216

7. Salazar, J., Okabe, K., Hirata, Y.: Path-following guidance using phantom sensation based vibrotactile cues around the wrist. IEEE Robot. Autom. Lett. 3(3), 2485–2492 (2018). https://doi.org/10.1109/LRA.2018.2810939

8. Nair, D., Stankaitis, G., Duback, S., Geoffrion, R., Jackson, J.B.: Handwriting correction system using wearable sleeve with optimal tactor configuration. In: 2021 18th International Conference on Ubiquitous Robots (UR), pp. 283–289 (2021). https://doi.org/10.1109/UR5 2253.2021.9494651

9. Bloomfield, A., Badler, N.I.: Virtual training via vibrotactile arrays. Presence: Teleoper. Virtual Environ. 17(2), 103–120 (2008). https://doi.org/10.1162/pres.17.2.103

10. Van Der Linden, J., Schoonderwaldt, E., Bird, J., Johnson, R.: MusicJacket—combining motion capture and vibrotactile feedback to teach violin bowing. IEEE Trans. Instrum. Meas. 60(1), 104–113 (2011). https://doi.org/10.1109/TIM.2010.2065770

11. Bark, K., et al.: Effects of vibrotactile feedback on human learning of arm motions. IEEE Trans. Neural Syst. Rehabil. Eng. 23(1), 51–63 (2015). https://doi.org/10.1109/TNSRE.2014.2327229

12. D'Amour, S., Harris, L.R.: Contralateral tactile masking between forearms. Exp. Brain Res. 232(3), 821–826 (2014). https://doi.org/10.1007/s00221-013-3791-y

13. Halfen, E.J., Magnotti, J.F., Rahman, M., Yau, J.M.: Principles of tactile search over the body. J. Neurophysiol. 123(5), 1955–1968 (2020). https://doi.org/10.1152/jn.00694.2019

14. Alluisi, E.A., Morgan, B.B., Hawkes, G.R.: Masking of cutaneous sensations in multiple stimulus presentations. Percept. Mot. Skills 20(1), 39–45 (1965). https://doi.org/10.2466/pms.1965.20.1.39

15. Shah, V.A., Casadio, M., Scheidt, R.A., Mrotek, L.A.: Spatial and temporal influences on discrimination of vibrotactile stimuli on the arm. Exp. Brain Res. 237(8), 2075–2086 (2019). https://doi.org/10.1007/s00221-019-05564-5

16. Yeganeh, N., Makarov, I., Kristjánsson, Á., Unnthorsson, R.: Discrimination accuracy of sequential versus simultaneous vibrotactile stimulation on the forearm. Appl. Sci. 14(1), 43 (2023). https://doi.org/10.3390/app14010043

17. Romo, R., Hernández, A., Zainos, A., Lemus, L., Brody, C.D.: Neuronal correlates of decision-making in secondary somatosensory cortex. Nat. Neurosci. 5(11), 1217–1225 (2002). https://doi.org/10.1038/nn950

18. Mahar, D.P., Mackenzie, B.D.: Masking, information integration, and tactile pattern perception: a comparison of the isolation and integration hypotheses. Perception 22(4), 483–496 (1993). https://doi.org/10.1068/p220483

19. Cholewiak, R.W., Collins, A.A.: Individual differences in the vibrotactile perception of a "simple" pattern set. Percept. Psychophys. 59(6), 850–866 (1997). https://doi.org/10.3758/BF03205503

20. Gallace, A., Spence, C.: The cognitive and neural correlates of "tactile consciousness": a multisensory perspective. Conscious. Cogn. 17(1), 370–407 (2008). https://doi.org/10.1016/j.concog.2007.01.005

21. Cholewiak, R.W., Craig, J.C.: Vibrotactile pattern recognition and discrimination at several body sites. Percept. Psychophys. **35**(6), 503–514 (1984). https://doi.org/10.3758/BF03205946
22. Lieberman, J., Breazeal, C.: TIKL: development of a wearable vibrotactile feedback suit for improved human motor learning. IEEE Trans. Rob. **23**(5), 919–926 (2007). https://doi.org/10.1109/TRO.2007.907481

MoveTouch: Robotic Motion Capturing System with Wearable Tactile Display to Achieve Safe HRI

Ali Alabbas[1], Miguel Altamirano Cabrera[1]([✉]), Mohamed Sayed[1],
Oussama Alyounes[1], Qian Liu[2], and Dzmitry Tsetserukou[1]

[1] Skolkovo Institute of Science and Technology (Skoltech), 121205 Moscow, Russia
{ali.alabbas,m.altamirano,mohamed.sayed,oussama.alyounes,
d.tsetserukou}@skoltech.ru
[2] Dalian University of Technology, China, Dalian 116024, China
qianliu@dlut.edu.cn

Abstract. The collaborative robot market is flourishing as there is a trend towards simplification, modularity, and increased flexibility on the production line. But when humans and robots are collaborating in a shared environment, the safety of humans should be a priority. We introduce a novel wearable robotic system to enhance safety during Human-Robot Interaction (HRI). The proposed wearable robot is designed to hold a fiducial marker and maintain its visibility to a motion capture system, which, in turn, localizes the user's hand with good accuracy and low latency and provides vibrotactile feedback to the user's wrist. The vibrotactile feedback guides the user's hand movement during collaborative tasks in order to increase safety and enhance collaboration efficiency. A user study was conducted to assess the recognition and discriminability of ten designed vibration patterns applied to the upper (dorsal) and the down (volar) parts of the user's wrist. The results show that the pattern recognition rate on the volar side was higher, with an average of 75.64% among all users. Four patterns with a high recognition rate were chosen to be incorporated into our system. A second experiment was carried out to evaluate users' response to the chosen patterns in real-world collaborative tasks. Results show that all participants responded to the patterns correctly, and the average response time for the patterns was between 0.24 and 2.41 s.

Keywords: Haptic feedback · Human Robot Interaction · Wearable devices · Motion Capture.

1 Introduction

The number of robots in the industry is increasing globally, reaching over 3.9 million robots in factories in 2022, according to the International Federation of Robotics (IFR) [10]. Robots are considered helpful associates for carrying

H. Kajimoto et al. (Eds.): EuroHaptics 2024, LNCS 14768, pp. 428–441, 2025.
https://doi.org/10.1007/978-3-031-70058-3_35

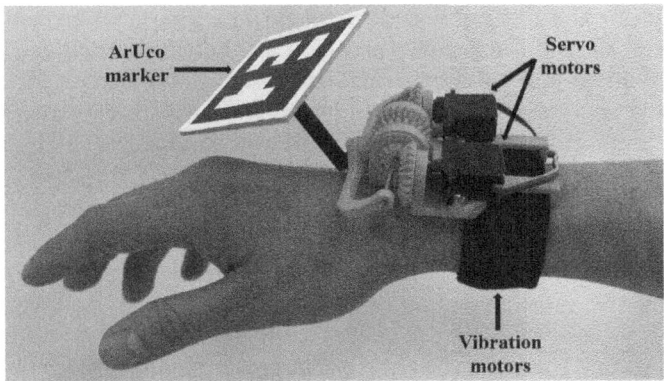

Fig. 1. MoveTouch, a novel wearable robot for position tracking and haptic feedback. An ArUco marker is located at the end effector to adjust its orientation, while five vibration motors deliver tactile feedback to the user.

out repetitive tasks, while human dexterity can be harnessed in the operation. However, robots are mostly being implemented in the industry as tools, not as companions for humans. To achieve collaboration between humans and robots, the safety of users has to be guaranteed [19].

Haptic feedback is considered one of the most efficient ways to guarantee safety in the application of Human-Robot Interaction (HRI) [16]. It works as an excellent notification system for humans [18].

Several studies have focused on enhancing safety in HRI scenarios through tactile devices [15]. Some have included tactile feedback in medical applications [7], virtual reality [4] or even in assisting blind people in indoor environments [11]. Tactile feedback can be applied to different parts of the human body, for example, to the fingertips [7], to the palm [5] or to the forearm [1,3,13].

In the applications where users need their hands to perform some tasks, providing tactile feedback to the forearm is considered a preferable option since it allows users to use their hands freely. Stanley et al. evaluated ten forms of tactile feedback to the human wrist using five wearable actuators (taper, dragger, squeezer, twister, and vibration) [17]. They showed that repeated taps on the subject's wrist on the side toward which they should turn enhanced the performance of the subjects. However, their work compared between different types of tactile feedback based on the suggested tactile actuators and did not evaluate different patterns of vibration. Chase et al. showed that, by training, it is possible to improve the signal identification in haptic feedback for novices [6].

This research work showed that tactile feedback can enhance the human response and help users perform some actions depending on the haptic feedback that they receive. However, they did not evaluate the different patterns in terms of how effectively users can recognize these patterns. Hong J. et al. developed a wrist haptic device to guide blind people's hands to perform different tasks [9]. They found, by using 4 and 8 different vibro motors, that single motor feedback was more efficient than interpolated feedback (using more than one motor at a time). However, they did not show the difference between giving feedback to the volar and the dorsal part of the forearm.

ArUcoGlide is a wearable robotic device that we previously developed to ensure that the motion capture system is always able to track the hand of the user [2]. The main idea of this device is to have a rotatable ArUco marker with two Degrees of Freedom (DOF) to maintain a specific angle with the camera and thus avoid any possible occlusion. In this work, we have enhanced the design of the ArUcoGlide to make it less bulky based on a differential gear train mechanism while ensuring the same performance and adding vibro motors to give tactile feedback to the user's wrist.

This paper aims to study the perception of humans for different tactile patterns exerted on the up (dorsal) and down (volar) parts of the wrist. The tactile feedback is applied through five vibro motors attached to a wearable band. A human study was conducted to compare between the applied patterns depending on the position (dorsal or volar) and the applied frequency. Following the results of the human study, we trained the users to perform some actions depending on the chosen tactile feedback patterns. A human-robot collaborative task was conducted to evaluate the system, including the new design of our motion capture system (ArUcoGlide).

2   MoveTouch Motion Capture System

The proposed system consists of the following components:

1. A wearable robot, dubbed MoveTouch, that is responsible for adjusting the orientation of an ArUco marker held at its end-effector and providing haptic feedback to the user. The movable ArUco marker ensures the visibility of the marker to the motion capture system, avoiding any occlusion due to the movement of the user's hand. The haptic feedback is provided to the user through five vibration motors controlled independently, which are able to generate different vibration patterns. The haptic system acts as a guidance system that tells the user about the optimal movement of their hand during Human-Robot Interaction (HRI) to avoid collisions.
2. A motion capture (mocap) system consisting of a computer and a single RGB camera that continuously captures a live video stream of the workspace that consists of a collaborative robot UR10 from Universal Robots and the wearable robot.

2.1   MoveTouch Wearable Robot

The primary goal of the MoveTouch wearable robot is to assist in determining the user's position and provide vibrotactile guidance, ensuring a secure HRI experience.

Mechanical Design. The proposed device is a two-degrees-of-freedom (2-DoFs) wearable robot, based on a differential gear train mechanism, that holds an ArUco marker at its end-effector with the ability to rotate this marker around two perpendicular axes through two servo motors [12,14]. Additionally, it includes an array of vibration motors to deliver vibration patterns to the user's wrist. The links, holders, and gears were designed and 3D-printed with PLA material; the 3D CAD model is shown in Fig. 1. The device maintains the visibility of the ArUco marker to the tracking system by continuously adjusting the motors' angular position to hold the marker in a fixed orientation with respect to the camera coordinate system.

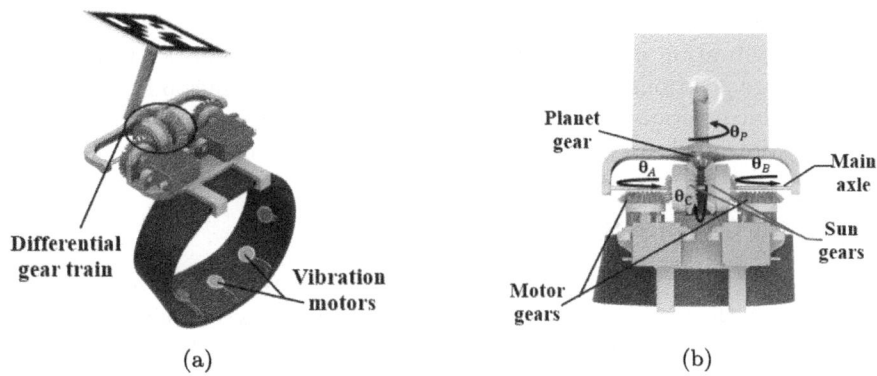

Fig. 2. MoveTouch design: 3D model perspectives (a) isometric view, (b) top view revealing the gear structure.

To achieve the rotation of the end-effector around two perpendicular axes using identical motors positioned side by side, we decided to use a differential gear train mechanism, as shown in Fig. 1(a). In this figure, the fixed base, the differential gear train, and the rotating Aruco holder are shown.

We denote $\Delta\theta_A$ and $\Delta\theta_B$ as the rotation angles of the first and second motors, respectively. The rotation angles that the ArUco marker should undergo around the lateral and longitudinal axes are $\Delta\theta_P$ and $\Delta\theta_C$ which can be seen in Fig. 1(b). These angles, $\Delta\theta_P$ and $\Delta\theta_C$, can be acquired from the motion capture system to ensure the ArUco marker maintains a fixed orientation relative to the camera coordinate system [2]. The equations for determining the required rotation angles of both motors, based on the desired rotation angles for the marker, are as follows:

$$\Delta\theta_A = n_a(n_s\Delta\theta_P + \Delta\theta_C) \tag{1}$$

$$\Delta\theta_B = n_b(n_s\Delta\theta_P - \Delta\theta_C), \tag{2}$$

where $n_a = n_b = 0.5$ are the gear ratio of the motor gears to the sun gears and $n_s = 0.5$ is the gear ratio of the sun gears to the planet gear. The rotation angles $\Delta\theta_A$ and $\Delta\theta_B$ are acquired via the motion capture system.

We chose to integrate five vibration motors within the system. These motors were chosen due to their small size. The selection of this number takes into consideration two key factors: 1) Enabling a uniform distribution of the vibration motors on one side of the user's wrist (either dorsal or volar), which typically falls within the range of 7.5–10 cm. 2) This number of vibration motors is sufficient for generating a diverse array of vibrotactile patterns to be employed in user experiments.

Electronic Design. The electronic setup includes an ESP32 microcontroller along with two Gotech GS-9025MG servo motors. The system is powered by a Li-Po 7.4 V battery connected through a DC/DC converter to power the microcontroller, the servo driver and the vibration motors.

Five coin vibration motors, capable of vibrating at frequencies ranging from 10 to 55 Hz, are positioned on the bracelet with a 2 cm separation between them.

The angular position of the servo motors and the activation of the vibration motors are controlled via Bluetooth from a base computer.

2.2   Motion Capture System

To ensure the safety of users, we need to track the real-time position of the operator within the working space. In this study, the interaction between one user wearing the MoveTouch robot on their wrist and a UR10 robotic manipulator was studied during collaborative task. Given that the MoveTouch marker is situated above the user's hand as shown in Fig. 1, we suggest that determining the marker's position enables us to locate the hand by applying a transformation from the marker to the center of the hand. In this paper, we will use the terms "user's hand position" and "Movetouch marker" interchangeably.

To achieve a safe interaction, we need to locate the user's hand in the robot coordinate system to make the robot avoid any collision with it. Our proposed mocap system is both cost-efficient and easy to install, utilizing an ArUco marker to track the operator's hand. The system comprises a base computer and an RGB webcam C930e from Logitech mounted on a stand that can be adjusted to capture different angles of the workspace, providing greater flexibility for the user. The basic task of the motion capture system is to transfer the user's hand position into the UR10 robot base coordinate system so that the robot can avoid collision with it. Since the MotionTouch marker, and thereby the user's hand position, can be estimated in the camera coordinate system [8], the transformation from the MotionTouch marker to the camera T_C^A can be known. By attaching another Aruco marker (we will call it the base marker) at a known transformation from the UR10 robot base, we can also estimate the position of the base of the robot in the camera coordinate system; thus, we can get the transformation T_C^B from the UR10 robot base to the camera coordinate system. As a result, we obtain the transformation between the user's hand and the UR10 robot base coordinate system T_B^A, and thus the position of the user's hand in the UR10 robot base coordinate system as follows:

$$T_B^A = (T_C^B)^{-1}T_C^A \tag{3}$$

Calculating the transformation from the base marker to the camera T_C^B is required only once before starting the experiments, as the camera will be in the same position throughout the whole experiment. However, if the camera's position or orientation is altered, the process needs to be repeated to derive the correct transformation matrix. Once the transformation matrix T_B^C is determined, we can utilize it to track the position of the MotionTouch marker that is attached to the user's hand, enabling us to locate it within the UR10 robot's coordinate system.

3   Vibrotactile Guidance System

The proposed wearable device, MoveTouch, provides vibrotactile guidance, facilitating human-robot collaboration. This innovative system harnesses tactile sensations to guide users in adjusting their hand position when the collaborative robot approaches and alerts them about potentially risky situations. The system transfers information about how the user should move their hands through vibration patterns exerted on the user's wrist. Ten vibration patterns have been chosen for testing in terms of recognition and discriminability on both the volar and dorsal parts of the user's hand

3.1   Patterns Design

Considering that our wearable robot is worn on the wrist of the user, it is logical to incorporate vibrotactile patterns in this specific region. Our initial step is to evaluate the user's ability to recognize vibrotactile patterns applied to their wrist and determine suitable frequencies for these patterns.

Five different patterns were designed using the five vibration motors, each with two different frequencies. The vibration patterns are illustrated in Fig. 3. In the first pattern, the vibration propagates sequentially through the five vibration patterns from the right side of the user to the left side. The second pattern is similar but the vibration propagation is from left to right. In the third pattern, vibration propagates symmetrically from the center to the outside. Initially, the central vibrator starts vibrating, followed by the simultaneous activation of the two neighboring vibrators. Subsequently, the vibration spreads to the two outermost vibrators. The same methodology for the fourth pattern but the direction is inversed to be from outside to the center. Finally, in the fifth pattern, all vibrators are activated together.

As for choosing the frequencies of the patterns, we established two levels: high and low. For the high-frequency patterns, each vibration motor was activated for 0.1 s, while for the low-frequency patterns, the activation period for each vibrator was 0.2 s. Consequently, the high-frequency rates were 2 Hz for the

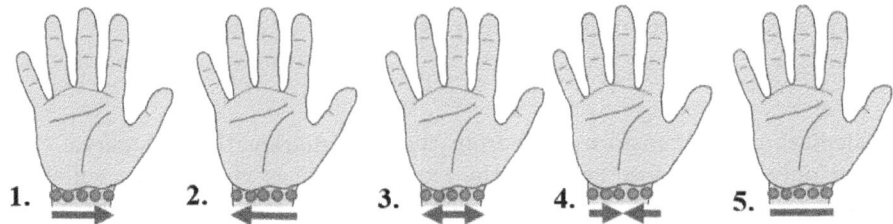

Fig. 3. The designed tactile patterns include: 1) Right-to-left propagation; 2) Left-to-right propagation; 3) Center-to-outside propagation; 4) Outside-to-center propagation; and 5) Simultaneous activation of all vibration motors.

first and second patterns; 3.3 Hz for the third and fourth patterns; and 10 Hz for the fifth pattern. The corresponding low-frequency rates were 1 Hz for the first and second patterns, 1.67 Hz for the third and fourth, and 5 Hz for the fifth pattern. We will label each pattern with H or L at the end to mention the frequency (e.g., 1H means the first pattern with a high frequency). The five patterns with two frequencies provide ten different patterns that we will utilize in our study.

In the following subsection, we present an experimental evaluation to assess the pattern recognition of the vibration patterns by users.

3.2   Pattern Recognition

In this experiment, we aimed to assess the ability of the participant to recognize and differentiate between the designed vibrotactile patterns. We invited 11 participants and asked them to tell us which haptic pattern they perceived. Their data was recorded and analyzed.

Subjects. Eleven participants, five women and six men, aged from 23 to 34 years (26 ± 3.05) took part in the experiment. All selected participants were right-handed. The participants were informed about the experiment and filled out the consent form.

User Study Procedure. Prior to the study, a training session was held to introduce the task, providing users with a detailed explanation of the patterns. Users were asked to wear the MoveTouch robot, and each pattern was rendered three times. A printed piece of paper displaying the patterns was placed in front of the participants throughout the entire training session.

During the study, each user was asked to wear the MoveTouch robot and sit in front of a PC with a graphical user interface (GUI) that allowed them to choose the pattern that they felt. An example of the user study of this experiment is shown in Fig. 4a.

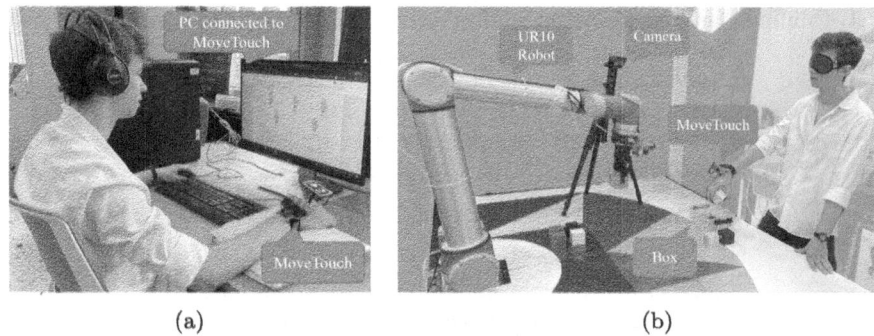

(a)   (b)

Fig. 4. The experimental setup: a) The pattern recognition experiment setup. b) The system evaluation in a real-world collaborative task setup.

To identify the most effective part form the wrist to perceive vibration (either dorsal or volar), participants were evaluated twice, wearing the device each time while the vibrators were either on the dorsal or volar part of their wrists. The participants were split into two groups: the first group tried the vibrators on the volar part first, followed by the dorsal part, while the second group followed the opposite sequence. Each of the 10 patterns was replicated five times in random order; thus, 50 patterns were provided to each participant in each evaluation. Participants were asked to wear headphones while playing white noise to prevent them from hearing the vibrations and engaged in the experiment through the GUI interface. They had the option to initiate the vibration pattern when they were ready and select from the displayed patterns the one they felt. During the experiment, participants were not allowed to repeat any pattern.

Results. The results for each evaluation (dorsal part and volar part) were analyzed and presented in a confusion matrix. In order to evaluate the statistical significance of the differences between the perceptions of the patterns with different frequencies, we analyzed the results using a single-factor repeated-measures ANOVA, with a chosen significance level of $\alpha < 0.05$. The open-source statistical packages Pingouin and Stats models were used for the statistical analysis.

The results of the user perception evaluation by rendering the patterns on the volar part of the wrist are summarized in the confusion matrix (see Table 1).

According to the ANOVA results, there is a statistically significant difference in the recognition rates for the different patterns on the volar part of the wrist: $F(9, 100) = 5.78, p = 2 \cdot 10^{-6}$. The ANOVA showed that the patterns and frequencies significantly influenced the perceptions of the users.

The paired t-tests with one-step Bonferroni correction showed statistically significant differences between the patterns 1H and 3H ($p = 4.95 \cdot 10^{-3} < 0.05$), 1L and 3H ($p = 1 \cdot 10^{-7} < 0.05$), 3H and 4L ($p = 7.3 \cdot 10^{-5} < 0.05$), 3H and

Table 1. Confusion matrix for actual and perceived patterns on the volar (down) part of the wrist

%	Answers (Predicted Class)									
	1H	1L	2H	2L	3H	3L	4H	4L	5H	5L
1H	0.80	0.11	0.00	0.00	0.05	0.00	0.02	0.02	0.00	0.00
1L	0.04	0.96	0.00	0.00	0.00	0.00	0.00	0.00	0.00	0.00
2H	0.02	0.00	0.69	0.15	0.05	0.02	0.04	0.02	0.00	0.02
2L	0.00	0.00	0.02	0.98	0.00	0.00	0.00	0.00	0.00	0.00
3H	0.02	0.00	0.00	0.00	0.38	0.05	0.42	0.07	0.05	0.00
3L	0.00	0.04	0.00	0.00	0.05	0.82	0.00	0.09	0.00	0.00
4H	0.07	0.00	0.00	0.00	0.13	0.02	0.64	0.07	0.07	0.00
4L	0.00	0.00	0.00	0.02	0.00	0.07	0.13	0.78	0.00	0.00
5H	0.00	0.00	0.00	0.00	0.11	0.00	0.02	0.00	0.82	0.05
5L	0.02	0.00	0.00	0.00	0.04	0.02	0.04	0.02	0.18	0.69

Patterns (row label, vertical)

5H ($p = 3.68 \cdot 10^{-2} < 0.05$), 2L and 3H ($p = 1 \cdot 10^{-7} < 0.05$), 2L and 4L ($p = 1.40 \cdot 10^{-2} < 0.05$), and 3H and 3L ($p = 2.24 \cdot 10^{-3} < 0.05$).

The results of the human perception evaluation by rendering the patterns on the dorsal part of the wrist are summarized in the second confusion matrix (see Table 2).

According to the ANOVA results, there is a statistically significant difference in the recognition rates for the different patterns on the dorsal part of the wrist: $F(9, 100) = 7.5724$, p $= 2.158 \cdot 10^{-8}$. The ANOVA showed that the patterns and frequencies significantly influenced the perceptions of the users.

The paired t-tests with one-step Bonferroni correction showed statistically significant differences between the patterns 1H and 3H ($p = 0.0026 < 0.05$), 1H and 4H ($p = 0.0176 < 0.05$), 1H and 4L ($p = 0.001 < 0.05$), 1L and 3H ($p = 0.0003 < 0.05$), 1L and 4H ($p = 0.003 < 0.05$), 1L and 4L ($p = 8 \cdot 10^{-5} < 0.05$), 2L and 3H ($p = 0.0001 < 0.05$), and 2L and 4H ($p = 0.002 < 0.05$).

From the two confusion matrices, the average recognition rate of all patterns on the dorsal part of the wrist was 66.6%, while on the volar side of the wrist, it was 75.64%.

Conclusion. We can notice that almost all the patterns are better recognized on the lower part of the wrist; thus, this position of the vibration motors was selected. These patterns include 1H, 1L, 2L, 3L, 4L, and 5H. However, to avoid user confusion, similar patterns like 1H and 1L are not included together. Instead, the pattern with the highest recognition, which is 1L, is selected. Consequently, patterns 1L, 2L, 3L, and 5H will used for further studies.

Table 2. Confusion matrix for actual and perceived patterns on the dorsal (up) part of the wrist

%	Answers (Predicted Class)									
	1H	1L	2H	2L	3H	3L	4H	4L	5H	5L
1H	0.80	0.11	0.00	0.00	0.05	0.00	0.02	0.02	0.00	0.00
1L	0.04	0.96	0.00	0.00	0.00	0.00	0.00	0.00	0.00	0.00
2H	0.02	0.00	0.69	0.15	0.05	0.02	0.04	0.02	0.00	0.02
2L	0.00	0.00	0.02	0.98	0.00	0.00	0.00	0.00	0.00	0.00
3H	0.02	0.00	0.00	0.00	0.38	0.05	0.42	0.07	0.05	0.00
3L	0.00	0.04	0.00	0.00	0.05	0.82	0.00	0.09	0.00	0.00
4H	0.07	0.00	0.00	0.00	0.13	0.02	0.64	0.07	0.07	0.00
4L	0.00	0.00	0.00	0.02	0.00	0.07	0.13	0.78	0.00	0.00
5H	0.00	0.00	0.00	0.00	0.11	0.00	0.02	0.00	0.82	0.05
5L	0.02	0.00	0.00	0.00	0.04	0.02	0.04	0.02	0.18	0.69

(row label: *Patterns*)

3.3 System Evaluation in a Collaborative Task

In this evaluation, the system was integrated into a real collaborative task to assess its effectiveness in enhancing the overall user safety and guiding users in a desired direction. The four selected patterns were integrated to direct users on hand movements as follows: Pattern 1L indicates moving the hand to the right, pattern 2L indicates moving the hand to the left, pattern 3L indicates moving the hand down, and pattern 5H indicates moving the hand back.

Subjects. Eight participants, comprising 5 men and 3 women, were randomly selected from the set of participants who took part in the first experiment.

Experimental Setup. The participants were asked to stand in front of a Siegmund welding table, where a UR10 robot was performing a pick-and-place task of cubes positioned at specific locations on the table. The participants wore the MoveTouch on their right wrist and utilized an eye cover to eliminate visual feedback about the location of the robot's Tool Center Point (TCP). Participants were also performing a pick-and-place task in the same box that the robot was using. A surveillance camera was placed on a tripod near the table, where it could monitor the entire workspace. Additionally, An observer was present during the experiment, poised to activate the emergency button to halt the robot in case of any unforeseen issues. The experimental setup is shown in Fig. 4(b).

Collaborative Task Procedure. Before the experiment starts, each participant underwent a training session covering the four patterns and the corresponding movements associated with each pattern.

The robot picked cubes from certain positions on the table and placed them in a box near the participant. The participants were asked initially to touch and familiarize themselves with the box to memorize its position. Throughout the experiment, participants were asked to pick cubes near the box and place them inside the box. The whole experiment was conducted while the user's eyes were covered.

During collaboration, when the robot's TCP is moving towards the participant's hand and gets to a distance closer than 40 cm (the activation area), the system activates a certain haptic pattern on the participant's wrist to make participants move their hand in response to each pattern as follows: 1) If pattern 1L is rendered, the participant should move their hand to the left. 2) When the pattern 2L is rendered, the participant should move the hand to the right. 3) For the pattern 3L, the movement should be downward. 4) For pattern 5H, the participant should move their hand backward.

For safety concerns, the robot was programmed to halt when its TCP came within a critical threshold (25 cm) from the participant's hand and continue when the participant's hand moved outside the critical area.

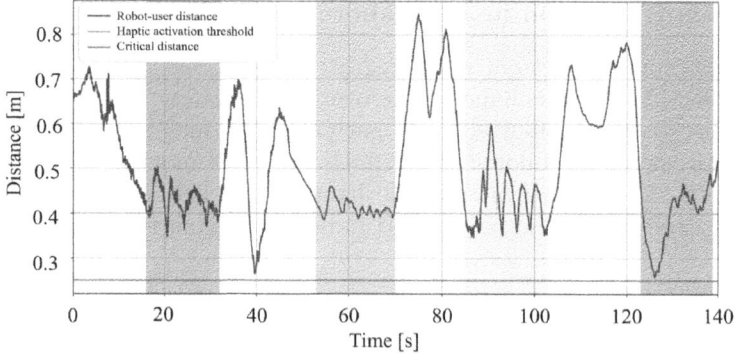

Fig. 5. The distance between the participant and the robot's TCP during one experiment. The red line is the critical distance, while the orange line represents the haptic activation distance. The four highlighted areas are the areas where the haptic patterns were activated.

Results. The distance between the participant's hand and the robot's TCP is illustrated in Fig. 5. We can observe that the critical distance was preserved throughout the experiment, which indicates safer collaboration, taking into account that the user was doing the task while eye-covered. All participants successfully responded to the haptic patterns, except for one instance where the participant responded to the pattern 5H by moving his hand downward instead of backward. We can see the effective response to the haptic patterns in the highlighted areas in Fig. 5. These highlighted areas correspond to moments when the robot approached the participant. For example, in the blue highlighted area, the participant attempted to continue the task and return his hand near the box many times, but the robot was still approaching, triggering the haptic pattern multiple times. Consequently, the participant moves their hand again, resulting in an oscillating distance between the robot and the user.

We measured the response time that the participants needed to get their hands out of the dangerous area. The time was measured between the moment the MoveTouch gives the haptic pattern and the moment users start to move their hands. The measured time for all users for each pattern can be seen in Fig. 6. We can see that the mean response times were 0.24 s, 0.61 s, 0.85 s, and 2.41 s for the first, second, fourth, and third patterns, respectively. We can see that the response time of the fourth pattern is the highest among all the patterns.

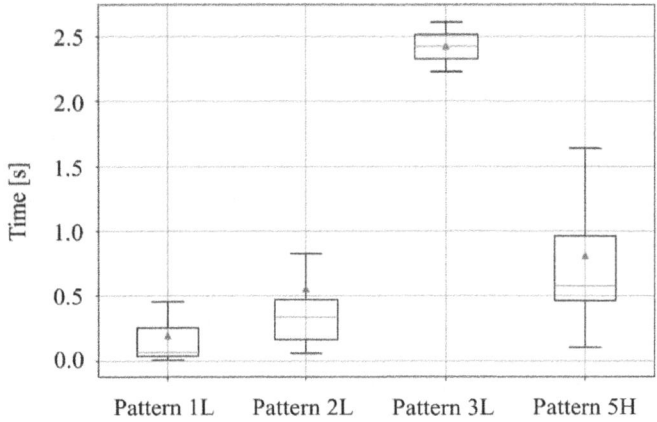

Fig. 6. The response time for participants to remove their hands for the four patterns during the collaborative task.

Conclusion. The critical distance between the eye-covered participant and the robot's TCP was preserved throughout the collaboration, indicating the effectiveness of the designed vibrotactile patterns in guiding user actions when rendered on the wrist. The patterns remained discernible even when the participant's attention was focused on performing another task. The measured perception time of each pattern can serve as a design parameter to establish an upper limit for the robot's speed during collaboration, to assure that the robot does not critically approach the user, and thus, enhance user safety during collaborative tasks.

4   Conclusion and Future Work

This research presents an innovative wearable robotic system featuring both motion capture and a haptic guidance system designed for human-robot interaction. The system includes a wearable 2-DoF robot responsible for adjusting a marker held at its end-effector, ensuring its constant visibility to the camera. The haptic guidance system incorporates five vibration motors fixed on the wearable robot wrist rubber band to render vibrotactile patterns at the user's wrist to guide their hand movement.

A total of ten haptic patterns were selected for experimentation on both the volar and dorsal parts of the wrist. A user study was conducted to evaluate the recognition and discriminability of these patterns. The average recognition rate of all patterns on the dorsal (up) part of the wrist was 66.6%, while on the volar (down) side of the wrist, it was 75.64%. In addition, almost all the vibration patterns were better recognized on the volar part of the wrist. Consequently, the decision was made to fix the vibration motors at the volar part of the wrist. Four patterns with high recognition rates were selected for further study.

Another experiment was made to evaluate the integration of the system into a real-world collaborative task incorporating the four selected patterns. We report that the critical distance had not been violated during the whole experiment and that the users were efficiently responding to the vibrotactile patterns. We also measured the perception time of each pattern and it was in the range 0.24–2.41 s.

For future work, we intend to expand the testing of the Movetouch robot across a broader user base and in diverse tasks to better evaluate the effectiveness of the proposed guidance system. We also aim to study how to enhance the recognition rate of the patterns by studying the effect of the pattern frequency. Additionally, we seek to minimize user response times to suggested patterns, considering that some patterns are perceived much faster than others.

Acknowledgements. The research reported in this publication was financially supported by the Russian Science Foundation grant No. 24-41-02039.

References

1. Ævarsson, E.A., Ásgeirsdóttir, T., Pind, F., Kristjánsson, Á., Unnthorsson, R.: Vibrotactile threshold measurements at the wrist using parallel vibration actuators. ACM Trans. Appl. Percept. (TAP) **19**(3), 1–11 (2022). https://doi.org/10.1145/3529259
2. Alabbas, A., Cabrera, M.A., Alyounes, O., Tsetserukou, D.: Arucoglide: a novel wearable robot for position tracking and haptic feedback to increase safety during human-robot interaction. In: 2023 IEEE 28th International Conference on Emerging Technologies and Factory Automation (ETFA), pp. 1–8 (2023). https://doi.org/10.1109/ETFA54631.2023.10275727
3. Altamirano Cabrera, M., Heredia, J., Tirado, J., Panov, V., Hagos, F., Tsetserukou, D.: Cohaptics: Development of human-robot collaborative system with forearm-worn haptic display to increase safety in future factories. In: 2021 IEEE 17th International Conference on Automation Science and Engineering (CASE), pp. 74–80 (2021). https://doi.org/10.1109/CASE49439.2021.9551579
4. Biswas, S., Visell, Y.: Haptic perception, mechanics, and material technologies for virtual reality. Adv. Funct. Mater. **31**(39), 2008186 (2021). https://doi.org/10.1002/adfm.202008186

5. Cabrera, M.A., Tirado, J., Heredia, J., Tsetserukou, D.: Linkglide-s: a wearable multi-contact tactile display aimed at rendering object softness at the palm with impedance control in VR and telemanipulation. In: 2022 IEEE 18th International Conference on Automation Science and Engineering (CASE), pp. 647–652. IEEE (2022). https://doi.org/10.1109/CASE49997.2022.9926451

6. Chase, E.D.Z., Israr, A., Preechayasomboon, P., Sykes, S., Gupta, A., Hartcher-O'Brien, J.: Learning vibes: Communication bandwidth of a single wrist-worn vibrotactile actuator. In: 2021 IEEE World Haptics Conference (WHC), pp. 421–426 (2021). https://doi.org/10.1109/WHC49131.2021.9517208

7. Enayati, N., De Momi, E., Ferrigno, G.: Haptics in robot-assisted surgery: challenges and benefits. IEEE Rev. Biomed. Eng. **9**, 49–65 (2016). https://doi.org/10.1109/RBME.2016.2538080

8. Garrido-Jurado, S., Muñoz-Salinas, R., Madrid-Cuevas, F.J., Marín-Jiménez, M.J.: Automatic generation and detection of highly reliable fiducial markers under occlusion. Pattern Recogn. **47**(6), 2280–2292 (2014). https://doi.org/10.1016/j.patcog.2014.01.005

9. Hong, J., Pradhan, A., Froehlich, J.E., Findlater, L.: Evaluating wrist-based haptic feedback for non-visual target finding and path tracing on a 2D surface. In: Proceedings of the 19th International ACM SIGACCESS Conference on Computers and Accessibility, pp. 210–219 (2017). https://doi.org/10.1145/3132525.3132538

10. IFR: World robotics industrial robots 2023, statistics, market analysis, forecasts and case studies. https://ifr.org

11. Khusro, S., Shah, B., Khan, I., Rahman, S.: Haptic feedback to assist blind people in indoor environment using vibration patterns. Sensors **22**(1), 361 (2022). https://doi.org/10.3390/s22010361

12. Krainev, A.: Dictionary-Reference Book on Mechanisms. Moscow (1987)

13. Moriyama, T., Kajimoto, H.: Wearable haptic device presenting sensations of fingertips to the forearm. IEEE Trans. Haptics **15**(1), 91–96 (2022). https://doi.org/10.1109/TOH.2022.3143663

14. Morozov, A., Angeles, J.: The design of an innovative large scale schÖnflies-motion generator. In: Proceedings of the Canadian Engineering Education Association (2011). https://doi.org/10.24908/pceea.v0i0.4013

15. Pacchierotti, C., Prattichizzo, D.: Cutaneous/tactile haptic feedback in robotic teleoperation: motivation, survey, and perspectives. IEEE Trans. Robot. (2023). https://doi.org/10.1109/TRO.2023.3344027

16. Seminara, L., Gastaldo, P., Watt, S.J., Valyear, K.F., Zuher, F., Mastrogiovanni, F.: Active haptic perception in robots: a review. Front. Neurorobot. **13**, 53 (2019). https://doi.org/10.3389/fnbot.2019.00053

17. Stanley, A.A., Kuchenbecker, K.J.: Evaluation of tactile feedback methods for wrist rotation guidance. IEEE Trans. Haptics **5**(3), 240–251 (2012). https://doi.org/10.1109/TOH.2012.33

18. Wang, Y., Millet, B., Smith, J.L.: Designing wearable vibrotactile notifications for information communication. Int. J. Hum. Comput. Stud. **89**, 24–34 (2016). https://doi.org/10.1016/j.ijhcs.2016.01.004

19. Zacharaki, A., Kostavelis, I., Gasteratos, A., Dokas, I.: Safety bounds in human robot interaction: a survey. Saf. Sci. **127**, 104667 (2020). https://doi.org/10.1016/j.ssci.2020.104667

Evaluation of HaptiComm-S for Replicating Tactile ASL Numbers: A Comparative Analysis of Direct and Mediated Modalities

Mounia Ziat[1]([✉])([ID]), Nurlan Kabdsyhev[2], Sven Topp[3], Basil Duvernoy[4], Jeraldine Milroy[5], and Zhanat Kappassov[1,2]

[1] XD, Bentley University, Waltham, MA, USA
mziat@bentley.edu
[2] ISSAI, Nazarbayev University, Astana, Kazakhstan
[3] University of Sydney, Sydney, Australia
[4] Center of Social and Affective Neuroscience, Linköping University, Linköping, Sweden
[5] University of New England, Armidale, NSW, Australia

Abstract. This research investigates the efficacy of HaptiComm-S, a haptic communication device designed to facilitate tactile communication for Deafblind individuals. The primary focus is on evaluating the device's capability to replicate the tactile American Sign Language (ASL) numbers 0 to 10. Participants performed under two distinct conditions: direct ASL signing and mediated ASL signing through two modalities (Tap and Tap-and-Hold). Our findings demonstrate significant differences in performance between the Direct and Mediated ASL modes. Direct ASL consistently exhibited higher accuracy compared to mediated conditions. Mediated ASL conditions were prone to perceptual errors in number identification. Notably, specific numbers, such as 4, 7, 8, and 9, posed challenges in the mediated conditions, often resulting in confusion among participants. These findings contribute valuable insights for the ongoing refinement in the design of haptic communication devices tailored to the needs of the Deafblind community.

Keywords: Tactile ASL · Haptic Communication · Deafblindness

1 Introduction

In the realm of assistive technology, ongoing research is focused on innovating communication methods for individuals with disabilities. Developing effective communication tools for deafblind individuals is particularly challenging, given the complex interplay of combined hearing and vision impairments. HaptiComm-S, an innovative haptic communication device, introduces a distinctive approach to tactile communication. Developed with the aim of exploring and broadening

H. Kajimoto et al. (Eds.): EuroHaptics 2024, LNCS 14768, pp. 442–448, 2025.
https://doi.org/10.1007/978-3-031-70058-3_36

the capabilities of haptic technology, HaptiComm-S focuses on simulating Tactile Fingerspelling (TFS) and Tactile American Sign Language (TASL). These modes of communication are essential for Deafblind individuals, involving specific hand-touch movements to convey letters and numbers. Our objective was to evaluate the efficacy of HaptiComm-S in conveying numbers from 0 to 10 to users' hands. To achieve this, an experiment was designed involving individuals without sensory impairments as test participants. The use of non-expert users in this initial phase served a dual purpose: firstly, to refine the technology functionally before introduction to the target audience, and secondly, to obtain a baseline performance of the device in a controlled setting.

2   Related Work

The development and refinement of assistive communication technologies for Deafblind individuals spanning three decades (refer to [11] for a comprehensive review), underscores a continual effort to improve autonomy and social inclusion. The focus on developing tactile communication devices aligns with the broader goal of enhancing interaction. Wearable Human-Machine Interfaces (HMIs) are gaining prominence, transitioning from traditional touch sensors to more sophisticated e-skin technologies [8].

Earlier studies extensively explored assistive devices for Deafblind individuals. Notable examples of device development include Dexter, a mechanical fingerspelling hand aiding communication and DB-HAND, a hardware/software system facilitating autonomous interaction [1,6]. Building upon these foundations, the Finger Braille Teaching System [5] demonstrated non-disabled individuals' ability to communicate with Deafblind individuals using Braille code. Similarly, Reed et al. [12] explored a tactile speech device translating phonemic codes through tactile stimulation, and Ozioko [10] presented a wearable tactile communication interface employing finger Braille, integrating actuation and sensing in the same location.

In recent years, there has been a substantial focus on developing wearable and glove-like technologies based on existing tactile communication languages and methods. The Mobile Lorm Glove, which translates the Lorm hand-touch alphabet into text messages [3], exemplifies this trend. Other wearable devices facilitating communication using tactile methods have explored the use of the British Deafblind manual alphabet and Braille [9], as well as the Malossi alphabet [1]. Additionally, Hirose and Amemiya [4] introduced Finger-Braille interfaces integrated with wearable computers, and Nicolau et al. [7] presented UbiBraille, a device leveraging Braille knowledge for reading electronic texts.

3   Experiment

3.1   Participants

Twenty participants (8 F, 12 M, mean age = 29.85, SD = 6.02) were recruited from Bentley University and duly compensated for their involvement. None reported hearing or visual deficits, hand injury, or nerve damage. The experimental procedure adhered to the Declaration of Helsinki guidelines, and the study protocol received approval from the Institutional Review Board of Bentley University.

3.2 Apparatus

Building upon the HaptiComm device [2], the HaptiComm-S was tailored for users with smaller hand sizes. The original design was scaled down by a factor of 0.85 and an increased height of 20 mm to reduce magnetic interference. It incorporates an array of twenty-four custom-made electrodynamic actuators, each meticulously positioned to deliver distinct tactile sensations. Users place their left hand on the device, enabling them to perceive tactile stimuli (Fig. 1).

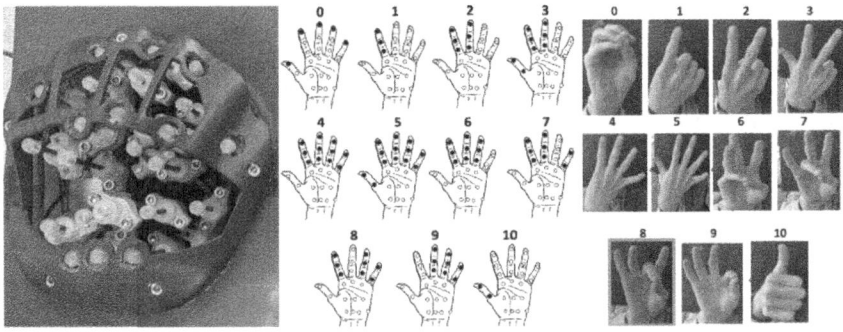

Fig. 1. Left) HaptiComm-S, Center) Mediated TASL as provided by the HaptiComm-S, and Right) ASL numbers.

3.3 Stimuli

We focused on TASL, a specific method of tactile communication where a deaf-blind individual discerns signs by placing a hand over that of the signer. This study evaluated the HaptiComm-S device's ability to transmit TASL numbers 0 to 10 under two conditions: Tap and Tap-and-Hold. In the Tap condition, participants experienced brief 40 ms taps, whereas in the Tap-and-Hold condition, the contacts lasted 200 ms. The control condition utilized direct TASL, where participants identified numbers by physically grasping the experimenter's hand to feel each fingering position. In the mediated conditions, the HaptiComm-S system replicated these positions by activating the corresponding actuators to simulate the sensation of the specific fingers used in direct TASL, delivering either short or long taps to mimic the actual finger positions. Figure 1 visually depicts the mediated TASL numbers as interpreted by the HaptiComm-S, in comparison to traditional TASL.

3.4   Procedure

Following informed consent, participants watched a video tutorial demonstrating ASL numbers 0 to 10, which was shared with them a day before the experiment. Upon arrival at the experimental site, we verified their comprehension by requesting that they manually sign the numbers as a prerequisite for commencing the experimental tasks.

The experiment consisted of three randomized blocks, each comprising 33 trials (3 × 11 numbers): 1) Direct ASL Signing, 2) Mediated ASL Signing with the Tap condition, and 3) Mediated ASL Signing with the Tap-and-Hold condition. In the Direct ASL block, participants identified numbers by holding the experimenter's hand with their left hand and releasing it immediately after providing their answer. In the Mediated ASL Signing blocks, participants initially familiarized themselves with numbers activated by the HaptiComm-S device, receiving feedback on their responses, and subsequently identified the numbers delivered on their left hand.

Numbers were presented in a randomized order within each block. To minimize visual and auditory distractions, participants wore blindfolds and noise-canceling headphones. A 5-min break occurred following each block before proceeding to the next. For each trial, participants verbally reported the perceived number before advancing to the next trial. Upon completing all blocks, participants assessed the difficulty level of the tasks and provided feedback on their overall experience.

4   Results

A two-way repeated measures ANOVA examined the impact of the ASL condition and Number on correct responses. Results revealed a significant main effect of Condition, $F(2, 38) = 98.38$, $p < .001$, $\eta_p^2 = .84$, indicating a substantial effect size. Similarly, a significant main effect of Numbers was observed, $F(10, 190) = 18.30$, $p < .001$, $\eta_p^2 = .49$, with a large effect size. Moreover, the interaction effect between Condition and Number was significant, $F(20, 380) = 9.61$, $p < .001$, $\eta_p^2 = .34$, suggesting that the effect of Condition on the dependent variable varied depending on the number presented.

Post hoc comparisons, with multiple comparisons, revealed significant performance differences across the three TASL conditions. In the Direct ASL condition, Participants consistently showed better performance with numbers 0 through 10 compared to the same numbers in both ASL Mediated conditions. Specifically, numbers 0, 4, 6, 7, 8, and 9 exhibited systematically higher performance in the Direct condition. Within the Mediated conditions, numbers 1, 2, 3, and 5 demonstrated higher performance relative to 7, 8, and 9 (Fig. 2).

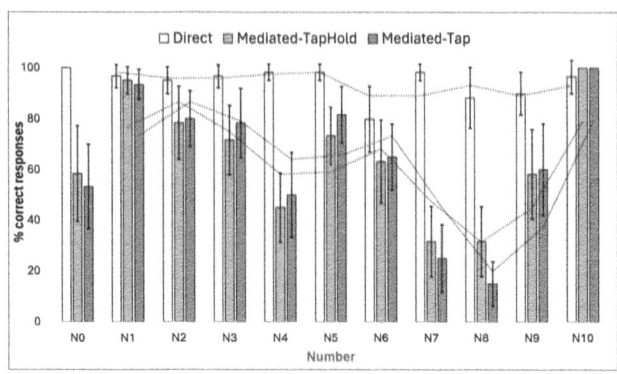

Fig. 2. Barplots for the three ASL conditions. Error bars represent 95% confidence intervals. The moving average is displayed for each condition.

4.1 Discussion and Conclusion

Our findings consistently highlight an advantage for Direct TASL across all numbers, underscoring the significant impact of the modality on performance. Within the Mediated conditions, performance was generally strong for numbers 0, 1, 2, 3, 6, and 10. However, accuracy declined in perceiving numbers 4, 7, 8, and 9, which were frequently misidentified. The confusion matrix for the direct ASL condition reveals high accuracy in participant responses, with correct identification rates predominantly ranging from 92% to 100%. This indicates effective communication of tactile ASL numbers by the human signer.

Contrastingly, the mediated ASL conditions face challenges with specific numbers. Notably, the number 4 in the Tap condition was often confused with 9 (22%), while in the Tap-and-Hold condition, it was frequently mistaken for 6 (13%), 7 (10%), 8 (10%), and 9 (12%). This pattern arises from the similarities in the mediated TASL for these numbers, where three fingers represent 6, 7, 8, and 9, and four fingers denote 4. Additionally, number 7 was consistently confused with 4 in both mediated conditions (42% and 35%), and the number 8 was frequently misidentified as 4, particularly in the Tap condition (55%) and to a lesser extent in the Tap-and-Hold condition (35%). The decision to maintain the natural hand configurations used in ASL for coding numbers on the HaptiComm-S device was intentional, aimed at minimizing the learning curve by leveraging pre-existing knowledge of ASL and to avoid introducing an entirely new signing method. Overall, the confusion matrices reveal perceptual errors, highlighting specific numbers more susceptible to misinterpretation in mediated conditions compared to direct human ASL (Fig. 3). This observation is corroborated by participants' evaluations of task difficulty on a 7-point scale (1-very easy, 7-very difficult). In the direct ASL condition, the average task difficulty was 1.65 (SD = 0.89), whereas for the mediated conditions, task difficulty had an average score of 4.85 (SD = 1.51). Most participants indicated the task with a human signer as easier than with the HaptiComm-S, reporting challenges in distinguishing between 7, 8, and 9 on the Hapticomm-S.

TH — RESPONSE

PRESENTED	0	1	2	3	4	5	6	7	8	9	10
0	58	0	0	3	3	17	13	2	2	2	0
1	2	96	2	0	0	2	0	0	0	0	1
2	0	0	78	7	5	0	7	2	2	0	0
3	2	0	10	72	0	2	7	5	2	2	0
4	0	2	0	5	45	3	13	10	10	12	0
5	3	0	0	3	7	73	3	0	3	7	0
6	0	0	2	7	12	2	63	5	3	7	0
7	0	0	5	0	42	3	10	32	2	7	0
8	0	2	2	2	30	2	8	7	32	17	0
9	2	0	0	3	12	2	10	12	2	58	0
10	0	0	0	0	0	0	0	0	0	0	100

T — RESPONSE

	0	1	2	3	4	5	6	7	8	9	10
0	53	0	0	8	3	22	7	0	3	3	0
1	0	93	2	2	0	0	2	0	0	0	2
2	0	2	80	5	2	0	3	3	3	2	0
3	0	0	8	78	0	7	3	2	0	2	0
4	0	0	0	3	50	5	12	7	2	22	0
5	0	0	0	3	3	82	8	2	0	2	0
6	0	0	3	5	8	2	65	3	2	12	0
7	0	0	5	5	35	12	7	25	5	7	0
8	0	0	2	2	55	0	12	7	15	8	0
9	0	0	0	3	10	0	12	3	12	60	0
10	0	0	0	0	0	0	0	0	0	0	100

D — RESPONSE

	0	1	2	3	4	5	6	7	8	9	10
0	100	0	0	0	0	0	0	0	0	0	0
1	2	97	0	0	0	0	0	0	0	0	2
2	0	0	95	0	0	0	0	0	3	0	2
3	0	0	0	97	0	0	0	2	2	0	0
4	0	0	0	0	98	0	0	0	0	2	0
5	0	0	0	0	2	98	0	0	0	0	0
6	0	0	0	0	10	0	80	2	0	8	0
7	0	0	0	0	0	0	0	98	2	0	0
8	0	0	0	0	0	0	2	10	88	0	0
9	0	0	0	3	0	0	7	0	0	90	0
10	0	3	0	0	0	0	0	0	0	0	97

Fig. 3. Confusion matrices showing the percentage of correct responses for the Tap-and-Hold (TH), Tap (T), and Direct (D) conditions.

Unlike letters, which can be contextually corrected within words, numbers demand precise delivery as they lack contextual cues for correction. Subsequent studies should explore alternative aspects of stimuli, such as the time interval between actuators or the spatial pattern of stimuli. Tactile ASL, encompassing both letters and numbers, necessitates incremental learning, practice, repetition, and memorization beyond mere familiarization. This research aims to contribute to assistive technology by developing innovative solutions to bridge communication gaps for individuals with complex sensory needs. The study's findings serve as a stepping stone towards more inclusive and accessible communication technologies, ultimately enhancing the quality of life for the Deafblind community.

Acknowledgments. This Research was supported by Google CS-ER.

References

1. Caporusso, N.: A wearable malossi alphabet interface for deafblind people. In: Conference on Advanced Visual Interfaces, pp. 445–448 (2008)
2. Duvernoy, B., Kappassov, Z., Topp, S., Milroy, J., Xiao, S., Lacôte, I., Abdikarimov, A., Hayward, V., Ziat, M.: Hapticomm: a touch-mediated communication device for deafblind individuals. IEEE Robot. Automat. Lett. **8**(4), 2014–2021 (2023)
3. Gollner, U., Bieling, T., Joost, G.: Mobile lorm glove: introducing a communication device for deaf-blind people. In: ACM TEI 0212, pp. 127–130 (2012)
4. Hirose, M., Amemiya, T.: Wearable finger-braille interface for navigation of deaf-blind in ubiquitous barrier-free space. In: HCII, vol. 4, pp. 1417–1421 (2003)
5. Matsuda, Y., Isomura, T., Sakuma, I., Kobayashi, E., Jimbo, Y., Arafune, T.: Finger braille teaching system for people who communicate with deafblind people. In: Conference on Mechatronics and Automation, pp. 3202–3207. IEEE (2007)
6. Meade, A.: Dexter–a finger-spelling hand for the deaf-blind. In: ICRA. IEEE (1987)
7. Nicolau, H., Guerreiro, J., Guerreiro, T., Carriço, L.: Ubibraille: designing and evaluating a vibrotactile braille-reading device. In: SIGACCESS, pp. 1–8 (2013)
8. Ozioko, O., Dahiya, R.: Smart tactile gloves for haptic interaction, communication, and rehabilitation. Adv. Intell. Syst. **4**(2), 2100091 (2022)

9. Ozioko, O., Hersh, M.: Development of a portable two-way communication and information device for deafblind people. In: Assistive Technology, pp. 518–525 (2015)
10. Ozioko, O., Karipoth, P., Hersh, M., Dahiya, R.: Wearable assistive tactile communication interface based on integrated touch sensors and actuators. IEEE Trans. Neur. Syst. Rehab. Eng. **28**(6), 1344–1352 (2020)
11. Reed, C.M., Tan, H.Z., Jones, L.A.: Haptic communication of language. IEEE Trans. Haptics **16**(2), 134–153 (2023)
12. Reed, C.M., et al.: A phonemic-based tactile display for speech communication. IEEE Trans. Haptics **12**(1), 2–17 (2018)

Memorable Vibration Pattern Design Based on Writing Pattern

Zi Ying Wong[1] ⓘ, Yongjae Yoo[2] ⓘ, and Sang-Youn Kim[1](✉) ⓘ

[1] Interaction Lab, Future Convergence Engineering, Korea University of Technology and Education, Cheonan-si, Republic of Korea
sykim@koreatech.ac.kr
[2] Department of Applied Artificial Intelligence, Hanyang University, Ansan-si, Republic of Korea
yongjaeyoo@hanyang.ac.kr

Abstract. In this paper, we presented memorable vibration patterns representing digit 0 to 9, which were designed based on writing patterns. Based on the collection of 50 participants' handwriting pattern of 10 digits we gathered, we designed two different types of vibration patterns to generate the digits: vibrotactile flows and discrete vibrotactile simulations. In the user study, we evaluated identifiability and learnability of the patterns we generated. First, participants successfully identified 69.4% of vibrotactile flow patterns and 77.5% with discrete vibrotactile simulations in their first session of 30 trials without training. The average recognition rate in their last session 30 trials increased to 83.6% for vibrotactile flows and 91.1% for discrete vibrotactile simulations after two sessions (60 trials), shows the ease of learning the vibration pattern. We also observed a lasting learning effect of both types of vibrotactile patterns in a delayed recall test was conducted 72–96 h after the first user study – 90.0% success rate for vibrotactile flows and 91.4% for discrete vibrations.

Keywords: Memorable Vibration Pattern · Digits · Writing Pattern

1 Introduction

Handwriting letters have been a most popular means to convey information to others in our life. In every moment, we write something–taking a memo, often with drawing something to memorize some information we need. Recent advances in technology make us doing such in a small form-factor mobile device, which impacted how we write and see, but the essentials of handwriting remained the same.

In mobile devices, there is another way to deliver information which has been underestimated—haptic functions. Due to many reasons, the information capability of mobile devices has been limited to convey a few different temporal patterns only in most devices, regardless of its size and price. One of the reasons would be the difficulty and inefficiency of learning tactile patterns, compared to those of audiovisual stimuli. In other words, most of the users do not have motivations to utilize tactile information delivery features.

H. Kajimoto et al. (Eds.): EuroHaptics 2024, LNCS 14768, pp. 449–463, 2025.
https://doi.org/10.1007/978-3-031-70058-3_37

To overcome this, many researchers studied how to effectively convey information using tactile patterns, including our efforts in the current work.

In this work, we suggested a method to design vibrotactile patterns which inspired from handwriting patterns of 10 digits (0 to 9). The vibrotactile patterns can be expressed using vibrotactile flows, continuously moving sensation of vibrotactile stimulation, and discrete vibrotactile stimulation, sequential vibrotactile stimulation with break in between each stimulation. We expected that using well-known spatiotemporal patterns can minimize the users' cognitive load to learning. The following of the paper consists of the literature survey, our approach, user studies, and discussions.

1.1 Related Work

The use of spatiotemporal patterns to convey information has been an essential research topic. Novich et al. [1] found that spatiotemporal patterns perform better than spatial and intensity encoded patterns in encoding information to the skin. Luzhnica et al. [2] proposed methods to optimize overlapped spatiotemporal patterns for alphabet and words rendering using a glove wearable. Seo and Choi [3] developed a rendering algorithm to create vibrotactile flow which is able to produce diverse perceptual effects. They extended the 1D vibrotactile flows to 2D vibrotactile by rendering vibrotactile flows along the edges of a mobile device using 4 actuators places at its corner [4]. Gong et al. [5] developed a handheld haptic device which able to create 3D vibrotactile flows.

While the exploration of spatiotemporal patterns continues, research on haptic communication using tactile devices have garnered attention. Devices and vibration patterns for communication in various forms have been proposed [6]. However, conveying information in a systematic manner to shorten learning time remain crucial to encourage users to use tactile information for communication.

Namely, Morse code is one of the methods that can convey information using vibration with simple vibration patterns. Vibrotactile Morse codes usually use short vibrations to represent dots and longer vibrations to represent dashes in Morse code [7–9]. Plaisier et al. [9] conducted a study and found that users can learn 15 to 24 vibration patterns in 30 min. Braille is also another common method to convey information via touch. Braille can be generated using a refreshable braille display, and many different approaches to display Braille on a smartphone or a wearable device have been proposed [10–12]. However, researchers found that it could be hard to learn Braille at a later age [13].

Liu et al. [14] proposed Vibrotactile Alphabets, which alphabets are generated by manipulating the duration and frequency of vibrations. Users recognized over 90% of words and symbols correctly after six to eight hours of training. Liao et al. [16] created a set of vibration patterns for alphabets and digits inspired by EdgeWrite [15] patterns. The vibration patterns are generated on a wrist-worn tactile display with 2×2 tactor array. Users recognized 88.6% and 85.9% of numeric and alphabetical patterns respectively after a 15-min learning session to learn the sequence of proposed vibration patterns and a training session in which the vibration patterns were displayed twice in random order. Cauchard et al. [17] proposed ActiVibe, which generate 1 to 10 on a smartwatch for communicating progress. They mentioned that 2 min of training is sufficient to understand and identify the set of vibrations.

A variety of vibration patterns that convert phonemes to words and express them in vibration on the forearm or arm were proposed in different works. Zhao *et al.* [18] proposed a phonemic approach in which each phoneme is paired with a unique vibro-tactile pattern based on articulation placement. The users were able to read 10 words with 60% accuracy, which gradually increased to 83% accuracy after three recognition-recall blocks with 30 trials in each block. De Vargas *et al.* [19] converted phenomes into vibrations based on frequency with two vibrotactile actuators attached to the forearm. Users were able to identify 65% of words correctly after 4.2 h of training.

1.2 Approach

First, we collected handwriting patterns of the 10 digit, 0 to 9 from 50 participants. Then we grouped the writing patterns based on the shape, starting point, direction, and sequence of strokes. As hardware, we made a mobile device mockup with six vibrotactile actuators. The six actuators were located on the left and right edges of the prototype (see Fig. 2a in Sect. 2.3). On this hardware setup, we designed two sets of vibration patterns to express the digits using vibrotactile flows and discrete vibrotactile simulations. Both types of patterns reflected how we write digits, which described in the Sect. 2.2. With this setup, we conducted a user study to observe the effectiveness of the patterns in information transfer, and their short- and long-term learnability.

The main contributions of this work are:

- We collected handwriting patterns of 0 to 9 from 50 participants and found generalized ways to write each number by following the majority did.
- We designed two types of spatiotemporal vibration patterns for the numbers, using vibrotactile flows and discrete vibrotactile simulations. The patterns were presented on a mobile device mockup with six actuators.
- We conducted two user studies; the first one with 13 participants and showed a rapid learning curve in the early-stage deployment of the patterns, and the second one of delayed recall test showed the patterns' long-term effects.

2 Vibration Pattern Design

2.1 Participants

We gathered handwriting patterns from 50 participants (26 female, age: 20−35 years, mean: 26.54, SD: 3.6) to design the vibration pattern for each digit. We recruited partic-ipants with different nationalities (27 Malaysian, 6 Korean, 5 Pakistanis, 4 Burmese, 4 Indian, 2 Russian, 1 Cambodian, 1 Uzbekistani). The participants are asked to record a video of themselves writing the digit 0 to 9 using a pen on a paper.

2.2 Writing Patterns Categorization

We grouped the collected writing patterns based on the shape, starting point, direction, and sequence of strokes. Figure 1 illustrates the collected writing patterns, with dots representing starting points, arrows indicating endpoints, and the number below each writing pattern representing the occurrence frequency of the writing pattern among the 50 participants.

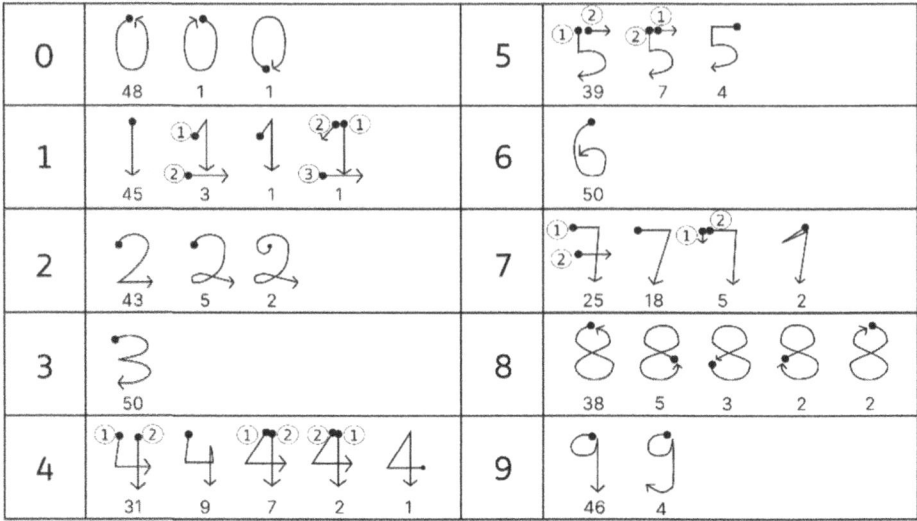

Fig. 1. Writing patterns collected from 50 participants. The dot indicates the start point and the arrow indicates the end point. The digit below each pattern is the frequency of occurrence.

2.3 Vibration Patterns Design

After grouping the writing patterns, we designed vibration patterns with six vibrotactile actuators arranged on the left and right, as shown in Fig. 2(a). We designed the vibration patterns to be generated using two different types of vibrotactile sensations, vibrotactile flows and discrete vibrotactile simulations. Figure 2(b) show the vibration pattern generated using vibrotactile flows and discrete vibrotactile simulations.

The following are factors we considered when designing the vibration patterns.

- **Occurrence Frequency of Writing Pattern.** In the collected writing patterns, most of the digits have a single writing pattern that occurred at more than 60% (Except digit 7). Hence, we designed the vibration patterns based on the writing pattern with the highest frequency of occurrence. However, we opted for the second-highest frequency pattern for digit 7 due to its simplicity.
- **Sharp Turning Point and New Starting Point.** Here, we define the point where a circular stroke ends and the immediately following next stroke starts as a sharp turning point. If a stroke initiates from a point different from where the preceding stroke ends, we identify this as a new starting point. Inspired by our observation that people often pause for a short moment at a sharp turning point and pause for a longer period when starting a new stroke when writing a digit, we added pause at the sharp turning points and the new starting points as one of the characteristics for certain digits (2, 3, 4, 5, 9) to facilitate recognition. We use two different pausing durations, 150 ms at a sharp turning point and 250 ms at a new starting point in both types of vibrotactile sensations. For discrete vibrotactile simulations generated vibration patterns, we repeated the vibration at sharp turning points after the pause to create

a more natural feeling, providing a sense of the ending of a circular stroke and the starting of a next stroke.

• **Different Starting Points among Digits.** To minimize confusion between digits with higher similarity, we designed the digits to have distinct starting points. For example, to reduce confusion between 0 and 6, the starting point of 0 is placed on the top left, while digit 6 starts from the top right. Another example is the digit 2, which has starting point at the middle left, setting it apart from other digits.

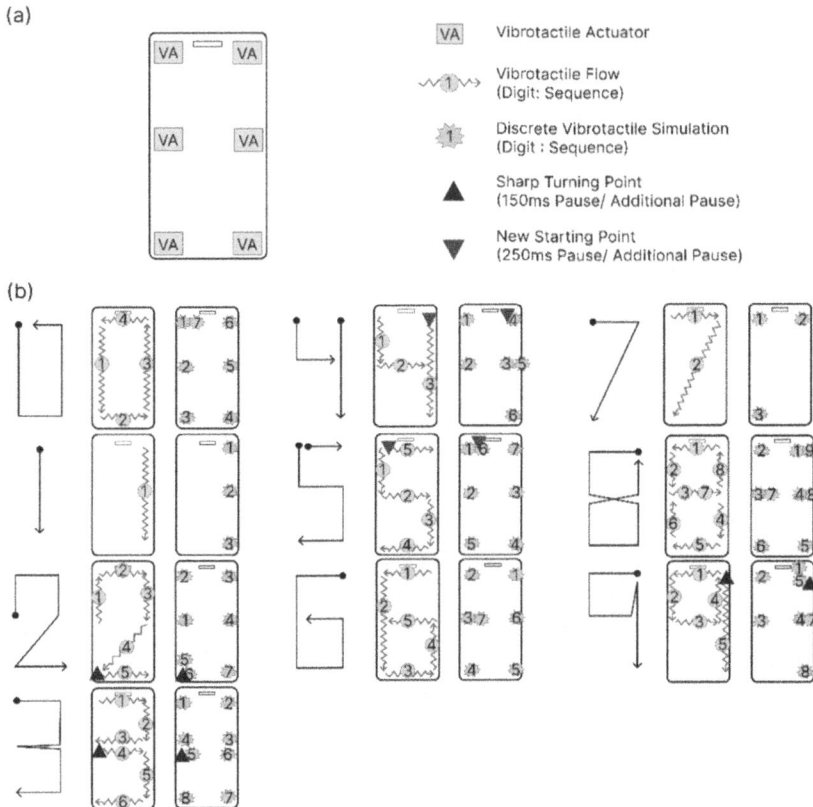

Fig. 2. (a) Arrangement of six vibrotactile actuators. (b) Vibration pattern generated using vibrotactile flows and discrete vibrotactile simulations.

3 User Study

3.1 Participants

We recruited 13 participants (7 female, age: 21−35 years, mean: 26.46, SD: 3.71) to participate in the user study from the university community. Our participants consist of individuals from various nationalities, including 6 Korean, 2 Indian, and one each from

Malaysia, Indonesia, Mongolia, Pakistan, and Tanzania. Among them, 2 participants are post-doctoral researchers at the university, 7 participants are graduate students, and 4 participants are undergraduate students. All participants reported themselves with no known sensory disorders or disabilities. The majority were right-handed, except for one participant who was ambidextrous. Each participant had at least 10 years of experience using a mobile phone and spend a minimum of 2 h using their mobile phone daily. All participants were paid KRW 10,000 per hour (about USD 7.50) after the experiment and an extra KRW 10,000 was paid to the two participants with the highest score.

3.2 Apparatus and Environment

We built a prototype (Fig. 3a) with a size of 138 mm * 65 mm *12m, which was selected based on the size of the iPhone SE. We used six linear resonant actuators (JAHWA JHV-10R1) in our prototype. The actuators were operated at 120Hz. An application running on an Android phone (Huawei Nova 2i) was developed to record participants' answers. Participants were asked to grab the prototype with their left hand and the Android phone was placed on a table in front of their right hand. There was no visual cue to participants while the experiment was conducted. During the study, the participants were asked to wear a soundproof earmuff to block out the noise caused by the vibrators.

(a)   (b)

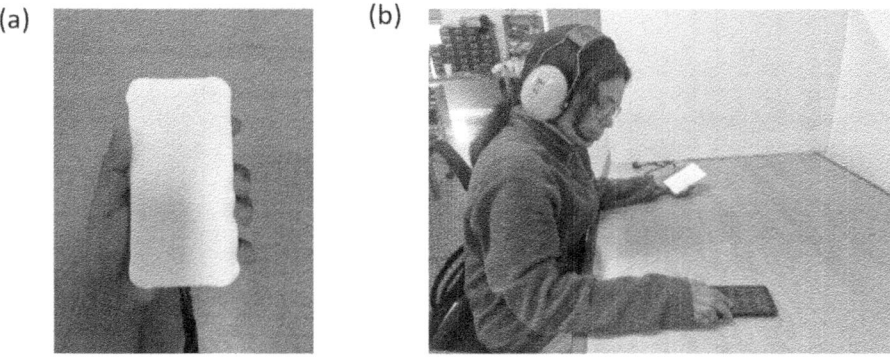

Fig. 3. (a) Prototype for user study. (b) Experiment setup. A participant wearing a soundproof earmuff holding the prototype in her left hand and selecting his answer on the mobile phone with his right hand.

3.3 Vibration Pattern Generation

To prevent participants from guessing the answer based on the total duration of the vibration, we controlled the total generation duration for each digit to be between 2900 ms–4000 ms.

The vibrotactile flows are generated by simultaneously driving two actuators (VA1 and VA2) with different amplitudes profile [13]. VA1 vibrates with decreasing amplitude across time while VA2 vibrates with increase amplitude across time as shown in Fig. 4(a).

When there is a sharp turning point or new starting point, a short pause is provided between two flows. Figure 4(b) shows generation of two flows with and without a pause.

For discrete vibrotactile simulations, we calculated the duration for each discrete vibrotactile simulation that fits the total duration to be between the total generation duration we controlled. Each discrete vibrotactile simulation is generated with a 70% duty cycle and additional pause time is added after the off time at the sharp turning points and new starting points. Figure 4(c) shows examples of generating digit 2 and digit 4 with discrete vibrotactile simulations.

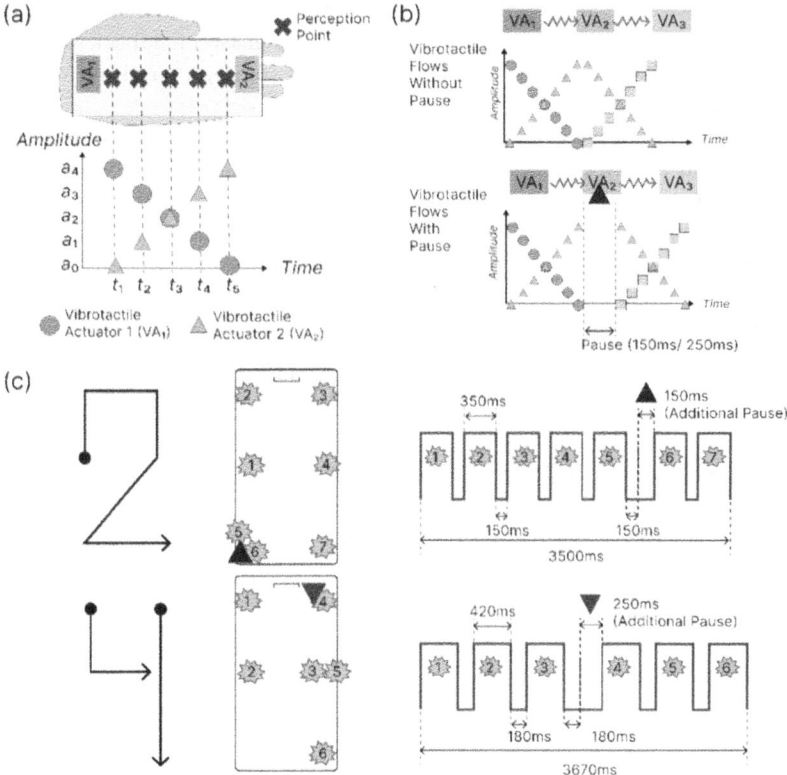

Fig. 4. (a) Generation of vibrotactile flow using phantom sensation. (b) *Top*: Two vibrotactile flows without pause in between. *Bottom*: Two vibrotactile flows with pause in between. (c)Generation of digit 2 and digit 4 with discrete vibrotactile simulations

3.4 Procedure

The user study consists of three sessions for each participant: a practice session for vibration flows generated pattern, a practice session for discrete vibrotactile simulations generated pattern, and a delayed recall test. Each session lasted for about 45 min. 7 participants did the practice session for the vibration flows generated pattern in the first session, while the remaining 6 participants did the practice session for discrete vibrotactile simulations generated pattern. The two practice sessions were conducted on the same day with a minimum 6-h gap between them. The delayed recall test took place three or four days later, depending on participant availability.

Before each practice session started, we verbally explained the vibration patterns and provided graphical illustrations of the patterns, which shows the sequence of the vibrotactile flows or the discrete vibrotactile simulations. There were three sets of trials in each practice session and each set consisted of 30 trials. On each trial, the participant perceived a randomly selected digit and then answered which digit was perceived by selecting their answer on an Android phone. The digit was generated only once for each trial, with no repetition allowed. We provided the correct-answer feedback to participants immediately after each trial. A 10-s break was provided after every trial and a five-minute break was given every 30 trials.

In the third session, we conducted a delay recall test to observe the ability of participants to recall the vibration patterns over time. The participants were tested with 30 trials for each type of vibrotactile sensation. A 10-s break was provided after every trial and a five-minute break was given after the test for the first vibrotactile sensation type. No revision to vibration patterns was provided before the session started and no correct-answer feedback was provided after participants selected their answers.

After the delayed recall test, we conducted a semi-structured interview with participants. We asked the participants to rate the following statements using a 5-Likert scale. Further questions were asked to collect participants' comments and opinions.

Table 1. 5-Likert scale statements in the interview (1-Strongly Disagreed; 5-Strongly Agreed)

	Statement
S1 (F/D)	The vibration pattern for all digits is easy to memorize. (Vibration Flow/ Discrete Vibrotactile Simulation)
S2 (F/D)	The vibration patterns are similar to writing patterns. (Vibration Flow/ Discrete Vibrotactile Simulation)
S3	The pause at sharp turning points and new starting points helps to read the vibrotactile flows generated digits better
S4	The pause is too short
S5	The additional pause at sharp turning points and new starting points helps to read the discrete vibration simulations generated digits better
S6	The additional pause is too short
S7	The repeated vibration at a sharp turning point helps to read the discrete vibrotactile simulations generated digits better
S8	The different starting points among the digits help to read the vibration better

4 Results

We excluded the result from P3 because of her low overall recognition rate. Her results were 30.3% to 36.9% lower than average for vibrotactile flows generated digits and 15.0% to 34.2% lower than average for discrete vibrotactile simulations generated digits throughout the practice sessions. P3 mentioned that she faced difficulty in identifying the exact location of vibrations, attributing it to challenges in grasping the prototype due to her small hand size.

4.1 Experiment Results

Table 2. Experiment results

| | Vibrotactile Flow | | | | Discrete Vibrotactile Stimulation | | | |
	Mean	Median	Std.	IQR	Mean	Median	Std.	IQR
Set 1	69.44%	68.33%	18.90%	40.00%	77.50%	75.00%	11.11%	18.33%
Set 2	83.61%	88.33%	14.94%	31.67%	88.33%	90.00%	9.05%	20.00%
Set 3	83.81%	90.00%	16.78%	29.17%	91.11%	91.67%	6.25%	13.33%
Delayed Recall	90.00%	96.67%	14.56%	15.00%	91.39%	96.67%	14.25%	10.00%

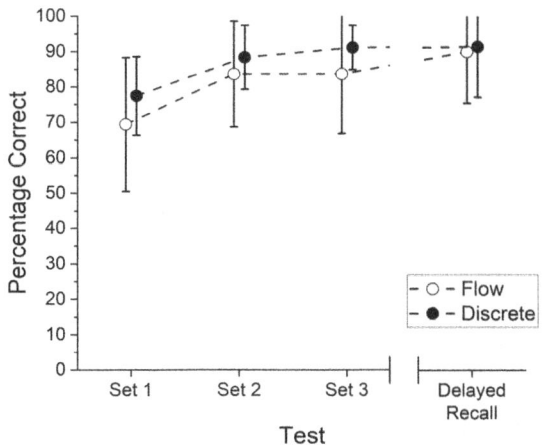

Fig. 5. Mean percentage correct scores. Error bars represent standard deviations.

Table 2 shows the results of our study while Fig. 5 shows the trend of mean percentage correct across digits and participants. The mean percentage correct for vibrotactile flows generated digits improved from 69.44% in the first set to 83.61% in the second set and maintained the same in the third set. For the discrete vibrotactile simulations generated digits, the mean percentage correct increased from 77.50% to 88.33% in the second set and gradually increased to 91.11% in the third set. The results of delayed recall test did not drop but improved to 90.00% for vibrotactile flows and 91.39% for discrete vibrotactile simulations.

We performed two-way repeated ANOVA on the percentage correct in each set to analyze the learning rate of the vibration patterns. The learning effect across the three sets of test was significant ($F(2, 22) = 25.978$, $p < 0.001$). There is no significant difference between types of vibrotactile sensations ($F(1, 22) = 3.219$, $p = 0.098$). The interaction term between the two factors was not significant ($F(2, 22) = 0.321$, $p = 0.728$).

Figure 6 shows the confusion matrix of results on the delayed recall test. For both vibrotactile flows and discrete vibrotactile simulations, digit 9 has the lowest percentage with 77.8% correct. Vibrotactile flows achieved the highest percentage correct for digit 2 at 97.2%, while discrete vibrotactile simulations performed equally well, with digit 0 and 6 both reaching the same highest accuracy of 97.2%.

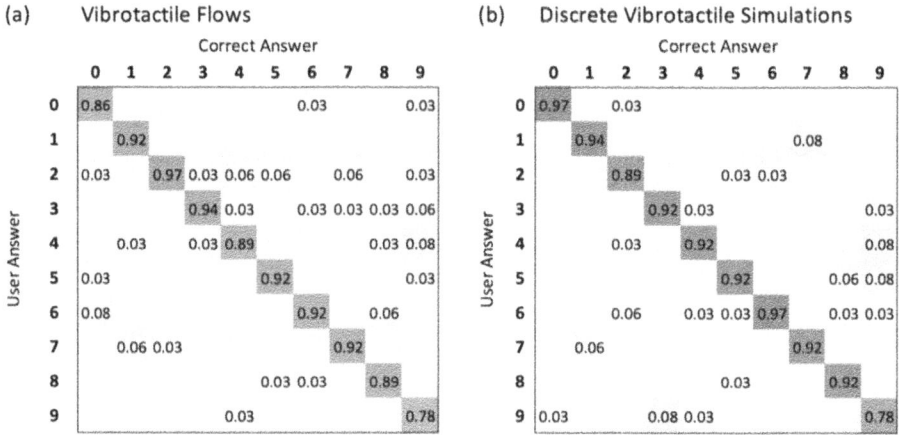

Fig. 6. (a) Confusion matrix for vibrotactile flows generated digits (b) Confusion matrix for discrete vibrotactile simulations generated digits

4.2 Post-experiment Interview

Figure 7 shows the rating results for statements in Table 1. For the vibrotactile flows generated vibration patterns, all participants agreed to the ease of memorizing (S1(F)) and the similarity of the vibration pattern to writing patterns (S2(F)). While for the discrete vibrotactile simulations generated vibration patterns, two participants remained neutral and one participant disagreed with the ease of memorizing (S1(D)) and the similarity of the vibration pattern to writing patterns (S2(D)).

Among twelve participants, nine found the pause between vibrotactile flows at sharp turning points and new starting (S3) helpful in distinguishing the digits, while three remained neutral. The additional pause between discrete vibrotactile simulations (S5) was generally deemed helpful by the majority of participants in distinguishing digits;

however, two participants held a different opinion. Most of the participants (ten out of twelve) disagreed with the statement stating that the pause between vibrotactile flows is too short (S4). Similarly, eleven participants disagreed with the statement regarding the additional pause between discrete vibrotactile simulations being insufficient (S6).

Participants hold different opinions on the helpfulness of the repeated vibration at sharp turning points for discrete vibrotactile simulations (S7) but the majority (nine out of twelve) stand on the positive side. All participants agreed that distinct starting points (S8) helped them to distinguish each digit better.

Besides the factors we considered when designing the vibration patterns, five participants mentioned that the ending points helped them to distinguish digits. While the distinct ending point is being helpful in the recognition of digits, high similarity endings cause confusion. For instance, a few participants also feedbacked that digit 4 and digit 9 are confusing because of their high similarity at the ending, which both have a pause and followed by vibration from top right to bottom right at the ending.

Participants also mentioned that the count of vibrations and duration of each vibration were helpful in reading discrete vibrotactile simulations generated digits. Digit 1 and 7 were created with only three discrete vibrotactile simulations which are less in count and longer in duration for each vibration compared to other digits, setting them apart from other digits. From the confusion matrix for distinct vibrotactile simulations in Fig. 6(b), we can observe that participants confused between digit 1 and 7 in some cases but never confused them with other digits.

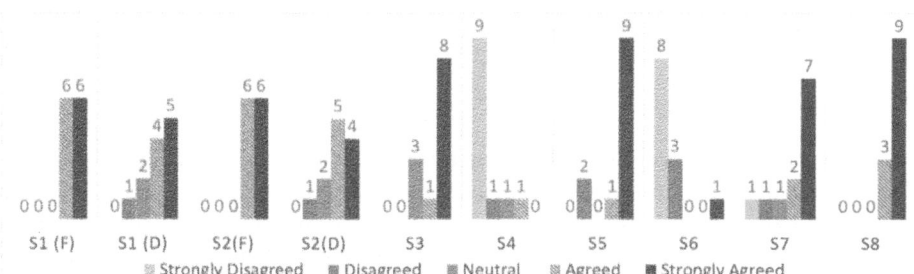

Fig. 7. Rating for statements in Table 1

5 Discussion

5.1 Confusion Between Certain Digits

In the post-experiment interview, we received comments on confusion between digit 4 and 9 due to their similarity at the ending (A pause and followed by vibration from the top right to the bottom right). Two participants felt confused when using the vibrotactile flows, another two participants felt confused when using the discrete vibrotactile simulations, and another three felt confused for both. There were also two participants mentioned that they got confused between 2 and 5 for vibrotactile flows due to the similarity in their shape and sequence. Both digits consist of a half-circular shape followed by a pause and a horizontal stroke from left to right.

Although the majority of participants did not think that the pause was too short, but we can observe that participants did not notice the difference in the pausing duration. Digit 4 and 5 have a longer pause for the new starting point (250ms) while digit 2 and 9 have a shorter pausing duration at the sharp turning point (150 ms). The pausing duration for sharp turning points and new starting points needs to be optimized to let users differentiate them easily so that users will be less confused.

5.2 Individual Variances in Recognition and Preference

Although in overall, the average recognition rates of discrete vibrotactile simulations generated digits are higher than the vibrotactile flows generated digits, but four participants had a higher top recognition rate for vibrotactile flows generated digits during practice sessions. During the delayed recall test, while other participants had recognition rates with lower than 7% difference between two types of vibrotactile simulations, P1 and P8 had higher recognition rates for discrete vibrotactile simulations generated digits with 26.6% and 20% difference respectively. On the other hand, P13 had a higher recognition rate for vibrotactile flows generated digits with a 30% difference.

In the post-experiment interview, the participants showed difference preferences over the types of vibrotactile simulations. Five participants rated the ease of memorizing vibrotactile patterns generated using vibrotactile flows higher than the discrete vibrotactile simulation, while two participants thought that discrete vibration patterns generated using vibrotactile simulations are slightly easier to memorize. The remaining five participants rated both the same.

5.3 Future Work

Validating Effectiveness of Special Characteristics. In this work, we added a few characteristics, such as sharp turning point and new starting point, in certain digits to enhance recognition. However, as outlined in Sect. 5.1, that some participants experience confusion when two digits share the similar or analogous special characteristics. This raises the question: did users really recognize the digit, or were they merely guessing based on the special characteristic? The latter scenario could potentially lead to more confusion, especially when more this work is extended to other characters, such as alphabets, where more characters may share the same characteristic. In future experiments, we will test the recognition rate by mixing fake digits with real digits. These fake digits will possess patterns containing the characteristics of the real digit but will differ from the real digits in other aspects. This experiment aims to determine whether users genuinely recognize the digit or merely guess based on specific characteristics.

Application in Real-World Situation. The results of user study showed that the proposed vibration patterns are easy to learn and memorize without requiring high effort from users. However, in real life situation, users usually read more than one digit at once and are surrounded by noise, for example, reading TAG number on a subway. An experiment to evaluate the performance of using the proposed vibrations patterns to read a series of digits in different environment need to be conducted.

Reducing Vibrotactile Actuators Count. In this work, our primary was to assess the learnability and retainability of vibration patterns designed based writing patterns, leading us to refrain from limiting the count of vibrotactile actuators during the vibration pattern design. However, future work could explore the feasibility of creating vibrotactile simulations with the same patterns using only four vibrotactile actuators at the edges, which can be achieved by generating simulations in the middle utilizing Phantom Sensation.

Implementing to Different Types of Devices. We tested the designed vibration patterns using a mobile phone-sized prototype, but the same vibration patterns can be easily implemented to different types of devices, including wrist-worn tactile displays, haptic suits, haptic chairs, and more. Future work could examine the adaptability of the vibration patterns across a spectrum of devices, varying in size and spatial location.

Extending to Alphabets and Other Characters. Currently, we designed the vibration patterns for digits based on writing patterns while the design of vibration patterns for alphabets is not included. However, future research aims to expand this work to include the generation of alphabet characters following writing patterns and potentially explore the feasibility of incorporating certain Asian characters, such as Korean, Japanese, or some Chinese letters. Research towards this direction is ongoing by us.

6   Conclusion

In this study, we proposed two set vibration patterns for digit 0 to 9, which were designed based on writing patterns using different types of vibrotactile simulations. The results confirmed that the proposed vibration patterns are able to convey digits haptically without requiring high effort to memorize and learn the vibration patterns. Writing patterns, as an intuitive and familiar element in our daily life, have a high potential to create intuitive and memorable vibration patterns.

Acknowledgments. This work was supported by Priority Research Centers Program through the National Research Foundation of Korea (NRF) funded by the Ministry of Education (NRF-2018R1A6A1A03025526), Institute of Information & communications Technology Planning & Evaluation (IITP) grant funded by the Korea government (MSIT) (No. 2022-0-01005, Development of non-wearable visual-tactile digital twin platform technology to provide various interpretation of digital objects) and Culture, Sports and Tourism R&D Program through the Korea Creative Content Agency (KOCCA) grant funded by the Ministry of Culture, Sports and Tourism (RS-2023-00226263).

References

1. Novich, S.D., Eagleman, D.M.: Using space and time to encode vibrotactile information: toward an estimate of the skin's achievable throughput. Exp. Brain Res. **233**, 2777–2788 (2015). https://doi.org/10.1007/s00221-015-4346-1

2. Luzhnica, G., Veas, E.: Optimising encoding for vibrotactile skin reading. In: Proceedings of the 2019 CHI Conference on Human Factors in Computing Systems, pp. 1–14 (2019). https://doi.org/10.1145/3290605.3300465
3. Seo, J., Choi, S.: Perceptual analysis of vibrotactile flows on a mobile device. IEEE Trans. Haptics **6**(4), 522–527 (2013). https://doi.org/10.1109/TOH.2013.24
4. Seo, J., Choi, S.: Edge flows: Improving information transmission in mobile devices using two-dimensional vibrotactile flows. In: 2015 IEEE World Haptics Conference (WHC), pp. 25–30 (2015). https://doi.org/10.1109/WHC.2015.7177686
5. Gong, Y., Wang, D., Guo, Q., Luo, H., Zhang, Y., Xiao, J.: Identification of vibrotactile flow patterns on a handheld haptic device. In: 2019 International Conference on Virtual Reality and Visualization (ICVRV), pp. 76–81 (2019). https://doi.org/10.1109/ICVRV47840.2019.00021
6. Reed, C.M., Tan, H.Z., Jones, L.A.: Haptic communication of language. IEEE Trans. Haptics **16**(2), 134–153 (2023). https://doi.org/10.1109/TOH.2023.3257539
7. Dhandapani, G., Ferguson, J., Freeman, E.: HapticLock: eyes-free authentication for mobile devices. In: Proceedings of the 2021 International Conference on Multimodal Interaction, pp.195–202 (2021). https://doi.org/10.1145/3462244.3481001
8. Varma, M., Watson, S., Chan, L., Peiris, R.: VibroAuth: authentication with haptics based non-visual, rearranged keypads to mitigate shoulder surfing attacks. In: Moallem, A. (ed.) HCII 2022. LNCS, vol. 13333, pp. 280–303. Springer, Cham (2022). https://doi.org/10.1007/978-3-031-05563-8_19
9. Plaisier, M.A., Vermeer, D.S., Kappers, A.M.: Learning the vibrotactile morse code alphabet. ACM Trans. Appl. Percept. (TAP) **17**(3), 1–10 (2020). https://doi.org/10.1145/3402935
10. Jayant, C., Acuario, C., Johnson, W., Hollier, J., Ladner, R.: V-braille: haptic braille perception using a touch-screen and vibration on mobile phones. In: Proceedings of the 12th International ACM SIGACCESS Conference on Computers and Accessibility, pp. 295–296 (2010). https://doi.org/10.1145/1878803.1878878
11. Nicolau, H., Montague, K., Guerreiro, T., Rodrigues, A., Hanson, V.L.: HoliBraille: multipoint vibrotactile feedback on mobile devices. In: Proceedings of the 12th International Web for All Conference, pp. 1–4 (2015). https://doi.org/10.1145/2745555.2746643
12. Dhar, A., Nittala, A., Yadav, K.: TactBack: VibroTactile braille output using smartphone and smartwatch for visually impaired. In: Proceedings of the 13th International Web for All Conference, pp. 1–2 (2016). https://doi.org/10.1145/2899475.2899514
13. Oshima, K., Arai, T., Ichihara, S., Nakano, Y.: Tactile sensitivity and braille reading in people with early blindness and late blindness. J. Vis. Impairment Blindness **108**(2), 122–131 (2014). https://doi.org/10.1177/0145482X1410800204
14. Liu, X., Dohler, M.: Vibrotactile alphabets: time and Frequency patterns to encode information. IEEE Trans. Haptics **14**(1), 161–173 (2020). https://doi.org/10.1109/TOH.2020.3005093
15. Wobbrock, J.O., Myers, B.A., Kembel, J.A.: EdgeWrite: a stylus-based text entry method designed for high accuracy and stability of motion. In: Proceedings of the 16th Annual ACM Symposium on User Interface Software and Technology, pp. 61–70 (2003). https://doi.org/10.1145/964696.964703
16. Liao, Y.C., Chen, Y.L., Lo, J.Y., Liang, R.H., Chan, L., Chen, B.Y.: EdgeVib: effective alphanumeric character output using a wrist-worn tactile display. In: Proceedings of the 29th Annual Symposium on User Interface Software and Technology, pp. 595–601 (2016). https://doi.org/10.1145/2984511.2984522
17. Cauchard, J. R., Cheng, J. L., Pietrzak, T., & Landay, J. A.: ActiVibe: design and evaluation of vibrations for progress monitoring. In: Proceedings of the 2016 CHI Conference on Human Factors in Computing Systems, pp. 3261–3271 (2016). https://doi.org/10.1145/2858036.2858046

18. Zhao, S., Israr, A., Lau, F., Abnousi, F.: Coding tactile symbols for phonemic communication. In: Proceedings of the 2018 CHI Conference on Human Factors in Computing Systems, pp. 1–13 (2018). https://doi.org/10.1145/3173574.3173966
19. de Vargas, M.F., Marino, D., Weill, A., Cooperstock, J.R.: Speaking haptically: from phonemes to phrases with a mobile haptic communication system. IEEE Trans. Haptics **14**(3), 479–490 (2021). https://doi.org/10.1109/TOH.2021.3054812
20. Kim, S.Y., Kim, J.O., Kim, K.Y.: Traveling vibrotactile wave-a new vibrotactile rendering method for mobile devices. IEEE Trans. Consum. Electron. **55**(3), 1032–1038 (2009). https://doi.org/10.1109/TCE.2009.5277952

Estimating Contact Force Rate Using Skin Deformation Cues

Bingxu Li[1(✉)], Gregory J. Gerling[2], and Tyler Cody[1]

[1] SmartHap, Charlottesville, VA, USA
bingxu.contact@gmail.com
[2] University of Virginia, Charlottesville, VA, USA

Abstract. Knowledge of contact force is important in various fields as it provides essential insights into the interactions between objects. Understanding the forces exerted during human-machine interactions is essential for optimizing performance and preventing discomfort or injury. Despite the significance of contact force, there are limited approaches for estimating it accurately when direct measurements are not available. Existing methods often come with constraints such as complexity, cost, and environmental limitations. This paper introduces a method for estimating contact force from skin deformation cues in bare-finger interactions using linear regression models. The statistical results from the linear models align with scientific studies, revealing a significant correlation between force and skin deformation that has a dependence on stimulus moduli. The developed models exhibit reliable estimations of force, with robustness to variances between trials and insensitivity to individual differences in skin properties. This approach offers an alternative method for estimating contact force and indicates the potential of models for estimating key variables associated with user experience.

Keywords: Tactile Feedback · Contact Force Prediction · Human-Machine Interaction

1 Introduction

Tactile force, relating to sense of touch, becomes a critical consideration in minimizing tissue damage in surgical procedures [4,22], facilitating responsive and adaptive capabilities in robotics and automation [16,17], and enhancing realism and immersion in virtual environments [16,19]. One way to measure force is to use stick-based tools designed with force sensors to provide feedback to the manipulator, commonly used in surgical robots and laparoscopic instrumentation for precision and safety control [3,26]. Although this approach is suitable for tasks demanding fine control, it requires additional equipment or accessories, involves higher learning curves during user training, and results in unnatural interactions. On the other hand, bare-finger interaction offers a natural and intuitive way to interact with objects by eliminating the need for additional tools

H. Kajimoto et al. (Eds.): EuroHaptics 2024, LNCS 14768, pp. 464–476, 2025.
https://doi.org/10.1007/978-3-031-70058-3_38

and enabling users to perform multiple actions simultaneously. Many applications benefit from utilizing bare-finger interactions to optimize user experiences. The key challenge is limited force feedback to the user, attributed to the absence of sensors on the contact surface and the non-linear characteristics of the natural skin. For example, particularly in virtual and augmented reality (VR/AR) environments where realism and immersion are primary for user experience, the measurement of force becomes crucial but constrained because users often interact with virtual objects without physical touch. One method is to apply physics-based models by implementing physics engines that simulate the dynamics and interactions between virtual objects, but it has limitations including sensitivity to model parameters, computational complexity, and simplified contact models [1,21,29]. Another method is to use myography techniques, such as surface electromyography(sEMG), which outputs electric activity of conductive materials upon physical contact [5]. This technology is frequently used in biomechanical settings [7,28] and human-computer interaction (HCI) applications to detect contact events [24,30]. However, this method has limitations including electrode wiring and implementation, low spatial resolution, and sensitivity to electric and magnetic interference [8,23]. In scenarios where the spaces or access is restricted, such as physical therapy, direct force assessment becomes impractical due to constraints of physical space and patient comfort. One common method is to use visual observation by tracking identified markers, carried out either by a dedicated hardware or a skilled individual [18,21], but it shows limitations in quantitative data and its high dependence on visual cues.

Thus, in bare-finger interactions, these limitations associated the current approaches highlight the need for complementary methods including predictive models that can also serve as an indirect, albeit virtual, measurement of force. In tactile sensing, contact force rate, along with skin deformation cues, contributes to the psychophysical perception of an object's compliance [11,25]. This relationship is influenced both by the material properties of the objects and dynamic aspects of the interaction [27]. In this paper, we construct multivariate linear models using skin deformation cues as predictors and biomechanical measurements from previously published studies in passive touch [14]. Our results show that the developed models can predict contact force rate using skin deformation cues with maximum error less than a detection threshold of 0.08 N/s, a detection threshold found in previous studies [15].

The remainder of this paper is structured as follows. Next, we describe the data collection, measurement, and modeling methods. Then, we present our results. The paper concludes with a discussion of the results.

2   Methods

In previous research [10], we employed a custom-built imaging system that facilitates direct visualization of skin surface in the form of a 3-D point cloud, while the skin udergoes indentation of contact. In previous studies [13,15], cues derived from the point cloud were used to quantify the spatial deformation of skin

surface. Biomechanical experiments were conducted across a group of participants, in which the relationship between contact force and skin deformation
with varying stimulus moduli was evaluated. In the following, the relevant details
of the imaging system and experiments are described and the data processing
and model construction methods used in this paper are given.

2.1 Participants

A total of 25 participants (mean age = 27, range = 21 - 30; 11 males and 14
females) were recruited in the experiments, which all fully completed. The experiments were approved by the local Institutional Review Board, and informed
consent was obtained from each participants. None had prior knowledge about
the experiments and their fingers were free of calluses and scars. All devices and
surfaces were sanitized after each use.

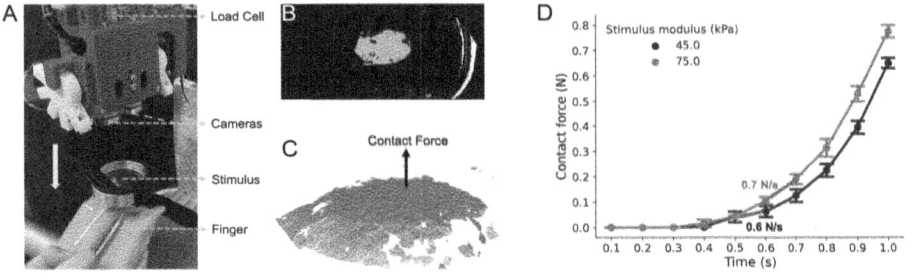

**Fig. 1. Experimental setup for observing skin deformation and measuring
contact force.** (A) Mechanical-electrical indenter that delivers elastic stimulus to the
fingertip. (B) One image of the fingertip captured by the right camera when indented
by a 45 kPa stimulus. (C) Construction of a 3-D point cloud from captured images
using disparity-mapping method. (D) Contact force is recorded simultaneously over
an indentation of 1 second. The rate of force is calculated as 0.6 and 0.7 N/s for the
45 and 75 kPa stimulus, respectively. The segment in red indicates the median value
of the intermediate change rates.

2.2 Apparatus

The apparatus [10] used for skin surface visualization, force measurement,
and biomechanical experiments consists of a custom-built, electrical-mechanical
motion indenter (ILS-100 MVTP, Newport, Irvine, CA, USA) paired with two
stereo cameras and a load cell installed on a cantilever, shown in Fig. 1A. Elastic stimuli are individually placed in a 3D printed rotatable housing, and each
can be delivered to the fingertip at controlled rate and displacement. A support
platform consists of a cylindrical, heavy duty aluminum rod attached with a
rectangular aluminum hand rest was designed to stabilize the participant's finger at a 30-° angle with respect the stimulus surface. The hand rest includes a

rigid plastic housing that holds the finger in position and helps minimizing subtle movements during experiments. An inclined 30-° contact is chosen because
of the high concentration of mechanoreceptors at the fingertip. The displacement
of stimulus is recorded by the motion controlled with resolution of 0.001 mm,
with force measured by the load cell (LCFD-5, Omegadyne, Sunbury, OH, USA)
at a frequency of 150 Hz, with a resolution of ±0.05 N. Images are captured by
the left and right cameras (Papalook PA150, Shenzhen Aoni Electronic Industry
Co., Guangdong, China) at 30 frames per second with maximum resolution of
1280×720 pixels, maintaining a manual focus during experiments. This custom-
built imaging system shows competitive advantages in capturing high spatial
resolution of the skin surface, facilitating direct visualization of skin dynamics
in bare-finger interactions, and providing empirical measurements to characterize
skin responses upon contact with elastic stimuli of varying compliance.

2.3   Skin Deformation Cues and Contact Force

Using the images collected by the cameras, a disparity-mapping method was
applied for generating a 3-D point cloud that represents the geometry of skin
surface, Fig. 1B–C. To quantitatively characterize skin geometry, in prior work
[14] we developed an ellipse fitting algorithm by fitting vertically stacked, uni-
orientated ellipses to the point cloud, to derive mechanical cues that describe
skin deformation over time. The skin deformation cues include contact area,
curvature, eccentricity, and penetration depth.

The change rates of contact force and skin deformation cues are calculated
and used in the statistical analysis, as rate-based cues describe both spatial and
temporal aspects of dynamic changes of the skin, and are found to be associated
with psychophysical responses [2,13]. In the experiments, contact force was measured by the load cell when the fingertip was making contact with the stimulus.
The change rate of force was determined as the median value of a sequence of
intermediate rates calculated at every 0.1 s time interval. Figure 1D shows the
change in force over an indentation by the 45 and 75 kPa stimulus, and their
force rate was calculated as 0.6 and 0.7 N/s, respectively.

2.4   Biomechanical Experiments

Biomechanical measurements of skin deformation were conducted in passive
touch. Two elastic stimuli were fabricated with moduli of 45 and 75 kPa. These
moduli were chosen to closely approximate a slightly higher modulus than that
of the skin at the fingertip [9]. During the experiments, the participants were
seated in a comfortable chair with elbows resting on the table. Their index finger
was stabilized and indented by the stimulus at 1.75 mm/s rate for 2 mm displacement. 3-D point clouds were generated at every 0.1 s during indentation,
followed by acquisition of the skin deformation cues. Each stimulus was indented
for 3 repeated repetitions. In total, there were 150 indentations conducted in this
experiments (25 participants, 2 stimuli, and 3 repetitions) and 1500 3-D point
clouds generated.

2.5   Data Processing and Model Construction

The 1500 3-D point clouds were processed into 1500 instances of skin deformation cues, and then further processed into 150 instances of change rate of cues—one instance for each indentation. The median change rates of contact area, curvature, eccentricity, and penetration depth are the 4 predictors (referred to genearlly as skin deformation cues in the following) and the median change rate of contact force is the predicted response. The instances were divided by stimuli into two 75 instance samples, and then further split into 70% subsamples of 52 instances for training and 30% subsamples of 23 instances for testing.

Three linear regression models were assessed in this study: Ordinary Least Squares (OLS), Lasso, and Ridge regression. The mathematical formula used for multivariate OLS regression is:

$$y = \beta_0 + \beta_1 x_1 + \ldots + \beta_4 x_4 + \epsilon \tag{1}$$

for multivariate Lasso regression is:

$$y = \beta_0 + \sum_{j=1}^{4} \beta_j x_j + \lambda \sum_{j=1}^{4} |\beta_j| \tag{2}$$

for multivariate Ridge regression is:

$$y = \beta_0 + \sum_{j=1}^{4} \beta_j x_j + \lambda \sum_{j=1}^{4} \beta_j^2 \tag{3}$$

where y is the dependent variable, $x_1, x_2, \ldots, x_4$ are the independent variables, and $\beta_0, \beta_1, \ldots, \beta_4$ are the regression coefficients. ϵ is the error term in OLS regression model that minimizes the sum of squared residuals (SSE) to estimate the coefficients. In Lasso and Ridge regression, the L_1 penalty $\sum_{j=1}^{4} |\beta_j|$ and the L_2 penalty $\sum_{j=1}^{p} \beta_j^2$ are used to decrease coefficients to zero or near-zero, respectively, and the regularization term λ is a penalty parameter that controls the decrement. To construct the model, the change rates of four skin deformation cues were considered as the independent variables to predict contact force rate which is the dependent variable. The entire dataset was normalized before regression fitting. The hyper-parameter λ for Ridge and Lasso regression at each stimulus were determined by the GridSearch method through optimizing r^2 score over a 10-fold cross-validation of their respective training subsamples. Subsequently, the linear models are constructed from the 52-instance training subsamples and tested by 23-instance testing subsamples.

3   Results

3.1   The Relationship Between Force and Skin Deformation Cues

The statistical correlations among the rate of change in skin deformation cues and their correlations with force rate were evaluated using the t-test for both the 45 and 75 kPa stimuli, as shown in Fig. 2A–B. The results indicate two major findings. First, the cues show statistical independence, as reflected in low correlation values. Second, the correlation between force and skin deformation cues varies with stimulus modulus. In particular, force rate exhibited high correlations with the rate of contact area (r = 0.76) and curvature (r = 0.7) for the 45 kPa stimulus, Fig. 2C; whereas it correlates with the rate of contact area (r = 0.71) and eccentricity (r = 0.77), for the 75 kPa stimulus, Fig. 2D.

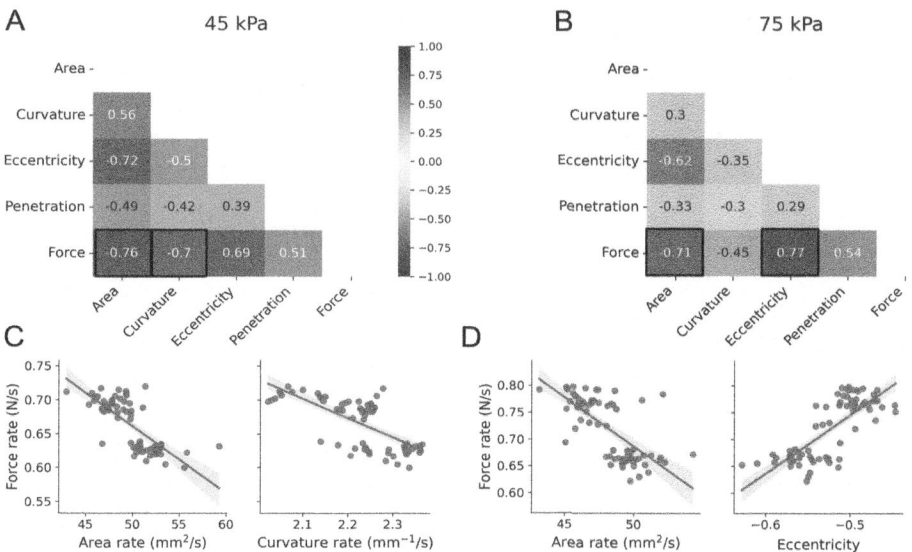

Fig. 2. Statistical correlations between the rate of skin deformation cues and force. The correlation values that higher than 0.7 are highlighted in a black frame for the (A) 45 kPa and (B) 75 kPa stimulus. Panel (C) and (D) present the overall data from 75 trials, indicating a high correlation between force rate with the rate of (C) contact area and curvature for the 45 kPa stimulus, and (D) contact area and eccentricity for the 75 kPa stimulus.

3.2 Constructing Regression Models for Force Prediction

The correlation analysis depicted in Fig. 2 suggests that the change rate of contract force can be predicted using the skin deformation cues. The prediction results from the linear models using all four skin deformation cues as predictors are presented in Fig. 3. The maximum error is calculated as the error mean plus two standard deviations. The results show that the OLS regression model yields the highest R^2 value of 0.701, with a maximum error of 0.058 N/s for the 45 kPa stimulus; whereas the Ridge regression model achieves R^2 value of 0.77, with a maximum error of 0.04 N/s for the 75 kPa stimulus. Notably, previous research found a force rate difference of 0.08 N/s between 45 and 75 kPa stimulus to not be differentiable by participants [13,15]. This implies that while errors were observed in the regression model, the maximum error is significantly below the detection threshold. Moreover, a higher prediction performance is observed for the stiffer stimulus (75 kPa). This is because the change rate of force is more associated with stimuli that are To validate the linearity of the predictive relationship between the skin deformation cues and the change rate of force, a second-order polynomial regression analysis was conducted, which revealed a low coefficient ($R^2 = 0.462$, mean error $= 0.11 \pm 0.06$) implying a predominantly linear relationship.

	45 kPa		75 kPa	
Model	R² score	Max error	R² score	Max error
OLS	0.701	0.057894	0.765	0.04914
Ridge	0.692	0.059701	0.773	0.041797
Lasso	0.700	0.057616	0.761	0.047278

Fig. 3. Results from the three regression models for force prediction The optimal parameter used for the Ridge and Lasso regression was determined using Grid search method by the highest R^2 score.

3.3 Significance of Skin Deformation Cues on Predicting Force

By using analyzing the Lasso regression models, we can compare the significance of each cue in predicting change rate of force. As part of the fitting process, Lasso regression increases and decreases the weight of predictors to optimize performance. The feature importance from the Lasso regression models are given in Fig. 4 and show that the change rates of contact area and curvature are primary cues used for predicting force rate for the 45 kPa stimulus, whereas the change rates of eccentricity and penetration depth are the primary cues for the 75 kPa stimulus. This aligns with the correlations found in Fig. 2.

To further validate the predictive significance of each cue, we conducted an analysis wherein the models were constructed by systematically excluding one cue at a time for force prediction. This approach allowed us to assess how much the model performance was affected by the excluded cue. The results are shown in Fig. 5. For the 45 kPa stimulus, the results show a significant drop in model outcomes when the change rates of curvature and contact area were excluded from constructing the model, Fig. 5A; whereas excluding the change rates of penetration depth and eccentricity resulted in a performance decrease in force prediction for the 75 kPa stimulus, Fig. 5B.

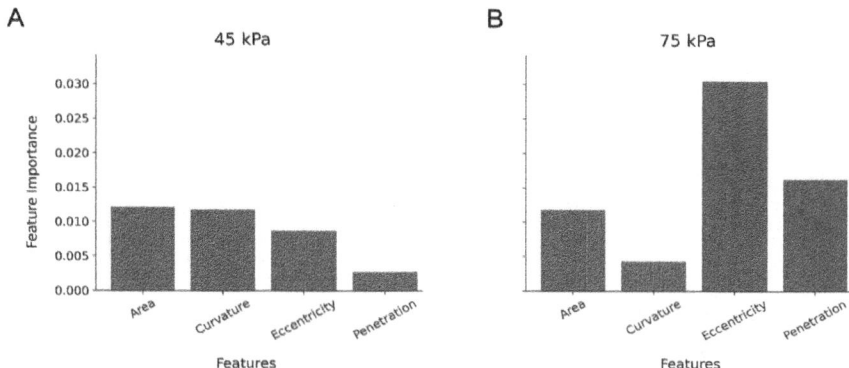

Fig. 4. Significance of skin deformation cues in predicting contact force using Lasso regression.

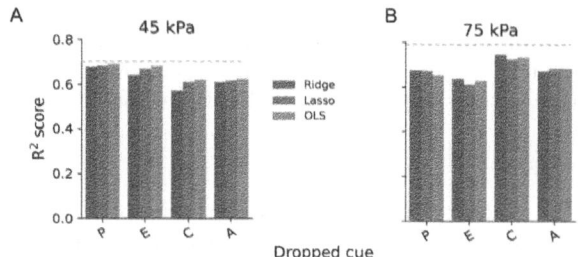

Fig. 5. Validation of the impact of each cue in force prediction by constructing the models with only three cues. The letter (P, E, C, A) on the x-axis indicates the corresponding cue that has been dropped. Letter P, E, C, A, represents: the change rate of penetration, eccentricity, curvature and area, respectively. The gray dash line indicates the R^2 value when the model uses all four cues for prediction.

3.4 Effects of Sample Size, Stimulus Modulus, and Individual Differences on Force Prediction

The predictive capacity of the models may be influenced by various factors, including the training sample size, the modulus of tested stimuli, variances between trials, and also individual differences in skin properties and between skin states.

To assess the impact of sample size, we calculated R^2 values using the regression models as the number of samples used for model training were incrementally increased, as shown in Fig. 6. The results indicate a minimum of 40 samples ($n = 40$) is necessary for the models to achieve stability and optimal performance. Note that errors less than 0.08 N/s are considered not differentiable [13,15]. We can observe that for the 45 kPa stimulus errors are below the detection threshold at $n = 29$, as shown in Fig. 6A, and for the 75 kPa stimulus at $n = 26$, as shown in Fig. 6B. Overall, the models exhibit reduced errors and demand a smaller sample size for enhanced performance in the case of the stiffer stimulus.

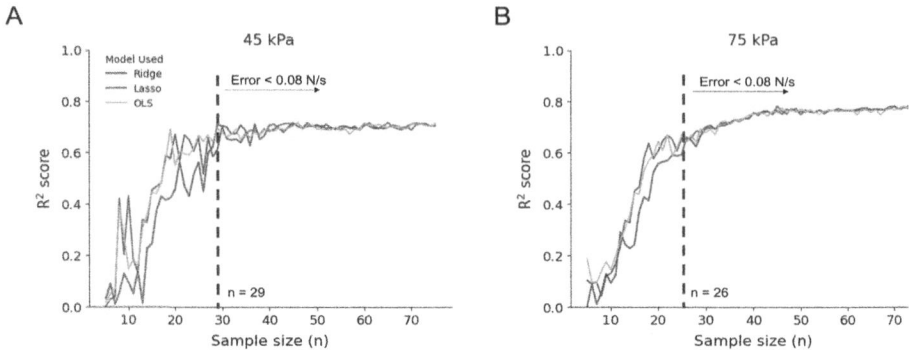

Fig. 6. Evaluation of the impact of sample size on model performance (n = 75). Starting at n = 5, the models use 70% for training and 30% for testing. The dash line indicates the sample size at which the error falls below the detection threshold.

Additionally, we evaluated the model's sensitivity to stimulus moduli by constructing a model using data from the 45 kPa stimulus and applied it to predict force rates for the 75 kPa stimulus, and vice versa, Fig. 7. The results revealed a notable decline in performance for cross model prediction, with errors exceeding the detection threshold of 0.08 N/s. This underscores the necessity for distinct models used in predicting force regarding the stimulus moduli.

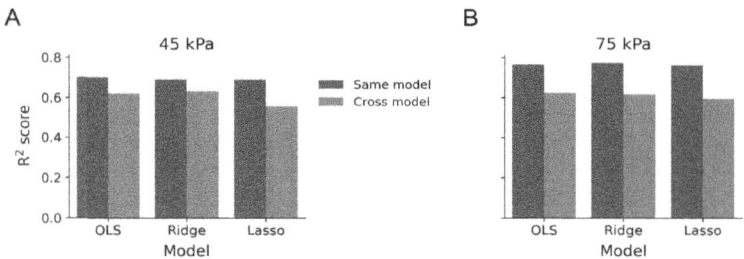

Fig. 7. Validation of the distinctiveness of model for force prediction. (A) The model is fit with the data of 45 kPa stimulus making predictions for both 45 (same model) and 75 kPa (cross model) stimuli. Similarly, in (B) the model is fit with the data of 75 kPa stimulus.

Lastly, we investigated the influence of individual differences and variations between trials on the model capability in force prediction. The data of two participants, one with a soft and another with a stiffer index finger, was chosen for the analysis, Fig. 8. Each participant underwent three experimental trials before and after skin modulation, in which their finger stiffness was manipulated by the application of hyaluronic acid at skin surface [13]. Regression models were used to predict force rate at each trial, which was compared with the actual values measured by the load cell, for the 45 and 75 kPa stimuli. The findings indicate consistent outcomes ($R^2 = 0.74, 0.75$ for Participant A and B for the 45 kPa, and $R^2 = 0.84, 0.82$ or Participant A and B for the 75 kPa, respectively) from the model for force prediction, with errors falling below the detection threshold (<0.03 N/s). This validates that the developed model is robust against variances between trials, and also insensitive to skin stiffness and changes in skin states.

4   Discussion

The main objective of this work was to develop models capable of predicting contact force using information from skin deformation cues. We constructed multivariate linear models for the change rate of contact force using the change rate of skin deformation cues as predictors. The models achieved a maximum error less than a detection threshold of 0.08 N/s. The threshold was determined through human-subject experiments conducted in [15], where the participants were asked to perceptually differentiate between the 45 and 75 kPa stimuli. Psychophysical results reveal that these two stimuli are not perceptually distinguishable ($<75\%$ correct rate). The observed difference in force rate was 0.08 N/s and the mean difference in force magnitude was 0.13 N at a stimulus indentation of 2 mm, is comparable with the just-noticeable difference (JND) of 0.1 N on the fingertips upon normal indentation [6,12]. Another advantage of the developed models is the requirement of a relatively small number of samples for model training, as shown in Fig. 6, as well as the models appear to be robust to variances between experimental trials and in skin properties, as shown in Fig. 8.

The observed predictive importance of each cue is aligned with scientific findings [15,20], showing that in tactile interaction, the change rate of contact area shows correlation with the change rate of contact force irrespective of stimulus moduli, whereas other cues exhibit such correlation depending on the stimulus moduli. This phenomena occurs because when indented by a soft stimulus, the finger tends to retain its shape which makes curvature stands out as a cue capturing spatial changes on skin surface; in contrast, when indented by a hard stimulus, the skin quickly flattens out at the contact surface, which makes eccentricity a stronger cue influencing contact force. This scientific understanding underscores and rationalizes the degradation in cross model performance shown in Fig. 7.

In future work, it is important to consider the means of measuring skin deformation cues. The experimental apparatus used in this paper is designed for scientific purposes, has high spatial and temporal resolution, and involves complex mechanical systems. Alternative approaches may be more practical for engineered systems.

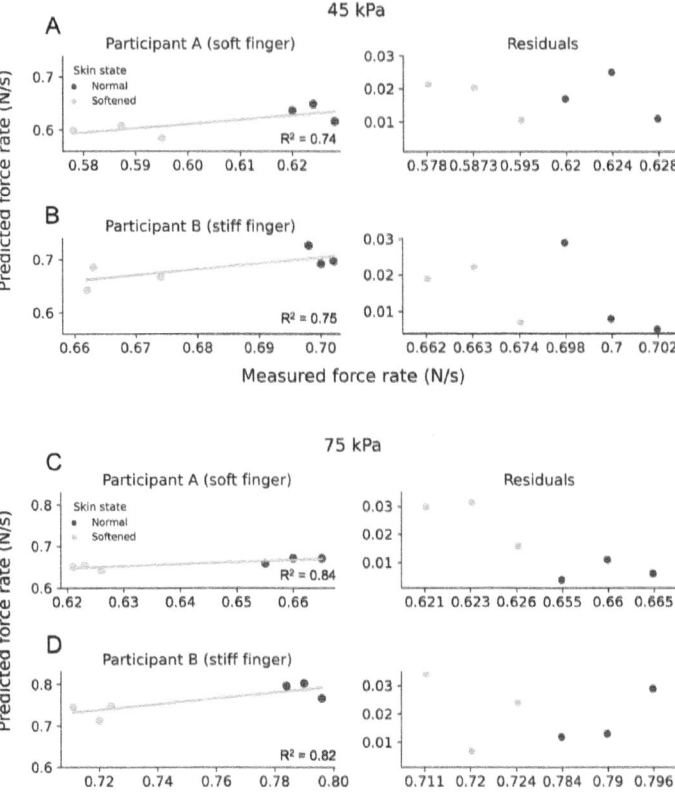

Fig. 8. Evaluating the effects of between-trial variances, skin stiffness and modulation of skin state, on force prediction capability. Participant A with soft finger (stiffness = 0.1 N/mm) and Participant B with stiff finger (stiffness = 0.15 N/mm) were selected for the analysis. Ridge regression models were used to predict force rate and compared with the actual value measured by the load cell. (A) and (B) indicate the prediction performance for the two participants before and after their skin was softened by hyaluronic acid, for the 45 kPa stimulus, and (C) and (D) for the 75 kPa stimulus. The error (residual) from each prediction is reflected correspondingly on the right side.

Acknowledgements. This work was supported in part by the National Science Foundation (IIS-1908115) and National Institutes of Health (R01NS105241, NCCIH R21AT011980, and U24AT011970).

Disclosure of Interests. The authors claimed to have no interest of conflict.

References

1. Abbiendi, G., collaboration, O., et al.: Tests of the standard model and constraints on new physics from measurements of fermion-pair production at 189-209 GeV at LEP (2003). arXiv preprint https://arxiv.org/abs/hep-ex/0309053
2. Ambrosi, G., Bicchi, A., De Rossi, D., Scilingo, E.P.: The role of contact area spread rate in haptic discrimination of softness. In: Proceedings 1999 IEEE International Conference on Robotics and Automation (Cat. No. 99CH36288C), vol. 1, pp. 305–310. IEEE (1999)
3. Autorino, R., Kaouk, J.H., Stolzenburg, J.U., Gill, I.S., Mottrie, A., Tewari, A., Cadeddu, J.A.: Current status and future directions of robotic single-site surgery: a systematic review. Eur. Urol. 63(2), 266–280 (2013)
4. Chen, E., Marcus, B.: Force feedback for surgical simulation. Proc. IEEE 86(3), 524–530 (1998). https://doi.org/10.1109/5.662877
5. Chowdhury, R.H., Reaz, M.B., Ali, M.A.B.M., Bakar, A.A., Chellappan, K., Chang, T.G.: Surface electromyography signal processing and classification techniques. Sensors 13(9), 12431–12466 (2013)
6. Dahiya, R.S., Metta, G., Valle, M., Sandini, G.: Tactile sensing-from humans to humanoids. IEEE Trans. Robot. 26(1), 1–20 (2009)
7. De Luca, C.J.: The use of surface electromyography in biomechanics. J. Appl. Biomech. 13(2), 135–163 (1997)
8. Disselhorst-Klug, C., Schmitz-Rode, T., Rau, G.: Surface electromyography and muscle force: limits in sEMG-force relationship and new approaches for applications. Clin. Biomech. 24(3), 225–235 (2009)
9. Gerling, G.J., Hauser, S.C., Soltis, B.R., Bowen, A.K., Fanta, K.D., Wang, Y.: A standard methodology to characterize the intrinsic material properties of compliant test stimuli. IEEE Trans. Haptics 11(4), 498–508 (2018). https://doi.org/10.1109/TOH.2018.2825396
10. Hauser, S.C., Gerling, G.J.: Imaging the 3-D deformation of the finger pad when interacting with compliant materials. In: 2018 IEEE Haptics Symposium (HAPTICS), pp. 7–13. IEEE (2018)
11. Hirota, K., Ujitoko, Y., Sakurai, S., Nojima, T.: Deformation matching: force computation based on deformation optimization. IEEE Trans. Haptics 15(2), 267–279 (2022). https://doi.org/10.1109/TOH.2022.3142053
12. LaMotte, R.H., Srinivasan, M.A.: Tactile discrimination of shape: responses of slowly adapting mechanoreceptor afferents to a step stroked across the monkey fingerpad. J. Neurosci. 7(6), 1655–1671 (1987)
13. Li, B., Gerling, G.J.: An individual's skin stiffness predicts their tactile discrimination of compliance. J. Physiol. 601, 5777 (2023)
14. Li, B., Hauser, S., Gerling, G.J.: Identifying 3-D spatiotemporal skin deformation cues evoked in interacting with compliant elastic surfaces. In: 2020 IEEE Haptics Symposium (HAPTICS), pp. 35–40 (2020). https://doi.org/10.1109/HAPTICS45997.2020.ras.HAP20.22.5a9b38d8
15. Li, B., Hauser, S.C., Gerling, G.J.: Faster indentation influences skin deformation to reduce tactile discriminability of compliant objects. IEEE Trans. Haptics 16(2), 215–227 (2023). https://doi.org/10.1109/TOH.2023.3253256
16. Magrini, E., Flacco, F., De Luca, A.: Estimation of contact forces using a virtual force sensor. In: 2014 IEEE/RSJ International Conference on Intelligent Robots and Systems, pp. 2126–2133 (2014). https://doi.org/10.1109/IROS.2014.6942848

17. Magrini, E., Flacco, F., De Luca, A.: Control of generalized contact motion and force in physical human-robot interaction. In: 2015 IEEE International Conference on Robotics and Automation (ICRA), pp. 2298–2304 (2015). https://doi.org/10.1109/ICRA.2015.7139504

18. Nakagaki, H., Kitagi, K., Ogasawara, T., Tsukune, H.: Study of insertion task of a flexible wire into a hole by using visual tracking observed by stereo vision. In: Proceedings of IEEE International Conference on Robotics and Automation, vol. 4, pp. 3209–3214. IEEE (1996)

19. Park, J., Khatib, O.: A haptic teleoperation approach based on contact force control. Int. J. Robot. Res. 25(5–6), 575–591 (2006)

20. Pawluk, D.T., Howe, R.D.: Dynamic Contact of the Human Fingerpad Against a Flat Surface (1999)

21. Pham, T.H., Kyriazis, N., Argyros, A.A., Kheddar, A.: Hand-object contact force estimation from markerless visual tracking. IEEE Trans. Pattern Anal. Mach. Intell. 40(12), 2883–2896 (2017)

22. Puangmali, P., Althoefer, K., Seneviratne, L.D., Murphy, D., Dasgupta, P.: State-of-the-art in force and tactile sensing for minimally invasive surgery. IEEE Sens. J. 8(4), 371–381 (2008). https://doi.org/10.1109/JSEN.2008.917481

23. Pylatiuk, C., Muller-Riederer, M., Kargov, A., Schulz, S., Schill, O., Reischl, M., Bretthauer, G.: Comparison of surface EMG monitoring electrodes for long-term use in rehabilitation device control. In: 2009 IEEE International Conference on Rehabilitation Robotics, pp. 300–304. IEEE (2009)

24. Qi, J., Jiang, G., Li, G., Sun, Y., Tao, B.: Intelligent human-computer interaction based on surface EMG gesture recognition. IEEE Access 7, 61378–61387 (2019)

25. Schorr, S.B., Okamura, A.M.: Three-dimensional skin deformation as force substitution: wearable device design and performance during haptic exploration of virtual environments. IEEE Trans. Haptics 10(3), 418–430 (2017)

26. Tholey, G., Desai, J.P., Castellanos, A.E.: Force feedback plays a significant role in minimally invasive surgery: results and analysis. Ann. Surg. 241(1), 102 (2005)

27. Van Kuilenburg, J., Masen, M.A., van der Heide, E.: A review of fingerpad contact mechanics and friction and how this affects tactile perception. Proc. Inst. Mech. Eng. Part J J. Eng. Tribol. 229(3), 243–258 (2015)

28. Vigotsky, A.D., Halperin, I., Lehman, G.J., Trajano, G.S., Vieira, T.M.: Interpreting signal amplitudes in surface electromyography studies in sport and rehabilitation sciences. Front. Physiol. 8, 985 (2018)

29. Xydas, N., Kao, I.: Modeling of contact mechanics and friction limit surfaces for soft fingers in robotics, with experimental results. Int. J. Robot. Res. 18(9), 941–950 (1999)

30. Zheng, M., Crouch, M.S., Eggleston, M.S.: Surface electromyography as a natural human-machine interface: a review. IEEE Sens. J. 22(10), 9198–9214 (2022)

Move or Be Moved: The Design of a Haptic-Tangible Manipulative for Paired Digital Education Interactives

Scott George Lambert[1]([✉]), Jennifer L. Tennison[1], Kyle Mitchell[1], Emily B. Moore[2], and Jenna L. Gorlewicz[1]

[1] Saint Louis University, Saint Louis, MO 63110, USA
scott.lambert@slu.edu
[2] University of Colorado Boulder, Boulder, CO 80309, USA

Abstract. One of the current limitations in digital educational experiences is the lack of touch. Touch is a critical component in the learning process and in creating inclusive educational experiences for sensorially diverse learners. From haptic devices to tangible user interfaces (TUI), a growing body of research is investigating ways to bring touch back into the digital world, yet many focus on a specific dimension (e.g. haptic feedback or kinesthetic manipulation) of touch. Learning, however, is a multi-dimensional touch experience - it is about moving and being moved. This work presents the Action Quad - a novel haptic-TUI design for teaching geometry (specifically quadrilaterals). The Action Quad is a multi-point-of-contact, reconfigurable tool that synergizes the affordances of both kinesthetic interaction and haptic feedback into a single form factor. We present findings from an initial user study ($N = 11$) investigating how sighted-hearing individuals approach, interact, and experience the Action Quad, and we present a case study with an individual with blindness. We share key takeaways from the design process and participant feedback on interactions with this novel haptic-TUI device, sharing design insights on an emerging area of research that could support a new class of educational learning tools rooted in touch.

Keywords: Haptics · Tangible User Interfaces · STEM Education · Inclusive Design

1 Introduction

There has been a rapid shift of educational content to the digital space—from online videos and instruction to interactive educational simulations [11]. This shift was heightened during the COVID-19 pandemic, which opened a new era of digital education offerings and opportunities, but also highlighted the challenges of the lack of interactive, hands-on learning experiences [23,34]. What once used to be manipulated and learned through movement and tangible interaction, is often now consumed visually and aurally on-screen. The lack of touch-based interactions is particularly detrimental for inclusion of sensorially diverse

© The Author(s), under exclusive license to Springer Nature Switzerland AG 2025
H. Kajimoto et al. (Eds.): EuroHaptics 2024, LNCS 14768, pp. 477–492, 2025.
https://doi.org/10.1007/978-3-031-70058-3_39

learners, such as those with blindness and low vision (BLV), where touch is a primary mode of information access [10]. Beyond access, however, touch is a key component in the learning process for all individuals; interactive, kinesthetic learning opportunities bring more of our sensory and motor functions into the cognition process, a well-known component of active learning and conceptual understanding in embodied cognition [2, 8].

Several efforts have focused on bringing touch back into the digital world, with two primary communities pioneering advancements: tangible user interfaces (TUIs) and haptics. Both haptics and TUI research aim to provide novel touch experiences, but do so through different means. TUIs are designed as physical representations of virtual objects, enabling users to control objects in the digital space by manipulating the tangible interface [28]. TUIs primarily serve as one-way, input devices, and often take on form factors similar to the virtual objects they represent. These interfaces can be challenging to design effectively, as noted by Shaer; it is vital to understand how a user approaches and makes sense of such a tool for effective design of physical form factor and haptic experiences [28].

TUIs, such as Media Blocks [31] or Block-Jam [21], bring enhanced interactivity to digital spaces through hands-on learning [17, 35], opportunities for collaboration [4, 7, 30], and accessibility benefits [16, 22]. Early work demonstrated the efficacy of TUIs in education applications, such as Programmable Bricks [25] which offers students a physical set of toy bricks that pair to a digital space, acting as handles or controllers for on-screen electronic objects. Recently, more efforts have been focused on TUIs for mathematics, such as a tabletop interactive space for trigonometry [32], a set of freeform objects for early-childhood numeracy conception [5], or 3D-printed tangibles for geometry learning [3]. These efforts have highlighted some of the benefits of tangible learning experiences in classrooms, especially for students with learning disabilities, and particularly in mathematics.

Haptic devices, on the other hand, are inherently focused on providing a user with touch feedback (primarily force or tactile), representative of interactions within a virtual or remote environment. Haptic devices often focus on the output to the user and have form factors that are wearable, graspable, and touchable [6] depending on the varied applications of use. The Phantom OMNI [19], a pioneering haptic device, was first introduced in 1994. This table-top device provides precise force feedback through a pen-like interface, allowing virtual objects to be felt with high resolution, and has been used in educational applications including biology [20] and physics [33]. Another educational haptic device is the Haptic Paddle [26]: this low-cost haptic tool has been used to demonstrate principles of dynamics, mechatronics, and controls [9, 18], and has shown effectiveness in improving student performance in dynamics laboratory courses [27]. Though not exhaustive, these examples represent the varying form factors, types of feedback, and contexts that haptic devices can provide within education. Unlike TUIs, which have a primary focus on manipulation, haptic devices tend to focus on the display and fidelity of feedback. Yet, learning is a compilation of all of these interactions - moving and being moved.

This work presents the first steps in designing a novel haptic-TUI: Action Quad (Fig. 1). Action Quad is a multi-point-of-contact, reconfigurable tool that promotes both kinesthetic interaction (moving) and haptic feedback (being moved), while maintaining a form factor that is representative of the object itself. Action Quad synergizes principles of haptics and TUIs, emphasizing the affordances of tactility as a sense-making modality. This intersection promotes the opportunity of providing dynamic haptic feedback [12] and enhanced manipulation [13] from a singular platform. Action Quad was designed for an educational context of learning geometry, specifically quadrilaterals, which dictated its initial form factor. This paper presents the design of Action Quad and a first usability assessment on how individuals approach and interact with this hybrid TUI-haptic interface.

In Sect. 2, we detail the hardware and software framework that supports this new interface. Through an initial user study ($N = 11$) presented in Sect. 3, we evaluate the usability of the Action Quad in two dimensions: how users approach and interact with the device, and how device performance impacts user assessment of the system. We discuss our findings in Sects. 4 and 5, highlighting user approach and exploration strategies, how users distinguish characteristics of the Action Quad using both local and global descriptions, and participant-assessed performance of the Action Quad. In Sect. 6, we present a case study with an individual with BLV, toward uncovering insights on how a user's approach, interac-

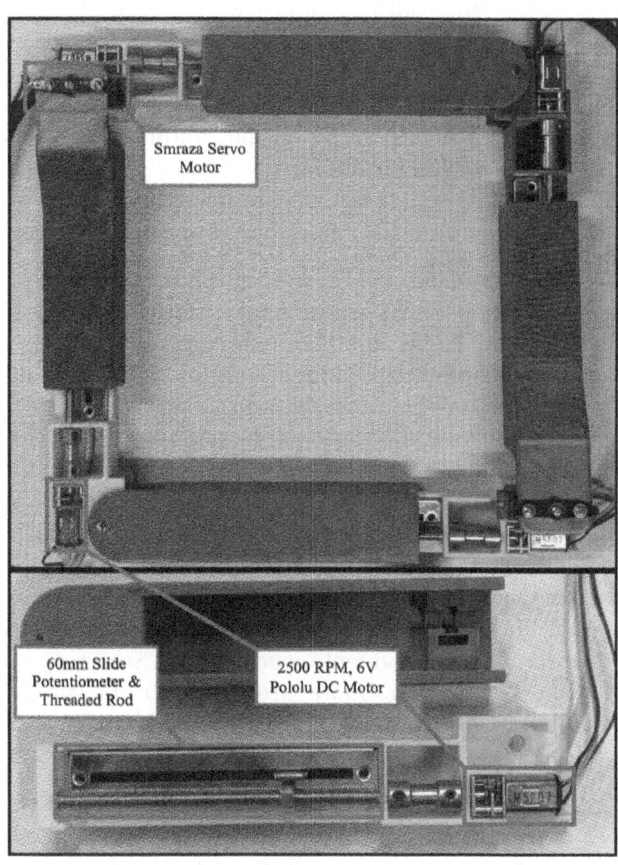

Fig. 1. Top: The Action Quad's 3D printed shell houses all components. Bottom: Each link contains a potentiometer, threaded rod with lead screw, coupler, and DC motor.

tions, and assessment of the system may vary from a nonvisual perspective. We conclude and share future work in Sect. 7. This work sets the stage for a new genre of educational tools—hybrid haptic-TUI systems—with an initial focus on how such systems can be designed to support usability and intuitive interaction with a broader goal of bringing touch back in multiple dimensions in digital education contexts.

2 System Design

Context: The Action Quad is part of a broader set of inclusively-designed geometry learning tools that consist of both hardware and software components [14,15], collectively called The Quad. The hardware consists of variants of a 3D printed tool—from a base design (Base Quad) that has no electronics and can be used as a standalone tool to a smart design (Smart Quad) that has sensors built in for length and angle tracking. The software consists of the *Quadrilateral* simulation (or "sim"), developed by the" PhET Interactive Simulations project [24,29]). The *Quadrilateral* sim can serve as a self-contained, accessible learning experience with interactive visual, auditory, and speech displays and multiple input options. For example, the Smart Quad can be paired with the *Quadrilateral* sim. The Action Quad represents the next constituent in this hardware lineage—providing the capability to move or be moved when paired to the sim.

Design Requirements: Working in the context of a pedagogical toolset for learning geometry, several requirements guided the Action Quad's design. The Action Quad needed to be small enough such that two hands could explore the entire top surface and large enough to allow for two-handed interactions along each link. A user must be able to interact with the tool in any location they choose to promote free kinesthetic explorations. The links and corners must actuate to reconfigure the Action Quad to a range of four-sided shapes: rectangles, trapezoids, parallelograms, squares, and rhombuses. The Action Quad needed to consist of readily-available, inexpensive parts to enable broad deployment, while keeping total cost under $100 to support adoption in educational settings. The Action Quad must be able to be commanded to any configuration in under 4 s and within 5% accuracy for both lengths and angles to provide timely responses between other components, such as the Smart Quad or the *Quadrilateral* simulation. The Action Quad setup must be powered via a standard wall outlet. Finally, the Action Quad must be able to receive input from multiple software interfaces, including the *Quadrilateral* sim or the Smart Quad, to offer flexibility in its use.

Design and Electronics: We employed a human-centered, iterative design approach in the creation of the Action Quad, drawing on findings from our previous Quad tool developments and frequent feedback from end users, including those with BLV. Specifically, prior to designing Action Quad, we iterated through over 10 design cycles and iterations across the Base and Smart Quad designs, determining the appropriate total footprint range (225 cm^2 to 625 cm^2), link

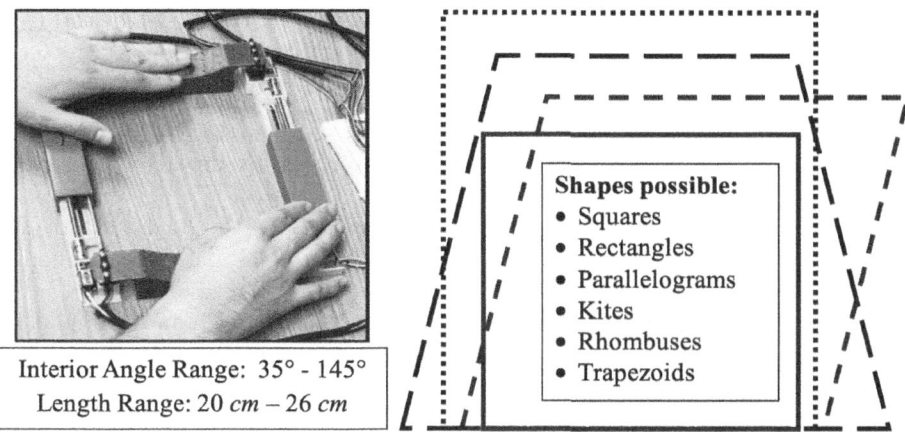

Fig. 2. The functionality of the Action Quad enables all convex four-sided shapes to be made with a range of sizes while enabling users to interact with the shapes using their hands

size (2 cm × 2 cm × 15 cm) and link geometry (rectangular vs. cylindrical). With Action Quad, our goal was to maintain these inherent form factor characteristics, while increasing capability. Initial designs employed motorized slide potentiometers in a 3D printed enclosure which were found to have insufficient strength for reliable, accurate actuation. The final Action Quad design (Fig. 1, top) consists of 4 links with 2 primary housing parts and 5 internal components per link, plus additional external components for a total of 31 parts. Aside from its internal components, it is entirely 3D printed and uses a lead-screw based mechanism which enables rapid and precise extension and retraction of the lengths.

The 4 links of Action Quad are connected at three corners via M3 screws for free rotation and at one corner via a housed servo motor (Smraza, SG90). Each link consists of two parts: a lower part that houses the DC motor (Pololu 15:1 Micro Metal Gearmotor LP 6 V @ 860RPM) coupled to a 100 mm, 0.5 mm pitch M5 threaded rod, and slide potentiometer (Mouser Electronics PTB6043-2010BPB103), and an upper part that houses an M5 nut that moves along the threaded rod as the motor spins (Fig. 1, bottom). A 60 mm slide potentiometer was implemented as a length sensor as our previous Quad tools had demonstrated its accuracy and ease of use.

The system is controlled by an Arduino MEGA with a L298N Motor Shield, powered by an external power supply, delivering 12 V @ 2 amps. It is connected via USB port to a PC to enable Serial connection to digital tools.

Assembly and Function: The Action Quad weighs 388 g and is capable of creating shapes as small as 20 cm × 20 cm and as large as 26 cm × 26 cm, with an angle range at each corner of 35° to 145°, enabling a wide range of quadrilaterals to be created (Fig. 2). The Action Quad housing for each link is 2.0 cm tall by 2.5 cm wide, 3D printed, and assembled in 8 h. Each link requires 5 soldered connections

(3 to the potentiometer, 2 to the DC motor) for wiring to the microcontroller, and installation of the motor, coupler, threaded rod, and potentiometer into the housing. The total cost of materials and components is approximately $65.

Communication and Control: The Action Quad can be controlled in two ways: 1) directly from the virtual *Quadrilateral* sim or 2) via the Smart Quad which transmits shape data via Bluetooth to the *Quadrilateral* sim. The sim connects to Action Quad through the p5.serialcontrol application [1], which runs on the user's computer. This app receives messages from the sim and relays the commanded configurations (four desired length values and one desired angle value) to the Arduino Mega, where the information is parsed, calibrated, and commanded to the Action Quad via a serial connection. The movement scheme for each length is position-based, using positional measurements from the potentiometer readings to reach desired configurations. A configuration was considered "reached" when the position was within 0.14 cm (2.35% of the full extension length) of the commanded position. This range was found to be the smallest tolerance for reliable movement results. The dual input for Action Quad enables a variety of use cases for a learner in a pedagogical context: 1) experiencing configurations of the Quad being produced in the sim; and 2) experiencing configurations of the Quad being produced by another tangible device, which can be controlled by the same or a different learner. While the studies in Sect. 4 focus specifically on the tangible interactions with the Action Quad in isolation and the Smart Quad to Action Quad, we detail the full integration of communication and control for completeness as it impacted the design of Action Quad.

Performance Validation: To evaluate the performance of the Action Quad, we ran a series of benchtop tests that measured the extension and retraction speeds, accuracies, and forces displayed. The Action Quad was powered by a 12 V external power supply connected to the motor shield, and was laid flat on a tabletop with 1 cm grid markings for length measurement. A PC was used to send configurations to the Arduino via serial communication. In the speed test, each length in the fully assembled device was moved through a series of lengths (20 cm, 26 cm, 20 cm, 26 cm, 20 cm) 3 times each to give 15 total trials, with time from initial position to final position being recorded. With the current design, each length individually can fully extend from a base value of 20 cm to an extended value of 26 cm in an average of 1.34 s ($SD = 0.1$), moving at a speed of 4.47 cm/s. When operating all four motors simultaneously, the Action Quad can extend from the home position (all lengths at 20 cm, all right angles, area $= 400$ cm^2) to fully extended (all lengths 26 cm, all right angles, area $= 676$ cm^2) in an average of 1.54 s ($SD = 0.1$). All rotation commands are stepped through over 450 ms to create smooth movements. Similarly, 3 trials were run to assess accuracy of reaching commanded configurations. Each length in Action Quad was moved through the same series of lengths as in the speed test. In each trial, the achieved length and angle measurements were compared against the desired length and angle values. Each length was accurate to within ± 0.22 cm. The corner actuation via servo motor was accurate to within $\pm 3°$.

Lastly, the forces for extension and retraction were measured using a Kistler Quartz Force Sensor (Type 9203) calibrated using a set of gram weights connected to a Kistler Charge Meter (Type 5010). In extension, one link of the Action Quad was constrained to the tabletop while the opposite end was extended into the force sensor. The maximum value before motor stall was recorded. Similarly in retraction, one link was constrained to the tabletop while the opposite end was connected via mounting screws to the force sensor and retracted, pulling the force sensor. The maximum force value before stall was recorded. This was repeated 10 times per link. Overall, the average maximum force during extension was $3\,\mathrm{N}$ ($SD = 0.13$) and $2\,\mathrm{N}$ ($SD = 0.24$) during retraction. Collectively, these results demonstrate the benchtop performance of the Action Quad—illustrating speeds, accuracies, and force displays that are appropriate for the educational context it is designed to work within.

3   Research Study

The Action Quad represents a novel interface with multiple points of contact and affordances of both a haptic device (in its ability to generate forces) and a tangible interface (in its ability to command configurations). This synergy of capability provides interesting usability and interaction questions, which motivate the study in this paper. For example, where, how, and why do users hold the device? How do they respond to "being moved" and "commanding movements"? How do users make sense of the configurations created? Here, we investigate these questions toward assessing the fundamental design and usability of the Action Quad. This study is a necessary first step to refining Action Quad for future assessment in educational contexts. To this end, we conducted a usability study to investigate how users ($N = 11$) approach interaction with the Action Quad and how its performance in isolation and in the Action Quad-Smart Quad system impacts the users' assessment of it. This study was conducted with 11 sighted participants (18–28 years). Additionally, as the Action Quad exists within a set of inclusive education tools, we wanted to garner inclusive perspectives on the system. To this end, we conducted a case study with an individual with blindness (22 years, diagnosis: Leber congenital amaurosis) (see Table 1 for participant demographics). To understand the Action Quad from both a usability and functionality perspective, we address the following research questions in this work:

1. In what ways do users approach Action Quad, and what interaction strategies are used to make sense of the Action Quad and its reconfigurable features?
2. How did Action Quad's performance affect users' assessment of the system?

To investigate the two research questions, the study was divided into three sections: 1) interactive; 2) observatory, and 3) comparative. In each section, participants completed a series of tasks and prompts that required them to explore the Action Quad and verbally comment about their interaction with the system. For the duration of the study (session length approximately 25 minutes), the experimenter sat across from the participant and controlled the Action Quad via a USB connection to a computer. The computer was running the PhET Simulation *Quadrilateral*. Data collected in each section includes participant comments and responses to feedback prompts specific to each task. All sections were video recorded for post-study analysis of interaction methods.

In the Interactive Section, we wanted to investigate how users organically approached the Action Quad. Here, participants were presented with a series of shape transformations wherein the experimenter configured the Action Quad from one shape to another using the connected computer. This task was designed to observe how participants chose to interact with the device during reconfigurations. Participants experienced 3 sets of the same shape transformations in the same order: 1) square-rectangle; 2) rectangle-parallelogram, and 3) parallelogram-square. Participants were not

Table 1. Participant Summary

#	Age	Sex	Dominant Hand
1	28	M	R
2	20	F	R
3	30	M	R
4	25	M	L
5	26	F	R
6	25	M	R
7	29	M	R
8	21	M	R
9	28	M	R
10	24	F	R
11	27	F	R

explicitly instructed on how to interact with the device during the 3 transformation sets, as we wanted to understand how users naturally approached the device. If the participant was reluctant to interact with the device during the first set of transformations, they were encouraged by the experimenter to hold the tool in specific locations (e.g. top/bottom, left/right sides) during the second set. After each set, participants were asked to provide feedback on their interaction with the Action Quad (e.g. "What changed on the device between transformations?","How did that change occur?"). Answers to the above questions coupled with participant interaction strategies were recorded for analysis. Participants also provided a rating of comfort with their interaction on a 7-point scale (1: uncomfortable, apprehensive, 7: comfortable, familiar). The data collected in the Interactive Section uncovered insights on hesitations or design limitations, addressing Research Question 1.

The Observatory Section was conducted to assess how well the Action Quad can create shapes and investigate if and how participants used shape properties to differentiate between shape configurations using the Action Quad. In this task, participants were presented with three sets of shape configurations on the Action Quad. These sets consisted of two shapes that highlighted a special property (two sets of parallel sides; equal shape area; presence of right angles, shown in Fig. 3) followed by a third shape without this shared property. Participants were instructed to identify a property (if any) shared by the first two shapes that distinguishes them from the third, final shape. The Action Quad was configured by the experimenter through the connected computer into the first shape and the participant was instructed to interact with the device in any way they chose (touch or observation). The Action Quad was configured into subsequent shapes in each set when the participant signaled they were ready. The participant's identification of the shared property of the shapes was recorded, as were

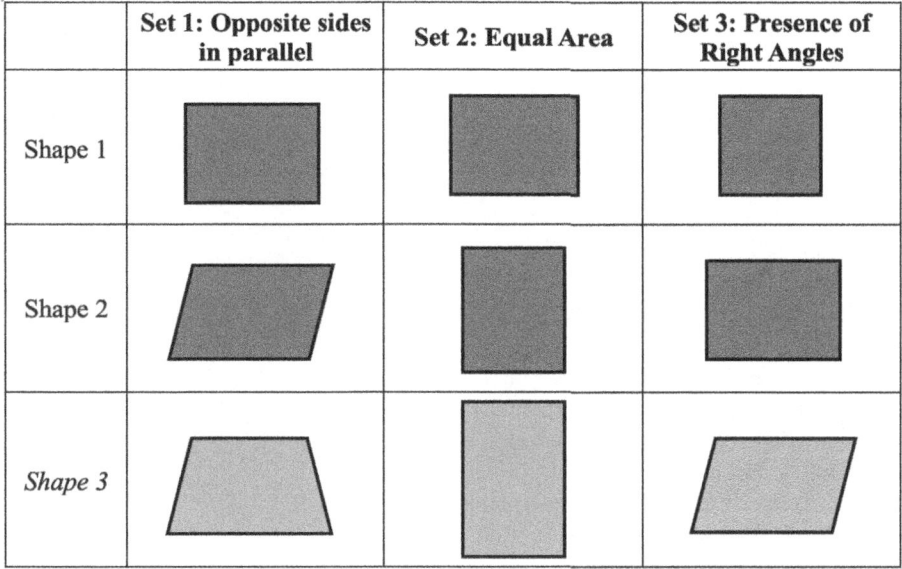

Fig. 3. The shape sets shown in the Observatory Section. Participants could ask to see the set of shapes in order as many times as desired.

their tangible interactions with the Action Quad. Data collected in this section informed how users made sense of the Action Quad as a device to display shapes, as well as highlighted the performance of the Action Quad in its ability to create notably different configurations.

Finally, in the Comparative Section, we assessed the Action Quad's performance in an initial tangible-tangible scenario and user's perception of its ability to replicate shape configurations they created. Participants utilized the Smart Quad tool to create their own configurations and represent them with the Action Quad (see Fig. 4). This task evaluated the participant's assessment of the system's accuracy. In this comparative task, participants were asked to change one parameter (a length or an angle) on the Smart Quad and to assess the accuracy of the Action Quad's replicated shape on a 7-point scale (1: completely inaccurate, 7: perfectly accurate). Then, participants were asked to create any configuration on the Smart Quad and assess the accuracy. Tangible interactions and participant ratings and comments were recorded. Participants did this task 3 times.

4 Results

The results are presented through a mixed-methods analysis approach and include both participant responses and ratings to the tasks within each study section and their associated video recording. Due to the low sample size of the study, we elected to use descriptive and summary analysis to better represent our findings.

Interactive Section. We analyzed video recordings of all participants observing key themes around: first interaction with Action Quad (with or without prompting); whether the device was held or kept flat; and how users talked about their interaction (e.g. global description of the shape created or local descriptions of the individual components of the shape). During the first set of transformations, only 1 of the 11 participants interacted with the Action Quad without prompting. During the second set of transformations, all participants interacted with the Action Quad (with prompting). Of these, 2 participants lifted the device off of the table, but 9 chose to place their hands flat along the top surface (Fig. 5). Nine participants commented on the increased tactile information when holding the opposite pair of sides during actuation (e.g. Holding the top/bottom during left/right actuation or vice versa. One participant noted, "If I hold the top and bottom [during top and bottom actuation], not really... Okay, now [holding the left and right] I do get the sense that... the shape got bigger").

The average comfort rating across all participants and interactions was a 6 out of 7 ($SD = 0.86$). To understand how participants discussed the movements of the Action Quad, we also analyzed the focus of their descriptions as global (directly mentioning shape by name) or local (commenting on the

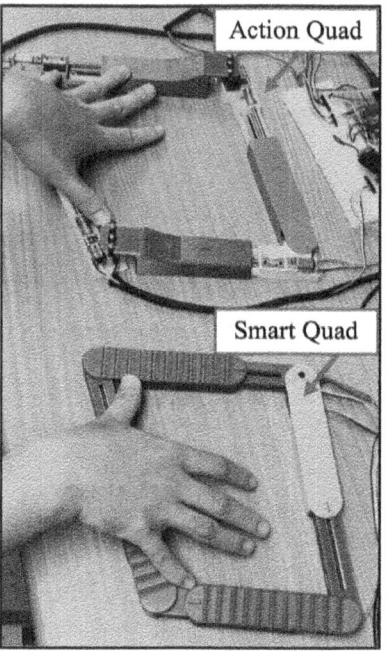

Fig. 4. In the Comparative Section, participants created a configuration on the Smart Quad (bottom) and compared the parameters with the Action Quad (top). Here, the participant in the case study compares the relative lengths of the Smart Quad (bottom) and Action Quad (top) by using their hands as a metric.

individual components of the tool such as link extension or corner rotation). Four participants described the device's movements by local changes (e.g. "The left got longer and the top right corner widened."), 3 described only the global changes (e.g. "It went from a rectangle to a parallelogram."), and 5 described both the parameters and shapes. Collectively, we observed that users were quite hesitant to interact with Action Quad, but after prompting, five participants commented on obtaining increased information about the transformation through touching the device: "Initially, visually... it felt like [the sides] pushed apart equally but this time I felt my left hand being pushed."

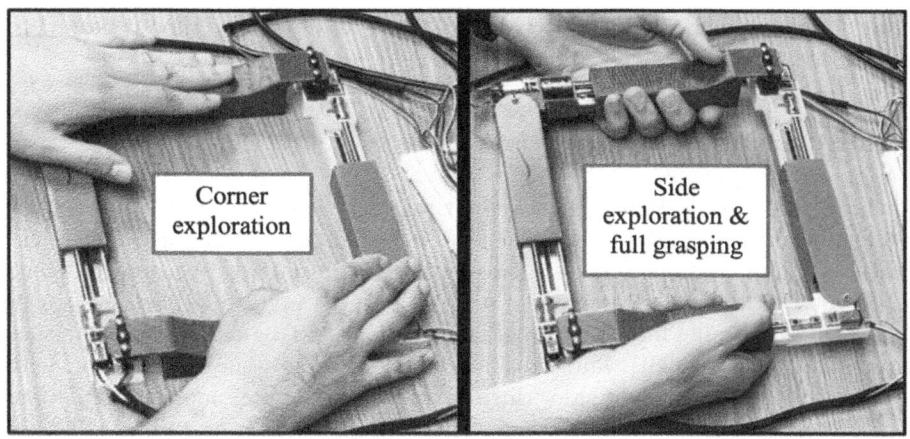

Fig. 5. Common exploration strategies from participants when interacting with the Action Quad. Most common was interaction flat along the top surface (shown on left, 9 out of 11 participants), while some participants chose to pick the device up and fully grasp the links (shown on right, 2 out of 11 participants).

Observatory Section. We analyzed participant responses regarding distinguishing features identified through 3 sets of shapes and used the video recordings to analyze participant tactile interactions, if any. In the first set of shapes, 8 of the 11 participants were able to correctly identify the number of parallel lines as the distinguishing feature (geometrically equivalent answers were accepted such as "opposite corners were equal angles" or "opposite sides were equal lengths"). In the second set, 5 participants correctly identified the difference in area between the first two shapes and the final shape, while 3 participants gave no answer. In the third set, all 11 participants correctly identified the presence of right angles in the first two shapes and not in the final shape. During these sets, 3 participants maintained touch throughout, while the remaining 8 opted to observe the shapes only.

Comparative Section. We used video recordings to analyze the participant-assessed accuracy of replicated shapes and the overall description of the input shape. When changing one parameter, the average perceived accuracy was a 6.0 out of 7 ($SD = 0.77$). Participants then created 3 custom configurations, with perceived accuracy averaging to 5.8 out of 7 ($SD = 1.38$). Configurations included rectangles (3 participants), trapezoids (5 participants), parallelograms (8 participants), while all 11 participants created non-standard convex quadrilaterals. Comments on the accuracies mainly highlighted angle measurements: "I gave it a perfect 90 [degree angle] to test that, and it nailed it", "The angle between the left side and bottom doesn't quite match what I made". A few comments were related to the Smart Quad - Action Quad system: 3 participants commented on the relative orientation to the table of the Action Quad compared to the

Smart Quad, and 2 commented on the overall footprint of the Action Quad being larger than the Smart Quad.

5 Discussion

Taken together, key takeaways from the three part study were 1) Approach and Comfort: Users were initially apprehensive about engaging via touch but quickly became familiar and often used a two-handed exploratory approach upon first use; 2) Interaction: Users were accurately able to distinguish key features of the Action Quad (parallel lines, right angles), used varied language to describe these features (local vs. global), and experienced increased tactile information when holding the opposite pair of sides during actuation; 3) Assessment: Participant-assessed replication accuracy of the Action quad was high, and all participants were eager to create non-standard geometries to probe the limits of the system. Additionally, during the third set of transformations, participants were eager to explore different interaction schemes, with some lifting the device off the tabletop, others exploring the corner joints to investigate where more significant haptic feedback might exist in movements. Interestingly, 5 participants remarked that they gained more information about the movements when holding the device compared to just visual observation, perhaps due to the forces displayed by each link when actuating. This illustrates the value add of haptics embedded into the tangible manipulative which will require future investigations to quantify. Participants described the tool using both local and global descriptions, which may suggest that the reconfigurable nature of the tool plays a role in interpretation of the Action Quad's configurations. Specifically, the use of local language ("the left side extended") suggests interpretations based on the actuation rather than the overall final state. Further investigation into how the reconfigurable nature of the tool and user focus on global versus local features affect mathematical understanding is needed.

The results from the observatory section illustrate the ability of the Action Quad to create and highlight significant geometric features such as parallelism or right angles between sides, but participants struggled to identify the difference in area. This could be due to a number of factors but is likely attributable to the limit of extension along each link; the links extended 3 cm, from 20 cm × 23 cm to 20 cm × 26 cm, equalling an area change of 460 cm^2 to 520 cm^2, respectively, a 13% increase in area. This suggests a need for a metric along each link to indicate length values, such as tactile indentations at regular intervals along each link.

The participant assessed accuracy in the comparative section was high. When commenting on inaccuracies, participant comments were themed around the angle accuracy more so than the length accuracy, especially with non-symmetric configurations and non-right angled shapes. This is likely due to the embedded actuation mechanisms which create a non-extendable section of each link. This limitation does not exist in the Smart Quad nor the *Quadrilateral* Simulation, and thus is hypothesized as a contributor to some of the inaccuracies. This difference in turn alters the angle relations between links. The wide range of input shapes from participants demonstrated their interest in understanding the affordances and limitations of the system and illustrated the system's ability to accommodate a wide variety of configurations. The resulting average perceived accuracy coupled with the results from the benchtop tests, provide initial validation of the Action Quad's performance.

6   Case Study: Individual with Blindness

The same user study was run with one participant with blindness (22 years; Leber congenital amaurosis) to garner initial impressions of the Action Quad and its functionality from someone with more familiarity with tactile exploration. The participant rated their average comfort with the device throughout the 3 sets of Interactive transformations as 5.75 out of 7 ($SD = 0.58$) and rated the accuracy of the replicated shapes across the four Comparative trials with an average of 5.25 out of 7 ($SD = 2.22$).

Overall, this participant had some surprising sense-making strategies that had not been used by the 11 sighted participants. As expected, this participant used the full extent of their hands immediately to approach and understand the device and its capabilities (see Fig. 4 for an example of tactile comparison strategy). They also compared the device to other objects available to them such as the table, for even greater understanding of the shapes in relation to the environment. This participant was also the only participant to remark on the ambient sounds of the device and incorporate sound into their sense-making strategy across the different study sections.

To determine the shape was growing larger, they used both tactile interactions ("There was some tactile feedback... I could tell that these two [sides] got larger") coupled with the duration of the mechanical sounds of the device as the Action Quad moved. They could also distinguish the sounds of the servo and DC motors and could map them to the device's angle and length movements, respectively. During the Observation Section, when asked to find the distinguishing feature of the shapes, this participant's initial sense-making strategy involved determining device-centric elements, such as the overall size of the device and its movement during transition, before eventually settling on shape-centric elements such as parallelism and relation between links as time and experience with the task increased. The participant also commented on aspects of the Action Quad that previously had not been mentioned, such as the speed :"The movement is pretty rapid, so it can be hard to follow that", and how tactile exploration can alter their understanding: "As it's mechanically moving, the position on the table changes, or when you touch it. So that loses some geometric integrity when exploring". Overall, the participant had positive feedback for the Action Quad: "It was really neat to watch it come alive".

7   Conclusion

In this paper, we present the design of Action Quad - a haptic-tangible interface to enable a full touch experience (moving and being moved) in the context of an inclusive toolset for geometry learning. We conducted initial user studies that investigate how users approach and interact with the device through a series of open-ended, exploratory tasks. This work offers important key takeaways in designing such interfaces including two-handed usage along the top surface, support for a wide range of shape sizes, and allowing for uninhibited touch interactions by minimizing exposed mechanisms. Future work on the mechanical aspects of the Action Quad will prioritize usage and user experience via enclosed hardware and a more seamless connection to The Quad system, allowing more in-depth evaluation of the educational applications of the multimodal, hardware-software tool, especially with younger student populations. Additional studies will also compare the Quad system with other methods of displaying shapes to evaluate these findings against other media, such as simulation alone. Research will also focus on how devices like Action Quad can be expanded to be more portable and flexible to span across numerous applications and pedagogical settings, enabling both individual and collaborative learning experiences that prioritize touch as a key interaction modality.

Acknowledgements. We thank Dr. Dor Abrahamson and Dr. Sofia Tancredi from the Embodied Design Research Laboratory at University of California Berkeley for bringing the idea of the Quad in an educational context to life and the PhET Interactive Simulations team including Brett Fiedler, Jesse Greenberg, and Taliesin Smith for bringing the virtual Quadrilateral sim to life. This study was funded by the National Science Foundation Award 1845490. This material is based on work supported by the National Science Foundation under DRL-1814220, DRL-1845490 (Gorlewicz), OISE-1927469, and IIS-2119303 (Moore).

References

1. Katsuya Endoh Aarón Montoya-Moraga Karl Fessel. p5.serialport. https://github.com/p5-serial/p5.serialport
2. Abrahamson, D., Lindgren, R.: Embodiment and Embodied Design, pp. 358–376. Cambridge University Press (2014). ISBN: 9781139519526. https://doi.org/10.1017/CBO9781139519526.022
3. Adusei, M., Lee, D.: "Clicks" appcessory for visually impaired children. In: Proceedings of the 2017 CHI Conference Extended Abstracts on Human Factors in Computing Systems, pp. 19–25 (2017)
4. Antle, A.N.,Droumeva, M., Ha, D.: Hands on what? Comparint children's mouse-based and tangible-based interaction. In: IDC (2009)
5. Beþevli, C., Göksun, T., Özcan, O.: An inquiry into the TUI design space for parent-child math engagement at home. In: Nordic Human-Computer Interaction Conference, pp. 1–12 (2022)
6. Culbertson, H., Schorr, S.B., Okamura, A.M.: Haptics: the present and future of artificial touch sensation. Annu. Rev. Control Robot. Auton. Syst. **1**, 385–409 (2018)
7. Fernaeus, Y., Tholander, J.: "Looking At the Computer but Doing It On Land": Children's Interactions in a Tangible Programming Space (2005)
8. Foglia, L., Wilson, R.A.: Embodied cognition. Wiley Interdisc. Rev. Cogn. Sci. **4**, 319–325 (2013). https://doi.org/10.1002/wcs.1226

9. Gorlewicz, J.L., Kratchman, L.B., Webster, R.J., III.: Haptic paddle enhancements and a formal assessment of student learning in system dynamics. Adv. Eng. Educ. **4**(2), n2 (2014)
10. Gorlewicz, J.L., et al.: Initial experiences using vibratory touchscreens to display graphical math concepts to students with visual impairments. J. Special Educ. Technol. **29**(2), 17–25 (2014)
11. Gorlewicz, J.L., et al.: The graphical access challenge for people with visual impairments: Positions and pathways forward. In: Interactive Multimedia Production and Digital Storytelling. IntechOpen (2018)
12. Ishii, A., Yasu, K.: FluxTangible: simple and dynamic haptic tangible with bumps and vibrations. In: Adjunct Proceedings of the 36th Annual ACM Symposium on User Interface Software and Technology, pp. 1–3 (2023)
13. Kildal, J.,: Kooboh: variable tangible properties in a handheld haptic-illusion box. In: Haptics: Perception, Devices, Mobility, and Communication: International Conference, EuroHaptics 2012, Tampere, Finland, June 13–15, 2012 Proceedings, Part II. Springer, pp. 191–194 (2012)
14. Lambert, S.G., et al.: Getting a grip on geometry: developing a tangible manipulative for inclusive quadrilateral learning. In: Giornale Italiano di Educazione Alla Salute Sport e Didattica Inclusiva (2022)
15. Lambert, S.G., et al.: A Tangible Manipulative for Inclusive Quadrilateral Learning (2022)
16. Marshall, P.: Do tangible interfaces enhance learning? In: TEI'07: First International Conference on Tangible and Embedded Interaction, pp. 163– 170 (2007). https://doi.org/10.1145/1226969.1227004
17. Marshall P., Rogers, Y., Hornecker, E.: Are Tangible Interfaces Really Any Better Than Other Kinds of Interfaces? (2003). http://www.cl.cam.ac.uk/conference/tangibleinterfaces/
18. Martinez, M.O., et al.: Open source, modular, customizable, 3-D printed kinesthetic haptic devices. In: 2017 IEEEWorld Haptics Conference,WHC2017, pp. 142–147 (2017). https://doi.org/10.1109/WHC.2017.7989891
19. Massie, T.H., Kenneth Salisbury, J., et al.: The phantom haptic interface: a device for probing virtual objects. In: Proceedings of the ASME Winter Annual Meeting, Symposium on Haptic Interfaces for Virtual Environment and Teleoperator Systems, vol. 55, pp. 295–300. Chicago, IL (1994)
20. Minogue, J., Jones, M.G.: Haptics in education: exploring an untapped sensory modality. Rev. Educ. Res. **76**(3), 317–348 (2006)
21. Newton-Dunn H., Nakano, H., Gibson, J.: Block jam: a tangible interface for interactive music. Journal of New Music Research 32.4 (2003), pp. 383–393.with Bumps and Vibrations. In: Adjunct Proceedings of the 36th Annual ACM Symposium on User Interface Software and Technology, pp. 1–3 (2023)
22. O'Malley, C., Fraser, D.S.: Literature Review in Learning with Tangible Technologies (2004)
23. Peek, N., et al.: Making at a distance: teaching hands-on courses during the pandemic. In: Extended Abstracts of the 2021CHI Conference on Human Factors in Computing Systems, pp. 1–5 (2021)
24. Perkins, K.: Transforming STEM learning at scale: PhET interactive simulations. Child. Educ. **96**(4), 42–49 (2020)
25. Resnick, A.L., et al.: I Programmable Bricks: Toys to Think with (1996)
26. Richard, C., Okamura, A.M., Cutkosky, M.R.: Getting a Feel for Dynamics: Using Haptix Interface Kits for Teaching Dynamics and Controls (997). http://cdr.stanford.edu/touch

27. Rose, C.G., et al.: Reflection on system dynamics principles improves student performance in haptic paddle labs. IEEE Trans. Educ. **61**(3), 245–252 (2018)
28. Shaer, O., Horneckerv, E.: Tangible user interfaces: past, present, and future directions. Found. Trends Hum. Comput. Interaction **3**, 1–137 (2009). https://doi.org/10.1561/1100000026
29. PhET Interactive Simulations. QuadrilateralKnuth: Computers and Typesetting. https://phet.colorado.edu/en/simulations/quadrilateral
30. Suzuki, H., Kato, H.: Interaction-Level Support for Collaborative Learning: AlgoBlock-An Open Programming Language (1995)
31. Ullmer, B., Ishii, H., Glas, D.: mediaBlocks: physical containers, transports, and controls for online media. In: Proceedings of the 25th Annual Conference on Computer Graphics and Interactive Techniques, pp. 379–386 (1998)
32. Urrutia, F.Z., Loyola, C.C., Marín, M.H.: A tangible user interface to facilitate learning of trigonometry. Int. J. Emerg. Technol. Learn. **14**, 152–164 (2019). https://doi.org/10.3991/ijet.v14i23.11433
33. Wiebe, E.N., et al.: Haptic feedback and students' learning about levers: unraveling the effect of simulated touch. Comput. Educ. **53**, 667–676 (2009). https://doi.org/10.1016/j.compedu.2009.04.004
34. Zha, S., et al.: Pandemic pedagogy in online hands-on learning for IT/IS courses. Commun. Assoc. Inf. Syst. **48**(1), 13 (2021)
35. Zuckerman, O., Gal-Oz, A.: To TUI or not to TUI: evaluating performance and preference in tangible vs. graphical user interfaces. Int. J. Hum Comput Stud. **71**, 803–820 (2013). https://doi.org/10.1016/j.ijhcs.2013.04.003

High–Fidelity Haptic Rendering Through Implicit Neural Force Representation

Christoforos Vlachos$^{(\boxtimes)}$ ⓘ and Konstantinos Moustakas ⓘ

Department of Electrical and Computer Engineering, University of Patras,
Rion-Patras, Greece
{chris.vlachos,moustakas}@ece.upatras.gr

Abstract. Recent research has demonstrated that neural networks using periodic nonlinearities may be used for implicit representation and reconstruction of continuous-time signals. Starting with a previously published network for representing the Signed Distance Function (SDF) of a mesh surface, we extend the concept and lay the foundation for introducing the additional representation of the Unit Normal Function (UNF). With the representation of these two functions at hand, we construct a penalty-based haptic rendering method. Our experiments suggest that this proposed method is able to handle very large meshes better than other competing alternatives, producing high-fidelity forces, free of discontinuities, by sampling a continuous implicit force function at the desired spatial accuracy.

Keywords: Haptic Rendering · Force Rendering · Implicit Representation · Neural Networks · Human–Computer Interaction

1 Introduction

Human sensory organs receive many different forms of input. Strengthening Human–Computer Interaction (HCI) can be definitely attained by mimicking these inputs to a sufficient degree. Although visual and auditory input to a person is nowadays considered mostly trivial, haptic input remains challenging. Consequently, the development of haptic devices, able to output haptic signals which are then sensed by a person, has attracted much attention. However, the haptic rendering methods that drive these devices feature computationally expensive operations [16]. Therefore, it raises no eyebrows how the importance of haptic rendering has been highlighted in many publications.

1.1 Related Work

There have been many attempts to categorize haptic rendering algorithms. One such categorization, which is presented in [27] is surface-based and volume-based techniques. This distinction, which is based on graphical rendering categorization, separates algorithms between those using polygonal representations

H. Kajimoto et al. (Eds.): EuroHaptics 2024, LNCS 14768, pp. 493–506, 2025.
https://doi.org/10.1007/978-3-031-70058-3_40

(explicit) or parametric representations (implicit) and those opting to represent meshes using voxels. Another way of classifying haptic algorithms, mentioned in the same article, is point-based vs. ray-based algorithms. This classification serves to differentiate between algorithms that take into account only the contact point of the haptic device, ignoring the rest of the probe, and those that model the entire probe as a ray before proceeding to calculate the response force. Finally, another grouping of haptic rendering techniques that may be encountered in recent publications [12] is the 3-degree-of-freedom (3DOF) vs. the 6-degree-of-freedom (6DOF) algorithms. 3DOF algorithms only account for the probe's position in 3D space, whereas 6DOF algorithms also consider the probe's rotation in determining the response force of the collision.

Similar ways of categorization can be carried out for haptic interfaces, but that falls beyond the scope of this paper. We refer the reader to [27] as well as [12] for obtaining further information regarding haptic devices.

In the last few years, a substantial assortment of haptic algorithms have surfaced. Since the god-object method was proposed in 1995 [30] and the voxel method in 1999 [14], numerous innovations have been made. A highly detailed account of the advancements in haptic rendering techniques in the last few years can be found in [11]. More recent methods have tackled the problem of continuous collision detection with deformable meshes [4], pseudo-rendering of constraints [13], haptic rendering of point clouds [29], leading to new and exciting fields of research.

Another domain that has been of much interest to the haptic rendering community has been data-driven haptic rendering. Introduced by Hover et al. in 2008 [6] and expanded by the same group in the following years [5,7,24,25]. Additional research on the area was performed by Choi's group [2,18,23] has elevated the subject and can nowadays be considered a mature field of research.

Despite those innovations, there is a weakness common to most haptic algorithms: their inability to handle very detailed meshes, with lots of vertices and triangles, without giving up either quality or speed of calculations. Moreover, interpolation-based methods commonly handle discontinuities badly, undesirably eliminating them where present or introducing them where absent. To resolve those issues, it was early on that the researchers' focus shifted to implicitly represented surfaces for haptic rendering [10,16,17,22]. Furthermore, recent work has focused on the use of neural networks to represent an object in 3D space [1,8,15,20]. Fusing these concepts has been the inspiration for our proposed method.

1.2 Motivation and Contribution

In their recent work [26], Sitzmann et al. have introduced the SInusoidal REpresentation Network (SIREN), a type of neural network which uses periodic activation functions in order to implicitly represent various types of signals. One such signal, demonstrated on that paper, is the Signed Distance Function (SDF) of a mesh surface.

In this paper, we show that it is possible to extend their proposed network to include computation of the Unit Normal Function (UNF) as well. This way, we arrive at an elegant solution for quickly and accurately approximating the response force generated by collision with the mesh, to be used in a haptic rendering context where high-fidelity discontinuity-free forces are necessary.

Specifically, our contribution is creating a haptic rendering method, which we call PANDIS (Periodic Activations for Normals and DIStances), that:

- is able to estimate the collision response force that should be generated and fed back to the user, in a predictable time.
- can be easily implemented and adapted to be used with any mesh. The haptic rendering method is mesh and resolution agnostic during run-time.
- does not contain unwanted discontinuities. Both the SDF and the UNF are implicitly represented continuous functions that can be sampled at the desired spatial accuracy. This implicitly generates smooth force fields and haptic interaction that leads to force feedback without discontinuities or the need for post-processing in the haptic rendering pipeline as is the case, for example, in the force shading method.

2 Methods

2.1 Formulation

Penalty-based algorithms model the response force between objects as a spring force whose magnitude increases as the objects penetrate deeper into one another [12]. Thus, to determine the collision response force, we will need to have knowledge of both the magnitude and the direction of the spring force. To acquire these, we will use two implicit functions, the Signed Distance Function (SDF) and the Unit Normal Function (UNF).

Given a space Ω and a boundary $\partial\Omega$ within that space, we can divide the space into an exterior region Ω^+ and an interior region Ω^-. A distance function $d(x)$ is an implicit function equal to the distance between x and the closest point on the boundary [19]. From this definition, it is apparent that on the boundary itself $d(x) = 0$, therefore $\partial\Omega$ represents the zero-level set of $d(x)$ and we can refer to it simply as Ω_0.

It should also be mentioned that, since d represents Euclidean distance, the following equation should hold:

$$|\nabla d| = 1 \tag{1}$$

This is intuitive, since moving twice as far away from the boundary, we would expect the distance function's value to double.

We can now adapt the definition of the (unsigned) distance function in order to define its signed counterpart. Whereas the unsigned distance function $d(x)$ is always non-negative, the Signed Distance Function (commonly abbreviated as SDF) $\phi(x)$ is positive in Ω^+ and negative in Ω^-, while maintaining its magnitude. In other words:

$$\phi(x) = \begin{cases} d(x) = 0 & x \in \partial\Omega \\ d(x) & x \in \Omega^+ \\ -d(x) & x \in \Omega^- \end{cases}$$

Similarly to the unsigned distance function, for the SDF it holds that:

$$|\nabla\phi| = 1 \qquad (2)$$

Regarding Eq. 1 and 2, it should be mentioned that distance functions are differentiable almost everywhere, but not completely. In the case of the unsigned distance function one quickly realizes that the boundary constitutes a local minimum of d and the gradient cannot be defined there [28]. Despite this being a problem only for the unsigned distance function (since SDFs are monotonic), there exist other points where both kinds of distance functions are not differentiable, specifically those where the closest boundary point is ambiguous (it is a "tie" between at least two points).

We also define the Unit Normal Function (UNF) $\psi(x)$ as the vector function that always points towards or away from the closest point of the boundary. For all points in Ω that the SDF is differentiable, $\psi = \nabla\phi$.

We are interested in determining an implicit approximation of the SDF of a mesh surface. We call this approximation $\Phi(x)$. An appropriate loss function for the training of a SIREN that is used to predict the SDF at a specific point in 3D space would then be [26]:

$$\begin{aligned} \mathcal{L}_{sdf} = & \int_\Omega |\,\|\nabla_x\Phi(x)\| - 1\,|dx \\ & + \int_{\Omega_0} (1 - \langle\nabla_x\Phi(x), n(x)\rangle)dx \\ & + \int_{\Omega_0} \|\Phi(x)\|dx \\ & + \int_{\Omega\setminus\Omega_0} \exp(-a \cdot |\Phi(x)|)dx \end{aligned} \qquad (3)$$

This loss function treats determining the SDF as a Boundary Value Problem (BVP). Ω represents the entire 3D space whereas Ω_0 represents only those points where the SDF is zero (the zero-level set). In the final integral, $a \gg 1$.

Since $\Omega \subseteq R^3$, the SDF is differentiable almost everywhere and its gradient satisfies the following Eikonal equation:

$$|\nabla_x\Phi(x)| = 1$$

Looking back at the loss function defined in Eq. 3, the first integral simply enforces the Eikonal equation on Ω.

The second integral ensures that, on the surface, the gradient of the SDF, $\nabla_x\Phi(x)$, and the surface normals, $n(x)$, point in the same direction. After all, on Ω_0 these two should be identical.

The third integral penalizes on-surface points that have an SDF that is not equal to zero, while the fourth penalizes off-surface points having an SDF close to zero.

While the SDF is crucial to calculating the magnitude of the response force used for haptic rendering, the direction is also needed. For that, we could use the same SIREN to implicitly represent the UNF of the mesh. Let's call this implicit representation $\Psi(x)$. It is important to note that, while $\Phi(x) \in R$, this is not the case for $\Psi(x)$, as $\Psi(x) \in R^3$.

Inspired by $\mathcal{L}_{sdf}$, we can begin to formulate our own loss function, to aid with the UNF prediction:

$$\mathcal{L}_{unf} = \int_{\Omega} (\|\Psi(x)\| - 1)dx$$
$$+ \int_{\Omega_0} (1 - \langle \Psi(x), n(x) \rangle)dx \qquad (4)$$
$$+ \int_{\Omega \setminus \Omega_0} (1 - \langle \Psi(x), \nabla_x \Phi(x) \rangle)dx$$

The first integral forces the UNF to have unit values, while the other two integrals have it pointing in the same direction as the mesh normals, where available, or the gradient of the SDF, in places where the normals are not defined. By carefully inspecting the two loss functions, it can be understood that, ideally, $\Psi(x) \equiv \nabla_x \Phi(x)$.

It should be noted that, instead of using some numerical method to differentiate the SDF, we opted to encode it as part of the same neural network. In this way, we arrive at an elegant solution where, adding barely any complexity, we are able to obtain both the SDF and the UNF with a single inference pass.

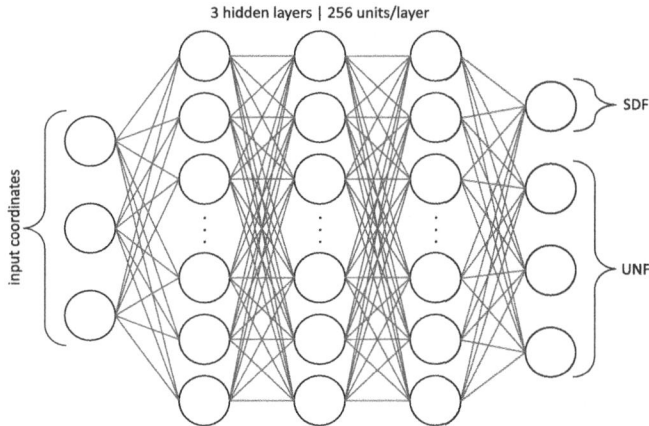

Fig. 1. The architecture of our proposed SIREN. The probe coordinates are entered in the Cartesian form. The first output is the SDF and the remaining 3 are the UNF.

A SIREN is now constructed, following the architecture shown in Fig. 1. That is, a Multi-Layer Perceptron (MLP) made up of one input layer, one output layer, as well as 3 hidden layers, each consisting of 256 sinusoidally-activated units. All layers are Fully Connected (FC). The 3D space coordinates of each mesh's vertices are input to the network and after passing through the three hidden layers, both $\Phi(x)$ and $\Psi(x)$ are output. For the training, we use the Adam optimizer and as a loss function a combination of the ones we defined previously:

$$\mathcal{L}_{total} = \mathcal{L}_{sdf} + \mathcal{L}_{unf} \tag{5}$$

One advantage of using a SIREN instead of an ordinary ReLU-based neural network for implicitly representing the SDF and the UNF is that the SIREN is able to represent these functions in such a way that their respective derivatives are also retained. Thus, we expect response forces that do not suffer from discontinuities as much as other haptic rendering algorithms.

It is moreover worth mentioning that each component (i.e., each integral) of the final loss function was given a weight, each discovered by trial and error and was validated experimentally. These are 50, 100, 3000, 100, 50, 100, and 100, respectively.

2.2   Training

All training took place on a system with an AMD Ryzen 5 3600X CPU and an NVIDIA GeForce RTX 2060 SUPER GPU with 8 GB GDDR6 VRAM. The installed memory was 32 GB DDR4 SDRAM. The operating system used was Windows 10, version 22H2.

Using PyTorch, we implemented the network we described and trained it on three different, quite large meshes, creating three models. These meshes were taken off the Stanford 3D Scanning Repository and are the Stanford Bunny, the Stanford Armadillo, and the Stanford Dragon. Each model was trained on its respective mesh for 10,000 epochs. The number of vertices and faces of each mesh, along with the training time (for 10,000 epochs), is displayed in Table 1.

Table 1. Number of vertices, faces, and training time of the meshes that were used for training.

Model	Vertices	Faces	Training Time
Bunny	35,947	69,451	09:29
Armadillo	172,947	345,944	47:11
Dragon	437,645	1,132,830	2:11:47

3   Results

To confirm the quality of our results, our testing was focused on the models of the Stanford Bunny, the Stanford Armadillo, and the Stanford Dragon. Firstly, we created a vector plot (also called a "quiver plot"), plotting the response force according to Hooke's Law as:

$$F = -kx \qquad (6)$$

where x is the SDF. The direction of the response force is indicated by the UNF. The generated vector plots along a planar section of each of the three meshes are shown in Fig. 2.

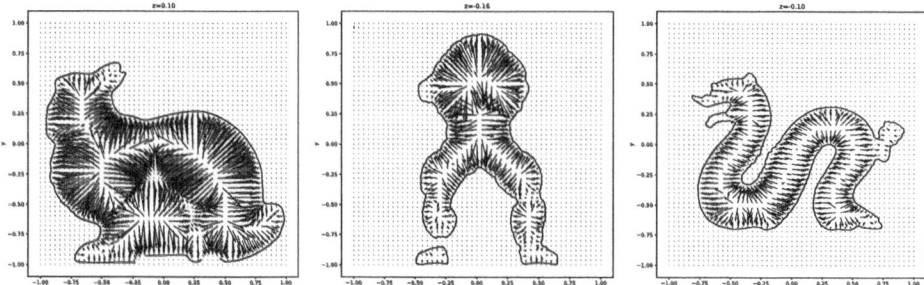

Fig. 2. Vector plot of the response force generated along a planar section of the (Left) Stanford Bunny, (Center) Stanford Armadillo, and (Right) Stanford Dragon.

After verifying that the model generates seemingly correct forces (2), we serialized the model, for use outside of PyTorch. We proceeded to write a C/C++ program using OpenHaptics® and LibTorch to enable utilizing a haptic rendering device to render the Stanford Armadillo using the forces produced and calculated by the neural network. Two such devices were used in our tests, the first being 3D SYSTEMS's flagship haptic device, the $Touch^{TM}$ (formerly the *Phantom Omni*), the other being the oldest and less accurate $Touch^{TM}$ *3D stylus*. Both devices as part of our experimental setup are visible in Fig. 3.

Although some steps have been taken towards standardizing the benchmarking of haptic rendering, they have either largely focused on the benchmarking of the haptic hardware itself, suffer from limitations, or ultimately failed to see widespread use [21]. It is because of this that we decided to focus our benchmarking on comparison with widespread, easily–implemented algorithms and objective statistical analyses.

We needed some comparison targets. To this extent, we implemented a few other algorithms for calculating the SDF and the UNF of a triangle mesh. First, we implemented a method for analytically calculating the above functions in

order to obtain ground-truth results (GT). Then, we ran that method on a
3D grid, effectively splitting the mesh's AABB into voxels. In this way, we can
report additional interpolated results. We implemented Nearest Neighbor(NN-i)
as well as Trilinear interpolation(Tri-i) options. It should be noted that trilinear
interpolation from a 3D grid could be considered State-of-the-Art, as featured
in recent publications [3,9]. We used two grid sizes: 32^3 and 64^3.

Table 2. Setup time (Top) and space (Bottom) required per method, per model. GT
stands for Ground Truth, NN-i for Nearest Neighbor Interpolation, Tri-i for Trilinear
Interpolation and PANDIS is the method we are proposing.

Model	GT	\multicolumn{4}{c}{Preprocessing/Training time}	PANDIS
		NN-i		Tri-i		
		32^3	64^3	32^3	64^3	
Bunny	0	07:38	58:52	07:38	58:52	09:29
Armadillo	0	40:21	5:25:00	40:21	5:25:00	47:11
Dragon	0	1:41:11	12:54:56	1:41:11	12:54:56	2:11:47

Model	GT	\multicolumn{4}{c}{Storage space}	PANDIS
		NN-i		Tri-i		
		32^3	64^3	32^3	64^3	
Bunny	0	384 KB	3 MB	384 KB	3 MB	800 KB
Armadillo	0	384 KB	3 MB	384 KB	3 MB	800 KB
Dragon	0	384 KB	3 MB	384 KB	3 MB	800 KB

3.1 Setup

The Analytical algorithm does not require any preparatory computations, since
it computes everything in real time. Thus it requires no time or space in advance
(Table 2). The story changes when it comes to the next algorithms. Both the
Nearest Neighbor and the Trilinear Interpolation methods need a precomputed
grid. Both the computation time and space change cubically as the grid size
increases, as long as we are talking about the same model. We observe that, for
a grid size of 32^3, the interpolation–based methods need preprocessing time that
is similar to the training time of PANDIS, while PANDIS falls behind when it
comes to storage requirements. That being said, increasing the grid size to 64^6
leaves the interpolation-based methods well outperformed by the SIREN–based
method on both regards. It should be noted that PANDIS, just as the others
chosen, is mesh-agnostic when it comes to the required storage space.

Table 3. Average computation time (in milliseconds) of response force. Times on top are from the $Touch^{TM}$, times below (inside parentheses) are from the $Touch^{TM}$ *3D stylus.*

Model	GT	NN-i		Tri-i		PANDIS
		32^3	64^3	32^3	64^3	
Bunny	13.851	<0.001	<0.001	<0.001	<0.001	0.745
	(14.053)	(<0.001)	(<0.001)	(<0.001)	(<0.001)	(0.741)
Armadillo	73.125	<0.001	<0.001	<0.001	<0.001	0.769
	(78.443)	(<0.001)	(<0.001)	(<0.001)	(<0.001)	(0.743)
Dragon	182.085	<0.001	<0.001	<0.001	<0.001	0.756
	(181.604)	(<0.001)	(<0.001)	(<0.001)	(<0.001)	(0.750)

3.2 Update Rate

Having finished preparing all the methods for testing, we measured the update rate capability of each one and report the average computation time, in milliseconds, for each method in Table 3. We reiterate that, for haptic rendering, the target rate is 1 kHz, which corresponds to a computation time of at most 1 ms.

From our testing, it is abundantly clear that the analytical method is too slow in dealing with large meshes. In contrast, the interpolation methods both respond very quickly. While PANDIS's speed is nowhere near that of the interpolation methods, it should be noted that it is still within target, 1 ms being the 97th percentile rank (i.e., less than 3% of the updates fall outside the target rate). Besides, we expect PANDIS to make up any lost ground when we look at the quality of the predicted forces.

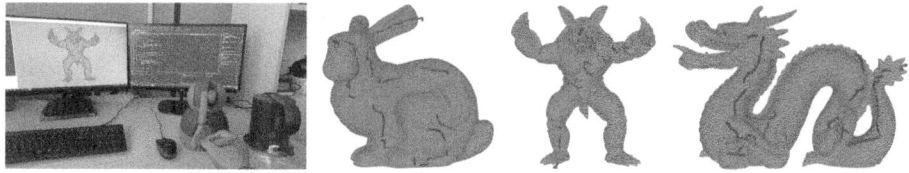

Fig. 3. (Leftmost) Our experimental setup, featuring the haptic devices used. (Left center – Rightmost) The synthetic trajectories used for testing the quality of the responses generated by each algorithm: (Left center) Bunny. (Right center) Armadillo. (Rightmost) Dragon

3.3 Quality

To measure the quality and accuracy of the forces produced, we traced and recorded a trajectory on the surface of each mesh (actually, penetrating the surface slightly, since we are testing penalty-based algorithms), making sure that the trajectory contained a variety of areas of the mesh's surface with vastly different gradients. The specific traces were generated using the state-of-the-art trilinear interpolation-based method and can be viewed in Fig. 3.

Table 4. Average error, as defined by Eq. 7, per method, per model.

Model	GT	NN-i		Tri-i		PANDIS
		32^3	64^3	32^3	64^3	
Bunny	0	0.086	0.031	0.057	0.026	0.006
Armadillo	0	0.139	0.081	0.099	0.046	0.022
Dragon	0	0.176	0.054	0.171	0.041	0.068

We then applied all methods to the synthetic trajectories we had collected. Defining an Average Error as

$$\text{Average Error}_i = \sum_j \frac{1}{j} \left(\|\text{force}_{i,j} - \text{force}_{\text{GT},j}\| \right), \tag{7}$$

where $i = \{\text{GT, NN-i, Tri-i, PANDIS}\}$, enables us to determine how each method performed, with respect to the accuracy of the predicted collision response force. These results can be seen in Table 4.

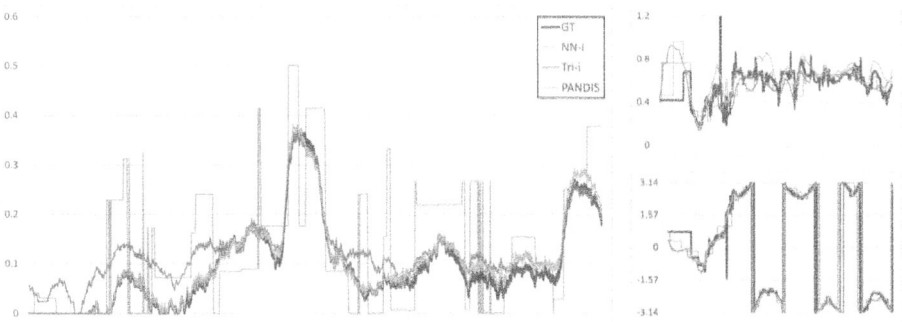

Fig. 4. Response force returned per method for a subset of the Armadillo's traced path, converted to spherical coordinates. (Left) r. (Top right) θ. (Bottom right) ϕ.

One can immediately notice that on both the Bunny and the Armadillo our method produces more accurate results, even on the increased grid resolution of 64^3. In the case of the Dragon, PANDIS loses—only barely—to the interpolation-based methods on a resolution of 64^3, albeit at the cost of significantly more preprocessing time and storage space.

Aiming to provide the reader with a better visual representation, we plot a subset of the response force produced by each algorithm in the case of the Armadillo model, in Fig. 4. It is evident that our proposed method tracks the Ground Truth results well, eliminating discontinuities present in other methods (such as nearest neighbor interpolation) and should be a serious consideration where a haptic representation of large meshes is required.

3.4 Theoretical Validation

In order to have a more complete picture of the quality and accuracy of the generated forces, we extended our testing to a sphere, whose SDF as well as the UNF can be defined in closed-form. Regarding a sphere located on the origin with radius r, it is well-known that its SDF is given by:

$$\phi(x, y, z) = \sqrt{x^2 + y^2 + z^2} - \mathrm{r}$$

Consequently, the UNF comes to:

$$\psi(x, y, z) = \nabla\phi(x, y, z) = \left(\frac{x}{\sqrt{x^2 + y^2 + z^2}}, \frac{y}{\sqrt{x^2 + y^2 + z^2}}, \frac{z}{\sqrt{x^2 + y^2 + z^2}} \right)$$

Since an implicit representation of the sphere was readily available, we were able to also extend our arsenal of methods to include the original implicit haptic rendering method of Salisbury and Tarr [22] as well as two other state-of-the-art methods of the same family [10,16]. We synthesized a trajectory on the sphere and the Ground Truth which we could trivially calculate from the closed-form SDF and UNF. We calculated the forces in the trajectory using all methods we had implemented, for more rigorous testing, and assessed the quality using the error defined in Equation 7. This led to the results of Table 5.

Table 5. Average error, as defined by Eq. 7, per method, on a sphere. For the interpolation-based methods (as well as [10]) a resolution of 100^3 was used, for [16] 642 support planes were created, and for PANDIS the underlying SIREN was trained on 2737 uniformly sampled points on the surface of the sphere.

Method	GT	NN-i	Tri-i	[22]	[10]	[16]	PANDIS
Error	0	0.097	0.082	0.083	0.367	0.296	0.162

One can see that PANDIS outperforms the other state-of-the-art methods ([10,16]) and produces results that are comparable to the remaining methods. It is obvious that interpolation-based methods play well with simple shapes, such as a sphere but, as previously proven, fail to keep that lead in complex meshes. On the other hand, the original implicit haptic rendering method of Salisbury and Tarr [22] does very well, but it should be noted that it only works on shapes whose implicit representation can be defined implicitly.

Therefore, we conclude that even playing from a losing position, on a scenario than is far from ideal for PANDIS (which shines brightest in the case of very complex meshes), our proposed method is able to keep up and even surpass other state-of-the-art methods.

4   Conclusion and Discussion

We have used a sinusoidally-activated neural network (SIREN) to create an implicit representation of both the Signed Distance Function (SDF) and the Unit Normal Function (UNF) with great accuracy. We have then proposed a new haptic rendering algorithm that makes use of this representation and succeeds in estimating the collision response force quickly, with high fidelity, and free of discontinuities. Furthermore, this method is easily scalable to very large meshes with no decrease in the update rate.

Although the aforementioned advances tackle critical and basic issues in haptic rendering, further advances have to take place in order to make machine learning based haptic rendering a game changer in haptic interaction. The basic current limitation is the fact that a deformation of the object would require a retraining of the SDF and UNF, making real-time deformation not feasible at the present stage. Moreover, both the SDF and UNF can be strictly theoretically defined only on closed manifold surfaces. Extension of the proposed approach solving these limitations through modern adaptive neural networks and utilization of pseudo-manifold surfaces are challenging directions of future work.

Acknowledgements. The research project is implemented in the framework of H.F.R.I call "Basic research Financing (Horizontal support of all Sciences)" under the National Recovery and Resilience Plan "Greece 2.0" funded by the European Union – NextGenerationEU (H.F.R.I. Project Number: 16469.).

Disclosure of Interests. The authors have no competing interests to declare that are relevant to the content of this article.

References

1. Atzmon, M., Lipman, Y.: Sal: Sign agnostic learning of shapes from raw data. In: Proceedings of the IEEE/CVF Conference on Computer Vision and Pattern Recognition, pp. 2565–2574 (2020). https://doi.org/10.1109/cvpr42600.2020.00264
2. Cha, H., Bhardwaj, A., Choi, S.: Data-driven haptic modeling and rendering of viscoelastic behavior using fractional derivatives. IEEE Access **10**, 130894–130907 (2022). https://doi.org/10.1109/ACCESS.2022.3230065
3. Corenthy, L., Otaduy, M.A., Pastor, L., Garcia, M.: Volume haptics with topology-consistent isosurfaces. IEEE Trans. Haptics **8**(4), 480–491 (2015). https://doi.org/10.1109/TOH.2015.2466239
4. Ding, H., Mitake, H., Hasegawa, S.: Continuous collision detection for virtual proxy haptic rendering of deformable triangular mesh models. IEEE Trans. Haptics **12**(4), 624–634 (2019). https://doi.org/10.1109/TOH.2019.2934104

5. Höver, R., Harders, M.: Measuring and incorporating slip in data-driven haptic rendering. In: 2010 IEEE Haptics Symposium, pp. 175–182. IEEE (2010). https://doi.org/10.1109/HAPTIC.2010.5444658

6. Hover, R., Harders, M., Székely, G.: Data-driven haptic rendering of visco-elastic effects. In: 2008 Symposium on Haptic Interfaces for Virtual Environment and Teleoperator Systems, pp. 201–208. IEEE (2008). https://doi.org/10.1109/HAPTICS.2008.4479943

7. Hover, R., Kosa, G., Szekely, G., Harders, M.: Data-driven haptic rendering-from viscous fluids to visco-elastic solids. IEEE Trans. Haptics **2**(1), 15–27 (2009). https://doi.org/10.1109/TOH.2009.2

8. Jiang, C., Sud, A., Makadia, A., Huang, J., Nießner, M., Funkhouser, T., et al.: Local implicit grid representations for 3D scenes. In: Proceedings of the IEEE/CVF Conference on Computer Vision and Pattern Recognition, pp. 6001–6010 (2020). https://doi.org/10.1109/cvpr42600.2020.00604

9. Kaluschke, M., Yin, M.S., Haddawy, P., Srimaneekarn, N., Saikaew, P., Zachmann, G.: A shared haptic virtual environment for dental surgical skill training. In: 2021 IEEE Conference on Virtual Reality and 3D User Interfaces Abstracts and Workshops (VRW), pp. 347–352 (2021). https://doi.org/10.1109/VRW52623.2021.00069

10. Kim, L., Sukhatme, G., Desbrun, M.: A haptic-rendering technique based on hybrid surface representation. IEEE Comput. Graph. Appl. **24**(2), 66–75 (2004). https://doi.org/10.1109/MCG.2004.1274064

11. Laycock, S.D., Day, A.: A survey of haptic rendering techniques. Comput. Graph. Forum **26**, 50–65 (2007). https://doi.org/10.1111/j.1467-8659.2007.00945.x

12. Lin, M.C., Otaduy, M.: Haptic rendering: foundations, algorithms, and applications. CRC Press (2008). https://doi.org/10.1201/b10636

13. Lobo, D., Otaduy, M.A.: Rendering of constraints with underactuated haptic devices. IEEE Trans. Haptics **13**(4), 699–708 (2020). https://doi.org/10.1109/TOH.2020.2981932

14. McNeely, W.A., Puterbaugh, K.D., Troy, J.J.: Six degree-of-freedom haptic rendering using voxel sampling. In: Proceedings of the 26th Annual Conference on Computer Graphics and Interactive Techniques, pp. 401—408. SIGGRAPH '99. ACM Press/Addison-Wesley Publishing Co. (1999). https://doi.org/10.1145/311535.311600

15. Mescheder, L., Oechsle, M., Niemeyer, M., Nowozin, S., Geiger, A.: Occupancy networks: Learning 3d reconstruction in function space. In: Proceedings of the IEEE/CVF conference on computer vision and pattern recognition, pp. 4460–4470 (2019). https://doi.org/10.1109/CVPR.2019.00459

16. Moustakas, K.: 6dof haptic rendering using distance maps over implicit representations. Multimed. Tools Appl. **75**(8), 4543–4557 (2016). https://doi.org/10.1007/s11042-015-2490-z

17. Moustakas, K., Tzovaras, D., Strintzis, M.G.: Sq-map: Efficient layered collision detection and haptic rendering. IEEE Trans. Vis. Comput. Graph. **13**(1), 80–93 (2006). https://doi.org/10.1109/TVCG.2007.20

18. Osgouei, R.H., Kim, J.R., Choi, S.: Data-driven texture modeling and rendering on electrovibration display. IEEE Trans. Haptics **13**(2), 298–311 (2019). https://doi.org/10.1109/TOH.2019.2932990

19. Osher, S., Fedkiw, R.: Signed distance functions. In: Level Set Methods and Dynamic Implicit Surfaces, pp. 17–22. Springer New York (2003). https://doi.org/10.1007/0-387-22746-6_2

20. Park, J.J., Florence, P., Straub, J., Newcombe, R., Lovegrove, S.: Deepsdf: Learning continuous signed distance functions for shape representation. In: Proceedings of the IEEE/CVF Conference on Computer Vision and Pattern Recognition, pp. 165–174 (2019). https://doi.org/10.1109/cvpr.2019.00025
21. Ruffaldi, E., Morris, D., Edmunds, T., Barbagli, F., Pai, D.K.: Standardized evaluation of haptic rendering systems. In: 2006 14th Symposium on Haptic Interfaces for Virtual Environment and Teleoperator Systems, pp. 225–232. IEEE (2006). https://doi.org/10.1109/haptic.2006.1627081
22. Salisbury, K., Tarr, C.: Haptic rendering of surfaces defined by implicit functions. In: ASME International Mechanical Engineering Congress and Exposition, vol. 18244, pp. 61–67. American Society of Mechanical Engineers (1997). https://doi.org/10.1115/IMECE1997-0378
23. Shin, S., Osgouei, R.H., Kim, K.D., Choi, S.: Data-driven modeling of isotropic haptic textures using frequency-decomposed neural networks. In: 2015 IEEE World Haptics Conference (WHC), pp. 131–138. IEEE (2015). https://doi.org/0.1109/WHC.2015.7177703
24. Sianov, A., Harders, M.: Data-driven haptics: Addressing inhomogeneities and computational formulation. In: 2013 World Haptics Conference (WHC), pp. 301–306. IEEE (2013). https://doi.org/10.1109/WHC.2013.6548425
25. Sianov, A., Harders, M.: Exploring feature-based learning for data-driven haptic rendering. IEEE Trans. Haptics **11**(3), 388–399 (2018). https://doi.org/10.1109/TOH.2018.2817483
26. Sitzmann, V., Martel, J., Bergman, A., Lindell, D., Wetzstein, G.: Implicit neural representations with periodic activation functions. Adv. Neural Inf. Process. Syst. **33**, 7462–7473 (2020)
27. Srinivasan, M.A., Basdogan, C.: Haptics in virtual environments: taxonomy, research status, and challenges. Comput. Graph. **21**(4), 393–404 (1997). https://doi.org/10.1016/s0097-8493(97)00030-7
28. Venkatesh, R., Sharma, S., Ghosh, A., Jeni, L., Singh, M.: Dude: deep unsigned distance embeddings for hi-fidelity representation of complex 3D surfaces (2020). https://doi.org/10.48550/arXiv.2011.02570
29. Zhu, L., Xiang, Y., Song, A.: Visible patches for haptic rendering of point clouds. IEEE Trans. Haptics **15**(3), 497–507 (2022). https://doi.org/10.1109/TOH.2022.3165119
30. Zilles, C., Salisbury, J.: A constraint-based god-object method for haptic display. In: Proceedings 1995 IEEE/RSJ International Conference on Intelligent Robots and Systems. Human Robot Interaction and Cooperative Robots, vol. 3, pp. 146–151. IEEE (1995). https://doi.org/10.1109/IROS.1995.525876

"It's Like Being on Stage": Staging an Improvisational Haptic-Installed Contemporary Dance Performance

Xuan Li(iD), Ximing Shen(iD), Youichi Kamiyama(iD), Danny Hynds(iD),
Arata Horie(iD), Sohei Wakisaka(iD), and Kouta Minamizawa(✉)(iD)

Keio University Graduate School of Media Design, Yokohama, Japan
kouta@kmd.keio.ac.jp

Abstract. There is increasing research exploring how to augment expressive movements in dance practices by using haptic technologies. Meanwhile, less is known about how the audience perceives such information. In this study, we explore the potential of using a haptic wristband to convey contemporary dancers' performative somatic information to the audience through real-time control of haptic feedback by a haptic DJ. We then evaluate audience members' expectations towards the haptic-enabled viewing dance in a public performance setting. Participants indicated satisfaction with the improvisational haptic dance viewing experience.

Keywords: Haptic · Wearable Device · Dance Performance

1 Introduction

Stage performance is an important part of the art form of dance that allows dancers to convey emotions and expressive movements [3]. Previous researchers have explored the potential of interactions between dancers and novel technologies, such as tracking dancer body movement to control drone swarms [15], or visualising the sound of muscle using mechanomyogram (MMG) data on stage [33]. However, the researchers also explored the audience's perception of these new performative experiences. Some performances brought audience members closer to the stage by improving the somatic experience of the performers using interactive systems in singing [14] and dancing [23]. Others took the reactions of the audience members to the dance stage by projecting their physiological data to the background [27], or incorporating posts on social networks and haptic feedback [17].

Audience experience can now be enhanced through multimodal sensory approaches, and haptic feedback has been explored in art and performances [9]. For example, utilising haptic interaction to enable a robot to partner stepping with expert dancers [6], and delivering the sensation of tap dance to hearing-impaired using haptic feedback [26]. While these works explored performers'

© The Author(s), under exclusive license to Springer Nature Switzerland AG 2025
H. Kajimoto et al. (Eds.): EuroHaptics 2024, LNCS 14768, pp. 507–518, 2025.
https://doi.org/10.1007/978-3-031-70058-3_41

evaluation and the interactive system architecture, less is known about the audience's experience on these haptic dance performances in the wild, and how they can potentially contribute to the future of digital media installations and the entertainment industry.

Inspired by these previous works, we set to explore the feasibility of adding a new figure, a haptic DJ, to convey a dancer's lived experience in dance to the audience using haptic feedback. To do so, we worked with a contemporary dance group to design the haptic choreography of a piece of their signature stage performance[2] using a haptic wristband installed with a VP2 Vibro-Transducer. We then staged the performance in a public theatre, and used the Arts Audience Experience Index (AAEI) [21] to evaluate the audience experience. We collected 131 valid responses from the audience members, and found overall satisfaction from the audience towards the haptic-enabled viewing experience. We summarise our three main contributions in this study: (1) we propose a haptic-installed dance performance to demonstrate the feasibility of using real-time improvisational haptic feedback to augment audience experience, (2) we conducted a public stage performance to identify the audience experience of our haptic-installed dance performance, (3) we found that the improvisational haptic performance is well received, providing a greater sense of presence, and new insights to the audience (Fig. 1).

Fig. 1. Haptic DJ during the actual performance

2   Related Works

2.1   Technology Explorations in Dance and Stage Performance

Previous dance research investigates the potential of interactive technology in learning and increasing the expression of somatic information. Researchers investigated how motion tracking and monitoring technologies such as accelerometers

[29], motion capture systems [31,35], and physiological sensors [18] can be used for the assessment of movement quality to improve the learning, practising, and performing experience of dancers. In parallel, using physiological data to augment dancers' somatic information during performance is another popular approach since dancers' physiological status and flow state show correlation [11]. For instance, using motion capture systems to capture dancers' movements in real-time to control background graphics [10], music [13], and virtual costumes [12]. Another branch of research used robotic manipulators to map dance movements [22,24]. In general, the technical implementations introduced above showcased how the act of capturing and quantifying the physical movements of dance benefits from the advancement of different interactive technologies. At the same time, there is less exploration on technical approach representing the expressiveness and dancer's intention.

2.2 Understanding the Audience Experience

There are growing voices about empowering the audience in art performances [1]. Audience participation in live performances has been found to be supportive of the promotion of social and mental health [20], and highly engaging performances are efficient in eliciting the feeling of empowerment for the audience [30]. Theodorou et al. [28] used video recording to assess the audience's physical movements during a performance, and proposed the potential of using stilled motion as an indicator of the audience's level of engagement. Cerratto-Pargman et al. [5] analysed the role of communication as a tool for audience participation in interactive performances, and discovered that expectations about the technology impact the audience's experience and interpretation of participation. In sum, these investigations demonstrate the increasingly important role of audience participation, as art performances are now transforming into more digital and interactive manners.

2.3 Haptic Experience in Art and Performance

Haptic technology has been employed in public art and performance-related research. For example, Vi et al. [34] installed a mid-air haptic experience inside a six-week art exhibition. Dima et al. [7] added tactile sensation to museum artefacts that are typically impossible to touch. Besides these explorations on enhancing the public exhibition experience, there are also systems created for dance performances. For example, McCormick et al. [16] developed a system for capturing dancers' movements using motion capture, and mapping it to haptics in real-time to the audience. Sasaki et al. [23] built an audio-haptic feedback system that can deliver the tap dancing sensation to groups of dancers, dancer students, and the audience. While most of these systems are designed to convey the physical movements of dancers to the audience, how to convey the expressive information of more vague artistic performances, such as contemporary dance, where a group of dancers usually perform different and improvisational movements at the same time, remains unknown. Therefore, we decided to incorporate a new figure, the haptic DJ, to reflect the dancers' embodied knowledge.

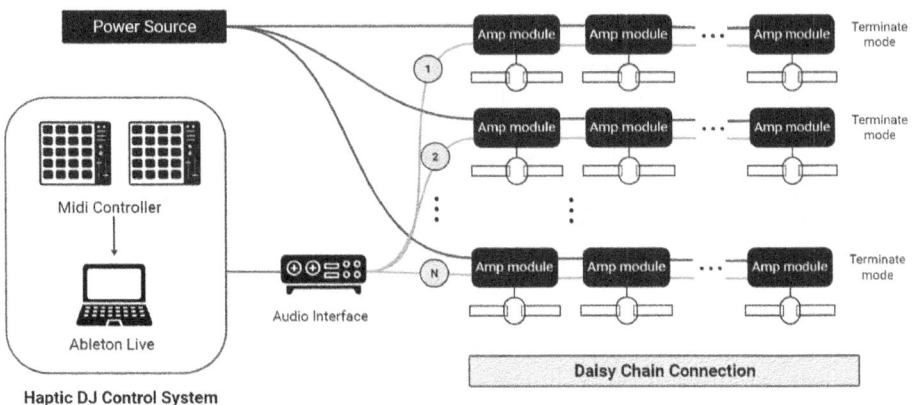

Fig. 2. Haptic system overview, including the Haptic DJ Control System and connection of the wearable device.

3 Designing an Improvisational Haptic-Installed Dance Performance

3.1 Haptic System Overview

Our system consists of (1) two midi keyboard controllers, (2) a laptop installed with the Ableton Live software, and (3) haptic wearable devices for the audience.

Haptic DJ Control System. The Haptic DJ prepared basic haptic pattern files such as different *sine waves* on Ableton before the performance, and mainly adjusted the volume to control the intensity of vibrations and the timing of the sample file.

Haptic Hardware and Connection Design. The haptic wearable device consists of two key components: (1) a VP2 Vibro-Transducer, and (2) a slap bracelet. We chose a wearable format and the slap bracelet so that the audience could easily adjust or remove the device by themselves at anytime.

We also designed a haptic amplifier unit that allows for daisy chaining connections (see Fig. 2), which features power and audio signal output for the wearable device. We then implemented an amplifier ratio adjustment function inside each haptic amplifier unit, to ensure the intensity of the haptic stimulation is consistent for each participant.

We highlighted that a potential hardware limitation for this system is the latency in the stimuli caused by the DJ's delayed response to the dancers' performance.

3.2 Haptic Choreography in Each Performance Scene

Contemporary dance is characterized by its improvisational nature and freedom of expression [8], which often deviates from the strict synchronization with

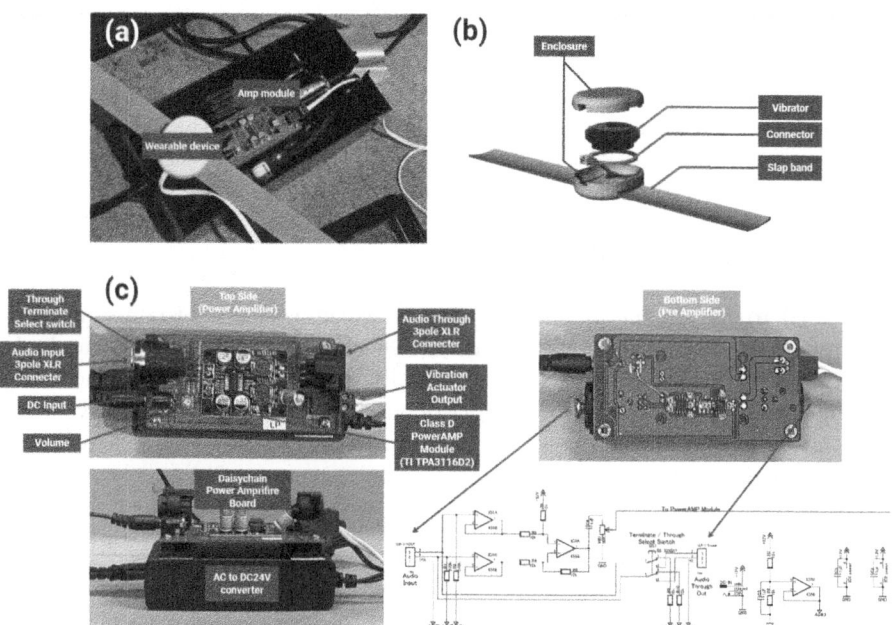

Fig. 3. Technical details of the haptic wearable system including (a) An overview photo of the haptic device setup for each participant, (b) 3D model of the wearable device, (c) Technical architecture of the Haptic Amp Module.

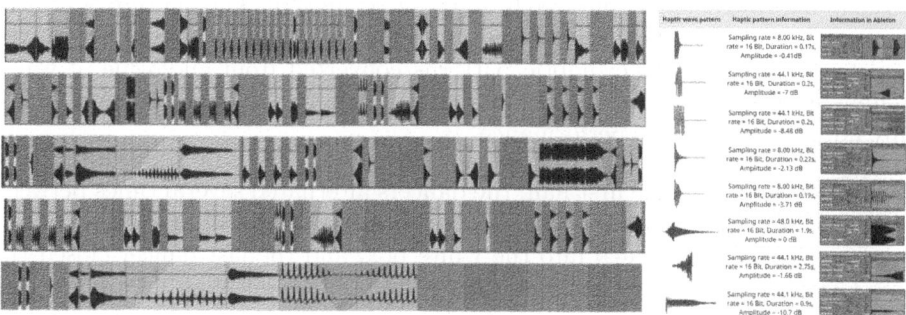

Fig. 4. Left: Haptic design in fully choreographed dance section. Right: Examples of haptic pattern profile in Ableton

music compared to other traditional forms of dance. In this performance, about 30% of the dance movements were predetermined (e.g., within a certain number of beats, the dancers need to move from one side of the stage to the other, but when and how to move was not decided). In line with the dance movements, we incorporated fully choreographed haptic sections and improvisational haptic sections into the performance. Here, we describe the haptic choreography of the dance performance in detail (Fig. 3):

Scene 1 (8 min). Four dancers entered the scene through a 2 m * 2 m * 2 m semi-closed white box, another dancer imitated the four dancers' movement from the outside. The four dancers pushed the box towards the outsider until they were covered by the white box. The haptic feedback of this scene was fully improvisational. *Sine waves* was the basic pattern that was used. We adjusted *sine wave* with variations in intensity to mirror the dancers' subtle shifts and sways. We applied short sensations when the dancers executed strong full-body movements like falling or striking.

Scene 2 (6 min). In the first half of the scene, dancers tried to eliminate each other, and two were able to escape. In the second half, one dancer was locked inside the box, and others acted mechanically around. One dancer then drove everyone else into the box. The imprisoned dancers put up an active dance inside the box. The last active dance part in the second half of this scene was fully choreographed (see Fig. 4).

Scene 3 (4 min). In this scene, one dancer lied on the ground and held the box on their abdomen, then gently rotated it. When they could not hold it anymore, the others came and supported to hold the box. This scene was fully improvisational. We applied subtle *sine waves* to reflect the rotation of the white box.

Scene 4 (6 min). In this scene, four dancers performed mirror-symmetric movements in two pairs, and tried to touch each other in reflection. This scene was fully improvisational, and free from music or rhythm. The dancers described this scene with metaphors related to the feeling of vibrancy in contrast to the previous scenes. Therefore, haptic patterns converted from audio file of water running or raining sound were used, whose parameters and intensities were manually adjusted in the live performance.

Scene 5 (6 min). In the first half, the dancers alternated several times between predetermined dance movements and improvisational dance segments. In the end, four dancers went back into the white box, and the last dancer closed the box. This scene includes a combination of fully choreographed haptic sections and improvisational haptic sections for the first part of the scene. At the end, random haptic patterns were applied to enhance the intense feeling that the performance has reached its climax rather than reflecting on any physical movements (Fig. 5).

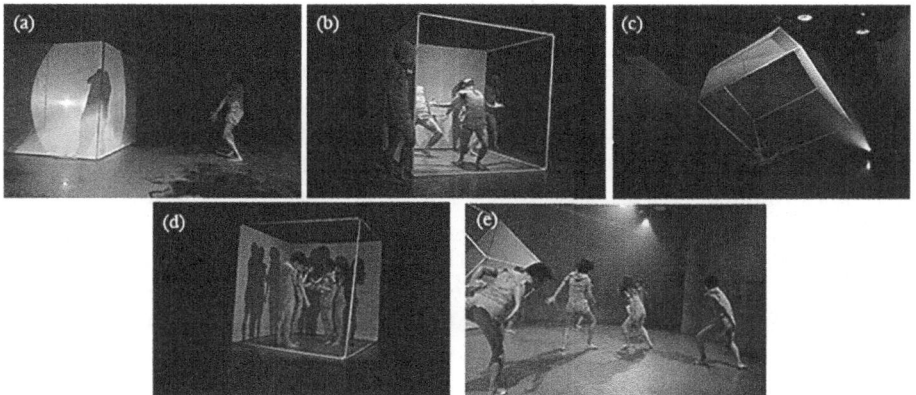

Fig. 5. Performance scenes. (a) Scene 1. (b) Scene 2. (c) Scene 3. (d) Scene 4. (e) Scene 5

4 Evaluating Audience Expectations in a Public Stage

4.1 Method

Questionnaire Design. We applied a within-subject approach to collect participants' subjective evaluations after watching the dance performance using the questionnaire. Previous works leveraged audience satisfaction to evaluate the efficacy of dance performances [19], and audience satisfaction towards the performance is shown to be affected by audience expectations [25] because expectation drives rational behaviour [4]. Inspired by these findings, we designed a study to evaluate audience satisfaction by measuring whether audience expectations towards the performance were met. We utilised the AAEI [21], an empirically validated scale [32], to measure the audience's experience related to their expectations. We modified the statements from previous investigation [32] to measure whether the probing items met the audience's expectations on three elements of AAEI.

Hypotheses. We constructed a priori hypotheses that our dance performance can achieve its proposed functionality of delivering haptic feedback to the audiences, and therefore is effective in inducing satisfying viewing experience on the emotional and cognitive levels. We constructed three hypotheses as follows:

- *H1*: The proposed dance performance meets the audience's expectation of a haptic-installed dance performance. (Authenticity)
- *H2*: The haptic-installed dance performance increases the audience's sense of engagement. (Collective-Engagement)
- *H3*: The haptic-installed dance performance enables the audience's to acquire new knowledge. (Knowledge)

4.2   Participants and Procedure

We received valid responses from 131 participants (seventy-eight female, forty-nine male, four unknown; average age = 43.08 years old, SD = 16.77 years old). Participants were professional dancers (5), amateur dancers with learning experience (50), and people with no experience in dancing (70). Participants' dance performance viewing experience was several times a month (50), once a month (25), several times a year (39), and never before (11).

The dance performance was staged three times in the same theatre in June 2023. Before each show, one staff member introduced the concept of the performance and the device on the stage. The audience gave informed consent on paper to this IRB-approved research before the show began. Participants watched the 30-minute show, and then answered the questionnaire.

4.3   Quantitative Results

We analysed our questionnaire responses in SPSS (see Table 1). Shapiro-Wilk tests were performed to check normality. We report descriptive statistics, as well as the results of Wilcoxon Signed Ranks Tests with p-value and effect size which indicates the statistical significance of the participants' fulfilled expectations. We used the baseline score (3) as hypothesised median for the Wilcoxon tests.

Table 1. Summary of the Hypotheses, Questions, and Quantitative Results.

Hypothesise	Statement	Mean (SD)	Significance
H1	*Q1*: The haptic dance performance experience met my expectations and I was very satisfied	3.85 (0.968)	$Z(129) = -7.169$, $p < 0.01$***,$d = -0.628$
H2	*Q2*: The tactile stimulation made me feel a greater sense of presence than I usually watch a dance performance	3.56 (1.121)	$Z(129) = -4.814$, $p < 0.01$***, $d = -0.421$
H2	*Q3*: I felt a connection to the dancers through the haptic device	3.39 (1.117)	$Z(129) = -3.520$, $p¡0.01$***,$d = -0.308$
H3	*Q4*: I felt that the tactile stimuli presented were consistent with the dancer's performative somatic information. (Note: Please answer by whether the timing matched the dancers' body movements, instead of the music)	3.67 (0.999)	$Z(129) = -6.148$, $p < 0.01$***, $d = -0.536$
H3	*Q5*: The haptic dance performance gave me new insights and inspiration about dance performance compared to the other performances I watched before	3.49 (0.998)	$Z(129) = -4.741$, $p < 0.01$***, $d = -0.414$

5   Discussion

In our work, we designed the new role of the haptic DJ bearing resemblance to that of the lighting designer in stage performances. Much like a lighting designer cues specific lighting effects to complement the choreography, the haptic DJ is responsible for timing and delivering the designed haptic patterns aligned with the dance movements in real-time. However, our work is specific to one contemporary dance piece with 5 male dancers, while a larger sample size and different dance genres would provide deeper insights into design and technology preferences.

Our audience expectation data suggests that overall the audience members were generally satisfied by the haptic-enabled viewing experience. For example, an increased sense of presence and the feeling of closer distance with the dancers were reported. The haptic sensation also provided them with enjoyable and multi-dimensional information to inspire their thinking and imagination. However, we did not compare their viewing experience without the haptic experience, or measure objective data such as the audience's biological data. A potential future direction is to conduct a comparison study between with haptic and without haptic audience viewing experience.

6   Conclusions

We proposed a haptic-installed contemporary dance performance with a new improvisational figure, the haptic DJ. During the public dance performance, we found that the haptic experience was well-received and earned overall positive feedback from the audience. The results we gathered are indicative of the possibility of using improvisational haptic feedback to connect the performers and audiences in the entertainment industry. We hope this study can encourage and assist future development in using haptic technology in live performance and installation settings.

Acknowledgements. This research is supported by JST Moonshot R&D Program "Cybernetic being" Project (Grant number JPMJMS2013). The performance was created in cooperation with Session House and Bushman, and supported by the Arts Council Tokyo.

Disclosure of Interests. The authors have no competing interests to declare that are relevant to the content of this article.

References

1. Bernstein, J.S.: Arts Marketing Insights: The Dynamics of Building and Retaining Performing Arts Audiences. John Wiley & Sons (2006)
2. Bushman: Zoumadefromfume (2023). https://sites.google.com/view/bushman-zoumadefromfume. Accessed 01 Sep 2023
3. Camurri, A., Lagerlöf, I., Volpe, G.: Recognizing emotion from dance movement: comparison of spectator recognition and automated techniques. Int. J. Hum. Comput. Stud. **59**(1), 213–225 (2003). https://doi.org/10.1016/S1071-5819(03)00050-8
4. Cardozo, R.N.: An experimental study of customer effort, expectation, and satisfaction. J. Mark. Res. (JMR) **2**(3), 244–249 (1965). https://doi.org/10.2307/3150182

5. Cerratto-Pargman, T., Rossitto, C., Barkhuus, L.: Understanding audience participation in an interactive theater performance. In: Proceedings of the 8th Nordic Conference on Human-Computer Interaction: Fun, Fast, Foundational, pp. 608–617. NordiCHI '14, Association for Computing Machinery, New York, NY, USA (2014). https://doi.org/10.1145/2639189.2641213

6. Chen, T.L., et al.: Evaluation by expert dancers of a robot that performs partnered stepping via haptic interaction. PLoS One **10**(5), e0125179 (2015). https://doi.org/10.1371/journal.pone.0125179

7. Dima, M., Hurcombe, L., Wright, M.: Touching the past: Haptic augmented reality for museum artefacts. In: Shumaker, R., Lackey, S. (eds.) Virtual, Augmented and Mixed Reality. Applications of Virtual and Augmented Reality. Lecture Notes in Computer Science, pp. 3–14. Springer International Publishing, Cham (2014). https://doi.org/10.1007/978-3-319-07464-1_1

8. Fink, A., Woschnjak, S.: Creativity and personality in professional dancers. Personal. Individ. Differ. **51**(6), 754–758 (2011). https://doi.org/10.1016/j.paid.2011.06.024

9. Hayes, L., Rajko, J.: Towards an aesthetics of touch. In: Proceedings of the 4th International Conference on Movement Computing. MOCO '17, Association for Computing Machinery, New York, NY, USA (2017). https://doi.org/10.1145/3077981.3078028

10. James, J., et al.: Movement-based interactive dance performance. In: Proceedings of the 14th ACM International Conference on Multimedia, pp. 470–480. MM '06, Association for Computing Machinery, New York, NY, USA (2006). https://doi.org/10.1145/1180639.1180733

11. Jaque, S.V., Thomson, P., Zaragoza, J., Werner, F., Podeszwa, J., Jacobs, K.: Creative flow and physiologic states in dancers during performance. Front. Psychol. **11**, 1000 (2020). https://doi.org/10.3389/fpsyg.2020.01000

12. Johnston, A.: Conceptualising interaction in live performance: reflections on "encoded". In: Proceedings of the 2nd International Workshop on Movement and Computing, pp. 60–67. MOCO '15, Association for Computing Machinery, New York, NY, USA (2015). https://doi.org/10.1145/2790994.2791003

13. Jung, D., Jensen, M.H., Laing, S., Mayall, J.: Cyclic.: an interactive performance combining dance, graphics, music and kinect-technology. In: Proceedings of the 13th International Conference of the NZ Chapter of the ACM's Special Interest Group on Human-Computer Interaction, pp. 36–43. CHINZ '12, Association for Computing Machinery, New York, NY, USA (2012). https://doi.org/10.1145/2379256.2379263

14. Kilic Afsar, et al.: Corsetto: a kinesthetic garment for designing, composing for, and experiencing an intersubjective haptic voice. In: Proceedings of the 2023 CHI Conference on Human Factors in Computing Systems, pp. 1–23. ACM, Hamburg Germany (Apr 2023). https://doi.org/10.1145/3544548.3581294

15. Kim, H., Landay, J.A.: Aeroquake: drone augmented dance. In: Proceedings of the 2018 Designing Interactive Systems Conference, pp. 691–701. DIS '18, Association for Computing Machinery, New York, NY, USA (2018). https://doi.org/10.1145/3196709.3196798

16. McCormick, J., Hossny, M., Fielding, M., Mullins, J., Vincent, J.B., Hossny, M., Vincs, K., Mohamed, S., Nahavandi, S., Creighton, D., Hutchison, S.: Feels like dancing: Motion capture-driven haptic interface as an added sensory experience for dance viewing. Leonardo **53**(1), 45–49 (2020). https://doi.org/10.1162/leon_a_01689

17. Moriaty, M., Sykes, L.: Deviced: Audience-dancer interaction via social media posts and wearable for haptic feedback. Wearable Technol. **3**, e3 (2022). https://doi.org/10.1017/wtc.2021.20
18. Niewiadomski, R., Mancini, M., Piana, S., Alborno, P., Volpe, G., Camurri, A.: Low-intrusive recognition of expressive movement qualities. In: Proceedings of the 19th ACM International Conference on Multimodal Interaction, pp. 230–237. ICMI '17, Association for Computing Machinery, New York, NY, USA (2017). https://doi.org/10.1145/3136755.3136757
19. Oppermann, L., Putschli, C., Brosda, C., Lobunets, O., Prioville, F.: The smart-phone project: an augmented dance performance. In: Proceedings of the 33rd Annual ACM Conference on Human Factors in Computing Systems, pp. 2569–2572. CHI '15, Association for Computing Machinery, New York, NY, USA (2015). https://doi.org/10.1145/2702123.2702538
20. Perkins, R., Mason-Bertrand, A., Tymoszuk, U., Spiro, N., Gee, K., Williamon, A.: Arts engagement supports social connectedness in adulthood: findings from the hearts survey. BMC Public Health **21**(1), 1208 (2021). https://doi.org/10.1186/s12889-021-11233-6
21. Radbourne, J., Johanson, K., Glow, H., White, T.: The audience experience: measuring quality in the performing arts. Int. J. Arts Manag. **11**, 16–29 (2009)
22. Rogel, A., Savery, R., Yang, N., Weinberg, G.: Robogroove: creating fluid motion for dancing robotic arms. In: Proceedings of the 8th International Conference on Movement and Computing. MOCO '22, Association for Computing Machinery, New York, NY, USA (2022). https://doi.org/10.1145/3537972.3537985
23. Sasaki, T., Okazaki, N., Yoshida, T., Balandra, A., Kashino, Z., Inami, M.: Sole-fultap: Augmenting Tap Dancing Experience Using a Floor-Type Impact Display (2023). arXiv:2304.00411
24. Saviano, G., Villani, A., Prattichizzo, D.: A PCA-based method to map aesthetic movements from dancer to robotic arm. In: 2023 IEEE International Conference on Advanced Robotics and Its Social Impacts (ARSO), pp. 71–77 (2023). https://doi.org/10.1109/ARSO56563.2023.10187492
25. Shepperd, J.A., Sweeny, K., Cherry, L.C.: Influencing audience satisfaction by manipulating expectations. Soc. Influ. **2**(2), 98–111 (2007). https://doi.org/10.1080/15534510601095772
26. Shibasaki, M., Kamiyama, Y., Minamizawa, K.: Designing a haptic feedback system for hearing-impaired to experience tap dance. In: Adjunct Proceedings of the 29th Annual ACM Symposium on User Interface Software and Technology, pp. 97–99. UIST '16 Adjunct, Association for Computing Machinery, New York, NY, USA (2016). https://doi.org/10.1145/2984751.2985716
27. Sugawa, M., et al.: Boiling mind: Amplifying the audience-performer connection through sonification and visualization of heart and electrodermal activities. In: Proceedings of the Fifteenth International Conference on Tangible, Embedded, and Embodied Interaction, pp. 1–10. TEI '21, Association for Computing Machinery, New York, NY, USA (2021). https://doi.org/10.1145/3430524.3440653
28. Theodorou, L., Healey, P.G.T., Smeraldi, F.: Exploring audience behaviour during contemporary dance performances. In: Proceedings of the 3rd International Symposium on Movement and Computing, pp. 1–7. MOCO '16, Association for Computing Machinery, New York, NY, USA (2016). https://doi.org/10.1145/2948910.2948928
29. Thiel, D.V., Quandt, J., Carter, S.J.L., Moyle, G.: Accelerometer based performance assessment of basic routines in classical ballet. Procedia Eng. **72**, 14–19 (2014). https://doi.org/10.1016/j.proeng.2014.06.006

30. Toelle, J., Sloboda, J.A.: The audience as artist? the audience's experience of participatory music. Music Sci. **25**(1), 67–91 (2021). https://doi.org/10.1177/1029864919844804

31. Trajkova, M., Cafaro, F.: Takes tutu to ballet: designing visual and verbal feedback for augmented mirrors. Proc. ACM Interact. Mob. Wearable Ubiquitous Technol. **2**(1), 38:1-38:30 (2018). https://doi.org/10.1145/3191770

32. Tung Au, W., Ho, G., Wing Chuen Chan, K.: An empirical investigation of the arts audience experience index. Empir. Stud. Arts **35**(1), 27–46 (2017). https://doi.org/10.1177/0276237415625259

33. Van Nort, D.: [Radical] Signals from life: from muscle sensing to embodied machine listening/learning within a large-scale performance piece. In: Proceedings of the 2nd International Workshop on Movement and Computing. p. 124-127. MOCO '15, Association for Computing Machinery, New York, NY, USA (2015). https://doi.org/10.1145/2790994.2791015

34. Vi, C.T., Ablart, D., Gatti, E., Velasco, C., Obrist, M.: Not just seeing, but also feeling art: mid-air haptic experiences integrated in a multisensory art exhibition. Int. J. Hum. Comput. Stud. **108**, 1–14 (2017). https://doi.org/10.1016/j.ijhcs.2017.06.004

35. Vincs, K., McCormick, J.: Touching space: Using motion capture and stereo projection to create a virtual haptics of dance. Leonardo **43**, 359–366 (2010). https://doi.org/10.1162/LEON_a_00009

Author Index

H. Kajimoto et al. (Eds.): EuroHaptics 2024, LNCS 14768, pp. 519–522, 2025.
https://doi.org/10.1007/978-3-031-70058-3

SPRINGER NATURE

GPSR Compliance

The European Union's (EU) General Product Safety Regulation (GPSR) is a set of rules that requires consumer products to be safe and our obligations to ensure this.

If you have any concerns about our products, you can contact us on ProductSafety@springernature.com

In case Publisher is established outside the EU, the EU authorized representative is:

Springer Nature Customer Service Center GmbH
Europaplatz 3
69115 Heidelberg, Germany

The manufacturer's authorised representative in the EU is Springer
Nature Customer Service Centre GmbH, Europaplatz 3, 69115 Heidelberg,
Germany. If you have any concerns regarding our products, please
contact ProductSafety@springernature.com

Printed and bound by CPI Group (UK) Ltd, Croydon, CR0 4YY

24/04/2026

02096358-0016